SO-ANU-726

THE LIBRARY
ST. MARY'S COLLEGE OF MARYLAND
ST. MARY'S CITY, MARYLAND 20686

Universitext

Craig Smoryński

Logical Number Theory I

An Introduction

With 13 Figures

Springer-Verlag
Berlin Heidelberg New York
London Paris Tokyo
Hong Kong Barcelona
Budapest

Craig Smoryński
429 S.Warwick
Westmont, IL 60559, USA

Cover picture: It does not seem supercrogatory to inform the reader that the cover illustration is derived from the frontispiece to the first edition of Christopher Marlowe's biography of Dr. Johannes Faustus, whose insightful remarks on logic are quoted in this book.

Mathematics Subject Classification (1980): 03-01, 03F30, 10B99, 10N05

ISBN 3-540-52236-0 Springer-Verlag Berlin Heidelberg New York
ISBN 0-387-52236-0 Springer-Verlag New York Berlin Heidelberg

Library of Congress Cataloging-in-Publication Data. Smoryński, C. Logical number theory: an introduction / Craig Smoryński. p. cm. – (Universitext) Includes bibliographical references and indexes. ISBN 3-540-52236-0 (Springer-Verlag Berlin: v. 1). – ISBN 0-387-52236-0 (Springer-Verlag New York: v. 1) 1. Number theory. 2. Logic, Symbolic and mathematical. I. Title. QA241.S614 1991 512'.7–dc20 90-25702

This work is subject to copyright. All rights are reserved, whether the whole or part of the material is concerned, specifically the rights of translation, reprinting, reuse of illustrations, recitation, broadcasting, reproduction on microfilms or in other ways, and storage in data banks. Duplication of this publication or parts thereof is only permitted under the provisions of the German Copyright Law of September 9, 1965, in its current version, and a copyright fee must always be paid. Violations fall under the prosecution act of the German Copyright Law.

© Springer-Verlag Berlin Heidelberg 1991
Printed in Germany

41/3140-543210 - Printed on acid-free paper

Preface

What would life be like without arithmetic, but a scene of horrors?

— *Rev. Sydney Smith*

It is with a mixture of trepidation and chutzpah that I offer a work bearing the audacious title of *Logical Number Theory*. Disadvantageous comparisons with the spectacularly successful *analytic* number theory are inevitable. It is a fact, however, that logicians have made their own studies of arithmetic, that, although not as deep or as fully developed as analytic number theory, there is a body of logical knowledge of the subject. The present work concerns itself with this fledgling logical study of number theory.

The present work, of which this book is the first of two volumes, is an *introductory* text. This means several things. First, it means that certain advanced and non-central topics have not been covered. Adequate expositions of some of these subjects already exist, and others are in preparation. Chapter VII of the second volume will offer references and recommendations to the reader who wishes to go further. For the present, let it suffice for me to say that these two volumes are intended to present the *core* of logical number theory and they do not go *much* beyond the basics. (I have, of course, indulged myself occasionally and included some items not clearly (or: clearly not) of central importance. I have done this whenever I felt the intruded topic to be of sufficient interest and brevity.)

A second characteristic of an introductory text should be a reasonably wide accessibility. It has been my intention that the present work be accessible to advanced undergraduate mathematics majors with minimal specific background knowledge as prerequisite. I presuppose that "mathematical maturity" which can no longer be presumed to be acquired from an American calculus course (the situation is better in Europe), a course in abstract algebra (always a good prerequisite for any serious logic course), and perhaps a little elementary number theory (if such was not covered in one's abstract algebra course) or a tiny bit of logic (say that available from the symbolic logic course offered by the philosophers). I emphasise that I do not wish to presuppose any amount of mathematical logic. For one thing, the arithmetical content of these two volumes ought to be of some interest to the general mathematician— who, in all probability, has had no prior contact with mathematical logic. To presuppose any logical background would thus lessen the book's potential usefulness. I do assume, however, that the reader has seen the connectives ($\neg$, $\wedge$, $\vee$, $\rightarrow$) and quantifiers ($\forall$, $\exists$), has an intuitive grasp of their meanings ("not", "and", "or", "implies" and "for all", "there exists", respectively), and will accept the linguistic condition that, if one wants to say, for example, that all objects x have property P, one writes this as

$$\forall x P(x),$$

and not as

$$P(x), \forall x,$$

this latter common usage having serious ambiguity if $P(x)$ is logically complex. (Think of the distinction between continuity and uniform continuity.)

I might mention at this point that the inclusion of logical material for the benefit of curious non-logicians has not *greatly* increased the size of the book. Because of the necessity, in the second volume, of verifying the derivability within formal systems of arithmetic of basic results of mathematical logic, all the basics have had to be discussed anyway. The addition of explanatory remarks for a wider audience did not take up much extra space. Indeed, the logical material that could be deleted on assumption of a prior logic course amounts to less than a single chapter.

The present work has proved somewhat longer than anticipated and is, thus, being split into two volumes. It is an entirely fortuitous circumstance that this, the first volume of the pair, is a complete work in itself. This volume could, in fact, be used as a slightly unorthodox textbook for an undergraduate introduction to mathematical logic. It covers almost everything I consider necessary for the "logical literacy" of the professional mathematician— the elements of recursion theory, the completeness theorem relating syntax and semantics, and the basic theorems on incompleteness and undecidability in their now traditional recursion theoretic formulations. What is missing is nonstandard analysis as well as some basic set theory, neither of which is a standard topic in introductory courses on mathematical logic anyway. The Table of Contents should give the potential instructor a hint of the extent of my unorthodoxy— the extra emphasis on arithmetic matters and the deëmphasis of pure logic: there is, for example, no section on the propositional calculus and truth tables. Such logical material (e.g., disjunctive normal form) is introduced as needed and omitted (e.g., duality) as it is not needed. Admitting such omissions, I must say, however, that I would not hesitate to use my book as an introductory logic textbook in a mathematics department. My book addresses the logical requisites of the general mathematician, not the future logician. (The latter can pick up whatever material I have overlooked in courses offered by the philosophy or informatics faculties, logicians not being as xenophobic as mathematicians in general.)

It was, of course, not my goal to write an introductory logic text, but, insofar as the present volume could serve as one, I offer a few suggestions on its use for such a purpose. First, given all the time in the world, I would cover the whole book. With a little less time I would omit the starred sections (in the case of I.9 omission is not completely advisable— skim it). More realistically, one could cover only Chapters I and III and have a solid core of logic. Pressed for time, a bare-bones course in logic could be given by I.3, I.7 - 12 (skimming I.9), III.2, III.3, and III.6 (plus III.7 or III.8 as time permits). All the central results would be covered.

As I said, my goal is not an introduction to logic, but to logical number theory. It would be rather pointless to use this volume in a course and not cover anything from Chapter II. The quickest route to the unsolvability of Hilbert's tenth problem is to be had by reading II.1, II.2, II.4, and II.5. The applications of II.6 and II.7 are, however, very nice and these sections ought to be at least minimally perused. Professional number theorists should read the whole of Chapter II; those with no prior background in recursion theory will need also I.7, I.8, I.10, I.11, and a bit of I.12. Number theorists who aren't familiar with the

Fueter-Pólya Theorem might also find I.4 - 5 to their liking. It presupposes III.2, but the material of III.4 may also be of some interest to the specialist.

Such dissections as above are compromises, distasteful to any author who, quite naturally, views his work as a whole. In the present case the whole consists of two volumes and I should like to say a few words about their overall plan. The simplest description is to be had by appeal to the unifying theme of undecidability: Chapter I introduces the theory of computability and offers abstract unsolvable problems; Chapter II shows that unsolvability is is not totally abstract by offering a concrete unsolvable problem. With the introduction of formal languages, Chapter III changes the nature of undecidability. Instead of offering so-called *mass problems* with no uniform effective solutions, it exhibits— *abstractly*— the existence of individual problems which cannot be decided on the basis of one's current set of axioms. It yields results like: "there are sentences which can neither be proven nor refuted", or, via Chapter II, "there are Diophantine equations possessing no solutions but for which there is no proof that they possess no solutions". Chapter IV begins the concretisation of such formal undecidability by presenting Gödel's original, deeper results. Not only can formal theories not decide everything, but there are specific assertions of intelligible content that cannot be decided. The first of these is the assertion of the consistency of a given formal theory. This is Gödel's Second Incompleteness Theorem and it is proven in Chapter IV of volume II. In Chapter VI, we will see the unprovability of specific combinatorial theorems in a strong formal theory of arithmetic. This proof uses nonstandard models of arithmetic, which are introduced and studied in Chapter V.

The above description is a gross oversimplification, but it will do for now as it illustrates the dichotomy between the two volumes: volume I covers the basics (at an advanced undergraduate level) with which all mathematicians ought to be partially familiar; and volume II covers more sophisticated material (at the first-year graduate level) with which all logicians ought to be partially familiar.

More years ago than should be clear from the final product, I was approached by the editors of the series *Perspectives in Mathematical Logic* to write a book on the "metamathematics of arithmetic". During the early stages of this work I received much financial support from the Heidelberg Academy of Sciences through the offices of Professor Doctor Gert H. Müller. Although the present book is not the same advanced monograph originally contracted for the *Perspectives*, it is the ultimate fruit of the seed planted by the editors of that series and I take the opportunity here to acknowledge both my conceptual and financial indebtedness.

While some of the writing in the earlier stages of the preparation of this work was performed in Heidelberg, most of the writing took place at the Rijksuniversiteit te Utrecht in the Netherlands under the hospitality of my dear friend Dirk van Dalen. This is particularly true of volume II, Chapters V and VI of which are based on the lecture notes of a course I gave in Utrecht during the Fall of 1978.

For their more direct involvement, I wish to thank my friend Friedemann Tuttas and the editorial staff of Springer-Heidelberg. The former proofread the typescript and made many corrections and suggestions for improvement; the latter have always been a pleasure to deal with.

Finally, I would like to express my gratitude to my friends Julia Robinson (now deceased) and Petr Hájek for their encouragement over the years.

Westmont, Illinois (USA)
June 1990

Craig Smoryński

Contents

Chapter I. Arithmetic Encoding

> Die Mathematiker sind eine Art Franzosen: redet man zu ihnen, so übersetzen sie es in ihre Sprache, und dann ist es sobald ganz etwas anderes.
>
> — *J. W. von Goethe*

1. Polynomials

The rare graduate student or advanced undergraduate who hasn't haunted the library and leafed through many an incomprehensible mathematical monograph and has, thus, not yet heard of (say) algebraic geometry or orthogonal polynomials or the unsolvability of Hilbert's tenth problem might well wonder why a book written for graduate students and advanced undergraduates should begin with something so simple as polynomials. Well, polynomials have had a long history and they form a recurring theme throughout all of mathematics.

Polynomials were not always what they are today. To the Hellenic Greeks (6th to 4th centuries B.C.), geometry was everything: terms of degree 1 denoted length, terms of degree 2 area, and terms of degree 3 volume. The Hellenic Greeks would not multiply four numbers together and they would not mix terms of different degrees. The Hellenistic Greeks (post-Alexander) got away from this: Heron's formula for the area of a triangle multiplies four numbers and Diophantus considered polynomials up to degree 6. All pre-Diophantine algebra that has been preserved is *rhetorical*; there are no variables, merely words. Diophantus made a tiny step toward symbolism by *syncopating*, or abbreviating, the names for his unknown, its square, and its cube, as well as the word for the copula. Thus, the polynomial equation

$$7x^4 + 3x^3 + 2x^2 + x = 17$$

would, if we were to use Diophantine notation, but modern English, read

$$s^q\ s^q\ 7\ c^u\ 3\ s^q\ 2\ x\ \ \text{eq}\ \ 17,$$

where "s^q" abbreviates "square", "c^u" abbreviates "cube", and "eq" abbreviates "equals". Diophantine notation is actually a little worse than this: Greek numeration was not nearly so pleasant as our modern decimal system; and it was better than this: he had a symbol for the minus sign to allow negative coëfficients and a notation to change powers to reciprocals. For example, he would write (modulo the change of language) $1/x^2$ as s^{qx}. If, nowadays, Diophantine notation seems laughable, we should remember that it was unequalled for centuries: Diophantus is conjectured to have lived around 250 AD, toward the end of the period of creativity of Greek mathematics. When the Arabs began to take command of the world of learning about half a millennium later, they reverted to the rhetorical approach. The Western European awakening did not see symbolism for a while.

In Cardano's *Ars Magna* (1545), for example, the language is again partially geometric. Cardano was able to handle non-homogeneous polynomials by padding his dimensions, and his student Ferrari was able to solve the general quartic, but he still illustrated the solution to cubic equations by drawing three dimensional figures and calculating volumes. Some achieved syncopation— *R* for *r*adix, *Rq* for square root, etc. The first great step toward symbolism in algebra was François Viète's *In artem analyticem isagoge* (1591), in which Viète introduced variables— vowels for unknowns and consonants for knowns. Moreover, unlike his contemporaries who used different notations for different powers of a given variable, Viète used only the given variable. He did not, however, have exponential notation and he wrote *A quad*, *A cub*, etc., for A^2, A^3, etc. Here we have a semi-syncopated and semi-symbolic algebra. With the additional symbolism for the operators $+$, $-$ (Stifel, 1544) and $=$ (Recorde, 1557), algebra was well on its way toward being symbolic and, if we ignore Newton's preference for xx over x^2, the next century saw not only a complete symbolisation of algebra, but, in fact, the current one.

The outcome of the Newton-Leibnitz controversy, i.e., the development of analysis on the European continent where Leibnitz's notation held sway and its stagnation in England where Newton's dots reigned supreme, affords us an object lesson in the importance of good notation. My brief precís on the development of algebraic notation should not be particularly helpful in understanding this point, but it should, at least, allow us to bring this point to bear on understanding the slow development of polynomials themselves. For example, polynomials of high degree or of a great many variables only arose quite recently if only because until quite recently one had no way of referring to them.

This introductory section will not pretend to summarise the history of polynomials. Such a task would be too immense. Even the simple question of what a polynomial is requires some thought. Bearing in mind the lack of notation, what was a polynomial to the ancients? There really wasn't such a thing. What existed at the time were word problems, which we would nowadays write as polynomial equations to be solved. The word problems of the Egyptians were linear; those of the Babylonians also included the quadratic. Diophantus actually wrote expressions we could accept as polynomials, and by the 16th century polynomials were expressions, parts of equations to be solved. In the 18th century the notion of function arose and polynomials became functions on the real numbers, eventually on the complex numbers. What are they today? Over a finite field they are not exactly functions, but they define functions. In abstract algebra, a polynomial can be defined to be an infinite sequence of elements of a given field, with all but finitely many entries in the sequence being 0. In very abstract algebra, a polynomial is an element of a free algebra over certain generators...

In this book, polynomials will be viewed in several ways. Informally, they will be understood to be whatever it was the reader took them to be before opening the book. Formally, they will be terms in a very strictly defined language. This language will be called the language of arithmetic and it will be our task to study this language and the mathematics associated with it. Let me phrase this differently: this is a book on logical number theory. Logic concerns itself not only with reasoning, but also with language— notations as it were. The goal of this book is to study the language of arithmetic and discover some of its strengths and weaknesses. So described, the sequel promises to be exceedingly dull. Thus, I hasten to say that my descriptive powers are hopelessly inadequate; the story I have to tell is truly fascinating and I think this will show through my turgid prose. Nonetheless, the reader will do well to keep such questions as notation, definability, and expressibility in mind.

A simple, yet (I hope) effective, illustration of the importance of notation is afforded by a comparison of the following four representations of the same polynomial:

$$x^4 - 7x^3 + 14x^2 - 8x \tag{1}$$

$$x(x-1)(x-2)(x-4) \tag{2}$$

$$x(-8 + x(14 + x(-7 + x))) \tag{3}$$

$$x(x-1)(x-2)(x-3) - x(x-1)(x-2) \tag{4}$$

The advantages and disadvantages of the first two of these are obvious; those of the last two are less so. Let us briefly go through some of these advantages and disadvantages. Representation (1) is best if we wish to differentiate or integrate the underlying function. It allows easy comparison with other polynomials written in the similar standard form. Representation (2) is not very pleasant to differentiate (unless we use logarithmic differentiation) or to integrate, but if we wish to know the zeroes of the polynomial, it is by far the best representation. The advantage of representation (3) over (1) is quite simple: it requires only three multiplications for an evaluation, whereas (1) requires six (or nine, if one repeats multiplications in calculating the series x^2, x^3, x^4). The superiority of (3) over (2) in this respect is not apparent because I chose a polynomial with such simple coëfficients in its factored form.

The special advantage afforded by representation (4) is not likely to be evident to anyone not at least partially familiar with the calculus of finite differences. This calculus is a pleasant discrete analogue to the differential calculus. The chief operation of this discrete calculus is the finite difference Δ defined by

$$(\Delta f)(x) = f(x+1) - f(x).$$

Just as in ordinary calculus, where there are simple formulae for differentiating certain functions, the calculus of finite differences has simple formulae for "differencing" certain functions. But these are not the same functions. The difference

$$\Delta x^n = (x+1)^n - x^n = \binom{n}{n-1}x^{n-1} + \ldots + \binom{n}{1}x + \binom{n}{0}$$

is hardly notationally compact. The appropriate replacement for x^n is $x^{(n)}$, defined by

$$x^{(n)} = x(x-1)\cdots(x-n+1).$$

For this function, the difference is easy to compute:

$$\begin{aligned}\Delta x^{(n)} &= (x+1)^{(n)} - x^{(n)}\\ &= x(x-1)\cdots(x-n+2) - x(x-1)\cdots(x-n+1)\\ &= x(x-1)\cdots(x-n+2)[(x+1) - (x-n+1)]\\ &= x^{(n-1)}n = nx^{(n-1)},\end{aligned}$$

nicely analogous to

$$\frac{d}{dx}x^n = nx^{n-1}.$$

Inverse to differencing is anti-differencing, the process of, given f, finding g such that $\Delta g = f$. As in ordinary calculus, two anti-differents differ by a constant (or, thinking of f, g as functions of a real variable and not a discrete one, they differ by a periodic function of period 1). Thus, as in the notation of the calculus,

$$\int f(x)\,dx = g(x) + C,$$

we have the notation,

$$\Delta^{-1}f(x) = g(x) + C \quad \text{(or: } g(x) + C(x), \text{ where } C(x) \text{ is periodic).}$$

The analogy is even better: just as the definite integral is a limit of sums, the definite inverse difference is a sum.

1.1. Theorem. *Let f be a function of natural numbers and let $\Delta g(x) = f(x)$. For any natural numbers a, b,*

$$\sum_{i=a}^{b} f(i) = \Delta^{-1}f(x)\Big|_a^{b+1} = g(b+1) - g(a).$$

Proof. Write $G(x) = \sum_{i=a}^{x-1} f(i)$ and observe

$$\Delta G(x) = \sum_{i=a}^{x} f(i) - \sum_{i=a}^{x-1} f(i) = f(x),$$

whence $G(x) = g(x) + C$. But

$$g(b+1) - g(a) = G(b+1) - G(a) = \sum_{i=a}^{b} f(i) - \sum_{i=a}^{a-1} f(i)$$

$$= \sum_{i=a}^{b} f(i),$$

where we implicitly assume $\sum_{i=c}^{d} f(i) = 0$ if $d < c$. QED

Getting back to our original polynomial, we see that its fourth representation can be abbreviated to

$$x^{(4)} - x^{(3)}. \tag{4'}$$

The advantage of this representation is now clear. Letting $P(x)$ denote the given polynomial, representation (4') immediately allows us to find

$$\Delta P(x) = 4x^{(3)} - 3x^{(2)} \quad \text{and} \quad \Delta^{-1}P(x) = x^{(5)}/5 - x^{(4)}/4 + C,$$

the latter very readily allowing us to find, say, $\sum_{i=1}^{15} P(i))$ without having to evaluate $P(x)$ 15 times.

It is something of an aside, but I cannot resist mentioning a little more about differencing and polynomials. The difference operator can be iterated, i.e. higher order differences can be taken. Let $P(x)$ denote any polynomial of degree n. We know from the calculus that $d^n/dx^n(P)$ is constant and $d^{n+1}/dx^{n+1}(P)$ is identically 0. It can similarly be shown (cf.

Exercise 4, below) that $\Delta^n P$ is constant and $\Delta^{n+1}P$ is identically 0. But then $(\Delta^{n-1}P)(k)$ can be determined from $(\Delta^{n-1}P)(0)$ using the constant $\Delta^n P$. Similarly, values of $\Delta^{n-2}P$ can be found from values of $\Delta^{n-1}P$. Etc. It turns out that the complete story of P is told by the values $P(0)$, $(\Delta P)(0)$, $(\Delta^2 P)(0), \ldots, (\Delta^{n-1}P)(0)$. If we take as our representative polynomial P the polynomial given by representations (1) - (4), above, and choose representation (4), we can find these differences readily:

$$\begin{aligned} \Delta P &= 4x^{(3)} - 3x^{(2)} & (\Delta P)(0) &= 0 \\ \Delta^2 P &= 12x^{(2)} - 6x^{(1)} & (\Delta^2 P)(0) &= 0 \\ \Delta^3 P &= 24x - 6 & (\Delta^3 P)(0) &= -6 \\ \Delta^4 P &= 24 \qquad ; & (\Delta^4 P)(0) &= 24. \end{aligned}$$

If we choose the more familiar representation (1), we have to calculate $P(0), P(1), \ldots, P(4)$ to determine $(\Delta P)(0) = P(1) - P(0)$, $(\Delta P)(1) = P(2) - P(1), \ldots, (\Delta P)(3)$, and then $(\Delta^2 P)(0) = (\Delta P)(1) - (\Delta P)(0)$, $(\Delta^2 P)(1)$, $(\Delta^2 P)(2)$, and then $(\Delta^3 P)(0)$,$(\Delta^3 P)(1)$, and finally $(\Delta^4 P)(0)$. However we do it, once we have calculated $P(0)$, $(\Delta P)(0)$, $(\Delta^2 P)(0)$, $(\Delta^3 P)(0)$, and $(\Delta^4 P)(0)$, we can make as large a table of values of P as we want. Figure 1.2 illustrates the start of this table. We can complete it column-by-column using the fact that successive rows list the differences of elements in successive columns. Entry i of column $j + 1$ is merely the sum of entries i and $i + 1$ of column j— except for the bottom row, where the entry is a constant 24. Figure 1.3 illustrates the completion of the second column. Figure 1.4 gives the complete table.

1.2. Figure. Beginning a Table.

x	0	1	2	3	4	5
$P(x)$	0					
$(\Delta P)(x)$	0					
$(\Delta^2 P)(x)$	0					
$(\Delta^3 P)(x)$	−6					
$(\Delta^4 P)(x)$	24					

1.3. Figure. Adding New Entries.

x	0		1	2	3	4	5
$P(x)$	0	→	0				
		↗					
$(\Delta P)(x)$	0	→	0				
		↗					
$(\Delta^2 P)(x)$	0	→	−6				
		↗					
$(\Delta^3 P)(x)$	−6	→	18				
		↗					
$(\Delta^4 P)(x)$	24	→	24				

1.4. Figure. The Final Table.

x	0	1	2	3	4	5
$P(x)$	0	0	0	−6	0	60
$(\Delta P)(x)$	0	0	−6	6	60	180
$(\Delta^2 P)(x)$	0	−6	12	54	120	210
$(\Delta^3 P)(x)$	−6	18	42	66	90	114
$(\Delta^4 P)(x)$	24	24	24	24	24	24

Constructing a table as we have just done can be a laborious process, but it has the tremendous advantage of requiring only additions. From the middle of the 17th century onward, mechanical devices capable of performing addition were available. Mechanical multiplication had, however, never been too reliable or, when reliable, not fully automatic. Around 1819, Charles Babbage started to think seriously about the problem of automatic computation. His thoughts ran as follows:

> I considered that a machine to execute the mere isolated operations of arithmetic, would be comparatively of little value, unless it were very easily set to do its work, and unless it executed not only accurately, but with great rapidity, whatever it was required to do.
>
> On the other hand, the method of differences supplied a general principle by which *all* Tables might be computed through limited intervals, by one uniform process. Again, the method of differences required the use of mechanism for Addition only.

In his *Passages From the Life of a Philosopher*, from which comes the above quote, Babbage explains further the mechanism he intended to use to construct his *Difference Engine*. Although he made a small working model of the Difference Engine, the main machine was never completed. One problem was that engineering techniques of the day were not advanced enough for the incredibly accurate machine Babbage wanted to build. Apparently Babbage was sufficiently capable to overcome this little difficulty, but not to overcome his own perfectionism. Not only did the development of the Difference Engine undergo numerous improved restarts, but by the 1830s he had had a bolder vision: Babbage's *Analytical Engine* would have been a far more general machine— but for speed, almost a modern computer. Although Babbage never completed either project, George Scheutz, a printer from Stockholm, and his son Edward constructed a fully operational difference engine which was displayed, with Babbage's blessing, at an exhibition in England in 1854 and, through Babbage's machinations, won the Scheutzes a gold medal at the Great Exposition of 1855 in Paris. In 1856, they sold their engine to the Dudley Observatory in Albany, New York, where it served faithfully for a number of years. (Analytical engines were also eventually constructed, but that is a different story.)

Exercises.

1. For $n \geq 1$, define

$$x^{|n|} = x(x+1)\cdots(x+n-1).$$

   i. Find $\Delta(1/x^{|n|})$
   ii. Find $\Delta^{-1}(1/x^{|n|})$, for $n > 1$
   iii. Find $\sum_{k=1}^{\infty} \frac{1}{k(k+1)}$.

2. Let a be a positive real number other than 1.
   i. Find Δa^x
   ii. Find $\Delta^{-1}a^x$.
3. Let $\mathcal{P}_n$ be the set of polynomials (in one variable) of degree at most n.
   i. Viewing $\mathcal{P}_n$ as a vector space over the reals, show that $\{1, x, x^2, ..., x^n\}$ and $\{1, x, x^{(2)}, ..., x^{(n)}\}$ are bases for $\mathcal{P}_n$.
   ii. Relative to a basis $B = \{e_1, ..., e_n\}$, let the column vector

$$[v]_B = \begin{bmatrix} a_1 \\ \cdot \\ \cdot \\ \cdot \\ a_n \end{bmatrix},$$

represent $v = a_1e_1 + ... + a_ne_n$. Let $n = 4$, and let $B_1 = \{1, x, x^2, x^3\}$ and $B_2 = \{1, x, x^{(2)}, x^{(3)}\}$ be bases of $\mathcal{P}_4$ as in part i, and find an invertible matrix A such that $A[v]_{B_1} = [v]_{B_2}$, for all $v \in \mathcal{P}_4$.
   iii. Use part ii to find $P(x) = \sum_{k=1}^{n} k^3$.
4. Prove: $\Delta^n P$ is constant if P is a polynomial (in one variable) of degree n. [For a thoroughly rigorous induction on n, you might first prove that ΔP has degree $n-1$.]
5. (Finite Taylor Theorem). Let P(x) be a polynomial of degree n. Show:

$$P(x) = \sum_{k=0}^{n} \frac{(\Delta^k P)(0)}{k!} x^{(k)}.$$

[Hint: By Exercise 3.i, $P(x) = \sum_{k=0}^{n} a_k x^{(k)}$.]
6. (Summation by Parts). Let f,g be functions.
   i. Show: $(\Delta fg)(x) = (\Delta f)(x)\cdot g(x+1) + f(x)\cdot(\Delta g)(x)$.
   ii. Show:

$$\sum_{k=0}^{n} f(k)g(k) = g(n)(\sum_{k=0}^{n} f(k))) - \sum_{k=0}^{n-1} [\Delta g(k)\cdot(\sum_{i=0}^{n} f(i))].$$

   iii. Given the obvious,

$$\sum_{k=0}^{n} 1 = n+1,$$

use ii to derive

$$\sum_{k=0}^{n} k = \frac{n(n+1)}{2}.$$

   iv. Use ii, iii to derive

$$\sum_{k=0}^{n} k^2 = \frac{n(2n+1)(n+1)}{6}.$$

   v. Find $\sum_{k=0}^{n} k\cdot 2^k$.

2. Sums of Powers

The standard examples of proofs by induction— standard in the didactic sense that they are invariably given when one is teaching or reviewing proof by induction for one's students— are proofs of the summation formulae:

$$\sum_{k=0}^{n} 1 = n + 1$$

$$\sum_{k=0}^{n} k = \frac{n(n+1)}{2}$$

$$\sum_{k=0}^{n} k^2 = \frac{n(2n+1)(n+1)}{6},$$

etc. What is not generally covered is a means of finding these formulae. Exercises 3 and 6 of the preceding section gave us two techniques for this, but not the best. Although it is a bit of a digression for us— in the sequel we will only need the formula for the sum of the first powers— I should like to present a recursive method of generating the next formula from a given one. We can always view this as an exercise in induction of a more sophisticated nature.

Before beginning, it would be nice to know such polynomials exist.

2.1. Theorem. *Let $n \geq 0$. There is a polynomial $P_n(x)$ of degree $n + 1$ such that, for all positive integers x,*

$$P_n(x) = \sum_{k=1}^{x} k^n.$$

(Remark: Note that I have given 1 rather than 0 as the lower bound for the summation. Only for $n = 0$ is there any difference and, in this case, we get $P_0(x) = x$, instead of $P_0(x) = x + 1 = \sum_{k=0}^{x} 1$, a marginally less attractive polynomial.)

Proof. By Exercise 3.i of the preceding section, we can write

$$x^n = a_n x^{(n)} + a_{n-1} x^{(n-1)} + \ldots + a_1 x + a_0.$$

But then

$$\Delta^{-1} x^n = \frac{a_n}{n+1} x^{(n+1)} + \frac{a_{n-1}}{n} x^{(n)} + \ldots + \frac{a_1}{2} x^{(2)} + a_0 x$$

and, by Theorem 1.1,

$$P_n(x) = \frac{a_n}{n+1}(x+1)^{(n+1)} + \frac{a_{n-1}}{n}(x+1)^{(n)} + \ldots + \frac{a_1}{2}(x+1)^{(2)} + a_0(x+1) - a_0,$$

which is clearly a polynomial of degree $n + 1$. (Exercise: Why is $a_n \neq 0$?) QED

Throughout the rest of this section, we will fix the notation P_n for the polynomial of Theorem 2.1. To figure out how to get from P_n to P_{n+1}, it is a good idea to list the first few such polynomials. The factored forms tend not to be too revealing:

$$P_0(x) = x$$

$$P_1(x) = \frac{x(x+1)}{2}$$

$$P_2(x) = \frac{x(2x+1)(x+1)}{6}$$

$$P_3(x) = \left[\frac{x(x+1)}{2}\right]^2.$$

Well, we do see x as a common factor and we might note that if we had summed from 0 to x, we would have had $P_0(x) = x + 1$ and $x + 1$ as a common factor. These are suggestive, but don't really seem to help much— what about the other factors? the denominators?

If we multiply these out and put our polynomials into standard notation we get the following.

2.2. Table of P_n's.

$$P_0(x) = x$$

$$P_1(x) = \frac{x^2}{2} + \frac{x}{2}$$

$$P_2(x) = \frac{x^3}{3} + \frac{x^2}{2} + \frac{x}{6}$$

$$P_3(x) = \frac{x^4}{4} + \frac{x^3}{2} + \frac{x^2}{4}.$$

Somewhere along the line, the thought of differentiating these might occur.

2.3. Table of Derivatives of P_n's.

$$P_0'(x) = 1$$

$$P_1'(x) = x + 1/2$$

$$P_2'(x) = x^2 + x + 1/6$$

$$P_3'(x) = x^3 + 3x^2/2 + x/2.$$

If one now compares P_0 with P_1', P_1 with P_2', and P_2 with P_3', one ought to notice a simple relationship.

2.4. Theorem. *Let* $n > 0$. *There is a constant* b_n *such that* $P_n'(x) = nP_{n-1}(x) + b_n$.

Proof. For $n = 1, 2,$ and 3 we can verify this by comparing Tables 2.2 and 2.3. The general proof is by induction, the *strong* form of induction: we do not conclude

$$P_{n+1}'(x) = (n+1)P_n(x) + b_{n+1} \tag{1}$$

from the hypothesis

$$P_n'(x) = nP_{n-1}(x) + b_n,$$

but from a whole collection of hypotheses,

$$P_k'(x) = kP_{k-1}(x) + b_k, \quad 1 \le k \le n. \tag{2}$$

Logically, this is equivalent to deriving (1) from the single, logically more complex hypothesis

$$\forall k[k \le n \ \& \ k \ne 0 \Rightarrow \forall x[P_k'(x) = kP_{k-1}(x) + b_k]],$$

i.e., this (or: any) instance of the strong form of induction reduces to a different instance of the weak form. Logical equivalence is, however, not psychological equivalence and we apply the strong form of induction as a principle in its own right, deferring further comment on variants of induction to volume II.

Before proceeding to the derivation of (1) from (2), we interpose a little lemmatic derivation. Let m be a positive integer and let k be a non-negative one. Then:

$$\begin{aligned}
P_{k+1}(m) &= \sum_{i=1}^{m} i^{k+1} = \sum_{i=0}^{m-1} (i+1)^{k+1} = \sum_{i=0}^{m-1} \sum_{j=0}^{k+1} \binom{k+1}{j} i^j \\
&= \sum_{j=0}^{k+1} \binom{k+1}{j} \sum_{i=0}^{m-1} i^j \\
&= m + \sum_{j=1}^{k+1} \binom{k+1}{j} P_j(m-1) \\
&= 1 + \sum_{j=0}^{k+1} \binom{k+1}{j} P_j(m-1).
\end{aligned} \tag{3}$$

Since (3) holds for all positive integral m, the equation is an identity:

$$P_{k+1}(x) = 1 + \sum_{j=0}^{k+1} \binom{k+1}{j} P_j(x-1). \tag{4}$$

Equation (4) is almost a basis for an inductive proof of (1) in that it relates P_{k+1} with $P_0, \ldots, P_k$. Unfortunately, P_{k+1} appears on both sides of the equation. If we bring both occurrences of P_{k+1} to the left, we get

$$P_{k+1}(x) - P_{k+1}(x-1) = 1 + \sum_{j=0}^{k} \binom{k+1}{j} P_j(x-1),$$

i.e.,

$$x^{k+1} = 1 + \sum_{j=0}^{k} \binom{k+1}{j} P_j(x-1). \tag{5}$$

Replacing k by $n + 1$ in (5) and solving for P_{n+1} we get

$$P_{n+1}(x-1) = \frac{1}{n+2}\left[x^{n+2} - 1 - \sum_{j=0}^{n} \binom{n+2}{j} P_j(x-1)\right]$$

whence replacement of $x - 1$ by x yields

$$P_{n+1}(x) = \frac{1}{n+2}\left[(x+1)^{n+2} - 1 - \sum_{j=0}^{n} \binom{n+2}{j} P_j(x)\right]. \tag{6}$$

With (6) we have P_{n+1} expressed in terms of $P_0, \ldots, P_n$. We can now assume (2) and differentiate:

$$P_{n+1}'(x) = \frac{1}{n+2}\left[(n+2)(x+1)^{n+1} - \sum_{j=0}^{n} \binom{n+2}{j} P_j'(x)\right]$$

$$= (x+1)^{n+1} - \frac{1}{n+2} P_0'(x) - \frac{1}{n+2}\sum_{j=1}^{n} \binom{n+2}{j} j\, P_{j-1}(x) + \text{constant}$$

$$= (x+1)^{n+1} - \frac{1}{n+2}\sum_{j=1}^{n} \binom{n+2}{j} j\, P_{j-1}(x) + \text{constant}$$

$$= (x+1)^{n+1} - \sum_{j=1}^{n} \binom{n+1}{j-1} P_{j-1}(x) + \text{constant} \tag{7}$$

$$= (x+1)^{n+1} - \sum_{j=0}^{n-1} \binom{n+1}{j} P_j(x) + \text{constant} \tag{8}$$

$$= (x+1)^{n+1} - 1 - \sum_{j=0}^{n-1} \binom{n+1}{j} P_j(x) + \text{constant}$$

$$= (n+1)P_n(x) + \text{constant}, \tag{9}$$

where (7) follows by the simple identity,

$$\binom{n+2}{j} j = \frac{(n+2)!}{(n+2-j)!\,j!} j = \frac{(n+2)(n+1)!}{(n+1-(j-1))!\,(j-1)!} = (n+2)\binom{n+1}{j-1},$$

(8) follows by a change in the index of summation, and (9) follows via (6) using n in place of $n + 1$. As equation (9) is of the desired form (1), we have completed the proof by induction. QED

Theorem 2.4 almost gives the passage from P_{n-1} to P_n. What remains is the determination of the constants b_n.

2.5. Corollary. *Let* $n > 0$. $b_n = 1 - \int_0^1 nP_{n-1}(t)\,dt$.

Proof. By the same argument used to prove Theorem 2.1, there is a polynomial $Q_n(x)$ such that, for all natural numbers m,

$$Q_n(m) = \sum_{k=0}^{m} k^n,$$

the sum now starting at 0. As $Q_n(m) = P_n(m)$ for all positive integral m, we have identity and

$$P_n(0) = Q_n(0) = \sum_{k=0}^{0} k^n = 0.$$

But then

$$P_n(x) = P_n(x) - P_n(0) = \int_0^x nP_{n-1}(t) + b_n\,dt,$$

by 2.4, whence

$$1 = P_n(1) = \int_0^1 nP_{n-1}(t)\,dt + b_n,$$

whence

$$b_n = 1 - \int_0^1 nP_{n-1}(t)dt.$$

QED

With 2.4 and 2.5 we have a complete determination of P_n from P_{n-1}. But why stop here? Perhaps a little extra work will yield a more efficient determination of the b_n's? Bearing in mind that $P_n(0) = 0$, we can iterate the application of Theorem 2.4 and create a short list of P_n's:

$P_0(x) = x$

$P_1(x) = \int(x + b_1)dx = x^2/2 + b_1x$

$P_2(x) = \int(x^2 + 2b_1x + b_2)dx = x^3/3 + b_1x^2 + b_2x$

$P_3(x) = \int(x^3 + 3b_1x^2 + 3b_2x + b_3)dx = x^4/4 + b_1x^3 + (3/2)b_2x^2 + b_3x$

$P_4(x) = x^5/5 + b_1x^4 + 2b_2x^3 + 2b_3x^2 + b_4x$

$P_5(x) = x^6/6 + b_1x^5 + (5/2)b_2x^4 + (10/3)b_3x^3 + (5/2)b_4x^2 + b_5x.$

The sequence of coëfficients of the b_kx^{n+1-k} terms are not as revealing as those inside the integrands, i.e. the sequence resulting if we multiply P_n by $n + 1$:

1			
1	2		
1	3	3	
1	4	6	4

etc.

These are just binomial coëfficients, the missing last term corresponding to the new b_{n+1}. This yields the immediate conjecture:

2.6. Theorem. *Let n > 0. Then*

$$P_n(x) = \frac{1}{n+1}\left[x^{n+1} + \binom{n+1}{n}b_1x^n + \ldots + \binom{n+1}{n+1-i}b_i x^{n-i+1} + \ldots + \binom{n+1}{1}b_n x\right].$$

Proof. By induction on n. For the basis, we have the above examples. For the induction step we will use Theorem 2.4. Assume

$$P_n(x) = \frac{1}{n+1}\left[x^{n+1} + \sum_{k=1}^{n} \binom{n+1}{n+1-k} b_k x^{n+1-k}\right]$$

and observe

$$(n+1)P_n(x) + b_{n+1} = x^{n+1} + \sum_{k=1}^{n} \binom{n+1}{n+1-k} b_k x^{n+1-k} + b_{n+1}$$

$$= x^{n+1} + \sum_{k=1}^{n+1} \binom{n+1}{n+1-k} b_k x^{n+1-k}.$$

Integrating both sides yields

$$P_{n+1}(x) = \frac{x^{n+2}}{n+2} + \sum_{k=1}^{n+1} \binom{n+1}{n+1-k} b_k \frac{x^{n+2-k}}{n+2-k}. \qquad (*)$$

But $\binom{n+1}{n+1-k}\frac{1}{n+2-k} = \frac{1}{n+2}\binom{n+2}{n+2-k}$, whence (*) yields

$$P_{n+1}(x) = \frac{1}{n+2}\left[x^{n+2} + \sum_{k=1}^{n+1} \binom{n+2}{n+2-k} b_k \frac{x^{n+2-k}}{n+2-k}\right]. \qquad \text{QED}$$

2.7. Corollary. *Let n > 0.*

i. $n = \sum_{k=1}^{n} \binom{n+1}{k} b_k$

ii. $b_n = \frac{1}{n+1}\left[n - \sum_{k=1}^{n-1} \binom{n+1}{k} b_k\right].$

Proof. i. By 2.6.

ii. By part i and the fact that $\binom{n+1}{n} = n$. QED

These numbers b_n, occasionally replaced by their absolute values or with their indices shifted, are called *Bernoulli numbers* after Jakob Bernoulli, who first discovered them and stated Theorem 2.6 in his *Ars Conjectandi* (1713). They pop up in old fashioned books on advanced calculus and in the calculus of finite differences as well as in analytic number theory. Thus, it pays to be aware of them and I hope the reader will forgive our seeming over-involvement in them. The exercises include a couple of other recursive relations involving the Bernoulli numbers. Separating us from these exciting exercises is the following table.

2.8. Table of Bernoulli Numbers.

$b_1 = 1/2$	$b_2 = 1/6$	$b_3 = 0$	$b_4 = -1/30$	$b_5 = 0$
$b_6 = 1/42$	$b_7 = 0$	$b_8 = -1/30$	$b_9 = 0$	$b_{10} = 5/66$.

The reader might enjoy extending this Table a bit to see if any pattern he notices continues.

Exercises.

1. i. Let $n > 0$. Show: $x + 1 \mid P_n$, i.e., $x + 1$ divides P_n.

   ii. Show: For $n > 0$, $\sum_{k=1}^{n} \binom{n+1}{k} b_k(-1)^{k+1} = 1$.

   [Hints: i. Let $Q_n(x) = \sum_{k=-1}^{x} k^n$ and compare $Q_n(-1)$ to $P_n(-1)$; ii. apply Theorem 2.6 to $P_n(-1)$.]

2. Show: For $n > 0$, $b_n = \frac{1}{n+1}\left[1 - \sum_{k=1}^{n-1} \binom{n}{k-1} b_k\right]$.

   [Hint: Use 2.7.i and the identity $\binom{n+1}{k} = \binom{n}{k} + \binom{n}{k-1}$.]

3. Show that $b_k = 0$ for $k > 1$ odd. [Hint: Apply Theorem 2.6 to $1 = P_n(1) + P_n(-1)$ for $n > 0$ to show, for even n,

$$\sum_{k=3,k \text{ odd}}^{n} \binom{n+1}{k} b_k = 0.$$

   Then use induction.]

3. The Cantor Pairing Function

Cantor invented the infinite in much the same way that Newton and Leibnitz invented the calculus, that is to say that, just as everyone was finding slopes and areas before Newton and Leibnitz systematised operations, plenty of people were working with the infinite before Cantor came along. There is a gross inexactness in this analogy: whereas most mathematicians prior to Newton and Leibnitz knew what they were about, the same can only be said of a handful of Cantor's precursors and contemporaries— e.g., Bernard Bolzano, Richard Dedekind, and Constantin Gutberlet. One of the earliest to accept the infinite and the fact that infinity behaves differently from finity was Galileo Galilei, familiarly called Galileo. In his *Dialogues Concerning the Two New Sciences* (1638), during the dialogue of the first day, he came to grips with the infinite. If we listen in we hear an amusing confusion. At one point, he discovers a paradox of measure— infinite sets can have measure 0— and misinterprets this. A bit later, he notes the one-one correspondence between positive integers and the squares thereof, but, with a little thought and the measure paradox draws the "wrong" conclusion. I cannot resist going a bit into these matters.

Galileo's paradox of measure— unrecognised as such by him— can be cast into modern terms as follows. Consider first the solid of revolution obtained by revolving the area below the straight line $y = 1 - x$ around the x-axis between the limits $x = r < 1$ and $x = 1$. It has volume

$$V_1(r) = \int_r^1 \pi(1 - x)^2\, dx.$$

Next, consider the volume of the solid obtained by revolving the area trapped between the curves $y = \sqrt{1 - (x - 1)^2}$ and $y = 1$ over the same interval. It has volume

$$V_2(r) = \int_r^1 \pi 1^2 - \pi(1 - (x - 1)^2)\, dx = \int_r^1 \pi(x - 1)^2\, dx.$$

Clearly $V_1(r) = V_2(r)$, for all r. But, as r tends to 1, V_1 becomes the volume of a point and V_2 that of a circle, i.e., both become 0. What does Galileo conclude? Answer: "It appears therefore that we may equate the circumference of a large circle to a single point... Hence in conformity with the preceding we may say that all circumferences of circles, however different, are equal to each other, and are equal to a single point."

A little later in the same dialogue, he notes the one-one correspondence between positive integers and squares, as well as the increasing sparseness of the squares: "... the larger the number to which we pass, the more we recede from infinity, because the greater the numbers the fewer are the squares contained in them; but the squares in infinity cannot be less than the totality of all the numbers, as we have just agreed; hence the approach to greater and greater numbers means a departure from infinity." Conclusion: "we are led to conclude that the attributes 'larger', 'smaller', and 'equal' have no place either in comparing infinite quantities with each other or in comparing infinite with finite quantities."

Galileo's *Dialogues* were written in the 1630s and appeared in 1638. By the third quarter of the 19th century, the understanding of the infinite had improved. The small end of the infinite spectrum— the infinitesimal— was being banished by the work of Cantor, Cauchy, Dedekind, Weierstrass, and others; the large end of this spectrum was beginning to achieve a certain amount of acceptance. A highly respectable example is afforded by Dedekind, who in his *Was sind und was sollen die Zahlen?* (written 1872 - 1878, published 1887), defined a set to be infinite just in case it could be put into one-one correspondence with a proper subset of itself. And, of course, there was Georg Cantor.

If Cantor was not the first mathematician to deal with the infinite or to have genuine insights thereinto, he was the first to devote himself wholeheartedly to it and to make fundamental distinctions; moreover, despite a growing interest in the subject on the part of others, he almost single-handedly developed a coherent theory of the infinite. For this achievement, he is nowadays deemed a great mathematician. In his own day, he was not so highly honoured: he taught in Halle, not in Berlin where a man of his calibre belonged, and was even pronounced a "Corrupter of Youth" (Verderber der Jugend) by his former friend and ultimate ideological opponent Leopold Kronecker. The error of Cantor's ways was that he was too original too early in his career and his mathematics was entirely too strange to the likes of the politically powerful Kronecker who, nowadays, is so overwhelmingly remembered for his opposition to Cantor that most mathematicians could not say what his positive contributions to mathematics were. History has proven an even harsher judge than this: Cantor was plagued by mental illness and even went to an asylum; in modern popular accounts, Kronecker is unfairly blamed for this.

What did Cantor do to arouse the opposition of Kronecker? To answer this one must digress to discuss the nature of mathematics. Obviously, I cannot do a competent job of such a discussion in the space available. It is, moreover, a twice-told tale, the second incarnation coming in the 1920s, with the final (?) outcome to be discussed in Chapter IV of volume II of this work. Thus, I shall here be most brief. Different mathematicians look for different things in mathematics. To some the search for TRUTH is all important. Two antithetical paths to truth are the *Secure* and the *Daring*. The Secure Path is the strait and narrow one associated with names like Leopold Kronecker, Luitzen Egbertus Jan Brouwer, Errett Bishop, and Morris Kline. Its followers declare truth to be that which can be established with no metaphysical speculation; they decry abstraction and are generally constructivists. The Daring Path is more in line with modern philosophies and is associated with names like Georg Cantor, David Hilbert, and René Thom. The Daring Path is not followed so much as found; its bold pathfinders make metaphysical assumptions, take risks, and invent new fields of mathematics. In short, the securists try to approximate truth from below, while the daredevils come down from above. The former may never reach the full truth, but they will not be in error; the latter will achieve a great deal more, but at the risk of seeing everything collapse when they derive $0 = 1$.

In one sense, the mathematical daredevil is a recent phenomenon, one of 19th and 20th century mathematics when the subject finally reached a level where abstraction was possible. One can, however, find parallels in prior centuries: the scholastics, clinging to their Aristotle and the holy scriptures, are the analogues of the security-minded constructivists and the men of daring were Galileo or even John Dee, one of the Illuminati of the spiritualists, whose Preface to the first English translation of Euclid's *Elements* contains a pitiable plea for understanding. To go on and liken Kronecker to the Spanish Inquisition is to overstate his power over Cantor as well as his malevolence, but it may not unfairly depict his attitude. The religious component of mathematical practice is not much talked about, but it is there. Kronecker felt the mathematician must stick to the integers which God made, and condemned everything else, especially Cantor's mathematics of infinity; Cantor, on the other hand, strove for the Church's approval of his work on the infinite.

But I find my brief digression lengthening and I should get back to the mathematics itself. I asked what Cantor did to arouse the ire of Kronecker, and the answer is that he created a theory of the infinite— with lots of *actual* infinities. Prior to the 19th century, mathematical infinity could be taken to be a *potential* infinity: an infinite set was never quite complete; it was an unfolding process— never ending, never endable. With Dedekind and Cantor this was no longer the case. Infinite sets existed as completed totalities. Moreover, under Cantor there were differents modes of infinity and different infinities.

Unlike Galileo, who was confused about different aspects of infinity and drew the conclusion that one could not compare size when infinity was involved, Cantor saw clearly that there were two separate aspects of infinity— ordinality and cardinality. This dichotomy leads to two types of infinite (or, as one likes to say: transfinite) numbers, namely, the ordinal and cardinal numbers. Both are studied in naïve and axiomatic set theory, and Cantor's ordinal numbers are of particular usefulness and interest in the advanced logical study of the arithmetic of the integers. Nonetheless, it is his infinite cardinal numbers that interest us here.

The basis of Cantor's theory of cardinality is the notion of one-one correspondence.

3.1. Definition. Let A, B be sets. A function $F{:}A \to B$ is a *one-one correspondence* between A and B if F is one-one and onto. If there is a one-one correspondence between A and B, we say A and B have the same *cardinality*.

If A and B possess the same cardinality, one may say also that A, B are *equi-cardinal* or even *equipollent*.

Note that Definition 3.1 does not say what a cardinal number is, but merely when two sets have the same cardinality whatever it may be. In the spirit of abstraction, one can postulate the existence of cardinal *numbers* representing these cardinalities, or one can look for specific representatives of given classes of equicardinal sets. The paradigm of this latter approach is the use of the set of natural numbers itself as the cardinality of the set.

3.2. Definition. The set of non-negative integers is called the set of *natural numbers* and is denoted by N or ω, thus:

$$N = \omega = \{x \in Z:\ x \geq 0\},$$

where Z denotes the set of integers. A set X is *denumerable* (or: *denumerably infinite*) if X is equicardinal with ω. X is *countable* if X is finite or denumerable.

If X is denumerable, it has the same cardinality as ω and we may write ω as the cardinal number of X. This is certainly done in most set theory books; if we do not formally do so here, it is simply because, with minor exception, ω is the only cardinal number that will interest us in this book.

There are three things Cantor accomplished with his cardinal numbers. In order of increasing importance to us, i.e., to the narrow goal of this book, they are:

i. he defined operations of addition, multiplication, and exponentiation on cardinal numbers;
ii. he showed that cardinality is non-trivial, i.e., that there are many distinct cardinal numbers; and
iii. despite this plurality, he calculated the cardinalities of various sets, i.e., he showed several sets to be denumerable and some sets to be equicardinal with the set of real numbers.

The first of these is easily dispensed with. Given two cardinal numbers μ and ν, choose two sets A, B of these respective cardinalities and define $\mu + \nu$, $\mu \cdot \nu$, and μ^{ν} to be the cardinal numbers of the sets

$$(A \times \{0\}) \cup (B \times \{1\})$$

$$A \times B = \{(a,b): a \in A \ \& \ b \in B\}$$

$$A^B = \{F: F \text{ is a function from } B \text{ to } A\},$$

respectively. It is fairly easy to show these operations to be well-defined and to derive a few of their simpler properties (e.g., commutativity and associativity of + and ·). (Cf. the Exercises, below.)

Cantor gave several proofs of the non-triviality of the notion of cardinality. The most famous of these is a proof of the uncountability of the set of real numbers, or, indeed, of the unit interval [0,1].

3.3. Theorem. *The closed unit interval is non-denumerable.*

Proof. Cantor's original proof is outlined in the Exercises (cf. Exercise 4). Here I shall present his later and more popular version.

Every real number $\boldsymbol{r} \in [0,1]$ can be written as an infinite decimal $\boldsymbol{r} = r_0 r_1 r_2 \ldots$. Moreover, if we agree never, except in the case $1 = .999...$, to allow an infinite string of successive 9's, this decimal expansion is unique, i.e., [0,1] is in one-one correspondence with those infinite sequences $(r_0, r_1, \ldots)$ of digits which are not of the form $(r_0, r_1, \ldots, r_{k-1}, 9, 9, \ldots)$, except for the $(9, 9, 9, \ldots)$ sequence which represents 1.

Suppose we had a one-one correspondence F of ω with these sequences:

$$\begin{aligned} r_0 &= F(0) = (r_{00}, r_{01}, r_{02}, \ldots) \\ r_1 &= F(1) = (r_{10}, r_{11}, r_{12}, \ldots) \\ r_2 &= F(2) = (r_{20}, r_{21}, r_{22}, \ldots) \\ &\vdots \end{aligned}$$

Such a correspondence cannot exist because the range of F omits the sequence

$$\boldsymbol{r} = (r_{00} \pm 1, r_{11} \pm 1, r_{22} \pm 1, \ldots),$$

obtained by going down the *diagonal* $r_{00}, r_{11}, r_{22}, \ldots$ and altering each entry by adding or subtracting 1, the choice being made to avoid getting a 9, e.g. by choosing +1 if $r_{ii}+1$ is not 9 and −1 otherwise.

QED

If I have been a bit sketchy in this last proof, it is partly because I am sure the reader has seen this before and partly because I want to elaborate outside the confines of a proof. The argument given is called a *diagonal argument*, or *Cantor's Diagonal Argument*, although it actually originated with Paul du Bois-Reymond. Be that as it may, the crucial thing is its procedure. Given a denumerable list of denumerable sequences, one writes down (or, imagines doing so) an infinite matrix, the rows of which are these sequences, and then goes down the diagonal to produce a new sequence not on the list by taking these diagonal elements and changing them. The new sequence differs from the 0th sequence in the 0th position, the 1st sequence in the 1st position, etc. This diagonal construction is very powerful and we will encounter it in various incarnations in the sequel.

As I said, however, Cantor's contribution of most immediate interest to us is his calculation of the cardinalities of specific sets. Given the notion of cardinality, this is the obvious sort of thing to do and yet it was revolutionary. Galileo's one-one correspondence of the natural numbers with their squares had been paradoxical, leading him more-or-less to the dismissal of infinite cardinality; Cantor went further and accepted what he found. In "Ein Beitrag zur Mannigfaltigkeitslehre", published in 1878, Cantor established a one-one correspondence between the real numbers in the unit interval and the pairs thereof in the unit square $[0,1] \times [0,1]$, thereby raising for the first time the problem of dimension: if the points in 1-dimensional space can be so corresponded with those in 2-dimensional space, are 1- and 2-dimensional spaces really different? When, later, Giuseppe Peano showed that one can even find a continuous function mapping $[0,1]$ onto $[0,1] \times [0,1]$, the problem became acute. The introduction of the topological notion of connectivity allows one to prove that such a function cannot be continuous, one-one, and onto: removing a single point (other than endpoints) from the interval disconnects it; no such removal disconnects the square. The student who has had the standard undergraduate introductory course in real analysis will see this immediately. That $[0,1] \times [0,1]$ does not bi-continuously one-one correspond with $[0,1] \times [0,1] \times [0,1]$ should not, however, be so obvious. The general problem of invariance of dimension was first solved by L.E.J. Brouwer, whom we shall meet again in Chapter IV of volume II, in 1911, and is usually taught in a graduate course on algebraic topology.

So we see just how revolutionary Cantor's cardinality calculations were. Not being so revolutionary myself, I prefer to discuss a more discrete analogue to Cantor's real pairing function, namely his pairing function on the natural numbers. Those curious readers who would like to see a real pairing function are referred to the Exercises where they can find a number of such calculations.

Cantor worked with positive integers and pairs thereof, and, in his paper already cited, simply noted along the way that

$$P(x,y) = x + \frac{(x+y-1)(x+y-2)}{2}$$

maps the set of pairs of positive integers one-one onto the set of positive integers. Adjusting for our choice of the natural numbers as a universe of discourse, this function becomes

$$P(x,y) = \left[(x+1) + \frac{((x+1)+(y+1)-1)((x+1)+(y+1)-2)}{2}\right] - 1$$

$$= x + \frac{(x+y+1)(x+y)}{2} = \frac{(x+y)^2+3x+y}{2}.$$

Let us declare our interest in this function formally:

3.4. Definition. The function

$$\langle x,y\rangle = \frac{(x+y)^2+3x+y}{2}$$

is called the *Cantor pairing function.*

In the sequel, we shall drop the eponymous adjective and refer to $\langle\cdot,\cdot\rangle$ simply as *the* pairing function. Some justification for the "the" will be given in the next two sections; justification for the words "pairing function" is given by the following.

3.5. Theorem. *$\langle\cdot,\cdot\rangle$ is a one-one correspondence from $\omega \times \omega$ onto ω.*

The most heuristic proof of Theorem 3.5 proceeds by first exhibiting such a one-one correspondence and then calculating this correspondence and discovering that it is the function we are calling the pairing function. There are two natural such correspondences— or, rather, two ideas behind such correspondences and two implementations of each. To describe these, we introduce some unnecessary, but pleasant, terminology.

3.6. Definition. A point $(x_0, \ldots, x_{n-1})$ in R^n, R being the set of reals, is a *lattice point* if $x_0, \ldots, x_{n-1}$ are all integral. In other words, for Z the set of integers, the points in Z^n are called lattice points.

The use of the letters Z and R, in some type-face or other, to denote the sets of integers and reals, respectively, is standard and is a convention that will be adhered to in this book.

The set $\omega \times \omega$ consists of the lattice points of the first quadrant (axes included). There are two natural enumerations of these lattice points— the *box enumeration* and the *diagonal enumeration.* The box enumeration proceeds by listing the pairs on the sides of progressively larger squares in, say, a clockwise march around the semi-perimeters. Thus, it lists

$$(0,0),\ (0,1),\ (1,1),\ (1,0),\ (0,2),\ (1,2),\ (2,2),\ (2,1),\ (2,0),\ \ldots$$

(cf. Figure 3.7.i). The diagonal enumeration proceeds by listing the entries on the progressively larger diagonals, say, going down from left to right (cf. Figure 3.7.ii). Thus, it lists

$$(0,0),\ (0,1),\ (1,0),\ (0,2),\ (1,1),\ (2,0),\ (0,3),\ \ldots$$

3.7. Figure. The Box and Diagonal Enumerations.

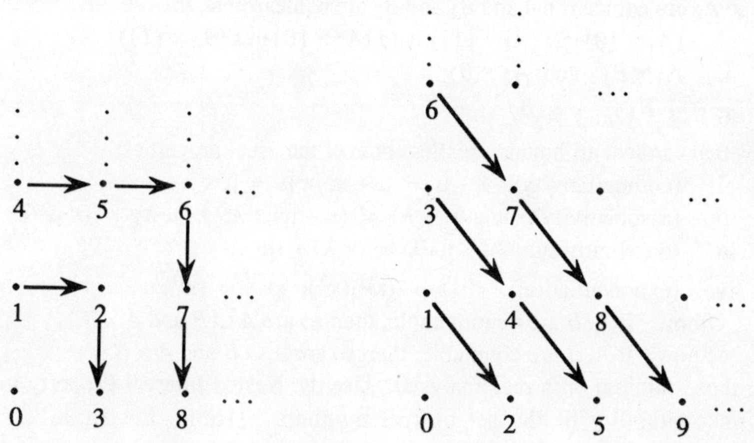

i. Box Enumeration ii. Diagonal Enumeration

The diagonal enumeration from ω to ω × ω is the function inverse to the Cantor pairing function. To see this, let (x_0,y_0) be a lattice point in the first quadrant and let us consider how many points precede it in the enumeration. First, (x_0,y_0) lies on the line with equation $x + y = x_0 + y_0$. Hence, before reaching this line one goes through

$$\sum_{k=0}^{x_0+y_0-1} (k+1) = \sum_{k=0}^{x_0+y_0-1} k + \sum_{k=0}^{x_0+y_0-1} 1$$

$$= \frac{(x_0+y_0-1)(x_0+y_0)}{2} + (x_0+y_0)$$

$$= \frac{(x_0+y_0+1)(x_0+y_0)}{2}$$

points, since there are $k+1$ points on each line $x+y = k$. Now, having reached the diagonal on which (x_0, y_0) lies, to get to (x_0, y_0) we must pass through $(0, x_0+y_0)$, $(1, x_0+y_0-1)$, ..., (x_0-1, y_0+1). There are thus x_0 points preceding (x_0, y_0) in this part of the enumeration, whence the total number of points preceding (x_0, y_0) in the whole enumeration is

$$\frac{(x_0+y_0+1)(x_0+y_0)}{2} + x_0 = \frac{(x_0+y_0)^2 + 3x_0 + y_0}{2}$$

$$= \langle x_0, y_0 \rangle.$$

Since we start our counting at 0, $\langle x_0, y_0 \rangle$ is the natural number mapped onto (x_0, y_0) by the diagonal enumeration and the Cantor pairing function is indeed the inverse to this enumeration. Hence the Cantor pairing function is a one-one correspondence between ω × ω and ω and Theorem 3.5 is established.

Exercises.

1. Prove the well-definedness of the cardinal arithmetic operations, i.e., prove that if A_1 and A_2 are equicardinal and B_1 and B_2 are equicardinal, then so are
   i. $(A_1 \times \{0\}) \cup (B_1 \times \{1\})$ and $(A_2 \times \{0\}) \cup (B_2 \times \{1\})$
   ii. $A_1 \times B_1$ and $A_2 \times B_2$
   iii. $A_1{}^{B_1}$ and $A_2{}^{B_2}$.
2. Show that cardinal arithmetic satisfies some of the usual properties:
   i. (commutativity). $\kappa + \mu = \mu + \kappa$; $\kappa \cdot \mu = \mu \cdot \kappa$
   ii. (associativity). $\kappa + (\mu + \lambda) = (\kappa + \mu) + \lambda$; $\kappa \cdot (\mu \cdot \lambda) = (\kappa \cdot \mu) \cdot \lambda$
   iii. (distributivity). $(\kappa + \mu) \cdot \lambda = (\kappa \cdot \lambda) + (\mu \cdot \lambda)$
   iv. (exponentiation). $\kappa^{\mu+\lambda} = (\kappa^{\mu})(\kappa^{\lambda})$; $\kappa^{\mu\lambda} = (\kappa^{\mu})^{\lambda}$.
3. i. Show: If A, B are denumerable, then so are $A \cup B$ and $A \times B$.
   ii. Show: If A, B are countable, then so are $A \cup B$ and $A \times B$.
4. (For those familiar with real analysis). Use the Nested Interval Property to show the uncountability of the set of real numbers. [Hint: Let $r_0, r_1, \ldots$ be an enumeration of the reals. Successively choose $I_0 \supseteq I_1 \supseteq \ldots$ so that each I_n is a closed interval of length $1/3^n$ and $r_n \notin I_n$.]
5. (Cantor-Schroeder-Bernstein Theorem). We write $\kappa \le \lambda$ if for some sets A, B of cardinalities κ, λ, respectively, there is a one-one function $F{:}A \to B$.
   i. Let $G{:}A \to B$, $H{:}B \to A$ be one-one. Let $a \in A$. Define the sequences a_n, b_n as follows:
   $$a_0 = a, \qquad b_0 = G(a_0)$$
   $$a_{n+1} = H(b_n), \qquad b_{n+1} = G(a_{n+1}).$$
   Further, define a_{-n}, b_{-n} by:
   $$b_{-n-1} = H^{-1}(a_{-n}), \text{ provided it exists}$$
   $$a_{-n-1} = G^{-1}(b_{-n}), \text{ provided it exists.}$$
   With all this, define the *trace* of a to be the set
   $$Tr(a) = \{a_n, b_n{:}\ n \in Z \text{ and } a_n, b_n \text{ exist}\}.$$
   What can the traces look like?
   ii. Given G, H as in part i, show that there is a one-one correspondence $F{:}A \to B$. Conclude: $\kappa \le \lambda$ & $\lambda \le \kappa \Rightarrow \kappa = \lambda$.
6. Use the Cantor-Schroeder-Bernstein Theorem (Exercise 5) to show the following:
   i. $2^{\omega} = \omega^{\omega} = (\omega^{\omega})^{\omega}$
   ii. $\mathrm{card}(Q) = \mathrm{card}(\omega \times \omega) = \omega$, where Q is the set of rational numbers and $\mathrm{card}(X)$ denotes the cardinality of a set X
   iii. $\mathrm{card}(R) = \mathrm{card}(R \times R)$
   iv. $\mathrm{card}(R) = \omega^{\omega} = 2^{\omega}$
   v. $\omega^{<\omega} = \{$finite sequences of natural numbers$\}$ has cardinality ω.
7. Show that there are exactly 2^{ω} pairing functions mapping $\omega \times \omega$ one-one onto ω. [Hint: Use Exercise 5.]
8. In some courses in number theory, one is introduced to Farey fractions, the enumeration of all rationals in [0,1] given by:

   0/1, 1/1, 1/2, 1/3, 2/3, 1/4, 3/4, ...

This enumeration proceeds by listing in order all fractions with denominator 1, then those with denominator 2 which do not reduce to fractions earlier in the list, etc. If we add the fictitious 0/0 and reïnstate the unreduced fractions, we get

0/0, 0/1, 1/1, 0/2, 1/2, 2/2, 0/3, 1/3, 2/3, 3/3, ...

Identifying y/x with (x,y), this enumerates the lattice points of the sector $\{(x,y)\colon\ 0 \le y \le x\}$. Find the inverse $P(x,y)$ mapping this sector onto ω:

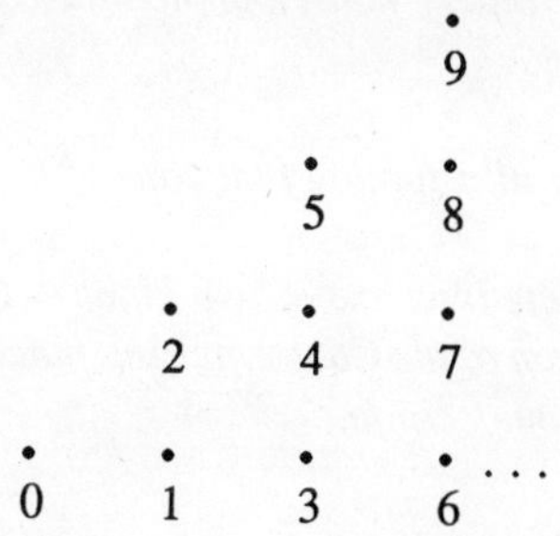

9. Show that

$$P(x,y) = \max\{x^2,y^2\} + x + \begin{cases} 0, & x \le y \\ x - y, & y < x \end{cases}$$

is inverse to the box enumeration of Figure 3.7.i.

10. One can iterate the pairing function to obtain tripling, quadrupling, ... general n-tupling functions, e.g., $\langle x, y, z \rangle = \langle x, \langle y, z\rangle\rangle$. It will not have escaped the reader's notice that the Cantor pairing function is a polynomial, in fact, a quadratic. This iteration yields high degree polynomials, in our example a 4th degree tripling function. Find a cubic tripling polynomial. [Hint: Enumerate the lattice points of successive planes $x + y + z = constant$ in the first orthant.]

11. Let $P(x,y)$ be the Cantor pairing function. Show: $x, y \le P(x,y)$ for all natural numbers x, y.

4. The Fueter-Pólya Theorem, I

The reader who has not lost sight of the landscape of polynomials during our little excursion into the Infinite will have noted that the Cantor pairing function is a quadratic polynomial. If he has faithfully worked on Exercise 10 of the previous section, he will have constructed a cubic tripling function and, perhaps, surmised the existence of a 4th degree quadrupling function. As apparently first noticed by Thoralf Skolem, the construction of the quadratic and cubic pairing and tripling functions generalises. For each n, there is an n-th degree n-tupling function. In fact, by permuting the variables, one sees that there are $n!$ such functions. In Exercise 10 we also saw how, given the pairing function, we could generate other tripling functions of degree 4. This generalises: one can compose pairing, tripling, etc. functions to produce quite a variety of polynomial n-tupling functions for larger n. To date, these are the only polynomial n-tupling functions known;

in particular, the only known polynomial pairing functions are the Cantor pairing function and the polynomial obtained by permuting the variables in it.

4.1. Conjecture. *The only polynomials* P *with real coëfficients which map* $\omega\times\omega$ *one-one onto* ω *are the Cantor pairing function* $\langle x, y\rangle$ *and its flip* $\langle y, x\rangle$.

4.2. Conjecture. *For each n, there are only finitely many polynomials establishing a one-one correspondence between* ω^n *and* ω.

4.3. Conjecture. *For each n, there are exactly n! n-tupling functions of degree n.*

4.4. Conjecture. *The polynomial n-tupling functions are of two kinds— the n! ones of degree n obtained by generalising the construction of the Cantor pairing function and those built up from lower order tupling functions by composition.*

In all these conjectures, *n* is assumed at least 2.

As the reader may conclude from the label "Conjecture" affixed to each of these assertions, their truth or falsity is unknown. Why should we believe in these conjectures? Evidence against them includes the very nearly polynomial box enumeration (Exercise 9 of the preceding section) and the existence of 2^{ω} pairing functions (Exercise 7 of that section). Further negative evidence will be the amazing coding tricks involving polynomials that we will encounter in the next Chapter. Positive evidence is the partial success of work in the direction of Conjecture 4.1 and the fact that 4.2 - 4.4 are modest generalisations thereof.

It was a number theorist, Rudolf Fueter, who first questioned the unicity of the Cantor pairing function. In a letter to the analyst Georg Pólya, he related how he had come across Cantor's function in reading an analysis text, asked himself if there were any other quadratic pairing functions, and proved (the proof being included in the letter) that there were none which mapped the origin (0,0) to 0, i.e., there was no enumeration of $\omega \times \omega$ with quadratic inverse that started at the origin— other than Cantor's function and its flip. Pólya responded with a new proof which did not need the extra restriction that (0,0) map to 0. They published this correspondence in a paper entitled "Rationale Abzählung der Gitterpunkte" ("Rational enumeration of lattice points") in 1923. Let me state their result formally:

4.5. Fueter-Pólya Theorem. *Let* $P(x, y)$ *be a quadratic polynomial with real coëfficients such that, restricted to* $\omega \times \omega$, P *maps* $\omega \times \omega$ *one-one onto* ω. *Then* P *is either the Cantor pairing function or its flip, i.e., either*

i. $\forall xy\ P(x, y) = \langle x, y\rangle$; *or*

ii. $\forall xy\ P(x, y) = \langle y, x\rangle$.

Elementary number theory abounds in simple results which can be stated comprehensibly to a layman and yet require (or: *seem* to require) advanced concepts for their proofs. The Fueter-Pólya Theorem is such an example from elementary logical number theory. Since one of the goals of logical number theory is the delineation, necessarily inexact, of those results which are accessible to more-or-less elementary techniques and those which are not, I have deemed it instructive to consider this result. Elementary methods simply don't go very far and the known proofs use quite advanced concepts.

We shall begin by trying to give a direct proof of the Fueter-Pólya Theorem. To this end, let $P(X,Y)$ be given and write P in the form

$$P(X, Y) = AX^{(2)} + BXY + CY^{(2)} + DX + EY + F,$$

with real coëfficients A, B, C, D, E, F and $X^{(2)} = X(X-1)$, $Y^{(2)} = Y(Y-1)$ as in section 1, above. For convenience, I use capital X, Y to denote the variables.

4.6. Lemma. *$2A, B, 2C, D, E, F$ are integers, i.e.*

$$P(X, Y) = \frac{a}{2}X^{(2)} + bXY + \frac{c}{2}Y^{(2)} + dX + eY + f, \qquad (*)$$

where a, b, c, d, e, f are integers.

Proof. $F = P(0,0)$ is integral.

$D = P(1,0) - P(0,0) = \Delta_X P(0,0)$ is integral.

$E = P(0,1) - P(0,0) = \Delta_Y P(0,0)$ is integral.

$B + D = P(1,1) - P(0,1) = \Delta_X P(0,1)$ is integral, whence B is integral.

$2A + D = P(2,0) - P(1,0) = \Delta_X P(1,0)$ is integral, whence $2A$ is integral.

$2C + E = P(0,2) - P(0,1) = \Delta_Y P(0,1)$ is integral, whence $2C$ is integral.

QED

Henceforth, we will write P in the form (*) of Lemma 4.6.

The alert reader will have noticed that I sneaked some undefined notation into the proof of Lemma 4.6. The partial difference operators Δ_X and Δ_Y relate to the difference operator Δ just as the partial differential operators D_x and D_y relate to D:

$$\Delta_X P(X_0, Y_0) = P(X_0 + 1, Y_0) - P(X_0, Y_0)$$

$$\Delta_Y P(X_0, Y_0) = P(X_0, Y_0 + 1) - P(X_0, Y_0).$$

Note that

$$\Delta_X P(X, Y) = aX + bY + d \qquad (1)$$

$$\Delta_Y P(X, Y) = bX + cY + e. \qquad (2)$$

4.7. Lemma. *$P(X, Y)$ is monotone in each variable iff $\Delta_X P$ and $\Delta_Y P$ are always non-negative.*

This is fairly obvious and I omit the proof. The real question is whether or not $\Delta_X P$ and $\Delta_Y P$ are always non-negative. By (1) the former would be the case if a, b, d were non-negative and by (2) the latter if b, c, e were non-negative. What is easy to establish is the following.

4.8. Lemma. $a, c > 0$.

Proof. Suppose, by way of contradiction, that a were negative. From some value x of X onward, we would have $\Delta_X P(x, 0)$ negative, whence a strictly descending sequence

P(x, 0), P(x + 1, 0), P(x + 2, 0), ... of non-negative integers, a contradiction. Thus, $a \geq 0$.

Similarly, $c \geq 0$.

Now suppose $a = 0$:

$$P(X, Y) = bXY + \frac{c}{2}Y(Y-1) + dX + eY + f.$$

Specialising $X = 0, Y = 0$ in turn yields

$$P(0, Y) = \frac{c}{2}Y(Y-1) + eY + f$$

$$P(X, 0) = dX + f.$$

Since P(X, 0) must be one-one, $d \neq 0$. Moreover, since P(X, 0) must map into ω, d cannot be negative. Thus $d > 0$. Choose a large even multiple of d to plug into the first of these:

$$P(0, 2kd) = d[kc(2kd-1) + 2ke] + f.$$

Since P(0,Y) must map one-one into ω and $d > 0$, for large k, $kc(2kd - 1) + 2ke$ is positive. Thus, for large enough positive k, both kd and $kc(2kd - 1) + 2ke$ are positive and we have

$$P(0, 2kd) = P(kc(2kd-1) + 2ke, 0),$$

contradicting the one-one-ness of P on ω^2.

Similarly, $c \neq 0$. QED

[It may be worth remarking that the first part of the proof of Lemma 4.8 is, essentially, an induction. The value of such a remark will not be apparent until the next volume, however.]

The strategy of our attempted elementary proof of the Fueter-Pólya Theorem will be to show that the function is increasing in each variable and, hence, its enumeration must proceed up or down successive diagonals. The tactics will differ slightly. We do this for a few steps until we have enough values of P to determine the coëfficients. To take a first step in this direction, we must make two assumptions.

4.9. Assumptions. *We assume*

i. P(0,0) = 0

ii. $b > 0$.

Since P(0,0) = 0, we must have P(0,1) and P(1,0) positive. But, by (1) and (2),

$$d = \Delta_X P(0,0) = P(1,0) - P(0,0) = P(1,0)$$

$$e = \Delta_Y P(0,0) = P(0,1) - P(0,0) = P(0,1).$$

Thus, we have $a, b, c, d, e \geq 0$ by 4.8, 4.9.ii, and this observation. Hence, Δ_XP, Δ_YP are always non-negative and P is an increasing function in each variable.

Our overall strategy is now to determine successively where P(x,y) = 1, P(x,y) = 2, ... until we have enough values to determine the coëfficients a, b, c, d, e ($f = 0$ by 4.9.i). The main tool we have for this determination is the fact that P is strictly increasing in each variable (increasing as we've just established, strictly so because P is one-one). Thus, if we mark the lattice points in the first quadrant which P maps to 0, 1, ..., n and ask where $n+1$ occurs, it will be a left-most or bottom-most such point, i.e. the possible candidates (x_0, y_0) for $P(x_0, y_0) = n+1$ are precisely those which have no neighbour $(x_0 - 1, y_0)$ or $(x_0, y_0 - 1)$ which has not already been assigned a value from $\{0, 1, ..., n\}$ (or, eventually, an even higher value). Stated more dynamically, the possible candidates (x_0, y_0) for $P(x_0, y_0) = n+1$ are the last ones that cannot be reached from lattice points in the quadrant by sliding leftward or downward without hitting a lattice point which maps into $\{0, 1, ..., n\}$. For $n = 0$, there are only two lattice points from which we can slide no further— (0,1) and (1,0).

4.10. Assumption - Remark. *We assume*

$$P(0,1) = 1.$$

This assumption will allow us to deduce

$$P(x,y) = \langle x,y \rangle,$$

for all x,y $\in \omega$. *The alternate assumption,*

$$P(1,0) = 1,$$

leads to the alternate conclusion,

$$P(x,y) = \langle y,x \rangle,$$

for all x,y $\in \omega$. *Hence, this assumption is a* case *assumption and not a restrictive one like* 4.9.

From this assumption we conclude

$$e = P(0,1) = 1. \tag{3}$$

By our sliding trick, there are two candidates for $P(x_0, y_0) = 2$, namely (0,2) and (1,0) (cf. Figure 4.11).

4.11. Figure.

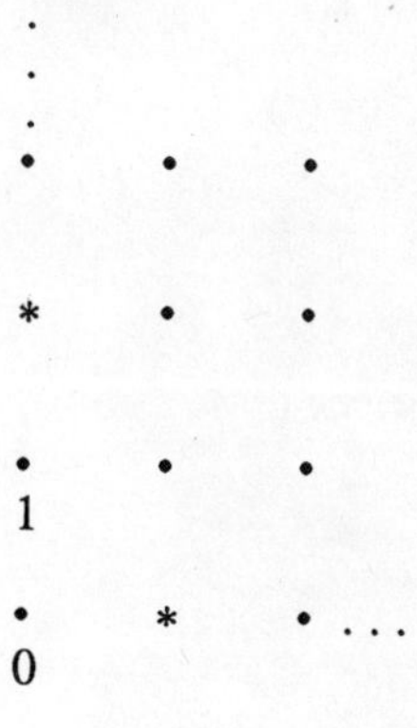

We can readily rule out (0,2). By (3), (2) reads

$$\Delta_Y P(X, Y) = bX + cY + 1.$$

Assuming P(0,2) = 2, this readily yields $c = 0$, contrary to Lemma 4.8. Hence, P(1,0) = 2 and

$$2 = P(1,0) = d. \tag{4}$$

If the reader now extends Figure 4.11, he will see that we have counted down the 0- and 1-diagonals. He will not be surprised to read that the next diagonal will finish the proof. What may be surprising, however, is that the next value of P offers the most difficulty. There are three candidates (x_0, y_0) for $P(x_0, y_0) = 3$. These are

$$(0,2), \ (1,1), \ \text{and} \ (2,0).$$

Ruling out (2,0) is, as with our determination that P(1,0) = 2, not too hard, but *ad hoc*:

4.12. Claim. P(2,0) ≠ 3.

Proof. By (1) and (4),

$$\begin{aligned} P(2,0) &= P(1,0) + \Delta_X P(1,0) = 2 + a\cdot 1 + b\cdot 0 + 2 \\ &= 4 + a > 3, \text{ by Lemma 4.8.} \end{aligned}$$

QED

4.13. Claim. P(1, 1) ≠ 3.

Proof. This is the tedious part. First, observe that (2), (4) yield

$$P(1,1) = P(1,0) + \Delta_Y P(1,0) = 2 + b + c\cdot 0 + 1 = 3 + b,$$

whence P(1, 1) = 3 implies $b = 0$ and

$$\Delta_X P(X, Y) = aX + 2 \tag{*}$$

$$\Delta_Y P(X, Y) = cY + 1. \tag{**}$$

Asking where P could possibly take the value 4 gives two possible answers: (0, 2) and (2, 0).

Case 1. P(2, 0) = 4. If this were the case, we would have

$$\begin{aligned} 4 = P(2,0) &= P(1,0) + \Delta_X P(1,0) = 2 + a + 2, \text{ by (*)} \\ &= 4 + a, \end{aligned}$$

whence $a = 0$, contradicting Lemma 4.8.

Case 2. P(0, 2) = 4. If this were the case, we would have

$$4 = P(0,2) = P(0,1) + \Delta_Y P(0,1) = 1 + c + 1, \text{ by (**)}$$

whence $c = 2$. But then

$$\begin{aligned} P(0,Y) &= Y^{(2)} + Y = Y^2 \\ P(1,Y) &= Y^{(2)} + 2 + Y = Y^2 + 2, \end{aligned}$$

yielding the values of the following figure.

4.14. Figure.

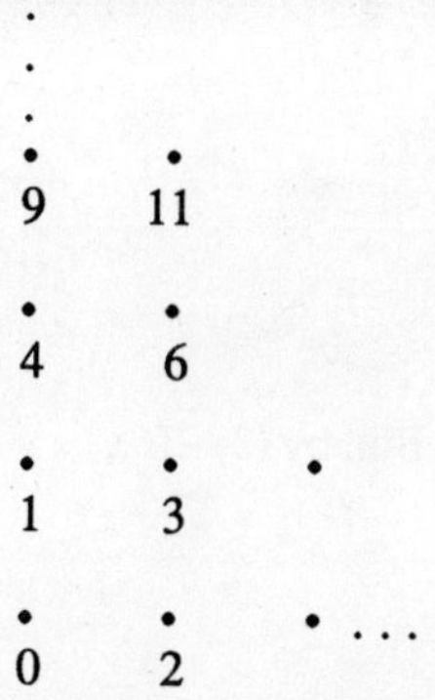

The sliding argument shows $P(2,0) = 5$. Formula (*) yields

$$\Delta_X P(1,0) = a+2,$$

but $\Delta_X P(1,0) = P(2,0) - P(1,0) = 5 - 2 = 3$, whence $a = 1$. Thus,

$$P(X, Y) = \frac{1}{2}X^{(2)} + Y^{(2)} + 2X + Y$$

and, in particular,

$$P(2, Y) = 1 + Y^2 + 4 = Y^2 + 5$$

and $P(2,2) = 9 = P(0,3)$, and $P(X, Y)$ is not one-one. QED

It follows that $P(0,2) = 3$. The correct picture so far is now Figure 4.15:

4.15. Figure.

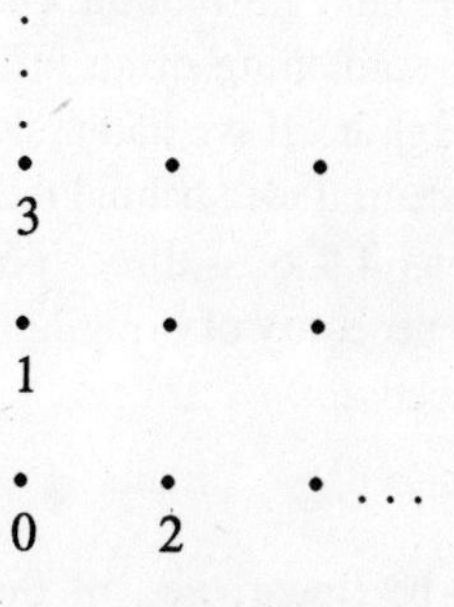

The value $P(0, 2) = 3$ yields several interesting things. First, by (2), (3),

$$\Delta_Y P(0, 1) = c + 1 = 3 - 1$$

entails

$$c = 1. \qquad (5)$$

This also yields

$$P(0,3) = P(0,2) + \Delta_Y P(0,2) = 3 + b{\cdot}0 + 1{\cdot}2 + 1 = 6.$$

With the sliding trick, this narrows down the possible pre-images of 4 to (1,1) and (2,0). Letting $Y = 0$, the argument given in Case 1 of the proof of the latest Claim applies and we have $P(2,0) \neq 4$. Hence $P(1,1) = 4$. But then

$$\begin{aligned} \Delta_X P(0,1) &= a\cdot 0 + b\cdot 1 + 2 = b + 2, \text{ by (1), (4)} \\ &= P(1,1) - P(0,1) = 4 - 1 = 3, \end{aligned}$$

whence

$$b = 1. \tag{6}$$

We have only to determine where 5 occurs to determine a. But, by (3) - (6),

$$P(0, Y) = \frac{1}{2}Y^{(2)} + Y \geq 6$$

for values of $Y > 2$ and

$$P(1, Y) = \frac{1}{2}Y^2 + \frac{3}{2}Y + 2 \geq 7$$

for values of $Y \geq 2$. Thus, the only candidate is (2,0). Hence $P(2,0) = 5$ and

$$\begin{aligned} \Delta_X P(1,0) &= a\cdot 1 + 2, \text{ by (2), (3)} \\ &= P(2,0) - P(1,0) = 5 - 2 = 3, \end{aligned}$$

whence

$$a = 1. \tag{7}$$

Putting (3) - (7) together, we have

$$P(X, Y) = \frac{1}{2}X^{(2)} + XY + \frac{1}{2}Y^{(2)} + 2X + Y,$$

which the reader can readily verify is the Cantor pairing function.

The proof just given has a clear overall strategy, muddled perhaps by an execution consisting of unmemorable *ad hoc* tricks. Because of this latter, as well as the limitations of generality imposed by the necessity of assuming 4.9 and the heavy use of the quadratic nature of P, the reader might even consider the proof to be something of an infliction. There is, however, a good, if pædagogical, reason to go through it. If we hadn't done so, we would not have seen the narrowness of applicability of the central idea behind the proof, i.e., we would not have seen the restriction to Assumptions 4.9 of a direct proof by enumeration. Without this, we would not fully appreciate the necessity of bringing to bear some fairly high-powered tools— as will be done in the next section.

Exercises.

1. By Lemma 4.8, a pairing function $P(X,Y)$ cannot be linear, i.e., of the form $dX + eY + f$. Give a shorter proof of this fact.
2. We call $P(X,Y) = \frac{a}{2}X^{(2)} + bXY + \frac{c}{2}Y^{(2)} + dX + eY + f$ *parabolic* if $b^2 - ac = 0$, i.e., if the graphs $P(X, Y) = constant$ are parabolas. Show that the assumption $b > 0$ can be dropped from our proof of a weakened Fueter-Pólya Theorem if P is assumed parabolic. [Hints: b cannot be 0. Write

$$P(X, Y) = \frac{1}{2a}(aX + bY)^2 + \frac{d'}{2}X + \frac{e'}{2}Y + f,$$

choose $X = -bt$, $Y = at$, and show P is not one-one.]

*5. The Fueter-Pólya Theorem, II

Of course, the Fueter-Pólya Theorem is interesting in its own right and I am justified by this interest in having gone this far. I wish in the present section to take the reader a bit farther by discussing a full proof of the Theorem. I say "discussing" instead of "presenting" because I will omit two steps from the proof, which steps will be described shortly. Now, it might strike the reader as odd that I am giving a complete, elementary proof of a weak version of the Theorem, and a partial, advanced proof of the full Theorem. The reasons for not giving the full proof are that: i. a full proof would require us to prove some deep results of analytic number theory and would take us too far afield, ii. a partial proof provides sufficient illustration of the main pædagogical and philosophical point to be made here by the Fueter-Pólya Theorem, and iii. a partial proof allows me to refer the reader to the literature for a complete proof— more anon.

The pædagogical and philosophical point to be made is a simple one: there is no way of knowing in advance what concepts or methods are appropriate for the solution to a given problem. In pre-calculus mathematics this is not the case: problems and methods match up rather closely. In calculus, one gets a hint of this when one discovers one must introduce logarithms and inverse trigonometric functions to find anti-derivatives for simple rational functions like $F(x) = 1/x$ and $F(x) = 1/(1+x^2)$. Elementary number theory abounds with readily stated results the intelligible proofs of which require seemingly irrelevant tools like complex integration or Fourier analysis. The Fueter-Pólya Theorem is an example of the non-predictability of the necessary tools for a given task. As such, it is a weak example in that it is a theorem for which i. the obviously relevant tools only seem to provide a partial result, ii. some apparently irrelevant tools can be applied to obtain the full result, and iii. all known proofs of the full result use such seemingly irrelevant tools; but it is not a theorem for which we have in any strict sense shown that something non-obvious must be done to prove it. In 1930, Kurt Gödel proved that there were such theorems, i.e., theorems which demonstrably required more than what seems relevant, and in the late 1970s logicians started to obtain results related to actual mathematical practice for which this is the case. We shall discuss versions of Gödel's results in Chapter III, below, and the newer theorems in Chapter VII of the volume II. For now, we have only the task of using powerful tools to prove the Fueter-Pólya Theorem.

The overall strategy of the proof is fairly simple. For any natural number n, the set

$$X_n = \{(x, y) \in R^2\colon\ 0 \le x \ \&\ 0 \le y \ \&\ \mathrm{P}(x, y) \le n\}$$

has exactly $n + 1$ lattice points. As n gets larger and larger, this area is asymptotic to the area A_n of the set

$$Y_n = \{(x, y) \in R^2\colon\ 0 \le x \ \&\ 0 \le y \ \&\ \mathrm{P}_2(x, y) \le n\},$$

where P_2 is the *homogeneous* quadratic part

$$\mathrm{P}_2(x, y) = \frac{a}{2}X^2 + bXY + \frac{c}{2}Y^2$$

of

$$P(X, Y) = \frac{a}{2}X^{(2)} + bXY + \frac{c}{2}Y^{(2)} + dX + eY + f.$$

The homogeneity of P_2 will guarantee that A_n/n is constant and we can determine this by simple calculus. A crucial result we will leave unproven asserts that the areas of X_n and Y_n approximate the number of lattice points in these sets, whence the constant ratio A_n/n measures the density of the range of P in ω. For P to be one-one and onto, this density must be 1. We can now look at the values of A_n/n in the cases that P_2 is elliptic, parabolic, or hyperbolic, and, although we do not know the actual values, we can conclude by their forms that A_n/n is transcendental except in the parabolic case. In this case, we quickly determine the values of a, b, c and the remaining coëfficients.

As the reader might guess, the second part omitted from the proof is the transcendence proof. The first omission— the closeness of area to number of lattice points— is an easily rectifiable matter requiring the addition of only a few pages of proofs; the latter omission simply cannot be overcome without too great a digression. However, if I cannot include these proofs, I can offer references for them. The transcendence result— Lindemann's Theorem— can be found in any number of textbooks. Perhaps the most accessible proof is in Lang's graduate *Algebra*. At the undergraduate level, Chapter IX of Lang's *Complex Analysis* gives, if not the full proof, at least the full flavour of the proof. The approximation of the number of lattice points in a decent region by the area of the region is certainly not in so standard a textbook as either of those of Lang, and I cannot say if it is in any text at all, but there is, nonetheless, a very good reference. The first of two papers under the title "Polynomial indexing of integer lattice points" (cf. the Reading List at the end of the chapter) by John Lew and Arnold Rosenberg is a model research paper (or: of what a research paper ought to be): it is well-written, fully documented, and fairly complete (the reader will have to go to one additional source, but only one). The student who has not yet begun to wean himself from textbooks, i.e., to read journal articles, could find no better introduction to the use of a research library than with this paper. (Incidentally, I have based the ensuing account on this paper.)

As I announced two paragraphs back, our first step is to look at the homogeneous part P_2 of a given pairing function P. As we shall be interested in homogeneous equations in the next Chapter, I might as well be fairly general in my introduction.

5.1. Definitions. A polynomial in the variables $X_1, ..., X_n$ of the form $aX_1^{i_1} \cdots X_n^{i_n}$ is called a *monomial* of *degree* $i_1 + ... + i_n$ (it being assumed each $i_j \in \omega$). By the distributive, associative, and commutative laws, any polynomial $Q(X_1, ..., X_n)$ can be written as a sum of monomials,

$$Q(X_1, ..., X_n) = \Sigma a_{i_1 \ldots i_n} X_1^{i_1} \cdots X_n^{i_n}.$$

Q is called a *homogeneous polynomial*, or a *homogeneous form*, or simply a *form*, if all the degrees $i_1 + ... + i_n$ are the same. This common degree is called the *degree* of the form.

For example, $X^2 + 2XY + 3Y^2$, $X^3 - XY^2$, and $2X - 7Y$ are forms, but $X^2 + 2X + Y^2$ and $X^2 + 2X + 1$ are not.

The fact that each monomial has the same degree has some powerful consequences for forms (and partially explains why we will replace P by P_2). The most immediate one is the following:

5.2. Theorem. *Let* $Q(X_1, ..., X_n)$ *be a polynomial of degree d (i.e. the degree of the highest term is d). Then: Q is homogeneous iff, for every real number t,*

$$Q(tX_1, ..., tX_n) = t^d Q(X_1, ..., X_n).$$

Proof. Assume Q is homogeneous of some degree d, say,

$$Q = \Sigma a_{i_1...i_n} X_1^{i_1} \cdots X_n^{i_n}, \text{ each } i_1 + ... + i_n = d.$$

Observe,

$$\begin{aligned} Q(tX_1, ..., tX_n) &= \Sigma a_{i_1...i_n} (tX_1)^{i_1} \cdots (tX_n)^{i_n} \\ &= \Sigma a_{i_1...i_n} t^{i_1+...+i_n} X_1^{i_1} \cdots X_n^{i_n} \\ &= t^d \Sigma a_{i_1...i_n} X_1^{i_1} \cdots X_n^{i_n} = t^d Q(X_1, ..., X_n). \end{aligned}$$

To prove the converse, write Q in the form

$$Q(X_1, ..., X_n) = \sum_{k=0}^{d} Q_k(X_1, \ldots, X_n),$$

where each $Q_k(X_1, ..., X_n)$ is homogeneous of degree k. By what has just been proven,

$$Q(tX_1, ..., tX_n) = \Sigma t^k Q_k(X_1, ..., X_n).$$

Subtracting the right side from the left, we get

$$\sum_{i_1 + ... + i_n < d} a_{i_1...i_n} X_1^{i_1} \cdots X_n^{i_n} (t^d - t^{i_1+...+i_n}) = 0.$$

Replacing t by, say, 2, we get

$$\sum_{i_1 + ... + i_n < d} c_{i_1...i_n} a_{i_1...i_n} X_1^{i_1} \cdots X_n^{i_n} = 0 \qquad (*)$$

with $c_{i_1...i_n} = 2^d - 2^{i_1+...+i_n} \neq 0$. But (*) can hold only if each coëfficient $c_{i_1...i_n} a_{i_1...i_n}$ is 0, i.e., if each $a_{i_1...i_n}$ for $i_1 + ... + i_n < d$ is 0 (cf. Exercise 1, below). Thus, no non-zero term of Q has degree less than d and Q is homogeneous of degree d. QED

5.3. Corollary. *If* $Q(X_1, ..., X_n)$ *is a form and* $Q(x_1, ..., x_n) = 0$, *then* $Q(tx_1, ..., t x_n) = 0$ *for any real number t.*

5.4. Corollary. *If* $Q(X, Y)$ *is a form of degree d and* $x \neq 0$, *then* $Q(x, y) = x^d Q(1, y/x)$.

Unlike ordinary polynomials, forms always have a zero: $Q(0, ..., 0) = 0$ for any form Q. Thus, when talking about zeroes of a form, we are interested only in *non-trivial* zeroes, i.e. zeroes other than (0,...,0). By Corollary 5.3, if $(x_1, ..., x_n)$ is a zero, every point $(tx_1, ..., tx_n)$ on the line passing through $(x_1, ..., x_n)$ and the origin is a zero of the given form. We, of course, will only be interested in lines in the plane and then only those parts of these lines lying in the first quadrant:

5.5. Definitions. Let (x, y) be any point in the first quadrant (axes included) other than the origin (0,0). By the *ray* determined by (x, y) we mean the set

$$\omega(x, y) = \{(tx, ty)\colon t \text{ is a positive real number}\}.$$

A ray $\omega(x, y)$ is called *rational* (*algebraic*) if $x = 0$ or the *slope* y/x of the ray is a rational (respectively, algebraic) number. (Recall that a real number is *algebraic* if it is a root of a polynomial with rational coëfficients.) Finally, a ray that is not rational (not algebraic) is called *irrational* (respectively, *transcendental*).

We may now begin in earnest to prove the Fueter-Pólya Theorem. To this end, assume $P(X, Y)$ is a quadratic pairing function. In the last section we proved, without extra assumption that

$$P(X, Y) = \frac{a}{2}X^{(2)} + bXY + \frac{c}{2}Y^{(2)} + dX + eY + f,$$

with a, b, c, d, e, f integral and a, c positive. P will be held fixed throughout the rest of this section, as will its homogeneous part,

$$P_2(X, Y) = \frac{a}{2}X^2 + bXY + \frac{c}{2}Y^2.$$

Our first lemma concerns P_2.

5.6. Lemma. *The set of zeroes of* P_2 *consists of the origin and at most* 2 (*possibly no*) *algebraic rays.*

Proof. Let $P_2(x, y) = 0$, with $(x, y) \neq 0$. Observe that x cannot be 0 since

$$P_2(0, y) = \frac{c}{2}y^2 = 0 \Rightarrow y = 0$$

by the positivity of c. Hence y/x is defined and by Corollary 5.4 we have

$$P_2(x, y) = 0 \Rightarrow x^2P_2(1, y/x) = 0$$
$$\Rightarrow P_2(1, y/x) = 0.$$

Thus y/x is a root of the polynomial

$$Q(Z) = \frac{a}{2} + bZ + \frac{c}{2}Z^2.$$

This has at most two roots,whence there are at most two slopes y/x of rays on which P_2 is 0. QED

Our next step is to show that these rays cannot be rational. If b were non-negative, this would be an easy matter (cf. Exercise 2, below). Unfortunately, we do not yet know that b is non-negative. Hence the following lemma:

5.7. Lemma. $P_2(x, y) > 0$ *on all rational rays.*

Proof. By contradiction. Let p,q be positive integers such that P_2 is 0 on the ray $\omega(p, q)$. (Exercise: Why don't we have to consider the possibility that P_2 be negative?)

Multiplying P out, we can write

$$P(X, Y) = \frac{a}{2}X^2 + bXY + \frac{c}{2}Y^2 + \frac{d'}{2}X + \frac{e'}{2}Y + f,$$

where $d'/2 = d - a/2$ and $e'/2 = e - c/2$. On $\omega(p,q)$, the quadratic terms vanish:

$$P(pt, qt) = \frac{d'}{2}pt + \frac{e'}{2}qt + f = \frac{d'p + e'q}{2}t + f.$$

Let $m = (d'p + e'q)/2$. m is an integer because $m = \Delta_t Q(t)$, where $Q(t) = P(pt, qt)$. Moreover, m is positive since $Q(t) = mt + f$ assumes infinitely many positive values as t ranges over ω.

Thus, the range of $Q(t) = P(pt, qt)$ for $t \in \omega$ consists of all natural numbers of the form $mt + f$. We shall get a contradiction by showing that P assumes some such value at a point not on the ray $\omega(p, q)$. To this end,look at

$$P(rm, sm) = \left[\frac{a}{2}r(rm - 1) + brsm + \frac{c}{2}s(sm - 1) + dr + es\right]m + f.$$

We have but to choose $r,s \in \omega$ so that, for

$$t_0 = \frac{a}{2}r(rm - 1) + brsm + \frac{c}{2}s(sm - 1) + dr + es,$$

i. t_0 is integral
ii. t_0 is positive, and
iii. $(r, s) \notin \omega(p, q)$.

The first condition is met by choosing r, s even. As to the second condition, $P(rm, sm) = t_0m + f$ will have the same sign as t_0 for all but finitely many values of (r, s). Thus, any large enough even $(r, s) \notin \omega(p, q)$ will do and $P(rm, sm) = P(pt_0, qt_0)$ with $(rm, sm) \neq (pt_0, qt_0)$, our desired contradiction.

QED

5.8. Corollary. $P_2(x, y) \geq 0$ *for all real* $x, y \geq 0$.

The Corollary follows from 5.7 by continuity and the density of the rational rays in the first quadrant.

A couple of quick comments might be in order. First, what we have proven so far, i.e., Lemmas 5.6 and 5.7, generalises to pairing functions of arbitrary degree. The argument in 5.7 becomes more sophisticated, but it is still rather elementary. The appeal to continuity in justifying the Corollary augers correctly: higher powered tools will quickly be brought into play. The next lemma will use the rational-irrational distinction, in the lemma following that we will use calculus in an essential way (as opposed to its formal, notational use in section 2, above), and ultimately we will use the algebraic-transcendental distinction.

To motivate our next lemma, let us consider the problem of finding the area A_n of the set

$$Y_n = \{(x, y) \in R^2 \colon 0 \leq x \ \& \ 0 \leq y \ \& \ P_2(x, y) \leq n\}.$$

The trick is to write the curve $P_2(x, y) = n$ in polar coördinates. Letting $x = r\cos\theta$, $y = r\sin\theta$, for $(x, y) \neq (0, 0)$ in the first quadrant, the equation

$$\frac{a}{2}x^2 + bxy + \frac{c}{2}y^2 = n$$

becomes

$$\frac{a}{2}r^2\cos^2\theta + br^2\cos\theta\sin\theta + \frac{c}{2}r^2\sin^2\theta = n,$$

whence

$$r = \frac{\sqrt{n}}{\sqrt{\frac{a}{2}\cos^2\theta + b\cos\theta\sin\theta + \frac{c}{2}\sin^2\theta}}. \tag{1}$$

Conversely, if θ is such that

$$\frac{a}{2}\cos^2\theta + b\cos\theta\sin\theta + \frac{c}{2}\sin^2\theta \neq 0,$$

we can use (1) to define r and then obtain $x = r\cos\theta$, $y = r\sin\theta$ satisfying $P_2(x,y) = n$. By Lemma 5.6, there are at most 2 angles θ for which we can fail to find such an r. Thus, our task of finding A_n is that of finding the area of a region that looks something like that in Figure 5.9.

5.9. Figure. A Hypothetical $r = f(\theta)$.

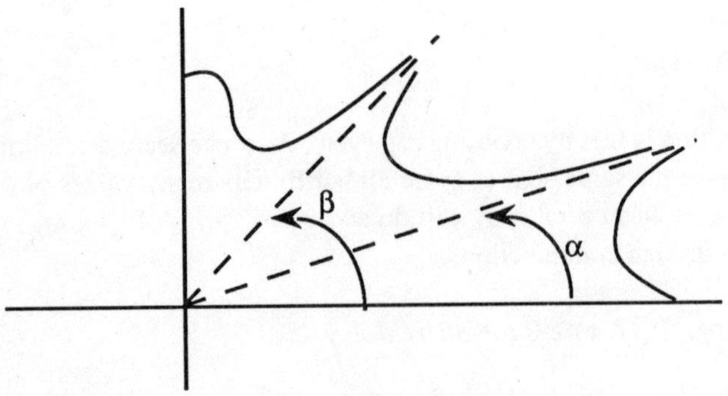

The area A_n is readily determined by the techniques of the calculus:

$$A_n = \int_0^{\alpha} r^2/2\,d\theta + \int_{\alpha}^{\beta} r^2/2\,d\theta + \int_{\beta}^{\pi/2} r^2/2\,d\theta = \int_0^{\pi/2} r^2/2\,d\theta$$

$$= \frac{1}{2}\int_0^{\pi/2} \frac{n\,d\theta}{\frac{a}{2}\cos^2\theta + b\cos\theta\sin\theta + \frac{c}{2}\sin^2\theta},$$

whence

$$A_n/n = \int_0^{\pi/2} \frac{d\theta}{a\cos^2\theta + 2b\cos\theta\sin\theta + c\sin^2\theta}. \tag{2}$$

[Note that A_n/n is constant. This is actually a consequence of Theorem 5.2— cf. Exercise 4, below.]

To evaluate the integral (2), the substitution $t = \tan\theta$ is natural, but $t = \sqrt{(c/a)}\tan\theta$ will be most convenient: Then $dt = \sqrt{(c/a)}\dfrac{d\theta}{\cos^2\theta}$ and

$$\begin{aligned}\int\frac{d\theta}{—} &= \int\frac{\sqrt{a/c}\,\cos^2\theta\,d\theta}{a\cos^2\theta + 2b\cos\theta\sin\theta + c\sin^2\theta}\\ &= \int\frac{\sqrt{a/c}\,dt}{a + 2b\sqrt{a/c}\,t + \frac{ac}{c}t^2}\\ &= \frac{\sqrt{a/c}}{a}\int\frac{dt}{1 + 2b/\sqrt{ac}\,t + t^2}\\ &= \frac{\gamma}{b}\int\frac{dt}{1 + 2\gamma t + t^2},\end{aligned} \tag{3}$$

where

$$\gamma = b/\sqrt{ac}. \tag{4}$$

Now, the temptation is to use (2) and (3) to write

$$A_n/n = \frac{\gamma}{b}\int_0^{\infty}\frac{dt}{1 + 2\gamma t + t^2} \tag{5}$$

since $\theta = 0$ corresponds to $t = 0$ and $\theta = \pi/2$ to $t = \infty$. However, such is overlooking the possible angles $\theta = \alpha, \beta$ at which P_2 is 0. Such angles correspond to zeroes of $1 + 2\gamma t + t^2$, whence we will establish (5) once we show $1 + 2\gamma t + t^2$ has no positive real roots. If $\gamma > -1$, we have, for $t > 0$,

$$\begin{aligned}1 + 2\gamma t + t^2 &> 1 - 2t + t^2\\ &> (1 - t)^2\\ &> 0,\end{aligned}$$

whence (5) will hold once we've proven the following.

5.10. Lemma. $\gamma > -1$, *i.e.*, $b > -\sqrt{ac}$.

Proof. Look at

$$Q(X, Y) = 2P_2(X,Y) = aX^2 + 2bXY + cY^2.$$

We can rewrite Q as

$$\begin{aligned} Q(X,Y) &= (\sqrt{a}X - \sqrt{c}Y)^2 + 2\sqrt{ac}XY + 2bXY \\ &= (\sqrt{a}X - \sqrt{c}Y)^2 + 2XY(b + \sqrt{ac}). \end{aligned}$$

Now

$$Q(\sqrt{c}, \sqrt{a}) = 2\sqrt{ac}(b + \sqrt{ac}),$$

which has the same sign as $b + \sqrt{ac}$. However, we know from Corollary 5.8 that $Q(\sqrt{c}, \sqrt{a}) \geq 0$ and it suffices to show $b + \sqrt{ac} \neq 0$.

That $b + \sqrt{ac} \neq 0$ is easy: If $b + \sqrt{ac} = 0$, then $\sqrt{ac} = -b$ is rational. However, if $b + \sqrt{ac} = 0$, then

$$P_2(\sqrt{c}, \sqrt{a}) = Q(\sqrt{c}, \sqrt{a}) = 0$$

and, by 5.7, $\sqrt{a}/\sqrt{c}$ is irrational. However, $\sqrt{a}/\sqrt{c} = \sqrt{ac}/c = -b/c$ is rational, a contradiction. QED

By this lemma,

$$A_n/n = \frac{\gamma}{b}\int_0^\infty \frac{dt}{1 + 2\gamma t + t^2}, \tag{5}$$

with the denominator of the integrand never being 0 and the integral thus clearly convergent. The task at hand is, thus, to calculate this integral. There are three cases: $-1 < \gamma < 1$ (the curves $P_2(x, y) = n$ being hyperbolas), $\gamma = 1$ (the parabolic case), and $\gamma > 1$ (the elliptic case).

5.11. Lemma. *For each* $n > 0$,

i. *if* $\gamma = 1$, $A_n/n = 1/b = 1/\sqrt{ac}$

ii. *if* $|\gamma| < 1$, *then* $\gamma = \cos\alpha$ *for some* $0 < \alpha < \pi$ *and*

$$A_n/n = \frac{\gamma}{b}\alpha\csc\alpha = \frac{1}{\sqrt{ac}}\alpha\csc\alpha$$

iii. *if* $\gamma > 1$, *then* $\gamma = \cosh\beta$ *for some* $\beta > 0$ *and*

$$A_n/n = \frac{\gamma}{b}\beta\,\mathrm{csch}\,\beta = \frac{1}{\sqrt{ac}}\beta\,\mathrm{csch}\,\beta.$$

Proof. i. The case $\gamma = 1$ is trivial:

$$A_n/n = \frac{\gamma}{b}\int_0^\infty \frac{dt}{(t+1)^2} = -\frac{\gamma}{b}\frac{1}{t+1}\Big|_0^\infty = 0 - -\frac{\gamma 1}{b 1} = \frac{\gamma}{b} = \frac{1}{\sqrt{ac}}.$$

ii. Let $\gamma = \cos\alpha$ and observe

$$\int_0^\infty \frac{dt}{1 + 2\gamma t + t^2} = \int_0^\infty \frac{dt}{(t+\gamma)^2 + 1 - \gamma^2}.$$

Let $t + \gamma = \sqrt{1-\gamma^2}\tan u$ be the usual trigonometric substitution:

$$\int \;=\; \int\frac{\sqrt{1-\gamma^2}\,\sec^2 u\,du}{(1-\gamma^2)(1+\tan^2 u)} = \frac{1}{\sqrt{1-\gamma^2}}\int du$$

$$= \frac{1}{\sqrt{1-\gamma^2}}\tan^{-1}\frac{t+\gamma}{\sqrt{1-\gamma^2}},$$

whence

$$\int_0^\infty \frac{dt}{1+2\gamma t+t^2} = \frac{1}{\sqrt{1-\gamma^2}}\tan^{-1}\frac{t+\gamma}{\sqrt{1-\gamma^2}}\Bigg|_0^\infty$$

$$= \frac{1}{\sqrt{1-\gamma^2}}\left[\frac{\pi}{2} - \tan^{-1}\frac{\gamma}{\sqrt{1-\gamma^2}}\right]$$

$$= \frac{1}{\sqrt{1-\gamma^2}}(\pi/2 - \tan^{-1}\cot\alpha)$$

$$= \frac{1}{\sqrt{1-\gamma^2}}(\pi/2 - (\pi/2 - \alpha))$$

$$= \frac{\alpha}{\sqrt{1-\gamma^2}} = \alpha\csc\alpha.$$

Multiplying the integral by γ/b (cf. (5) above) yields the result.

iii. $\gamma > 1$. In this case, $1 + 2\gamma t + t^2$ has 2 roots, $-\gamma \pm \sqrt{\gamma^2 - 1}$. Letting $t_1 > t_2$ be these roots, we get

$$\int\frac{dt}{1+2\gamma t+t^2} = \int\frac{dt}{(t-t_1)(t-t_2)}$$

$$= \frac{1}{2\sqrt{\gamma^2-1}}\left(\int\frac{dt}{t-t_1} - \int\frac{dt}{t-t_2}\right)$$

$$= \frac{1}{2\sqrt{\gamma^2-1}}\log\left|\frac{t-t_1}{t-t_2}\right|,$$

the definite integral thus becoming

$$\int_0^\infty \ldots = \frac{-1}{2\sqrt{\gamma^2-1}}\log|t_1/t_2|.$$

Now let $\gamma = \cosh\beta$ and observe that

$$\sqrt{\gamma^2 - 1} = \sinh\beta \qquad (*)$$

and

$$-\gamma \pm \sqrt{\gamma^2-1} = -\cosh\beta \pm \sinh\beta,$$

whence

$$t_1/t_2 = (-\cosh\beta + \sinh\beta)/(-\cosh\beta - \sinh\beta)$$

$$= (-\cosh\beta + \sinh\beta)^2$$
$$= e^{-2\beta}. \qquad (**)$$

Putting (*) and (**) together, we get

$$\int_0^\infty \ldots = -1/2(\operatorname{csch}\beta)(-2\beta) = \beta\cosh\beta.$$

Again, multiplying by the factor γ/b yields the desired formula. QED

Think of a large square of size $n \times n$. Its area is n^2. But how many lattice points does it contain? If its sides contain lattice points, it should contain $(n+1)^2$ such points. If the sides contain no lattice points, it could have something of the order of $4n$ fewer such. In any event, the area approximates the number of lattice points and these numbers, area and number of lattice points are asymptotically 1. It takes a bit more work to establish such an asymptotic result for a more complicated region like

$$Y_n = \{(x, y)\colon\ 0 \le x\ \&\ 0 \le y\ \&\ P_2(x, y) \le n\}.$$

Moreover, some work is required to relate Y_n, for large n, to

$$X_n = \{(x, y)\colon\ 0 \le x\ \&\ 0 \le y\ \&\ P(x, y) \le n\},$$

which has exactly $n + 1$ lattice points. Nonetheless, the expected relation, that A_n is asymptotically equal to $n + 1$, hence to n, is true.

5.12. Proposition. $A_n/n = 1$.

I refer the reader to the paper of Lew and Rosenberg for the proof. As the result is moderately plausible, I don't really feel the need to include the proof here. Besides, the next step is the one I am anxious to discuss.

5.13. Lemma. *If* $\gamma \neq 1$, A_n/n *is transcendental, i.e.*, (by 5.12) $\gamma = 1$.

Proof. I shall assume the reader sufficiently familiar with complex numbers to be comfortable with such identities as

$$\sinh(ix) = i\sin x, \quad \text{for } i^2 = -1.$$

I shall also invoke *Lindemann's Theorem*: if θ is an algebraic number other than 0, then e^θ is transcendental.

Let $\gamma \neq 1$. Write $\sigma = A_n/n$, choose β as in 5.11.iii if $\gamma > 1$ and let $\beta = i\alpha$ as in 5.11.ii if $|\gamma| < 1$. In both cases, Lemma 5.11 yields

$$\sigma = \frac{1}{\sqrt{ac}}\beta\operatorname{csch}\beta = \frac{1}{\sqrt{ac}}\beta(\gamma^2 - 1)^{-1/2},$$

whence

$$\beta = \sigma(ac)^{1/2}(\gamma^2 - 1)^{1/2}.$$

If σ were algebraic, then β would be a product of algebraic numbers, and it can be shown that the product of algebraic numbers is algebraic (Exercise 5, below). Thus, if σ were algebraic, then β would be algebraic. By Lindemann's Theorem, we will get a contradiction to the assumption that σ is algebraic by showing e^β to be algebraic.

Now, $\gamma = b/\sqrt{ac}$ is algebraic, but γ also equals $\cosh \beta$. Let $Q(X) = \sum a_i X^i$ be such that $Q(\cosh \beta) = Q(\frac{e^\beta + e^{-\beta}}{2}) = 0$. Expanding the expression for $Q(\frac{e^\beta + e^{-\beta}}{2})$ yields an equation

$$\sum_{k=-m_1}^{m_2} c_k e^{k\beta} = 0, \quad m_1, m_2 \geq 0.$$

Multiplying by $e^{m_1\beta}$ yields

$$\sum_{k=0}^{m_1+m_2} c_{k-m_1} (e^\beta)^k = 0,$$

demonstrating e^β to be algebraic, the desired contradiction. QED

The rest is elementary, but detailed.

5.14. Corollary. $a = b = c = 1$.

For, $1 = A_n/n = \gamma/b = 1/b = 1/\sqrt{ac}$ with a, c positive integers.

Thus, we have so far

$$P(X,Y) = \frac{1}{2}X^{(2)} + XY + \frac{1}{2}Y^{(2)} + dX + eY + f.$$

Now, $d \neq e$ since otherwise $P(X, Y) = P(Y, X)$ and P would not be one-one. Replacing $P(X, Y)$ by $P(Y, X)$ if necessary, we can assume $d > e$.

The best way of proving that P is the Cantor pairing function would be to show that P increases as one goes down the diagonals and as one jumps from the bottom of one diagonal to the top of the next. To aid us in doing this, we write P as much as possible as a function of the diagonal $X + Y$, bearing in mind the ultimate form we expect P to be in:

$$\begin{aligned} P(X, Y) &= \frac{1}{2}(X + Y)(X + Y + 1) + (d - 1)X + (e - 1)Y + f \\ &= \frac{1}{2}(X + Y)(X + Y + 1) + g(X + Y) + hX + f, \qquad (6) \end{aligned}$$

where $g = e - 1$, $h = d - e$, and g, h, f are integral.

On a given diagonal $X + Y = \textit{constant}$, the move down from the point (x,y) to the point $(x + 1, y - 1)$ will increment the value of P by h, as a glance at (6) will show, i.e., $P(x + 1, y - 1) - P(x, y) = h > 0$. Thus, P does increase in travelling down the lattice points of the diagonals. Jumping from the bottom of one diagonal to the top of the next yields

$$P(0, n + 1) - P(n, 0) = g + n + 1 - nh = g + 1 + n(1 - h). \qquad (7)$$

Since $h > 0$, we must show $h = 1$ to have any chance of this expression being positive. The usual *ad hoc* trick will yield this.

5.15. Lemma. $h = 1$.

Proof. Suppose instead that $h > 1$. Then, for large positive integral m, $hm - g$ and $(h-1)m - g - 1$ are positive integers. But a quick calculation shows $P(m, (h-1)m - g - 1) = P(0, hm - g)$, whence P is not one-one, a contradiction. QED

We are just about done. Since $h = 1$, (7) tells us that $P(0, n+1) - P(n, 0) = g + 1$, i.e. as we jump from the bottom of one diagonal to the top of the next the increase is $g + 1$. Consider what happens if g is negative and we pass from the diagonal $X + Y = n$ to $X + Y = n + 1$. The value of $P(n, 0)$ drops by $g + 1$ to $P(0, n+1)$ and then increases by 1 each step along the diagonal $P(1, n), P(2, n-1),\ldots$ For $n > |g|$, we reach the same value $P(n, 0)$ before exiting the diagonal, contrary to the one-one-ness of P on ω^2. [In fact, $P(n, 0) = P(-g-1, n+g+2)$.] Thus, $g \geq 0$.

Thus, we have

$$P(X, Y) = \frac{1}{2}(X+Y)(X+Y+1) + g(X+Y) + X + f$$

with $g, f \geq 0$. Moreover, we know that $P(0,0)$ is f, that as we go from the bottom of one diagonal to the top of the next we increase the value of P by $g + 1$, and that we increase this value by 1 each step down a diagonal. Thus, the smallest value of P on ω^2 occurs at $(0,0)$ and, since P must map onto ω, $P(0,0) = 0$. Thus, $f = 0$. Finally, to determine g notice that if we drop the g-term, we obtain

$$Q(X, Y) = \frac{1}{2}(X+Y)(X+Y+1) + X = \frac{1}{2}((X+Y)^2 + 3X + Y),$$

the Cantor pairing function. Since $P(x,y) \geq Q(x,y)$ and Q is onto, we must have $g = 0$ or P would skip values. Thus, P is the Cantor pairing function.

[Two small comments: This is not the only place where we used the fact that P is onto. The other was when we blithely asserted that the set $X_n = \{(x,y)\colon\ 0 \leq x\ \&\ 0 \leq y\ \&\ P(x,y) \leq n\}$ contained exactly $n+1$ lattice points, which fact was central to the unpresented proof of Proposition 5.12. By the way, some steps back we assumed $d > e$. The opposite assumption that $e > d$ would yield the other Cantor pairing function $P(Y, X)$.]

Exercises.

1. Show that the set of monomials $\{X_1^{i_1}\cdots X_n^{i_n} : i_1,\ldots,i_n \in \omega\}$ is linearly independent. [Hint: Use induction on $n \geq 1$. Write a polynomial $Q(X_1,\ldots,X_n)$ in the form $\sum Q_k(X_1, \ldots, X_{n-1})X_n^k$).]
2. For $P(X, Y) = \frac{a}{2}X^{(2)} + bXY + \frac{c}{2}Y^{(2)} + dX + eY + f$, suppose $b \geq 0$ and show:

$$\forall \varepsilon > 0\, \exists r > 0\, \forall x,y > r\left[|P(x,y)| < (1+\varepsilon)|P_2(x,y)|\right].$$

   Use this to give a simpler proof of Lemma 5.7 in this case. [Hint: $P(x,y)$ can be 0 for only one lattice point $(x, y) \in \omega^2$.]
3. The proof of Lemma 5.7 asserting $P_2(x,y) > 0$ on all rational rays only showed $P_2(x,y) \neq 0$. Why could we assume $P_2(x,y) \geq 0$?
4. Show by direct appeal to Theorem 5.2 that $A_n = n\cdot A$, where A is the area of the set $\{(x,y)\colon\ 0 \leq x\ \&\ 0 \leq y\ \&\ P_2(x,y) \leq 1\}$.

5. Let C denote the field of complex numbers and let F be a subfield. Let α be algebraic over F, i.e., $P(\alpha) = 0$ for some polynomial
$$P(X) = a_0 + a_1X + \dots + a_nX^n$$
with coëfficients in F. Define
$$F[\alpha] = \{b_0 + b_1\alpha + \dots + b_{n-1}\alpha^{n-1}:\ b_0,\dots,b_{n-1} \in F\}.$$
i. Show: $F[\alpha]$ is a field

ii. Show: If n is the minimum degree of a polynomial over F of which α is a root, then $F[\alpha]$ is an n-dimensional vector space over F.

iii. Let β be algebraic of degree m over $F[\alpha]$. Show: $(F[\alpha])[\beta]$ is a vector space of degree mn over F.

iv. Show: If α, β are algebraic over Q, so are $\alpha\cdot\beta$ and $\alpha + \beta$.

6. The Chinese Remainder Theorem

In this section we shall make a slight digression— a digression which, in the next section, we will see is not so great a digression after all. Our new topic for discussion is the Chinese Remainder Theorem.

The Chinese Remainder Theorem can be likened to the fabulous elephant described by the seven blind men: one grasped the elephant by the tail and said it was very like a rope, another touched the side and likened the beast to a wall, ... The Chinese Remainder Theorem has been an important tool in astronomical calculations and in religious observance (what day does Easter fall on?); it has been a source for mathematical puzzles; it has been abstracted in algebra to a theorem on the isomorphism of one homomorphic image of a ring of a given type to a product of two homomorphic images of the ring (indeed, in American colleges this can be the only version of the Chinese Remainder Theorem the student may see); it has been applied by computer scientists to obtain *multiple precision*; and, somewhere along the way, it has been used in logic as a means of coding finite sequences.

To state and discuss the Chinese Remainder Theorem, we must recall a few notions from an elementary number theory or a beginning abstract algebra course, namely: congruence and relative primality.

6.1. Definition. Let a, b, m be integers with $m \neq 0$. We say a is *congruent to b modulo m*, written $a \equiv b \pmod m$, if m divides $a - b$, written $m \mid a - b$. (We say $a \mid b$ iff $b = ac$ for some integer c.)

The key properties of congruence are its preservation of each of + and $\cdot$.

6.2. Lemma. *For all a, b, c, d, m, with $m \neq 0$,*

i. $a \equiv c \pmod m$ & $b \equiv d \pmod m \Rightarrow a + b \equiv c + d \pmod m$

ii. $a \equiv c \pmod m$ & $b \equiv d \pmod m \Rightarrow a \cdot b \equiv c \cdot d \pmod m$.

I shant prove Lemma 6.2 as I expect the reader already to be familiar with this. If I am wrong about the reader's memory, I have supplied him with a little exercise.

6.3. Definition. Integers a, b are *relatively prime* if their only common divisors are ± 1. For a list $a_0, \ldots, a_{n-1}$ of $n > 2$ integers, we say that they are *relatively prime* if again ± 1 are the only integers dividing all the a_i's, and that they are *pairwise relatively prime* (or: *relatively prime in pairs*) if each pair a_i, a_j for $i \neq j$ is relatively prime.

A quick example: 4, 6, 9 are relatively prime since the divisors of 4 are precisely ± 1, ± 2, ± 4 and of these only ± 1 divide 9. 4, 6, 9 are not pairwise relatively prime, however, since 4, 6 share 2 as a divisor or since 6, 9 share 3. In talking about divisors, it is customary to mention only the positive ones and I have most surely committed a *faux pas* in admitting the negative ones in Definition 6.3, but I promise to watch my step in the sequel.

We can now state the Chinese Remainder Theorem.

6.4. Chinese Remainder Theorem. *Let $m_0, \ldots, m_{n-1}$ be pairwise relatively prime positive integers. Let, further, integers $a_0, \ldots, a_{n-1}$ be given. There is an integer x such that, for each i,*

$$x \equiv a_i \pmod{m_i}. \tag{1}$$

Quick Proof. Recall from elementary number theory or abstract algebra the Euler ϕ-function: for positive integral m,

$$\phi(m) = \text{the number of positive integers} \leq m \text{ relatively prime to } m.$$

The set of positive integers $\leq m$ relatively prime to m forms a group under multiplication modulo m. This group has order $\phi(m)$, whence, for a relatively prime to m,

$$a^{\phi(m)} \equiv 1 \pmod{m}. \tag{2}$$

Using this fact, we get an explicit formula for x:

$$x = a_0\left(\frac{M}{m_0}\right)^{\phi(m_0)} + a_1\left(\frac{M}{m_1}\right)^{\phi(m_1)} + \ldots + a_{n-1}\left(\frac{M}{m_{n-1}}\right)^{\phi(m_{n-1})}, \tag{3}$$

where $M = m_0 \cdots m_{n-1}$.

To see that (3) works, simply note that m_i, M/m_i are relatively prime, whence

$$\left(\frac{M}{m_i}\right)^{\phi(m_i)} \equiv 1 \pmod{m_i}$$

by (2). But, if $i \neq j$,

$$\frac{M}{m_j} \equiv 0 \pmod{m_i}.$$

Thus,

$$\begin{aligned} x &\equiv a_0 \cdot 0 + \ldots + a_{i-1} \cdot 0 + a_i \cdot 1 + \ldots + a_{n-1} \cdot 0 \pmod{m_i} \\ &\equiv a_i \pmod{m_i}. \end{aligned}$$

QED

The number x given by (3) is quite large. For example, if we choose $m_0 = 2, m_1 = 3, m_2 = 5, a_0 = 1, a_1 = 2, a_2 = 4$, we get $M = 30$ and

$$x = 1\cdot 15^1 + 2\cdot 10^2 + 4\cdot 6^4 = 5399.$$

However, we could also take x to be 29, which is quite a bit smaller than 5399.

6.5. Chinese Remainder Theorem. (Sharper Formulation). *Let $m_0, ..., m_{n-1}$ be pairwise relatively prime positive integers, and let $a_0, ..., a_{n-1}$ be n arbitrary integers. Then: for any integer a, there is a unique integer x satisfying*
i. $a \le x < a + m_0 \cdots m_{n-1}$
ii. $x \equiv a_i \pmod{m_i}$, *for each i.*

Proof. Rather than reduce this result to 6.4, we prove it anew. Without loss of generality, we can assume $a = 0$.

Let $0 \le x, y < m_0 \cdots m_{n-1}$ and consider the sequences,

$$(\mathrm{Rem}(x, m_0), ..., \mathrm{Rem}(x, m_{n-1}))$$

$$(\mathrm{Rem}(y, m_0), ..., \mathrm{Rem}(y, m_{n-1})),$$

where, for $m > 0$,

$$\mathrm{Rem}(z, m) = \text{the remainder of } z \text{ after division by } m.$$

If these sequences are identical, then each $m_i \mid x - y$. The fact that the m_i's are pairwise relatively prime then tells us that $m_0 \cdots m_{n-1} \mid x - y$. Since $0 \le x, y < m_0 \cdots m_{n-1}$, this implies $x = y$. Hence, distinct natural numbers $< m_0 \cdots m_{n-1}$ yield distinct sequences of remainders.

There are $m_0 \cdots m_{n-1}$ natural numbers below $m_0 \cdots m_{n-1}$ and the same number of sequences $(b_0, ..., b_{n-1})$ with $0 \le b_i < m_i$ for each i. Hence each sequence is the sequence of remainders of a unique natural number $x < m_0 \cdots m_{n-1}$. Choose x for the sequence given by $b_i = \mathrm{Rem}(a_i, m_i)$.

QED

Getting back to our example with $m_0 = 2, m_1 = 3, m_2 = 5, a_0 = 1, a_1 = 2, a_2 = 4$, we see that, in addition to the solution $x = 5399$ to

$$x \equiv a_i \pmod{m_i},\ i = 0, 1, 2,$$

there must be a solution with $0 \le x < 2\cdot 3\cdot 5 = 30$. Thus, we see explained the size of the new solution $x = 29$. But what is the relation between the two solutions?

6.6. Corollary. *Let $m_0, ..., m_{n-1}$ be pairwise relatively prime positive integers, let $a_0, ..., a_{n-1}$ be given, and suppose*

$$x_0 \equiv a_i \pmod{m_i}, \text{ for each } i.$$

Then, for any integer x, the following are equivalent:
i. $x \equiv x_0 \pmod{m_0 \cdots m_{n-1}}$
ii. $x \equiv a_i \pmod{m_i}$, *for each i.*

Proof. Let x be given.

i ⇒ ii. This is fairly straightforward:

$$x \equiv x_0 \pmod{m_0 \cdots m_{n-1}} \Rightarrow x \equiv x_0 \pmod{m_i}$$
$$\Rightarrow x \equiv a_i \pmod{m_i},$$

for each i.

ii ⇒ i. By 6.5, there is a unique y satisfying the inequalities,

$$x \le y < x + m_0 \cdots m_{n-1},$$

and all the congruences,

$$y \equiv a_i \pmod{m_i}.$$

It was established in the first step of the proof of 6.5 that

$$y \equiv x_0 \pmod{m_0 \cdots m_{n-1}}.$$

If ii holds, uniqueness yields $x = y$, whence i. QED

The reflexion of this Corollary in our ongoing example is: Rem(5399, 30) = 29.

We now have two proofs of the Chinese Remainder Theorem, each providing its own extra information. The first proof gives us an explicit formula for the quantity sought (well— explicit up to the Euler ϕ-function); the second proof gives us some bounds. For the purposes of this book, the (nearly) explicit formula offers no particular advantage. The use we wish to put the Chinese Remainder Theorem to in the immediately following section requires merely the existence of the solution x. The bounds offered in 6.5 will only assume significance in the next chapter. However, there are other applications of the Chinese Remainder Theorem for which the extra information given by 6.4 and 6.5 is not enough and a *third* proof is used! We will eventually need this third proof ourselves (Chapter IV of volume II), so we might as well digress a bit to consider some applications for which Theorems 6.4 and 6.5 and their proofs are inadequate. The uninterested reader can jump ahead to the next section.

The Chinese Remainder Theorem gets its name partly from the following problem from the *Sun-tzu suan ching* from the 4th century AD:

> There is an unknown number of things. When counted in threes, they leave a remainder of two; when counted by fives, they leave a remainder of three; and when counted by seven, they leave a remainder of two. Find the number of things.

This sort of puzzle, common enough in the ancient writings, is not particularly exciting, but it does bring us to the heart of the matter. The problem says "find the number", not "prove the number exists". The formula of 6.4 gives

$$x_0 = 2 \cdot (35)^{\phi(35)} + 3 \cdot (21)^{\phi(21)} + 2 \cdot (15)^{\phi(15)}.$$

Assuming we know enough elementary number theory to calculate $\phi(35) = 24$, $\phi(21) = 12$, and $\phi(15) = 8$, we see

$$x_0 = 2 \cdot 35^{24} + 3 \cdot 21^{12} + 2 \cdot 15^{8},$$

a rather large number which we will want, by 6.5, to reduce modulo $3 \cdot 5 \cdot 7 = 105$. We could carry out the reductions as we perform the multiplications, thus guaranteeing we never have to multiply more than three-digit numbers, but we nevertheless have to perform

44 multiplications. The task of performing 44 multiplications of one-, two-, and three-digit numbers would not have daunted any 18th or 19th century mathematician, who would instantly spot some short cuts. For example, since $35^{24} = 35^{16} \cdot 35^{8}$, we have but to square 35, reduce, square that, etc., a few times— only 4 multiplications (not counting the work in reducing modulo 105). Using this trick of writing the exponent as a sum of powers of 2 and repeatedly squaring (which trick, in a very liberal sense, goes back to the ancient Egyptians), we can readily see that we can take

$$x_0 = 2 \cdot 70 + 3 \cdot 21 + 2 \cdot 15 = 233$$

and then reduce modulo 105 to get a final answer,

$$x = 23.$$

It is amusing to note that Sun Tsŭ (or Sun-tzu, or Suentse, or ...) obtains the value precisely by taking $x_0 = 2 \cdot 70 + 3 \cdot 21 + 2 \cdot 15$ and then reducing.

What made the above computation so easy was a combination of things. One was the Egyptian trick to reduce the number of multiplications, one was the reduction modulo 105 to keep the sizes of the factors small, and one was my having skipped the actual computation of $\phi(x)$. In general, the computation of $\phi(x)$ depends on a factorisation of x— and factorisation is a notoriously difficult task.

The numbers in Sun Tsŭ's problem are small enough that the factorisation of the m_i's— whence the calculation of the $\phi(m_i)$'s— is no problem. Hence, in this case 6.4 offers a practicable algorithm. What about 6.5? The obvious algorithm is to try $n = 0, 1, ..., 104$, taking remainders modulo 3, 5, 7 until the desired pattern appears. This is a small enough task that one can do it on a home computer easily; doing it by hand, with the possibility of over 100 triples of remainders to search out, might be moderately daunting. A later Chinese example dramatically shows up the computational shortcomings of 6.4 and 6.5.

Around the end of the first millennium AD, Chinese scholars devoted a lot of time to calculating the *Grand Cycle*, or, actually, the number of years having passed since the Grand Cycle, which would have been the last time the winter solstice fell exactly at midnight on the first day of the 11th month, which day also being the first day of a 60 day cycle. The "month" referred to is the synodic month of roughly $29\frac{1}{2}$ days. The 60 day cycle is evidently a more secular period. In the 13th century *Shu-shu chiu chang*, Ch'in Chiu-shao solved the problem in the following manner: Let

a = length of a tropical year

b = length of a synodic month

R_1 = day number of the 60-day cycle on which the winter solstice falls

R_2 = day number of the 11th month on which the winter solstice falls

N = number of years since the Grand Cycle.

Then one has but to solve

$$aN \equiv R_1 \pmod{60}$$
$$\equiv R_2 \pmod{b}.$$

The values chosen for a, b are

$$a = 365\frac{4108}{16900}, \qquad b = 29\frac{8967}{16900}.$$

As these are not whole numbers, one must multiply by the denominators to get

$$\begin{aligned} 6172608N &\equiv 16900R_1 \pmod{1014000} \\ &\equiv 16377R_2 \pmod{499067}. \end{aligned}$$

In problem 12 of his book, Ch'in Chiu-shao chose values of R_1, R_2 yielding the congruences

$$\begin{aligned} 6172608N &\equiv 193440 \pmod{1014000} \\ &\equiv 16377 \pmod{499067}. \end{aligned}$$

For the sake of discussion, let us ignore the factor 6172608 and consider the congruences,

$$x \equiv 193440 \pmod{1014000} \tag{4}$$

$$\equiv 16377 \pmod{499067}. \tag{5}$$

Obviously one does not want to apply the search supplied by 6.5 to this problem (there are 506,053,938,000 numbers n through the pairs of the remainders of which one would have to search!). The method of 6.4 offers a little more hope: although large, 1014000 factors easily:

$$1014000 = 2^4 \cdot 3 \cdot 5^3 \cdot 13^2.$$

Elementary number theory tells us that

$$\phi(1014000) = \phi(2^4) \cdot \phi(3) \cdot \phi(5^3) \cdot \phi(13^2)$$

and that $\phi(p^k) = p^k - p^{k-1}$ for any prime p, whence

$$\phi(1014000) = (16-8)(3-1)(125-25)(169-13) = 249600.$$

The smaller number, 499067, is more difficult to factor, but my computer assures me it is prime. Thus,

$$\phi(499067) = 499066,$$

whence we can solve (4) and (5) by taking

$$x_0 = 193440(499067)^{249600} + 16377(1014000)^{499066},$$

reducing modulo 506,053,938,000 as we perform our multiplications. I leave this calculation as an exercise to the reader.

Even without having seen the word "computer" mentioned a couple of times above, it will have occurred to the reader to let a computer do the work for him. He might discover, however, that his computer doesn't handle as many significant digits as occur in this problem. What is he to do? The answer (or, perhaps: the name of the problem) is *multiple precision.* A computation dealing with (say) 15 digit numbers on a computer handling only 10 digit numbers cannot normally be done without the loss of significant digits, and one must resort to multiple precision methods to carry out the computations without the loss. The obvious approach is to split the large numbers into several blocks of digits and deal with these smaller blocks. For example, to multiply two 8 digit numbers, say,

$$12345678 \text{ and } 87654321,$$

one can split each into two 4 digit numbers,

$$\begin{array}{c|c} 1234 & 5678 \\ \hline A & B \end{array} \qquad \begin{array}{c|c} 8765 & 4321 \\ \hline C & D \end{array}$$

and find

$$AC = 1081 \mid 6010$$
$$AD = 533 \mid 2114$$
$$BC = 4976 \mid 7670$$
$$BD = 2453 \mid 4638.$$

The product will be a 16 digit number, the last 4 digits of which will be the last 4 of BD; the next 4 will be the sum of the last 4 of BC and AD and the carry— the first 4 digits of BD; any excess will join the last 4 digits of AC and the carries— the first 4 digits of AD and BC; and, finally, any excess joins the first 4 digits of AC to yield the full number:

$$
\begin{array}{cccc}
 & & 2453 & 4638 \\
 & 4976 & 7670 & \\
 & 533 & 2114 & \\
1081 & 6010 & & \\
\hline
1082 & {}^{1}1520 & {}^{1}2237 & 4638
\end{array}
$$

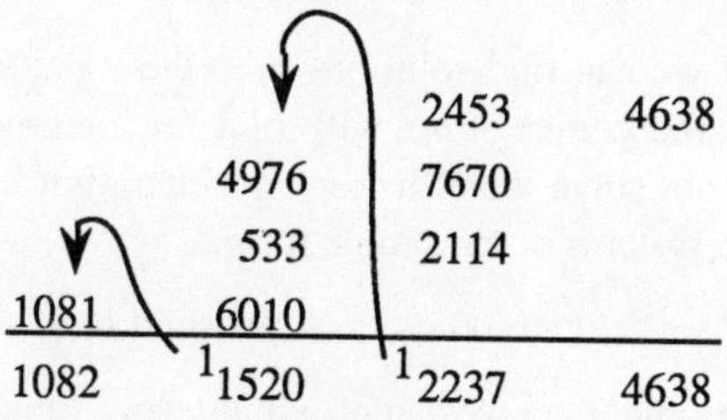

This is simple enough— it is the usual multiplication algorithm performed in base 10000.

If all of one's calculations are going to be additions and multiplications, and if there are going to be a lot of them, one can use the Chinese Remainder Theorem for multiple precision to avoid the carrying operation and the keeping track of the column number in which intermediate products belong. The idea is very simple. One first chooses a large product of small relatively prime factors, $m = m_0 \cdots m_{n-1}$. In practical considerations there is always such an upper bound to the numbers one wants to use, whence we cannot computationally distinguish between arithmetic in the ring of integers Z and that in the reduced ring $Z/(mZ)$ of integers modulo m. But the Chinese Remainder Theorem and Lemma 6.2 assert that the map

$$a \mapsto (\mathrm{Rem}(a, m_0), \ldots, \mathrm{Rem}(a, m_{n-1}))$$

is an isomorphism of $Z/(mZ)$ with $Z/(m_0 Z) \times \ldots \times Z/(m_{n-1} Z)$. Each addition or multiplication in Z is, for large enough m, such an operation in $Z/(mZ)$ and becomes n such operations on small numbers in the various $Z/(m_i Z)$'s. If n is small enough, this is not too much extra work and is often less as one is dealing with small numbers and as one has not to keep track of the carrying operations. Of course, there are two places in all of this where the congruences are no help. The first is the encoding step, where one must transform the possibly large numbers a into their n-tuples $(\mathrm{Rem}(a, m_0), \ldots, \mathrm{Rem}(a, m_{n-1}))$; the second is the decoding step where one must apply an efficient form of the Chinese Remainder Theorem to solve the congruences,

$$x \equiv a_i \pmod{m_i} \qquad (*)$$

to replace the n small numbers by the single large number they encode. Ignoring the problem of multiple precision in the encoding and decoding steps, the chief algorithmic problem offered is to solve the simultaneous congruences (*) efficiently. It is to this problem we next turn our attention.

If we let $b_i = (M/m_i)^{\phi(m_i)}$, where $M = m_0 \cdots m_{n-1}$, formula (3) becomes

$$x = a_0 b_0 + \ldots + a_{n-1} b_{n-1}. \qquad (6)$$

These particular values of b_i satisfy two congruences,

$$b_i \equiv 1 \pmod{m_i}, \quad b_i \equiv 0 \pmod{M/m_i}, \qquad (7)$$

and these congruences are the only properties of the b_i's used in proving, in the course of the proof of Theorem 6.4, that

$$x \equiv a_i \pmod{m_i},$$

for x as in (6). Hence, if we can find solutions b_i to the n pairs of congruences (7), the number x given by (6) for this choice of b_i's will solve the congruences (1).

This last remark does not solve the efficiency problem, but it does afford a reduction. If we can efficiently solve systems of the simple form,

$$x \equiv 1 \pmod{a}, \; x \equiv 0 \pmod{b}, \qquad (8)$$

where a, b are relatively prime, then we can efficiently solve any system (1). To this end, note that x is a solution to (8) iff there are integers k, h such that

$$x = ak + 1, \quad x = bh,$$

i.e., iff $bh = ak + 1$. However, even the reader who hasn't taken a course in elementary number theory will have learned in his abstract algebra course that the set of positive integers of the form $bh - ak$ has the greatest common divisor of a, b as minimum element, whence, for a, b relatively prime, he knows h, k to exist so that $bh - ak = 1$. In such a course, he may or may not have seen an algorithm for finding h, k. There is a simple such algorithm.

Let a, b be given relatively prime positive integers. Suppose $a > b$ and apply the Division Algorithm learned in grade school to obtain

$$a = bq_1 + r_1, \quad 0 \le r_1 < b.$$

The exhortation to repeat this is known as the Euclidean Algorithm:

$$b = q_2 r_1 + r_2, \; 0 < r_2 < r_1$$

$$r_1 = q_3 r_2 + r_3, \; 0 < r_3 < r_2$$

$$\vdots$$

$$r_l = q_{l+2} r_{l+1}.$$

For example, let $a = 52, b = 11$:

$$52 = 4\cdot 11 + 8 \tag{9}$$
$$11 = 1\cdot 8 + 3 \tag{10}$$
$$8 = 2\cdot 3 + 2 \tag{11}$$
$$3 = 1\cdot 2 + 1 \tag{12}$$
$$2 = 2\cdot 1.$$

To solve (8), one backs this procedure up (ignoring the very last equation):

$$3 - 1\cdot 2 = 1, \text{ by (12)}$$
$$3 - 1\cdot(8 - 2\cdot 3) = 1, \text{ by (11)}$$
$$3\cdot 3 - 1\cdot 8 = 1, \text{ by simplification}$$
$$3\cdot(11 - 1\cdot 8) - 1\cdot 8 = 1, \text{ by (10)}$$
$$3\cdot 11 - 4\cdot 8 = 1, \text{ by simplification}$$
$$3\cdot 11 - 4\cdot(52 - 4\cdot 11) = 1, \text{ by (9)}$$
$$19\cdot 11 - 4\cdot 52 = 1, \text{ by simplification.}$$

Hence a solution to (8) for our given values of a, b is given by

$$x = 19\cdot 11 = 4\cdot 52 + 1 = 209.$$

That this method offers a great computational advantage will be clear when the reader applies it to the pair (4), (5) of congruences arising from Ch'in Chiu-shao's problem (cf. Exercise 3, below).

Exercises.

1. (Chinese Remainder Theorem for arbitrary moduli). Let $m_0, ..., m_{n-1}$ be positive integers, let $a_0, ..., a_{n-1}$ be arbitrary integers, and let d_{ij} be the greatest common divisor of the pair m_i, m_j. Show: The following are equivalent:
   i. There is an x such that, for all i, $x \equiv a_i \pmod{m_i}$
   ii. For all $i \neq j$, $a_i \equiv a_j \pmod{d_{ij}}$.
2. Perform the calculations in the solution given in the text to Sun Tsŭ's problem:
   $x \equiv 2 \pmod 3$
   $x \equiv 3 \pmod 5$
   $x \equiv 2 \pmod 7$.
3. i. Solve: $x \equiv 193440 \pmod{1014000}$
   $x \equiv 16377 \pmod{499067}$.
   ii. Solve: $6172608N \equiv x \pmod{1014000} \equiv x \pmod{499067}$,

   for x as in part i.

4. i. Show: There is no one-one function $f:\omega^2 \to \omega$ which preserves +, i.e., such that, for all $x, y, a, b \in \omega$, $f(x+a, y+b) = f(x, y) + f(a, b)$.

ii. Show: There is no one-one function $f:\omega^2 \to \omega$ which preserves + in each variable, i.e., such that, for all $x, y, a, b \in \omega$,

$$f(x+a, y) = f(x, y) + f(a, y) \text{ and } f(x, y+b) = f(x, y) + f(x, b).$$

iii. Show: There is a pairing function f on $(\omega^+)^2$ (where $\omega^+ = \omega - \{0\}$) which preserves multiplication: For all $x, y, a, b \in \omega$,

$$f(xa, yb) = f(x, y) \cdot f(a, b).$$

iv. Is there a one-one function $f:(\omega^+)^2 \to \omega^+$ which preserves multiplication in each argument, i.e., such that, for all $x, y, a, b \in \omega^+$,

$$f(xa, y) = f(x, y) \cdot f(a, y) \text{ and } f(x, yb) = f(x, y) \cdot f(x, b)?$$

[Hints: i - ii. $f(0,1) \cdot f(1,0)$ has at least two pre-images; iii. use the Fundamental Theorem of Arithmetic and a one-one correspondence of the set P of prime numbers with each of two disjoint subsets P_1, P_2 of P, the union of which is P; iv. what is the pre-image of 1?]

7. The β-Function and Other Encoding Schemes

As stated (perhaps: barely hinted) in the last section, the Chinese Remainder Theorem allows one to code large numbers by sequences of smaller ones. In 1930, Kurt Gödel had already drawn the opposite conclusion: if, for $i = 0, \ldots, n-1$,

$$a_i = \text{Rem}(x_0, m_i),$$

then $x_0, m_0, \ldots, m_{n-1}$ code the sequence $a_0, \ldots, a_{n-1}$. The use of $n+1$ numbers to code n numbers might not strike one as particularly clever or economical, but we must not overlook the possibility that there may be more structure to the sequence $m_0, \ldots, m_{n-1}$ of moduli than to an arbitrary sequence $a_0, \ldots, a_{n-1}$ of numbers.

7.1. Lemma. *Let $n > 0$, $d = n!$ Then the numbers $1 + d, \ldots, 1 + (i+1)d, \ldots, 1 + (n+1)d$ are pairwise relatively prime.*

Proof. Suppose p is a prime number such that

$$p \mid 1 + (i+1)d, \quad p \mid 1 + (j+1)d$$

for some $0 \le j \le i \le n$. Then p divides the difference: $p \mid (i-j)d$. Since p cannot divide d, it follows that $p \mid i-j$ and $p > n$. Since $0 \le i-j < n < p$, we have $i = j$. QED

7.2. Definition. (Gödel's β-Function). For arbitrary natural numbers c, d, i, define

$$\beta(c, d, i) = \text{Rem}(c, 1 + (i+1)d).$$

7.3. Theorem. *Let $a_0, ..., a_{n-1}$ be an arbitrary finite sequence of natural numbers. There are c, d such that, for $i = 0, ..., n-1$, $\beta(c, d, i) = a_i$.*

Proof. Let $a = \max\{a_0, ..., a_{n-1}\} + 1$ and $d = (an)!$ By Lemma 7.1, $1 + d, ..., 1 + nd$ are pairwise relatively prime. By the Chinese Remainder Theorem there is some $c \in \omega$ such that

$$c \equiv a_i \pmod{1 + (i+1)d},$$

for $i = 0, ..., n-1$. But

$$a_i \le a \le an \le d < 1 + (i+1)d,$$

whence

$$a_i = \text{Rem}(c, 1 + (i+1)d) = \beta(c, d, i), \text{ for } i = 0, ..., n-1. \qquad \text{QED}$$

Thus, using Gödel's β-function we can code arbitrary finite sequences of natural numbers by pairs of natural numbers (whence, via the pairing function, by single numbers). How nice a result is this? The coding is clearly not one-one— many pairs c, d code $a_0, ..., a_{n-1}$. [The reader showed, back in section 3, Exercise 6.v, that there is a one-one correspondence between the sets of natural numbers and of finite sequences thereof.] The length of a sequence coded via c, d is not clearly defined: which sequence is coded by c, d is not uniquely determined. And, although computationally the decoding (i.e., finding a_i from c, d, i) is simple, the encoding (at least the algorithm given) is horrid: the number d is *immense*, whence, except in the case where $c = a_0 = a_1 = ... = a_{n-1}$ and consequently $c < d$— c is even larger. Nonetheless, there is something pleasing about the use of the β-function as a means of encoding finite sequences. For one thing, it is an extremely simply defined function. It is, in a sense to be made clear three sections hence, the simplest such coding scheme known. Moreover, it was the historically first *arithmetic* coding scheme, a fact I cannot really elaborate on until the adjective "arithmetic" is precisely defined— as will be done in section 10, below.

What I have been aiming at in this chapter, as is clear from the title if not yet from the text, is an indication of the expressive power of arithmetic. Thus, after the first two introductory sections, we examined the Cantor pairing function— and then devoted two sections to explaining that it is the simplest pairing function possible, at least according to one definition of simplicity. And, of course, the last section was really a dilatory preparation for the encoding just given. For the rest of the present section, I propose to look at a few additional number-theoretic coding schemes. How do we judge among them, i.e., by what criteria do we declare one to be simpler than or preferable to another? Criteria for simplicity would seem to be computability and definability. Alas, once we get to dealing with sequences, we are dealing with exponential growth and none of the schemes is really computationally feasible. However, these codings are computable in theory, if not in practice, and we will still have the definability criterion. The rest of the Chapter will devote itself to the study of the notions of theoretical computability and definability, and the strong interrelationship between the two.

For now, however, we have more mundane matters to deal with. What other methods of encoding does elementary number theory offer us? In the same paper in which Gödel introduced the β-function, he made heavy use of the Fundamental Theorem of Arithmetic.

7.4. Coding Sequences as Exponents. Let $p_0 = 2, p_1 = 3, \ldots$ be the sequence of primes. To any non-empty sequence $(a_0, \ldots, a_{n-1})$ of natural numbers we associate the code,

$$p_0^{a_0+1} \cdots p_{n-1}^{a_{n-1}+1};$$

to the empty sequence we can arbitrarily assign the number 0, or, perhaps, 1.

There are a couple of things to remark about this coding. The first is that, if we didn't increase the exponents by 1, we would give identical codes to $(a_0, \ldots, a_{n-1})$ and $(a_0, \ldots, a_{n-1}, 0, \ldots, 0)$. Second, note that, because we perform this increase, numbers which skip primes in their factorisations are not codes of sequences.

7.5. *b*-adic Coding of Sequences. Given $(a_0, \ldots, a_{n-1})$, choose $b > a_0, \ldots, a_{n-1}$ and assign $(a_0, \ldots, a_{n-1})$ the b-code

$$a_0 + a_1 b + a_2 b^2 + \ldots + a_{n-1} b^{n-1}.$$

For fixed b, every non-empty sequence, however long, of natural numbers below b has a unique b-code. To handle the empty sequence, we could again increase the coëfficients by 1, assign the empty sequence the value 0, and toss out those sequences in which $b-1$ appears. In doing so, we lose the one-one correspondence as, e.g., $1 + b^2$ no longer is a code.

7.6. Dyadic Coding of Finite Sets. In Recursion Theory, one defines the finite set D_x with *canonical index* x by:
$D_0 = \{\ \}$
$D_x = \{x_0, x_1, \ldots, x_{n-1}\}$, where $x = 2^{x_0} + 2^{x_1} + \ldots + 2^{x_{n-1}}$ and $x_0 < x_1 < \ldots < x_{n-1}$.

This gives a one-one correspondence between ω and the collection of all finite sets of natural numbers.

7.7. Dyadic Coding of Strictly Increasing Finite Sequences. If $(s_0, \ldots, s_{n-1})$ is a strictly increasing sequence of natural numbers, we can identify it with the set $\{s_0, \ldots, s_{n-1}\}$ and assign it the code assigned in 7.6. Conversely, each natural number can be uniquely decoded to such a sequence.

7.8. Dyadic Coding of Finite Sequences. Assign the empty sequence the number 0. If $(a_0, \ldots, a_{n-1})$ is a non-empty sequence of natural numbers, associate with it the strictly increasing sequence $(a_0, a_0 + a_1 + 1, \ldots, a_0 + \ldots + a_i + i, \ldots, a_0 + \ldots + a_{n-1} + n-1)$ and, thus, the code

$$2^{a_0} + 2^{a_0+a_1+1} + \ldots + 2^{a_0+\ldots+a_i+i} + \ldots + 2^{a_0+\ldots+a_{n-1}+n-1}.$$

The encoding given by 7.8 establishes a one-one correspondence between ω and the set of all finite sequences of natural numbers.

7.9. Coding Finite Sequences by Iterating the Pairing Function. First, define n-tupling functions:

$$\langle x_0 \rangle_1 = x_0$$
$$\langle x_0, x_1 \rangle_2 = \langle x_0, x_1 \rangle$$
$$\langle x_0, ..., x_n \rangle_{n+1} = \langle \langle x_0, ..., x_{n-1} \rangle_n, x_n \rangle.$$

Since each natural number codes a unique n-tuple for each n (*Exercise*: Prove this by induction on n.), a likely way of coding arbitrary finite sequences is to think of a given number as a pair, the first entry giving the length of the sequence being coded:

$$(x_0, ..., x_{n-1}) \mapsto \langle n, \langle x_0, ..., x_{n-1} \rangle_n \rangle.$$

To handle the empty sequence and make the encoding a one-one correspondence between ω and the set $\omega^{<\omega}$ of all finite sequences of natural numbers, we shift things slightly:

$$(\,) \mapsto 0$$
$$(x_0, ..., x_{n-1}) \mapsto \langle n-1, \langle x_0, ..., x_{n-1} \rangle_n \rangle + 1.$$

The big question is: what are we going to do with this plethora of coding schemes? In the next two sections, we will put them to two different uses. For now, let me spout a few generalities.

Let me begin with the pairing function. As I mentioned in section 3, Cantor's primary interest had been the pairing function on R. He had sought to prove such a thing couldn't exist, his response on finding such a function being "I see it, but I don't believe it." Experience in analysis had taught him that n-dimensional space was different from m-dimensional space for $m \neq n$, and he had just proven otherwise. However, his map was not (and, indeed, could not be) continuous. Hence, one cannot use the pairing function on R to reduce questions about continuous functions of, say, two variables to questions about continuous functions of one variable. On ω, however, we usually don't have any topology and for most purposes we can identify functions F on ω^n with associated functions $\check{F}$ on ω: simply define

$$\check{F}(\langle x_0, ..., x_{n-1} \rangle_n) = F(x_0, ..., x_{n-1}).$$

Of course, with questions of linearity, the algebraic invariance of dimensions from linear algebra blocks this (cf. also Exercise 4 of section 6, above). However, for more logical questions like definability and effectiveness, this identification will eventually allow us, whenever we find it convenient, to deal with unary functions.

And what about the encoding of finite sequences? As one rarely uses functions of a variable number of arguments, a similar motivation for this latter encoding would not be convincing, particularly in light of the fact that several of our encodings were not one-one correspondences. However, we can compare any one of these encodings of the finite sequences of natural numbers with the encoding in naïve set theory of functions by sets of ordered pairs. Just as the ultra-simple concepts of set and membership, through such encoding tricks, engender the full range of mathematical practice, the encoding of finite functions in number theory augers an astounding breadth of theory. In this and the next volume we shall get a glimpse of how wide a subject number theory truly is. In the present volume we shall see how devious and complex the expressive power of arithmetic is, and obtain thereby a partial explanation of the necessity of the use of devious methods of proof in number theory— as, for example, the genuine use of the calculus in section 5, above. As I say, this explanation will be partial: it won't explain why, in any given circumstance, we must transcend the methods obviously appropriate to the problem at hand; it will merely explain that we will often have to do so.

On a less philosophical note, let me make the rather obvious remark that a coding scheme depends not only on the success of the encoding, but also on that of its decoding: if x codes a sequence $(x_0, ..., x_{n-1})$, we need not only to go from $(x_0, ..., x_{n-1})$ to x, but also from x to $(x_0, ..., x_{n-1})$— in particular, from x and i, $0 \le i < n$, to x_i. Decoding via Gödel's β-function has already been explained. The reader should try his hand at the other decoding problems before moving on to the next section.

Exercises.

1. Find a code for the sequence (1, 2, 3) in each coding scheme 7.3, 7.4, 7.5, 7.8, and 7.9.
2. Modify 7.8 by mapping

$$( ) \mapsto 0$$
$$(a_0, ..., a_{n-1}) \mapsto 2^{a_0} + 2^{a_0+a_1} + ... + 2^{a_0+...+a_{n-1}}.$$

Show: Restricted to positive integers, this yields a one-one correspondence of the set of finite sequences of positive integers with the set of positive integers.

3. For $x, y, n \in \omega$, suppose $n = \langle x, y\rangle$.

i. Define $A(n) = [\sqrt{2n}]$, where $[\cdot]$ denotes the greatest integer function. Show:

$$A(n) = \begin{cases} x + y, & y \ge x \\ x + y - 1, & y < x \end{cases}$$

ii. Define

$$B(n) = \begin{cases} A(n), & 2n - A(n)^2 \ge A(n) \\ A(n) - 1, & 2n - A(n)^2 < A(n). \end{cases}$$

Show: $B(n) = x + y$.

iii. Show: $x = n - \dfrac{B(n)(B(n)+1)}{2}$

$y = B(n) - x$.

iv. Find x, y for $n = 64$, $n = 438$, and $n = 14285$.

4. As in Exercise 3, show how to decode $(x_0, ..., x_{n-1})$ for schemes 7.5, 7.6, and 7.8 using functions like $[a/b]$, $+$, $\cdot$, Rem(a,b), and $\dot{-}$ defined by

$$x \dot{-} y = \begin{cases} x - y, & x \ge y \\ 0, & x < y. \end{cases}$$

8. Primitive Recursion

In this section we will take a first look at computability by studying the class of those functions computable by the simplest— most primitive— sort of recursion. These functions are called the primitive recursive functions.

The computation (or: definition) of functions by recursion is hardly new. The Euclidean algorithm, encountered two sections back, is a very old example. More famous, if not nearly so old, is the function enumerating the Fibonacci numbers of Fibonacci (aka Leonardo of Pisa) from the 12th century:

$$F(0) = 0, \quad F(1) = 1$$
$$F(n+2) = F(n) + F(n+1).$$

These recursions, however, are logically sophisticated. There is a primitive form, first formally annunciated by Richard Dedekind (1870s) in his *Was sind und was sollen die Zahlen?*:

8.1. Theorem on Recursive Definition. *Let X be a set, G a function from X to X, and x_0 an arbitrary element of X. There is a (unique) function $F:\omega \rightarrow X$ satisfying*
i. $F(0) = x_0$
ii. $F(n+1) = G(F(n))$, *for all* $n \in \omega$.

I shall not prove this here. Once one understands induction, the truth of the Theorem is clear. Dedekind had to prove this Theorem because he had to show his set theoretic characterisation of the set of natural numbers would account for all the usual properties of the set of natural numbers. In Chapter IV of volume II, we will have to prove a version of this theorem to verify that the proof can be carried out within a given formal theory. For now, however, we can simply accept the result as obvious.

Dedekind's immediate use of definition by recursion was to show that the natural numbers could be characterised uniquely up to isomorphism by its inductive property. He then went on to observe that addition and multiplication could be defined inductively first by defining, for each $a \in \omega$, the auxiliary functions

$$\begin{cases} A_a(0) = a \\ A_a(b+1) = A_a(b) + 1 \end{cases}$$

$$\begin{cases} M_a(0) = a \\ M_a(b+1) = A_a(M_a(b)) \end{cases}$$

and then by defining

$$a + b = A_a(b)$$
$$a \cdot b = M_a(b).$$

If one allows parameters in 8.1, the recursive definitions of addition and multiplication simplify:

$$\begin{cases} a + 0 = a \\ a + (b + 1) = (a + b) + 1 \end{cases}$$

$$\begin{cases} a \cdot 0 = 0 \\ a \cdot (b + 1) = (a \cdot b) + a . \end{cases}$$

Dedekind gave these recursions and proved, by induction, a few basic properties of addition and multiplication such as the commutative, associative, and distributive laws. We will encounter these proofs in Chapter IV of volume II, when we discuss the so-called Peano axioms for arithmetic.

In 1919, Thoralf Skolem went much further than Dedekind in developing elementary number theory on a recursive basis. In his paper (published in 1923), the German title of which is longer than this chapter, he restricted his recursions to functions from natural numbers to natural numbers (i.e., he took X to be ω in 8.1), but he did allow parameters. Skolem's goal was philosophical: he didn't particularly care for the use of quantifiers and wanted to show how the "recursive mode of thought" could account for a significant fragment of number theory. Although we will eventually see that no such restriction on methods or language will yield everything, it is worth noting that he got as far as elementary prime number theory, recursively generating primes and inductively proving properties of them. Much of what follows in this section was begun by Skolem in this long-named paper.

I did not mention Skolem, however, with the purpose of giving him his due credit (his work is often credited to others), but rather to make the seemingly small point that he explicitly allowed parameters and restricted his recursions to functions from natural numbers to natural numbers. By 1925 Wilhelm Ackermann had shown (and in 1928 he published proof) that more functions were definable by recursion if one allowed auxiliary recursive definitions which did not have this restriction on X. In defining the class of primitive recursive functions, we will have to be exceedingly precise. If we do not actually reach the point of splitting hairs, we at least go a shade beyond dotting the i's and crossing the t's. Such a level of precision was first attained by Kurt Gödel in his famous paper of 1931, from which mine we have already unearthed the β-function, from which we will presently extract much more, and which we will not exhaust until we have finished with Chapter IV of the next volume. Because of his precise definition of the primitive recursive functions and the use to which he put them, some authors erroneously credit Gödel with having first introduced this class of functions.

8.2. Definition. Let $k > 0$. A function $F{:}\omega^k \rightarrow \omega$ of natural numbers is *primitive recursive* if it can be defined after finitely many steps by means of the following rules:

F1. $Z(x) = 0$ *Zero*

F2. $S(x) = x + 1$ *Successor*

F3^n_i. $P^n_i(x_0, \ldots, x_{n-1}) = x_i \quad (0 \le i < n)$ *Projection*

$F4^n_m$. $F(x_0, ..., x_{n-1}) = G(H_0(x_0,...,x_{n-1}), ..., H_{m-1}(x_0,...,x_{n-1}))$ *Composition*

$F5^n$. $\begin{cases} F(0,x_1,\ldots,x_n) = G(x_1,\ldots,x_n) \\ F(x+1,x_1,\ldots,x_n) = H(F(x,x_1,\ldots,x_n),x_1,\ldots,x_n) \end{cases}$ *Primitive Recursion*

where $m > 0$ in $F4^n_m$, and $n > 0$ in $F3^n_i$, $F4^n_m$, and $F5^n$.

The functions Z, S, and P^n_i are the initial functions from which all others are to be generated by composition and recursion. Note that through compositions, the successor and the zero functions will allow the construction of all the unary constant functions. With these, the projection functions, and composition, we get the closure of the class of primitive recursive functions under such mild explicit definitions as are effected by plugging in constants, permuting variables, adding dummy variables, and even diagonalising—identifying two or more variables. E.g., if G is a primitive recursive 5-ary function, the function

$$F(x_0, x_1, x_2) = G(0, x_0, x_0, x_2, x_1)$$

is also primitive recursive. For:

$$F(x_0, x_1, x_2) = G(Z(P^3_0(\boldsymbol{x})), P^3_0(\boldsymbol{x}), P^3_0(\boldsymbol{x}), P^3_2(\boldsymbol{x}), P^3_1(\boldsymbol{x})),$$

where $\boldsymbol{x}$ abbreviates the sequence x_0, x_1, x_2.

8.3. Examples. *The following functions are primitive recursive:*

i. $K^n_k(x_0, ..., x_{n-1}) = k$, *any* $k \in \omega$ *Constant*

ii. $A(x, y) = x + y$ *Addition*

iii. $M(x, y) = x \cdot y$ *Multiplication*

iv. $E(x, y) = x^y$ *Exponentiation*

v. $pd(x) = \begin{cases} x - 1, & x > 0 \\ 0, & x = 0 \end{cases}$ *Predecessor*

vi. $x \dot{-} y = \begin{cases} x - y, & x \geq y \\ 0, & x < y \end{cases}$ *Cut-Off Subtraction*

vii. $sg(x) = \begin{cases} 0, & x = 0 \\ 1, & x > 0 \end{cases}$ *Signum*

viii. $\overline{sg}(x) = \begin{cases} 1, & x = 0 \\ 0, & x > 0. \end{cases}$ *Signum Complement*

Proof. i. $K^n_k(x_0, ..., x_{n-1}) = S \cdots SZ(P^n_0(x_0, ..., x_{n-1}))$, where the number of compositions of S is k.

ii. The usual recursion for addition,

$$A(x, 0) = x$$
$$A(x, y+1) = A(x, y) + 1 = S(A(x, y)),$$

is on the *wrong* variable. The formal definition of primitive recursion (8.2.*F5*) requires recursion to be performed on the first variable of the function. We get around this either, since addition is commutative as well as associative, by doing the recursion on x,

$$A(0, y) = y$$
$$A(x+1, y) = S(A(x, y)),$$

or by first defining $B(y, x) = x + y$ by recursion on y,

$$B(0, x) = x$$
$$B(y+1, x) = S(B(y, x)),$$

and then defining $A(x, y) = B(P_1^2(x, y), P_0^2(x, y))$.

iii - iv. Left to the reader.

v. Observe: $pd(0) = 0$
$pd(x+1) = x.$

vi. Observe: $x \dot{-} 0 = x$
$x \dot{-} (y+1) = pd(x \dot{-} y).$

vii - viii. The signum function and its complement can be defined by recursion,

$$sg(0) = 0 \qquad \overline{sg}\,(0) = 1$$
$$sg(x+1) = 1; \qquad \overline{sg}\,(x+1) = 0,$$

or by cut-off subtraction,

$$\overline{sg}\,(x) = 1 \dot{-} x, \qquad sg(x) = 1 \dot{-} \overline{sg}\,(x).$$

QED

8.4. Lemma. (Definition-by-Cases). *Let G_1, G_2, H be primitive recursive n-ary functions, and define F by*

$$F(x_0, ..., x_{n-1}) = \begin{cases} G_1(x_0, ..., x_{n-1}), & H(x_0, ..., x_{n-1}) = 0 \\ G_2(x_0, ..., x_{n-1}), & H(x_0, ..., x_{n-1}) \neq 0. \end{cases}$$

Then: F is primitive recursive.

Proof. Letting $\boldsymbol{x}$ abbreviate $x_0, ..., x_{n-1}$, we see

$$F(\boldsymbol{x}) = G_1(\boldsymbol{x}) \cdot \overline{sg}\,(H(\boldsymbol{x})) + G_1(\boldsymbol{x}) \cdot sg(H(\boldsymbol{x})).$$

The closure of the class of primitive recursive functions under composition yields the primitive recursiveness of F. QED

The use of H in Lemma 8.4 suggests the following definition.

8.5. Definitions. A relation $R \subseteq \omega^n$ is *primitive recursive* if its *representing function*,

$$\chi_R(x) = \begin{cases} 0, & R(x) \text{ holds} \\ 1, & R(x) \text{ fails,} \end{cases}$$

is a primitive recursive function.

Warning. χ_R is not the usual *characteristic* function one comes across in mathematics. In recursion theory it is customary to reverse the rôles of 0 and 1 so that 0 represents truth and 1 falsehood. The reason is simple: 0 stands out among the natural numbers, particularly in definitions by recursion, and is thus most deserving of playing the rôle of truth.

8.6. Example. *The relation of equality is primitive recursive.*

Proof. Define

$$H(x,y) = |x - y| = (x \dot{-} y) + (y \dot{-} x),$$

so

$$H(x,y) = 0 \text{ iff } x = y.$$

Either apply Lemma 8.4 or observe,

$$\chi_=(x,y) = sg(H(x,y)).$$ QED

To apply Lemma 8.4 successfully, we will want a larger stock of primitive recursive relations. We can get quite a few by simply enumerating closure conditions. The following Lemma cites some easy ones.

8.7. Lemma. *Let* $T, U \subseteq \omega^n$ *be primitive recursive relations.*

i. *If* $G_1,...,G_n{:}\omega^m \to \omega$ *are primitive recursive functions, the relation*

$$R(x_0, ..., x_{m-1}): \quad T(G_1(x_0, ..., x_{m-1}), ..., G_n(x_0, ..., x_{m-1}))$$

is primitive recursive.

ii. *The relations*

$$R_\wedge(x_0, ..., x_{n-1}): \quad T(x_0, ..., x_{n-1}) \wedge U(x_0, ..., x_{n-1})$$
$$R_\vee(x_0, ..., x_{n-1}): \quad T(x_0, ..., x_{n-1}) \vee U(x_0, ..., x_{n-1})$$
$$R_\neg(x_0, ..., x_{n-1}): \quad \neg T(x_0, ..., x_{n-1})$$

are primitive recursive.

Proof. i. Observe, for $\boldsymbol{x} = x_0, ..., x_{n-1}$,

$$\chi_R(\boldsymbol{x}) = \chi_T(G_1(\boldsymbol{x}), ..., G_n(\boldsymbol{x})).$$

ii. Observe,

$$\chi_{R_\wedge}(\boldsymbol{x}) = sg(\chi_T(\boldsymbol{x}) + \chi_U(\boldsymbol{x}))$$
$$\chi_{R_\vee}(\boldsymbol{x}) = sg(\chi_T(\boldsymbol{x})\cdot\chi_U(\boldsymbol{x}))$$
$$\chi_{\neg R}(\boldsymbol{x}) = \overline{sg}(\chi_T(\boldsymbol{x})).$$ QED

I could have written $T \cap U$, $T \cup U$, and $\omega^n - T$ for $R_\wedge$, $R_\vee$, and $R_\neg$ in 8.7.ii instead of $T \wedge U$, $T \vee U$, and $\neg T$. However, using a mixture of 8.7.i and 8.7.ii (using the projection functions in 8.7.i), we readily see that we do not have to assume T, U to have the same variables. For example, if T is binary and U is ternary, $T(x_0, x_1) \wedge U(x_1, x_2, x_3)$ makes sense and is primitive recursive if T, U are, even though $T \cap U$ ought not to be defined:

$$(T \wedge U)(\boldsymbol{x}):\quad T(P^4_0(\boldsymbol{x}), P^4_1(\boldsymbol{x})) \wedge U(P^4_1(\boldsymbol{x}), P^4_2(\boldsymbol{x}), P^4_3(\boldsymbol{x})).$$

A second point about this proof is that the use of the sum and product to treat conjunction and disjunction generalises:

$$\chi_{\forall y \le x R(y,\boldsymbol{x})}(x, \boldsymbol{x}) = sg\Big(\sum_{y \le x} \chi_R(y, \boldsymbol{x})\Big)$$

$$\chi_{\exists y \le x R(y,\boldsymbol{x})}(x, \boldsymbol{x}) = \prod_{y \le x} \chi_R(y, \boldsymbol{x}),$$

where $\boldsymbol{x}$ abbreviates the sequence $x_1, \ldots, x_n$.

8.8. Lemma. i. *Let $G: \omega^{n+1} \to \omega$ be primitive recursive and let F_1, F_2 be defined by*

$$F_1(x, \boldsymbol{x}) = \sum_{y \le x} G(y, \boldsymbol{x}) = \sum_{y=0}^{x} G(y, \boldsymbol{x})$$

$$F_2(x, \boldsymbol{x}) = \prod_{y \le x} G(y, \boldsymbol{x}) = \prod_{y=0}^{x} G(y, \boldsymbol{x}),$$

where $\boldsymbol{x}$ abbreviates the sequence $x_1, \ldots, x_n$. Then: F_1, F_2 are primitive recursive.
ii. *Let $T \subseteq \omega^{n+1}$ be primitive recursive and define R_1, R_2 by*

$$R_1(x, \boldsymbol{x}):\ \forall y \le x\, T(y, \boldsymbol{x})$$

$$R_2(x, \boldsymbol{x}):\ \exists y \le x\, T(y, \boldsymbol{x}),$$

where $\boldsymbol{x}$ abbreviates the sequence $x_1, \ldots, x_n$. Then: R_1, R_2 are primitive recursive.

Proof. i. F_1 is defined by the recursion,

$$F_1(0, \boldsymbol{x}) = G(0, \boldsymbol{x})$$

$$F_1(x+1, \boldsymbol{x}) = F_1(x, \boldsymbol{x}) + G(x+1, \boldsymbol{x}).$$

F_2 is similarly defined.
ii. By part i and the remark preceding the lemma. QED

We are now in position to give some interesting primitive recursive relations.

8.9. Examples. *The following relations are primitive recursive:*

i. $x \mid y$: $\exists z \le y (y = x \cdot z)$

ii. *x is even*: $2 \mid x$

iii. *x is odd*: $2 \nmid x$, or $\exists y \le x (x = 2y + 1)$

iv. $x \le y$: $\exists z \le y(x = z)$

$x < y$: $\exists z \le y(z = x+1)$

v. $r = \mathrm{Rem}(x,y)$: $\exists q \le x(x = q \cdot y + r \wedge r < y)$

vi. $z = [x/y]$ (*z is the greatest integer in x/y*): $\exists r \le x((x = z \cdot y + (r \dot{-} 1)) \wedge y > 0$

vii. $z = \langle x, y\rangle$: $2z = (x+y)^2 + 3x + y$

viii. $x = \pi_1^2(z)$ (*x is the first coördinate of the pair coded by z*): $\exists y \le z(z = \langle x, y\rangle)$

$y = \pi_2^2(z)$: $\exists x \le z(z = \langle x, y\rangle)$

ix. $y \in D_x$ (*y is an exponent in the dyadic expansion of x*— cf. 7.6):

[Observe

$$y \in D_x \text{ iff } [x/2^y] \text{ is odd}$$
$$\text{iff } \exists z \le x(z = [x/2^y] \wedge z \text{ is odd}).]$$

In this last example, I went one step further than may have appeared necessary. By 8.7.i, given the primitive recursiveness of the relation "x is odd" and the primitive recursiveness of a function $F(x,y)$, we can conclude the primitive recursiveness of the relation "$F(x,y)$ is odd". So why did I not stop at "$[x/2^y]$ is odd"? The answer is that in 8.9 we have only shown the primitive recursiveness of the *graphs* of the functions $\mathrm{Rem}(\cdot,\cdot)$, $[\cdot,\cdot]$, $\langle\cdot,\cdot\rangle$, π_1^2 and π_2^2; we have not shown the functions themselves to be primitive recursive. There are functions with primitive recursive graphs, but which are not themselves primitive recursive.If, however, one can bound the value of such a function primitive recursively, the function will be primitive recursive.

8.10. Definition. (Bounded Minimisation). Let $R \subseteq \omega^{n+1}$ be given. By

$$F(x, \boldsymbol{x}) = \mu y < x\, R(y, \boldsymbol{x}),$$

we mean the function F assigning to x and $\boldsymbol{x} = x_1, ..., x_n$ the least $y < x$ such that $R(y, \boldsymbol{x})$ holds, if such a y exists, and the value x otherwise.

8.11. Lemma. *Let $R \subseteq \omega^{n+1}$ be primitive recursive and let*

$$F(x, \boldsymbol{x}) = \mu y < x\, R(y, \boldsymbol{x}).$$

Then: F is primitive recursive.

Proof. First define the auxiliary function,

$$F_1(x, \boldsymbol{x}) = \begin{cases} 0, & \exists y \le x\, R(y, \boldsymbol{x}) \\ 1, & \neg\exists y \le x\, R(y, \boldsymbol{x}), \end{cases}$$

by cases. Then observe that

$$F(x, \boldsymbol{x}) = \sum_{y<x} F_1(y, \boldsymbol{x}). \qquad \text{QED}$$

8.12. Corollary. *If $R \subseteq \omega^{n+1}$ is primitive recursive and $H:\omega^{n+1} \to \omega$ is primitive recursive, then*

$$F(x, \mathbf{x}) = \mu y < H(x, \mathbf{x})\, R(y, \mathbf{x}) \qquad (*)$$

is primitive recursive.

This function F is defined from that of 8.11 by simple composition— whence the result.

By 8.11 and 8.12, the functions occurring in 8.9 are primitive recursive:

$$\mathrm{Rem}(x, y) = \mu r < y\, [\exists q \le x(x = qy + r)]$$

$$[x/y] = \mu z < x + 1\, [x < (z+1)y]$$

$$\langle x, y\rangle = \mu z < (x+y)^2 + 3x + y + 1\, [2z = (x+y)^2 + 3x + y]$$

$$= [((x+y)^2 + 3x + y)/2]$$

$$\pi_1^2(z) = \mu x < z + 1\, [\exists y \le z(z = \langle x, y\rangle)]$$

$$\pi_2^2(z) = \mu y < z + 1\, [z = \langle \pi_1^2(z), y\rangle].$$

[It may be remarked that the first and second of these definitions give values for $y = 0$. The reader may amend these values through Definition-by-Cases.]

By these last three examples, the pairing function and its inverses or projections are primitive recursive. What about those functions dealing with the encoding of sequences? First, what are the appropriate functions? And, second, are these functions primitive recursive? Experience shows that the crucial functions are the length function,

$$lh(\,(x_0, \ldots, x_{n-1})\,) = n,$$

the projections,

$$(x_0, \ldots, x_{n-1}) \mapsto x_i,$$

sequence formation,

$$x \mapsto (x),$$

and concatenation,

$$(x_0, \ldots, x_{n-1}) * (y_0, \ldots, y_{m-1}) = (x_0, \ldots, x_{n-1}, y_0, \ldots, y_{m-1}).$$

As we had several coding schemes for finite sequences, we have several sets of problems— one for each such scheme. For the sake of definiteness, we shall use the dyadic coding of 7.8 as our standard.

8.13. Definitions. For $x \ne 0$, let $x = 2^{x_0} + \ldots + 2^{x_{n-1}}$, with $x_0 < \ldots < x_{n-1}$. Define the sequences S_0, S_x coded by $0, x$ to be

$$S_0 = (\,)$$

$$S_x = (x_0, x_0 - x_1 - 1, x_2 - x_1 - 1, \ldots, x_{n-1} - x_{n-2} - 1)$$

$$= (a_0, a_1, \ldots, a_{n-1}), \text{ say.}$$

We also define

$$lh(0) = 0, \qquad lh(x) = n,$$

$$(0)_i = 0, \text{ for all } i,$$

$$(x)_i = \begin{cases} x, & i \geq lh(x) \\ a_i, & i < lh(x), \end{cases} \quad \text{where } a_i = \begin{cases} x_0, & i = 0 \\ x_i - x_{i-1} - 1, & 0 < i < n, \end{cases}$$

and $\quad (0) = 2^0 = 1, \qquad (x) = 2^x.$

Finally, if $S_x = (a_0, a_1, ..., a_{n-1})$, $S_y = (b_0, b_1, ..., b_{m-1})$, we define

$$x{*}y = \text{the unique } z \text{ such that } S_z = (a_0, a_1, ..., a_{n-1}, b_0, b_1, ..., b_{m-1}).$$

8.14. Lemma. *The functions $lh(x)$, $(x)_y$, (x), and $x{*}y$ are primitive recursive.*

Proof. The first and third of these functions are easily seen to be primitive recursive. For the length function, observe

$$h(x) = \sum_{y \leq x} \overline{sg}\,(\chi_{y \in D_x}(y, x));$$

for sequence formation, simply note $(x) = 2^x$.

For projection, first find the y-th element of $D_x = \{x_0, ..., x_{n-1}\}$:

$$EL(x, 0) = \mu z < x\,[z \in D_x]$$

$$EL(x, y+1) = \mu z < x\,[z > EL(x, y) \wedge z \in D_x].$$

If $x = 2^{x_0} + ... + 2^{x_{n-1}}$, then

$$EL(x, i) = \begin{cases} x_i, & 0 \leq i < n \\ x, & n \leq i. \end{cases}$$

[Note that $EL(0,i) = 0$.] Thus:

$$(x)_0 = EL(x, 0)$$

$$(x)_{y+1} = EL(x, y+1) \dot{-} (EL(x, y) + 1).$$

Finally, suppose

$$S_x = (a_0, a_1, ..., a_{n-1}),\ S_y = (b_0, b_1, ..., b_{m-1}).$$

The code of $S_x * S_y = (a_0, a_1, ..., a_{n-1}, b_0, b_1, ..., b_{m-1})$ is

$$2^{x_0} + 2^{x_0+x_1+1} + ... + 2^{x_0+x_1+...+x_{n-1}+n-1} +$$

$$+ 2^{x_0+...+x_{n-1}+y_0+n} + ... + 2^{x_0+...+x_{n-1}+y_0+...+y_{m-1}+n+m-1},$$

which is

$$x + 2^{x_0+x_1+...+x_{n-1}+n}\left(2^{y_0} + ... + 2^{y_0+...+y_{m-1}+m-1}\right) =$$

$$= x + 2^{x_0+x_1+...+x_{n-1}+n} \cdot y = x + 2^{EL(x, lh(x) \dot{-} 1)+1} \cdot y.$$

Thus,

$$x*y = \begin{cases} 0, & x = 0 \wedge y = 0 \\ y, & x = 0 \wedge y \neq 0 \\ x, & x \neq 0 \wedge y = 0 \\ x + 2^{EL(x, lh(x) \dot{-} 1)+1} \cdot y, & x \neq 0 \wedge y \neq 0. \end{cases}$$

QED

The primitive recursiveness of these sequence operations greatly increases our ability to encode objects. Much of the sequel will be devoted to such actual encoding. For now, we will simply use these sequence operations to verify the closure of the class of primitive recursive functions under a couple of more exotic recursions.

8.15. Definition. Let $F:\omega^{n+1} \to \omega$ be given. The *course of values function* $\tilde{F}$ associated with F is defined by

$$\tilde{F}(x, \boldsymbol{x}) = (F(0, \boldsymbol{x}), F(1, \boldsymbol{x}), \ldots, F(x, \boldsymbol{x})),$$

the sequence of the first $x + 1$ values of F.

8.16. Theorem. (Course-of-Values Recursion). *Let G, H be primitive recursive and suppose F is defined by*

$$F(0, \boldsymbol{x}) = G(\boldsymbol{x})$$
$$F(x+1, \boldsymbol{x}) = H(\tilde{F}(0, \boldsymbol{x}), x, \boldsymbol{x}).$$

Then: F is primitive recursive.

Note that this recursion differs from the usual primitive recursion in that H uses $\tilde{F}(0,\boldsymbol{x})$— whence all of $F(0,\boldsymbol{x})$, ..., $F(x,\boldsymbol{x})$— and not merely $F(x,\boldsymbol{x})$. The definition of the Fibonacci sequence takes this form:

$$F(0) = 0$$

$$F(x+1) = \begin{cases} 1, & x = 0 \\ F(x-1) + F(x), & \text{otherwise,} \end{cases}$$

whence we can take

$$H(y, x) = \begin{cases} 1, & x \leq 1 \\ (y)_x + (y)_{x-1}, & x > 1. \end{cases}$$

Proof of 8.16. We first show $\tilde{F}$ to be primitive recursive by noting

$$\tilde{F}(0,\boldsymbol{x}) = (G(\boldsymbol{x})) = 2^{G(\boldsymbol{x})}$$
$$\tilde{F}(x+1,\boldsymbol{x}) = \tilde{F}(x,\boldsymbol{x}) * H(\tilde{F}(x,\boldsymbol{x}), x, \boldsymbol{x}).$$

[To be pedantic one last time, we have

$$\tilde{F}(0,\boldsymbol{x}) = G_1(\boldsymbol{x})$$

$$\widetilde{F}(x+1,\boldsymbol{x}) = H_1(\widetilde{F}(x,\boldsymbol{x}),x,\boldsymbol{x}),$$

where

$$G_1(\boldsymbol{x}) = 2^{G(\boldsymbol{x})}$$
$$H_1(y,x,\boldsymbol{x}) = y*(H(y,x,\boldsymbol{x})).]$$

To obtain the primitive recursiveness of F, simply note

$$F(x,\boldsymbol{x}) = (\widetilde{F}(x,\boldsymbol{x}))_x.$$

QED

We have already illustrated an application of Course-of-Values Recursion in citing the Fibonacci sequence. Closely related to the Fibonacci sequence are many Pell sequences, the exact significance of which we will learn in the next chapter. The Pell sequences we will be most interested in can be viewed as a pair of functions:

$$\begin{cases} X(0,a) = 1 \\ X(x+1,a) = aX(x,a) + (a^2-1)Y(x,a) \end{cases}$$
$$\begin{cases} Y(0,a) = 1 \\ Y(x+1,a) = X(x,a) + aY(x,a). \end{cases}$$

One way of proving the primitive recursiveness of these functions X, Y is to separate the two recursions completely. A little algebra yields:

$$\begin{cases} X(0,a) = 1 \\ X(1,a) = a \\ X(x+2,a) = 2aX(x+1,a) - X(x,a) \end{cases}$$
$$\begin{cases} Y(0,a) = 0 \\ Y(1,a) = 1 \\ Y(x+2,a) = 2aY(x+1,a) - Y(x,a). \end{cases}$$

We can then appeal to Course-of-Values Recursion to conclude the primitive recursiveness of X and Y. Such a strategy readily handles *linear* simultaneous recursions, but not the general case, for which we need the following.

8.17. Theorem. (Simultaneous Recursion). *Suppose G_1, G_2, H_1, H_2 are primitive recursive and F_1, F_2 are defined by*

$$\begin{cases} F_1(0,\boldsymbol{x}) = G_1(\boldsymbol{x}) \\ F_1(x+1,\boldsymbol{x}) = H_1(F_1(x,\boldsymbol{x}),F_2(x,\boldsymbol{x}),x,\boldsymbol{x}) \end{cases}$$
$$\begin{cases} F_2(0,\boldsymbol{x}) = G_2(\boldsymbol{x}) \\ F_2(x+1,\boldsymbol{x}) = H_2(F_1(x,\boldsymbol{x}),F_2(x,\boldsymbol{x}),x,\boldsymbol{x}). \end{cases}$$

Then: F_1 and F_2 are primitive recursive.

Proof. Define F by the primitive recursion:

$$F(0,\boldsymbol{x}) = (F_1(0,\boldsymbol{x}), F_2(0,\boldsymbol{x})) = (G_1(\boldsymbol{x}), G_2(\boldsymbol{x}))$$

$$F(x+1,\boldsymbol{x}) = (H_1((F(x,\boldsymbol{x}))_0, (F(x,\boldsymbol{x}))_1, x, \boldsymbol{x})), H_2((F(x,\boldsymbol{x}))_0, (F(x,\boldsymbol{x}))_1, x, \boldsymbol{x})).$$

A simple induction shows:

$$F(x, \boldsymbol{x}) = (F_1(x,\boldsymbol{x}), F_2(x,\boldsymbol{x})).$$ QED

For 8.17 we could have used the pairing function instead of the sequence operations in the proof, or we could have introduced a course-of-values for F and allowed simultaneous course-of-values recursion in the statement of the Theorem. There is quite a variety of forms of recursion under which one may test the closure of the class of primitive recursive functions, the test sometimes proving successful and sometimes not. A few positive cases are to be found in the Exercises; a major negative one occupies the next section. A major study of variations of these recursions was carried out in the 1930s by Rosza Péter and appeared in book form in 1950 (cf. the references to this section). The two closure results we have given are due to her.

Exercises.

1. Prove the primitive recursiveness of the function,

$$gcd(x, y) = \text{greatest common divisor of } x, y,$$

discussed in section 6, above.

2. Let a, b, c, d be positive integers and define the two sequences,

$$X(0) = 1 \qquad Y(0) = 0$$
$$X(n+1) = aX(n) + bY(n); \qquad Y(n+1) = cX(n) + dY(n).$$

Show: X, Y satisfy individual course-of-values recursions of the form,

$$Z(n+2) = eZ(n) + fZ(n+1),$$

for rational e, f.

3. Prove the primitive recursiveness of the length, projection, sequence formation, and concatenation functions for the b-adic scheme 7.5. These functions should include b among their arguments, e.g.,

$$length(x,b) = n$$

if

$$x = a_0 + a_1 b + \dots + a_{n-1} b^{n-1},$$

where $b > 1$, $a_{n-1} \neq 0$, and $0 \leq a_i < b$ for each i.

4. There are two schemes for obtaining a coding of finite sequences by iterating the pairing function. Left-associating (as in 7.9), one gets

$$\langle x_0, \dots, x_n \rangle_{n+1} = \langle \langle x_0, \dots, x_{n-1} \rangle_n, x_n \rangle$$

for one's (n+1)-tupling function; and, right-associating, one gets

$$\langle x_0, \dots, x_n \rangle_{n+1} = \langle x_0, \langle x_1, \dots, x_n \rangle_n \rangle.$$

The recursions for the projection functions $\pi(n, i, x) = \pi_i^n(x) = x_{i-1}$ for $x = \langle x_0, \dots, x_{n-1} \rangle$ satisfy

$$\pi(n+1,i,x) = \begin{cases} \pi(n,i,\pi_1^2(x)), & 1 \le i < n \\ \pi_2^2(x), & i = n+1 \end{cases}$$

assuming left-association, and

$$\pi(n+1,i,x) = \begin{cases} \pi_1^2(x), & i = 1 \\ \pi(n,i-1,\pi_2^2(x)), & 2 \le i \le n+1 \end{cases}$$

assuming right-association.

i. Show, by suitably generalising course-of-values recursion, that each recursion yields a primitive recursive function.

ii. Show the primitive recursiveness of the length, sequence formation, projection, and concatenation functions for the encoding scheme 7.9.

[*Remark.* The apparent necessity of appeal to course-of-values recursion— hence to another sequence encoding scheme— is a definite weakness. Is it one of the coding scheme or of the presentation of primitive recursion?]

5. Let $p_0, p_1, \ldots$ be the sequence 2, 3, ... of prime numbers.

i. Show the primitive recursiveness of the *set* of primes, i.e., of the representing function for the set of primes.

ii. (Euclid). Show: $p_{n+1} \le \left(\prod_{i=0}^{n} p_i\right) + 1.$

iii. Prove by induction: $p_n \le 2^{2^n}$.

iv. Prove the primitive recursiveness of the prime-enumerating function $P(n) = p_n$.

v. Prove the primitive recursiveness of the set of sequence numbers (i.e., codes of finite sequences), and of the length, projection, sequence formation, and concatenation functions of coding scheme 7.4.

[Hint: iii. Use ii and the formula, $\sum_{i=0}^{k} 2^i = 2^{k+1} - 1$.]

[*Remarks.* In logic, one traditionally appeals to ii to conclude $p_{n+1} \le p_n! + 1$, where $k!$ is defined as in Exercise 6, below. The estimate iii is due to Pólya, who proved that the numbers $2^{2^m} + 1$ and $2^{2^k} + 1$ are relatively prime for $k \ne m$ (exercise). Better bounds are possible. According to Bertrand's Postulate, proven by Chebyshev, for all $n > 0$ there is a prime number between n and $2n$. This yields the estimate $p_{n+1} \le 2^{n+2}$. A consequence of the Prime Number Theorem is the asymptotic formula $p_n \sim n \log n$. Thus, for large enough n, $p_{n+1} \le n^2$. The reader might enjoy comparing these estimates for $n \le 5$.]

6. The binomial coëfficients $\binom{n}{k}$ satisfy the recursion,

$$\binom{n}{0} = 1$$

$$\binom{n}{k+1} = \begin{cases} 0, & k+1 > n \\ \binom{n-1}{k} + \binom{n-1}{k+1}, & \text{otherwise} \end{cases}$$

i. Show, by suitably generalising course-of-values recursion, that $B(n,k) = \binom{n}{k}$ is a primitive recursive function.

ii. Observing that $\binom{n}{k} = \frac{n!}{k!(n-k)!}$ for $k \le n$, where

$$k! = \begin{cases} 1, & k = 0 \\ k(k-1)!, & k > 0, \end{cases}$$

is the factorial function, give a simpler proof of the primitive recursiveness of the binomial coëfficient function.

iii. Observing (or proving) $\binom{n}{k} \le 2^n$, show:

a. $\sum_{i=0}^{k} \binom{n}{i} 2^{in} = \mathrm{Rem}((1+2^n)^n, 2^{n(k+1)})$

b. $\binom{n}{k} = \left\lfloor \left(\sum_{i=0}^{k} \binom{n}{i} 2^{in}\right) / 2^{kn} \right\rfloor.$

Using these, give another proof of the primitive recursiveness of the binomial coëfficient function.

*9. Ackermann Functions

David Hilbert believed that constructive and classical, non-constructive mathematics would coïncide. To him, anything provable would ultimately yield to constructive methods and, in fact, the standard mathematical objects would all turn out to be constructible. In particular, he believed— and even believed he could show— every function from natural numbers to natural numbers was definable by some sort of recursion. That every such function is primitive recursive is quickly ruled out: there are, as the reader showed in Exercise 6 of section 3, uncountably many functions from ω to ω and, as we shall see in section 11, below, only countably many primitive recursive functions. In brief, one can diagonalise one's way out of the class of primitive recursive functions. What is remarkable is that the diagonalisation is itself effective and can be done by a recursion.

The obvious approach to such a diagonalisation is to begin with an enumeration of the primitive recursive functions, $F_0, F_1, \ldots$; show that the binary enumerating function,

$$F(n,x) = F_n(x),$$

can be defined by a recursion of a sort; and finally define a diagonal function,

$$F(x) = F_x(x) + 1,$$

which differs from each F_n at at least one argument. A less obvious, but in some sense more elementary, approach is to diagonalise on the growth rates of the primitive recursive functions: an application of primitive recursion can result in a function growing much more rapidly than the functions used in the definition. If one iterates the recursion— itself a recursive process— one will obtain a function embodying at once all primitive recursive rates of growth, whence its diagonal will outstrip all primitive recursive functions. This realisation is due, apparently independently, to two of Hilbert's students, Wilhelm Ackermann and Gabriel Sudan. Although they gave essentially the same recursion, Sudan worked with functions of transfinite ordinals and Ackermann with functions of natural numbers, whence Hilbert cited Ackermann and Ackermann's is the name associated with the resulting functions.

The idea behind Ackermann's function is simple. The primitive recursive function of addition,

$$F_0(x, y) = x + y,$$

grows more rapidly than the constant, successor, or projection functions (or, it will once we've explained how to compare growth rates of functions of differing numbers of arguments). A recursion on this, yielding multiplication,

$$F_1(x, 0) = 0$$
$$F_1(x, y+1) = F_0(x, F_1(x, y)) = x + F_1(x, y),$$

yields a function that will grow more rapidly than any primitive recursive function obtainable from a single primitive recursion from the initial constant, successor, or projection functions. Exponentiation,

$$F_2(x, 0) = 1$$
$$F_2(x, y+1) = F_1(x, F_2(x, y)) = x \cdot F_2(x, y),$$

grows more rapidly than anything obtainable by two primitive recursions; and iterated exponentiation,

$$F_3(x, 0) = x$$
$$F_3(x, y+1) = F_2(x, F_3(x, y)) = x^{F_3(x,y)},$$

will grow more rapidly than anything obtainable by three primitive recursions. Generally, one can define

$$F_{n+1}(x, 0) = x \text{ (say)}$$
$$F_{n+1}(x, y+1) = F_n(x, F_{n+1}(x, y)).$$

Letting

$$G(x, z) = \begin{cases} z, & z \leq 1 \\ x, & z > 1, \end{cases}$$

define the initial values of these recursions, we see that these functions satisfy the following recursion:

$$F_0(x, y) = x + y$$
$$F_{n+1}(x, 0) = G(x, n)$$
$$F_{n+1}(x, y+1) = F_n(x, F_{n+1}(x, y)),$$

whence the ternary function,

$$A(x, y, z) = F_z(x, y),$$

satisfies

$$\begin{aligned} A(x, y, 0) &= x + y \\ A(x, 0, z + 1) &= G(x, n) \\ A(x, y + 1, z + 1) &= A(x, A(x, y, z + 1), z). \end{aligned}$$

This function $A(x, y, z)$ is *the* Ackermann function, and it was shown by Ackermann not to be primitive recursive.

9.1. Theorem. *The function $A(x,x,x)$ grows more rapidly than any unary primitive recursive function, i.e., for any unary primitive recursive function $F(x)$ there is a natural number x_0 such that*

$$\forall x > x_0\ (F(x) < A(x,x,x)).$$

In the 1930s, Rosza Péter observed that the variable x plays no rôle in the recursion defining $A(x,y,z)$. Building on this observation, she proved an analogous result for the binary function P defined by the following simpler recursion:

$$\begin{aligned} P(0, y) &= y + 1 \\ P(x + 1, y) &= P(x, 1) \\ P(x + 1, y + 1) &= P(x, P(x + 1, y)). \end{aligned}$$

Because of the greater elegance of Péter's recursion equations, it is customary in expositions to treat the function P instead of A. It has similarly become customary to refer to P as the Ackermann function. We shall follow the first custom in proving in place of Theorem 9.1 an analogue to that Theorem for P. With respect to the second custom, I suggest that we not call P *the* Ackermann function but rather *an* Ackermann function, i.e., that we make explicit another implicit practice of calling any function constructed by such a diagonalisation on the growth rates of the primitive recursive functions an Ackermann function.

9.2. Theorem. *The function $P(x,x)$ grows more rapidly than any unary primitive recursive function, i.e., for any unary primitive recursive function $F(x)$ there is a natural number x_0 such that*

$$\forall x > x_0\ (F(x) < P(x,x)).$$

The construction of P, as well as the proof of Theorem 9.2, depends on a hierarchical property of the functions $P_n(y) = P(n,y)$. The result underlying Theorem 9.2 is the following.

9.3. Theorem. *Let F be an n-ary primitive recursive function. There is a constant m such that, for all $x_0, ..., x_{n-1}$,*

$$F(x_0, ..., x_{n-1}) < P(m, \max\{x_0, ..., x_{n-1}\}).$$

The reduction of Theorem 9.2 to Theorem 9.3 depends on the monotonicity of P in its first argument: Given a unary primitive recursive function F, choose m so that $F(x) < P(m,x)$ for all x. Letting $x_0 = m$, observe that, for all $x > x_0$, monotonicity will yield $F(x) < P(m,x) \leq P(x,x)$. To establish this monotonicity, we require to know that P is moderately large as a function of its second argument, i.e., $P(x,y) > y$ for all x, y. The following lemma collects together this result, the monotonicity results for each variable, and a strengthened form of monotonicity in the first argument.

9.4. Lemma. *For all* $x, y \in \omega$,

i. $P(x, y) > y$

ii. $P(x, y+1) > P(x, y)$

iii. $P(x+1, y) > P(x, y)$

iv. $P(x+1, y) \geq P(x, y+1)$, *with equality only for* $y = 0$.

Proof. The obvious method of proving anything about a function defined by recursion is induction; a double recursion will require a double induction. For convenience in handling the ensuing inductions, it is best to turn the first argument of P into in index, i.e., to define

$$P_x(y) = P(x, y).$$

The recursion defining P translates into the following recursions:

$$P_0(y) = y + 1$$

$$\begin{cases} P_{x+1}(0) = P_x(1) \\ P_{x+1}(y+1) = P_x(P_{x+1}(y)). \end{cases}$$

i. We wish to show, for each x, y, that $P_x(y) > y$. We do this by induction on x.

Basis. $x = 0$. $P_0(y) = y + 1 > y$.

Induction step. Assume that

$$P_x(y) > y, \text{ for all } y. \qquad (*)$$

We will show that $P_{x+1}(y) > y$, for all y, by a subsidiary induction on y.

Subsidiary basis. $y = 0$. Observe

$$\begin{aligned} P_{x+1}(0) &= P_x(1) \\ &> 1, \text{ by } (*) \\ &> 0. \end{aligned}$$

Subsidiary induction step. Observe

$$\begin{aligned} P_{x+1}(y+1) &= P_x(P_{x+1}(y)) \\ &> P_{x+1}(y), \text{ by } (*) \\ &\geq 1 + P_{x+1}(y) \\ &> 1 + y, \end{aligned}$$

by the subsidiary induction hypothesis. This completes the proof of the subsidiary induction step and therewith that of the induction step and part i of the Theorem.

ii. We wish to show $P_x(y+1) > P_x(y)$ for all y. The proof is by cases.
Case 1. $x = 0$. Simply observe: $P_0(y+1) = (y+1)+1 > y+1 = P_0(y)$.
Case 2. We look at P_{x+1}:

$$P_{x+1}(y+1) = P_x(P_{x+1}(y)) > P_{x+1}(y), \text{ by part i.}$$

iii. Since iv is stronger than iii, we will prove only iv.
iv. We wish to prove a strengthening of iii:

$$P_{x+1}(y) \geq P_x(y+1),$$

with equality only for $y = 0$. We prove this by induction on y.
Basis. $y = 0$. Observe: $P_{x+1}(0) = P_x(1)$.
Induction step. Observe,

$$\begin{aligned} P_{x+1}(y+1) &= P_x(P_{x+1}(y)) \\ &\geq P_x(P_x(y+1)), \text{ by ii and the induction hypothesis} \\ &> P_x(y+1), \text{ by i.} \end{aligned}$$

QED

Part iii of the Lemma is the monotonicity of P in its first argument and, as cited before stating the Lemma, yields the reduction of Theorem 9.2 to Theorem 9.3. We have but to prove Theorem 9.3. Lemma 9.4 gives us enough to begin this proof.
Proof of Theorem 9.3. The proof will be by induction on the inductive generation of the class of primitive recursive functions. The basis consists of the treatment of the initial functions (zero, successor, and projection) and the induction step handles closure under composition and primitive recursion.
Zero. By 9.4.i, for any $x \in \omega$,

$$P_0(x) > x \geq 0 = Z(x).$$

Projection. Observe that, for any $x_0, ..., x_{n-1} \in \omega$,

$$\begin{aligned} P_0(\max\{x_0, ..., x_{n-1}\}) &> \max\{x_0, ..., x_{n-1}\} \\ &> x_i = P_i^n(x_0, ..., x_{n-1}). \end{aligned}$$

Successor. Since $S(x) = x+1 = P_0(x)$, we cannot dominate S by P_0. But, 9.4.iii yields $P_1(x) > P_0(x) = S(x)$.
Composition. Let

$$F(x_0, ..., x_{n-1}) = G(H_0(x_0, ..., x_{n-1}), ..., H_{m-1}(x_0, ..., x_{n-1}))$$

and let $g, h_0, ..., h_{m-1}$ be numbers satisfying

$$P_g(\max\{y_0, ..., y_{m-1}\}) > G(y_0, ..., y_{m-1}) \quad (1)$$

$$P_{h_i}(\max\{x_0, ..., x_{n-1}\})) > H_i(x_0, ..., x_{n-1}) \quad (2)$$

for all $y_0, ..., y_{m-1}$ and all $x_0, ..., x_{n-1}$. Let $k = \max\{g, h_0, ..., h_{m-1}\}$. (1) and (2) yield through application of Lemma 9.4.iii the following:

$$P_k(\max\{y_0, ..., y_{m-1}\}) > G(y_0, ..., y_{m-1}) \quad (3)$$

$$P_k(\max\{x_0, ..., x_{n-1}\})) > H_i(x_0, ..., x_{n-1}). \quad (4)$$

Now, observe that, for any $x_0, ..., x_{n-1} \in \omega$,

$$\begin{aligned} F(x_0, ..., x_{n-1}) &= G(H_0(x_0, ..., x_{n-1}), ..., H_{m-1}(x_0, ..., x_{n-1})) \\ &< P_k(\max \{H_0(\boldsymbol{x}), ..., H_{m-1}(\boldsymbol{x})\}, \text{ by (3)} \\ &< P_k(P_k(\max \{x_0, ..., x_{n-1}\})), \end{aligned} \tag{5}$$

by (4) and 9.4.ii, where we have temporarily let $\boldsymbol{x}$ abbreviate $x_0, ..., x_{n-1}$. [*N.B.* We shall repeatedly use such an abbreviation without mention throughout the proof.] Noting that $P_k(\max \{x_0, ..., x_{n-1}\}) > 0$, and observing that

$$\begin{aligned} P_{k+1}(y + 1) &= P_k(P_{k+1}(y)) \\ &> P_k(P_k(y + 1)), \text{ by 9.4.iv and 9.4.ii,} \end{aligned}$$

we can conclude from (5) that

$$F(x_0, ..., x_{n-1}) < P_{k+1}(\max \{x_0, ..., x_{n-1}\})$$

for all $x_0, ..., x_{n-1} \in \omega$.

Primitive recursion. Let

$$\begin{aligned} F(0, \boldsymbol{x}) &= G(\boldsymbol{x}) \\ F(x + 1, \boldsymbol{x}) &= H(F(x, \boldsymbol{x}), x, \boldsymbol{x}). \end{aligned}$$

Suppose $g, h \in \omega$ are such that

$$G(x_0, ..., x_{n-1}) < P_g(\max \{x_0, ..., x_{n-1}\}) \tag{6}$$

$$H(y, x, x_0, ..., x_{n-1}) < P_h(\max \{y, x, x_0, ..., x_{n-1}\}) \tag{7}$$

for all $x, y, x_0, ..., x_{n-1} \in \omega$. Letting $k = \max \{g, h\}$, (6) and (7) quickly yield, for all natural numerical arguments,

$$G(x_0, ..., x_{n-1}) < P_k(\max \{x_0, ..., x_{n-1}\}) \tag{8}$$

$$H(y, x, x_0, ..., x_{n-1}) < P_k(\max \{y, x, x_0, ..., x_{n-1}\}). \tag{9}$$

We will attempt to show by induction on x that

$$F(x, x_0, ..., x_{n-1}) < P_{k+1}(\max \{x, x_0, ..., x_{n-1}\}).$$

Basis. Observe,

$$\begin{aligned} F(0, x_0, ..., x_{n-1}) &= G(x_0, ..., x_{n-1}) \\ &< P_k(\max \{x_0, ..., x_{n-1}\}), \text{ by (8)} \\ &< P_k(\max \{0, x_0, ..., x_{n-1}\}) \\ &< P_{k+1}(\max \{0, x_0, ..., x_{n-1}\}), \text{ by 9.4.iii.} \end{aligned}$$

Induction step. This step almost works:

$$\begin{aligned} F(x + 1, x_0, ..., x_{n-1}) &= H(F(x, \boldsymbol{x}), x, \boldsymbol{x}) \\ &< P_k(\max \{F(x, \boldsymbol{x}), x, \boldsymbol{x}\} \\ &< P_k(\max \{P_{k+1}(\max \{x, \boldsymbol{x}\}), x, \boldsymbol{x}\}), \end{aligned}$$

by induction hypothesis

$$< P_k(P_{k+1}(\max\ \{x, \boldsymbol{x}\})),\ \text{by 9.4.i}$$

$$< P_{k+1}(1 + \max\ \{x, \boldsymbol{x}\}),\ \ (10)$$

since $P_{k+1}(y + 1) = P_k(P_{k+1}(y))$. Now, (10) will yield

$$F(x + 1, x_0, ..., x_{n-1}) < P_{k+1}(\max\ \{x + 1, \boldsymbol{x}\}) \tag{11}$$

only when $x + 1 = \max\ \{x + 1, \boldsymbol{x}\}$, i.e., we can conclude (11) and finish the induction step only for large values of x or when $\boldsymbol{x} = x_0, ..., x_{n-1}$ is an empty list. Thus, the full induction will only hold in the case in which there are no parameters.

A complete handling of the primitive recursive case requires a lemma slightly stronger than Lemma 9.4.

9.5. Lemma. *For any* $x, y, z \in \omega$,
i. $P_{x+1}(y + z) = P_x^{(z)}(P_{x+1}(y))$
ii. $P_2(y) = 2y + 3$,
where $F^{(z)}$ *denotes the z-fold application of F:*

$$F^{(0)}(y) = y$$

$$F^{(z+1)}(y) = F(F^{(z)}(y)).$$

Proof. i. By induction on z.
Basis. $z = 0$. Observe

$$P_{x+1}(y + 0) = P_x^{(0)}(P_{x+1}(y)).$$

Induction step. Observe,

$$P_{x+1}(y + z + 1) = P_x(P_{x+1}(y + z)),\ \text{by the recurrence relation}$$

$$= P_x(P_x^{(z)}(P_{x+1}(y))),\ \text{by induction hypothesis}$$

$$= P_x^{(z+1)}(P_{x+1}(y)).$$

ii. Exercise. QED

Proof of Theorem 9.3, continued. Let

$$F(0, \boldsymbol{x}) = G(\boldsymbol{x})$$

$$F(x + 1, \boldsymbol{x}) = H(F(x, \boldsymbol{x}), x, \boldsymbol{x})$$

and choose k so that

$$G(\boldsymbol{x}) < P_k(\max\ \{x_0, ..., x_{n-1}\})$$

$$H(y, x, \boldsymbol{x}) < P_k(\max\ \{y, x, x_0, ..., x_{n-1}\})$$

for all $y, x, x_0, ..., x_{n-1} \in \omega$. We show by induction on x that

$$F(x, \boldsymbol{x}) < P_k^{(x+1)}(\max\ \{x_0, ..., x_{n-1}\}). \tag{12}$$

Basis. We have already seen this:

$$F(0, \boldsymbol{x}) = G(\boldsymbol{x}) < P_k(\max\ \{x_0, ..., x_{n-1}\}).$$

Induction step. Observe,

$$\begin{aligned} F(x+1,\boldsymbol{x}) &= H(F(x,\boldsymbol{x}), x, \boldsymbol{x}) \\ &< P_k(\max\{F(x,\boldsymbol{x}), x, \boldsymbol{x}\}) \\ &< P_k(\max\{P_k^{(x+1)}(\max\{\boldsymbol{x}\}), x, \boldsymbol{x}\}), \text{ by ind. hyp.} \\ &< P_k(P_k^{(x+1)}(\max\{\boldsymbol{x}\})), \text{ by 9.4.i} \\ &< P_k^{(x+2)}(\max\{\boldsymbol{x}\}). \end{aligned}$$

This completes the induction and establishes (12).

Once we have (12), we can apply 9.5 to show

$$F(x+1,\boldsymbol{x}) < P_{k+2}(\max\{x+1, x_0, \ldots, x_{n-1}\}).$$

(We have already shown a stronger inequality for $F(0,\boldsymbol{x})$.) To this end, observe

$$\begin{aligned} F(x+1,\boldsymbol{x}) &< P_k^{(x+2)}(\max\{\boldsymbol{x}\}), \text{ by (12)} \\ &< P_k^{(x+1)}(P_k(\max\{\boldsymbol{x}\})) \\ &< P_k^{(x+1)}(P_{k+1}(\max\{\boldsymbol{x}\})) \\ &< P_{k+1}(x+1+\max\{\boldsymbol{x}\}), \text{ by 9.5.i} \\ &< P_{k+1}(2\cdot\max\{x+1, \boldsymbol{x}\}) \\ &< P_{k+1}(P_{k+2}(\max\{x, x_0 \dot{-} 1, \ldots, x_{n-1} \dot{-} 1\}) \\ &< P_{k+2}(\max\{x, x_0 \dot{-} 1, \ldots, x_{n-1} \dot{-} 1\} + 1) \\ &< P_{k+2}(\max\{x+1, \boldsymbol{x}\}). \end{aligned}$$

QED

The above proof is rather long and not very insightful. Indeed, its excessive amount of detail obscures what insights it might offer. Upon reviewing it, however, one will find that, except in the case of primitive recursion with parameters, the proof is straightforward. For this reason, some authors prefer first to prove that the class of primitive recursive unary functions can be generated from a broader class of initial functions by composition and a parameter-free recursion scheme. They can then give the simpler proof not requiring Lemma 9.5. Such an approach has the advantage of simplifying the details of the proof that one's Ackermann function dominates all primitive recursive functions. It has the disadvantage of requiring one to prove an extra characterisation result of both limited interest and limited applicability, of creating the impression that the domination result depends on such a characterisation, and of not forcing one to look more closely at the recursion underlying P.

The approach we have taken in proving Theorem 9.3 is a compromise and is unsatisfying precisely because it is a compromise. The older treatments, e.g., that in Péter's book, do not introduce any abstraction: A and P are viewed as functions of several variables, not as sequences of functions. The higher order of these functions is mentioned only to motivate their definitions, but is not invoked in carrying out the details of the proofs. The treatment given here mentions the sequence of functions and in Lemma 9.5 actually goes a bit beyond using them merely as a notational device. It does, however, not go far enough.

9.6. Theorem. i. $P_0(y) = y + 1$
ii. $P_{x+1}(y) = P_x^{(y+1)}(1)$.

Proof. i is immediate.
ii. Apply 9.5.ii:

$$\begin{aligned} P_{x+1}(y) &= P_x^{(y)}(P_{x+1}(0)) \\ &= P_x^{(y)}(P_x(1)), \text{ by the recursion equation for } P \\ &= P_x^{(y+1)}(1). \end{aligned}$$

QED

Of course, we could prove Theorem 9.6.ii by a direct induction on y and thereby bypass 9.5, which is perhaps a less well-motivated result, and which is an immediate consequence of 9.6. With Theorem 9.6, Lemma 9.5 is obviated and the entire proof can be viewed with a bit more insight. Indeed, the reader ought to use Lemma 9.4 and Theorem 9.6 as the basis for another exposition of the proof of Theorem 9.3. Will the resulting exposition be clearer? Will the details of the proof change?

A full acceptance of the sequence of functions, which is itself a function from ω to the class of functions from ω to ω, leads to a third definition of an Ackermann function. Define the sequence $G_0, G_1, \ldots$ of numerical functions as follows:

$$\begin{aligned} G_0(y) &= y + 1 \\ G_{x+1}(y) &= G_x^{(y+1)}(y). \end{aligned}$$

Further, define $G(x, y) = G_x(y)$.

9.7. Theorem. *G is an Ackermann function, i.e., for any unary primitive recursive function $F(x)$ there is a natural number x_0 such that*

$$\forall x > x_0 \,(F(x) < G(x,x)).$$

The proof of Theorem 9.7 is left to the reader as an exercise.

As I mentioned at the beginning of this section, the Ackermann function was first introduced for the purpose of showing that the diagonalisation leading out of the class of primitive recursive functions could itself be given by a recursion— in fact, by a double recursion. To Hilbert, this was not so much an indication of the complexity of the class of numerical functions as a demonstration of the strength of recursion, whence support for his conjecture that all numerical functions would be definable by some— possibly transfinite— recursion. In the 1930s, the emphasis changed. Rigorous definitions of the notion of a computable function turned the Ackermann function into an example of a computable but not primitive recursive function. It had changed from a positive result to a counterexample. As such, it was an initial counterexample in a whole hierarchy of functions generated by multiple recursions. Eventually, the emphasis shifted again and the individual functions of the sequence of functions embodied by the Ackermann function was seen to give rise to a hierarchy of the primitive recursive functions. It is not the Ackermann function, but the individual primitive recursive functions making it up that are currently viewed as most important. The book, *Subrecursion; Functions and Hierarchies*, by H.E. Rose, tells this story nicely.

In the present book, Ackermann functions will primarily play a pædagogical rôle. There are at least three pædagogical points that can be made using the Ackermann function. The historical point, only hinted at in the previous paragraph, is that perspectives change, as do the uses one wishes to put a tool to. This point is also demonstrated by the history alluded to in the previous section, of the primitive recursive functions themselves: Skolem used them for his foundation of mathematics; Gödel used them as mere coding tools; and soon they formed a primitive notion of computable function, an explication shown not to be adequate by Ackermann's function. Mathematics is not the static subject that textbooks make it appear to be.

The second pædagogical point is one that shall only be made by reference. Although developed for logical purposes, Ackermann functions have begun to pop up outside of logic in such diverse areas as combinatorial number theory and analysis. In fact, since 1977 super-Ackermann functions in combinatorics have become quite common; Chapter VI of volume II will be devoted to this subject.

The best-known combinatorial result connected with Ackermann's function has, for a number of years, been a theorem of B.L. van der Waerden, a 20th century Dutch mathematician. Until recently, it has been *de rigueur* to mention van der Waerden's Theorem (as it is called) in connection with the Ackermann function because the bounds given by all known proofs were, on analysis, Ackermann functions. Recently, however, Saharon Shelah has found a new proof giving primitive recursive bounds, and mention of van der Waerden's Theorem becomes almost irrelevant, and, perhaps, a bit embarrassing. Nonetheless, I shall mention it here:

9.8. Theorem. *Let m, c be positive integers. There is a number n so large that, if we partition the first n natural numbers into c classes, at least one of these classes will contain an arithmetic progression of length m.*

Although Theorem 9.8 can now be provided with a primitive recursive bound (for n as a function of m and c), it still remains the best-known result connected— if only traditionally— with the Ackermann function. Moreover, it still affords a good, well-documented, example of the extraction of bounds from pure existence proofs. This is a topic from advanced logic and will not be touched on in either volume of this book, but it is worth being aware of, however subliminally.

My third pædagogical point will not be completely made until Chapter VI of volume II. This point is the necessity of abstraction and the unpredictability of what is needed to solve a problem. Such unpredictability is, perhaps, best illustrated by the proof of the Fueter-Pólya Theorem in section 5, above, where such apparent irrelevancies as Lindemann's Theorem and the integral calculus played key rôles. The abstraction appealed to in our discussion of Ackermann functions, that is the viewing of the binary numerical function $P(x,y)$ as a function from number x to *functions* P_x, is not nearly as "irrelevant" and, moreover, can be formally avoided (although, heuristically speaking, one would be hard pressed to explain where the recursion came from and what suggested Lemma 9.5 if one did not think of this sequence). However, when we attempt to prove something a bit more straightforward about an Ackermann function like P— namely, that it is well-defined— the Ackermann function becomes more impressive: the fact that we do not know how to prove the Fueter-Pólya Theorem without using moderately powerful and abstract tools does not mean that it cannot be done (any more than the fact that, for about 5 decades, one couldn't prove van der Waerden's Theorem without the use of very rapidly growing functions meant that the bounds couldn't be improved); we can prove (and will do so in Chapter VI of volume II) that the Ackermann function, the totality of which is an easy application of Dedekind's Theorem 8.1, cannot be proven to be well-defined by the same restricted methods we will be able to apply to the primitive recursive functions.

Exercises.

1. Define $F_0, F_1, \ldots$ as the sections of the Ackermann function,

$$F_n(x, y) = A(x, y, n).$$

   i. Show: $F_1(x, y) = xy$
   ii. Show: $F_2(x, y) = x^y$
   iii. Show: $F_2(x, y) = x^{x^{\cdot^{\cdot^{x}}}}\Big\}y$.

2. i. Show: $P_1(y) = y + 2$.
   ii. Show: $P_2(y) = 2y + 3$.

3. i. Show that the following recursion fails to define a function:

$$F(0, 0) = 0$$
$$F(x + 1, y) = F(y, x + 1)$$
$$F(x, y + 1) = F(x + 1, y).$$

   ii. Show that no function satisfies the following recursion:

$$F(0,0) = 0$$
$$F(x + 1, y) = F(x, y + 1) + 1$$
$$F(x, y + 1) = F(x + 1, y) + 1.$$

   iii. Prove, by appeal to Theorem 8.1, that the recursion equations for $P(x, y)$ yield a well-defined function on $\omega \times \omega$.

4. Prove by giving a variation on the proof of Theorem 9.3 that, for each n-ary primitive recursive function F, there is a number k so large that, for all $x_0, \ldots, x_{n-1} \in \omega$,

$$F(x_0, \ldots, x_{n-1}) < P_k(x_0 + \ldots + x_{n-1}).$$

5. Prove Theorem 9.7.
   i. Do this by mimicking the proof of Theorem 9.2.
   ii. Do this by reduction to Theorem 9.3 by showing
      a. $G_0(y) = P_0(y)$
      b. $G_x(y) > P_x(y)$ for all $x, y > 0$.

   [*Warning*. ii. The exact analogue to Theorem 9.3 does not hold because the G_x's all start out at 1.]

6. The function $G(x, y) = G_x(y)$ of Theorem 9.7 is not directly defined by a multiple recursion. Find a function H defined by a double recursion over the primitive recursive functions and such that G is explicitly definable from H.

10. Arithmetic Relations

Till now we have been considering a rather disparate collection of odds and ends— polynomials, finite sequences, primitive recursive computability and non-primitive recursive computability. What ties these together is the notion of arithmetic definability, to which we now turn.

Definability sounds like a very logical notion and indeed it is. Nonetheless, the most spectacular results on definability predate mathematical logic. Lindemann's transcendence result on the undefinability of e and π by simple equations is probably the best known and most obvious example of a result on definability. Galois theory, which the reader might not yet have encountered in his study of mathematics, is another, albeit less obvious, example: Galois theory asserts, in a special case, the equivalence of a syntactic notion of definability (an element of one number field is one of finitely many solutions to an algebraic equation over a smaller number field) with a semantic one (what cannot be changed in a larger structure under an automorphism which makes no change in a smaller one?). And, coëval with mathematical logic, there is one example the reader has definitely seen— the Fueter-Pólya Theorem.

As is particularly evident from Lindemann's Theorem and the Fueter-Pólya Theorem, negative results on definability are hard to prove and can depend on deep and powerful tools. Specific positive results, like the existence of the Cantor pairing function or the Gödel β-function, often require *ad hoc* tricks. It is the possibility of such tricks that makes undefinability results so difficult to establish. Fortunately for us, we will mainly concern ourselves with general positive results. Moreover, we have in fact already done all the tricky work in establishing the arithmetic nature of everything we have thus far considered. Alas, for the more refined positive results of the next chapter, we will require an arsenal of apparently new *ad hoc* tricks.

As just hinted, the task of the present section is to verify the arithmetic definability of the various concepts we have already introduced. This entails, of course, the necessity of explicating the term "arithmetic definability". In Chapter III, below, an excessively formal definition of the arithmetic language and its consequent formal definition of arithmetic definability will be given; for now, I hope to get by with a merely moderately formal definition— or, set of definitions: we will encounter no fewer than three distinct notions of arithmetic definability— arithmetic definability, Σ_1-definability, and Diophantine definability. The first and third of these notions are quite natural. Arithmetic definability will (eventually) just mean definability in an obviously arithmetic language; and Diophantine definability will be a restricted notion obviously tied to Diophantine equations, which equations will be studied in the next chapter. The middle-mentioned notion of Σ_1-definability is not so natural. It is a proof-generated concept— the arithmetic definitions of primitive recursive functions just happen to be Σ_1. Such happenstance would, in itself, hardly justify more than a footnote. In the next two sections, in which we discuss computability in general, we will see that the Σ_1-definable relations are precisely those that are *semi-computable*.

A preliminary definition is that of a polynomial relation.

10.1. Definition. A relation $R \subseteq \omega^n$ is a *polynomial relation* if there is a polynomial $P(X_0, ..., X_{n-1})$ with integral coëfficients such that, for all $x_0, ..., x_{n-1} \in \omega$,

$$R(x_0, ..., x_{n-1}) \text{ iff } P(x_0, ..., x_{n-1}) = 0.$$

[Let me remark quickly that the requirement that P have integral coëfficients can easily be relaxed to allow rational coëfficients: if P has rational coëfficients and d is the least common denominator of these coëfficients, then $Q = d \cdot P$ has, on simplifying, integral coëfficients and defines the same polynomial relation as P.]

Polynomial relations form a rather narrow class and they do not have a great many closure properties. They do, however, form a basis from which one can generate the broader classes of Diophantine relations, Σ_1-relations, and arithmetic relations by specifying additional closure properties. Closure under existential quantification yields the Diophantine relations.

10.2. Definition. A relation $R \subseteq \omega^n$ is a *Diophantine relation* if there is a polynomial $P(X_0, ..., X_{n-1}, Y_0, ..., Y_{m-1})$ with integral coëfficients such that, for all $x_0, ..., x_{n-1} \in \omega$,

$$R(x_0, ..., x_{n-1}) \text{ iff } \exists y_0 ... y_{m-1} \in \omega[P(x_0, ..., x_{n-1}, y_0, ..., y_{m-1}) = 0].$$

In other words, $R \subseteq \omega^n$ is Diophantine if, for some m and some polynomial relation $T \subseteq \omega^{m+n}$ we have, for all $x_0, ..., x_{n-1} \in \omega$,

$$R(x_0, ..., x_{n-1}) \text{ iff } \exists y_0 ... y_{m-1} \in \omega T(x_0, ..., x_{n-1}, y_0, ..., y_{m-1}).$$

The class of Diophantine relations will be extensively studied in the next chapter. For now, we simply note its existence.

The classes of Σ_1–relations and arithmetic relations are not so directly definable. Each class is inductively generated from the class of polynomial relations by closing under some natural logical operations. Two of these logical operations need to be defined formally before we can define definability.

10.3. Definitions. Let S, T be numerical relations, say $S \subseteq \omega^m$ and $T \subseteq \omega^k$. Let $n > m, k$ and suppose $i_0, ..., i_{m-1}$ ($i_r \neq i_s$ for $r \neq s$) and $j_0, ..., j_{k-1}$ ($j_r \neq j_s$ for $r \neq s$) are sequences from $\{0, ..., n-1\}$. The relation

$$R(x_0, ..., x_{n-1}): \; S(x_{i_0}, ..., x_{i_{m-1}}) \wedge T(x_{j_0}, ..., x_{j_{k-1}})$$

is called a *conjunction* of S and T; the relation

$$R(x_0, ..., x_{n-1}): \; S(x_{i_0}, ..., x_{i_{m-1}}) \vee T(x_{j_0}, ..., x_{j_{k-1}})$$

is called a *disjunction* of S and T.

The need for such a definition lies in the fact that we want to conjoin and disjoin relations of different arities. Cf. Lemma 8.7 and its succeeding discussion to review how we handled conjunctions and disjunctions in discussing primitive recursive relations. (Incidentally, the importation of the technique of 8.7 into the present context shows that the requirements that the i_r's be distinct and the j_r's be distinct are æsthetic rather than restrictive.)

10.4. Definition. The class of *arithmetically definable relations*, or *arithmetic relations*, is the smallest class of relations of natural numbers containing the polynomial relations and closed under the logical operations of conjunction, disjunction, negation, and quantification (existential and universal), i.e., a relation $R \subseteq \omega^n$ is an arithmetic relation if it can be obtained by finitely many applications of the following rules:

i. if S is a polynomial relation, then S is arithmetic

ii. if S, T are arithmetic relations and U is a conjunction of S, T or a disjunction of S, T, then U is an arithmetic relation

iii. if $S \subseteq \omega^n$ is an arithmetic relation, then its complement $\omega^n - S$ is an arithmetic relation

iv. if $S \subseteq \omega^{n+1}$ is an $(n+1)$-ary arithmetic relation and $0 \leq i \leq n$, then

$$T(x_0, ..., x_{i-1}, x_{i+1}, ..., x_n): \ \exists x_i \in \omega\ S(x_0, ..., x_n)$$

$$U(x_0, ..., x_{i-1}, x_{i+1}, ..., x_n): \ \forall x_i \in \omega\ S(x_0, ..., x_n)$$

are arithmetic relations.

This is not the definition of "definability" as "definability in an arithmetic language" as described above. The exhibition of a relation as arithmetic one will, however, be achieved by writing down an arithmetic formula defining the relation. Such a formula in a natural way describes the steps used in generating the relation from polynomial relations. For now, we do not need to be very explicit about this.

The Σ_1-definable relations are generated by dropping the rule allowing closure under negation and requiring universal quantifiers to be bounded in the last clause.

10.5. Definition. A relation $R \subseteq \omega^n$ is a Σ_1-relation if it can be obtained by finitely many applications of the following rules:

i. if S is a polynomial relation, then S is a Σ_1-relation

ii. if S, T are Σ_1-relations and U is a conjunction of S, T or a disjunction of S, T, then U is a Σ_1-relation

iii. if $S \subseteq \omega^n$ is an $(n+1)$-ary Σ_1-relation and $0 \leq i \leq n$, then

$$T(x_0, ..., x_{i-1}, x_{i+1}, ..., x_n): \ \exists x_i \in \omega\ S(x_0, ..., x_n)$$

is a Σ_1-relation

iv. if $S \subseteq \omega^{n+1}$ is an $(n+1)$-ary Σ_1-relation and $0 \leq i, j \leq n$ with $i \neq j$, then

$$T(x_0, ..., x_{i-1}, x_{i+1}, ..., x_n): \ \exists x_i \leq x_j\ S(x_0, ..., x_n)$$

$$U(x_0, ..., x_{i-1}, x_{i+1}, ..., x_n): \ \forall x_i \leq x_j\ S(x_0, ..., x_n)$$

are Σ_1-relations.

It has already been stated that the concept of Σ_1-definability is a proof-generated one. As this is not much of an explanation of the concept, I should devote a few words to the significance of Σ_1-definability. This significance can, in fact, be summed up in one word: verifiability. One can verify that a sequence of numbers stands in a polynomial relation simply by calculating the polynomial and checking if the value is 0. The closure conditions 10.5.ii-iv are precisely those which clearly preserve verifiability of membership. If, for example, one can effectively verify membership in $R \subseteq \omega^{n+1}$,

$$S(x_0, ..., x_{n-1}) \Leftrightarrow \exists x_n \leq x_j \; R(x_0, ..., x_n),$$

and $(x_0, ..., x_{n-1}) \in S$, one can verify this by finding $k \leq x_j$ such that $R(x_0, ..., x_{n-1}, k)$, and verifying this. To verify, say, $\forall x_n \leq x_j \; R(x_0, ..., x_n)$ for $x_0, ..., x_{n-1}$, one verifies in turn $R(x_0, ..., x_{n-1}, 0), R(x_0, ..., x_{n-1}, 1), ..., R(x_0, ..., x_{n-1}, x_j)$. Moreover, the closure conditions not included are precisely those which do not *clearly* preserve such verifiability. For example, even given the verifiability of R, a direct verification of $\forall x_n R(x_0, \ldots, x_n)$ would seem to require the infinitely many verifications of $R(x_0, ..., x_{n-1}, 0)$, $R(x_0, ..., x_{n-1}, 1), ...$— and there is no reason to believe that this can be accomplished in finitely many steps (say, by some *uniform* verification). Finally, let me note that in the next section we will have a formal counterpart to verifiability (or, as stated earlier: semi-computability) and we will prove the verifiable relations to be precisely the Σ_1-relations.

For now, our goal is more modest:

10.6. Theorem. *Let $F: \omega^n \to \omega$ be primitive recursive. The graph of F,*

$$R(x_0, ..., x_{n-1}, y): F(x_0, ..., x_{n-1}) = y,$$

is a Σ_1-relation.

Proof. By induction on the number of steps in the inductive generation of F.
Basis. F is the zero, the successor, or a projection function. Observe:

$$Z(x) = y \;\text{ iff }\; 0 \cdot x + y = 0$$

$$S(x) = y \;\text{ iff }\; y - (x + 1) = 0$$

$$P_i^n(x_0, ..., x_{n-1}) = y \;\text{ iff }\; y - x_i = 0.$$

Thus, the graphs of the initial functions are polynomial relations.
Induction step. Either F is the composition of primitive recursive functions which already have Σ_1-definitions or F is defined by primitive recursion from such functions.

Suppose first that

$$F(x_0, ..., x_{n-1}) = G(H_0(x_0, ..., x_{n-1}), ..., H_{m-1}(x_0, ..., x_{n-1})),$$

where the graph of G is a Σ_1-relation R_G and the graphs of $H_0, ..., H_{m-1}$ are Σ_1-relations $R_0, ..., R_{m-1}$, respectively. Observe:

$$F(x_0, ..., x_{n-1}) = y \;\text{ iff }\; \exists z_0 ... z_{m-1} [G(z_0, ..., z_{m-1}) = y \;\wedge \bigwedge_{i=0}^{m-1} H_i(x_0, ..., x_{n-1}) = z_i \,]$$

$$\text{iff } \exists z_0...z_{m-1}[R_G(z_0, ..., z_{m-1}) \wedge \bigwedge_{i=0}^{m-1} R_i(x_0, ..., x_{n-1}, z_i)].$$

Next, consider the case of primitive recursion:

$$F(0, x_0, ..., x_{n-1}) = G(x_0, ..., x_{n-1})$$
$$F(x + 1, x_0, ..., x_{n-1}) = H(F(x, x_0, ..., x_{n-1}), x, x_0, ..., x_{n-1}).$$

This case is the crucial one, so I must take care not to present it too quickly.

Consider the sequence of values,

$$\begin{aligned} y_0 &= G(x_0, ..., x_{n-1}) &&= F(0, x_0, ..., x_{n-1}) \\ y_1 &= H(y_0, 0, x_0, ..., x_{n-1}) &&= F(1, x_0, ..., x_{n-1}) \\ y_2 &= H(y_1, 1, x_0, ..., x_{n-1}) &&= F(2, x_0, ..., x_{n-1}) \\ &\vdots && \vdots \\ y_x &= H(y_{x-1}, x-1, x_0, ..., x_{n-1}) &&= F(x, x_0, ..., x_{n-1}). \end{aligned}$$

If we can arithmetically define the projection function for some sequence encoding scheme, we can define the graph of F as follows:

$$\exists \text{ sequence } s \ \{(s)_0 = G(x_0, ..., x_{n-1}) \wedge (s)_x = y \wedge$$
$$\wedge \ \forall z \le x \ [(s)_{z+1} = H((s)_z, z, x_0, ..., x_{n-1})]\},$$

(where $(s)_i$ is the i-th element of s, as in 8.13). Looking back at the coding schemes discussed in section 7, exactly one is clearly arithmetically definable— the β-function:

$$\begin{aligned} \beta(c, d, x) = y \ &\text{iff } \ \mathrm{Rem}(c, 1 + (x + 1)d) = y \\ &\text{iff } \ \exists z \, [c = (1 + (x + 1)d)z + y \ \wedge \ y < 1 + (x + 1)d] \\ &\text{iff } \ \exists zw \, [c = (1 + (x + 1)d)z + y \ \wedge \\ &\qquad \wedge \ y + w + 1 = 1 + (x + 1)d]. \end{aligned}$$

We can now obtain a Σ_1-definition of the graph of F:

$$\exists cd \ \{\exists y_0 \, [\beta(c, d, 0) = y_0 \ \wedge \ R_G(x_0, ..., x_{n-1}, y_0) \ \wedge$$
$$\wedge \ \forall z {\le} x \ \exists y_0 y_1 [\beta(c,d,z) = y_0 \wedge \beta(c,d,z+1) = y_1 \wedge R_H(y_0, z, x_0, ..., x_{n-1}, y_1)] \wedge$$
$$\wedge \ \beta(c,d,x) = y,$$

where R_G and R_H are the Σ_1-graphs of G, H, respectively. QED

The most important thing about the above proof is the use of an arithmetically definable sequence encoding scheme to allow us to code the recursion by saying we have a sequence beginning with the correct value at 0 and using the recursion to carry us from one value to the next. We can then read off the value at x by projecting the sequence onto its x-entry. Can we use a similar trick to define arithmetically the graph of the Ackermann function? I.e., how general is this trick? Another question: to what extent are we tied to the β-function? By the Theorem, the other coding schemes of section 7, being primitive recursive, are also Σ_1-definable. But can we see this directly, i.e., without recourse to Theorem 10.6 and the β-function?

Answering the above questions are two directions we can follow. A third is the improvement of the Σ_1-presentations of the graphs. Observe that the constructions of the proof never used the closure of the class of Σ_1-relations under disjunction and bounded existential quantification. Can one eliminate more? The answer is yes: in the next chapter, we will show that every Σ_1-relation is, in fact, a Diophantine relation.

All three of these directions deserve to be followed, and so they shall be. Since I wish to do the least with it, I shall start with the question of the necessity of the β-function.

By Theorem 10.6, all of the coding schemes of section 7 are Σ_1-definable. However, each of these definitions produced presupposes the β-function. A powerful result of the next chapter can be applied to give other Σ_1-definitions of these coding schemes, but these definitions are even more roundabout than those given by Theorem 10.6. The question is, then, not if this or that coding scheme is necessary, but which ones can be simply and directly defined. Unfortunately, "simply" and "directly" are not particularly clear concepts and, until they have been clarified, there is no hope of proving any negative results. All that can be said of a negative nature is that none of the other coding schemes of section 7 have any known direct arithmetic definitions and that all known Diophantine definitions of such schemes rely, if not on the β-function itself, at least on the Chinese Remainder Theorem.

There are, however, positive results. Although one doesn't know any direct arithmetic definition of the functions used in the coding scheme based on the b-adic expansion (scheme 7.5) for arbitrary bases b, prime bases are more amenable. In a monograph published in 1961, Raymond Smullyan showed how a suitable coding scheme based on the dyadic expansion could be used to prove Theorem 10.6 without any reference to the Chinese Remainder Theorem at all. Perhaps a decade later Saul Kripke streamlined the b-adic approach to Theorem 10.6 by using p-adic expansions for *variable* primes p. A variant of Kripke's approach is given in the Exercises at the end of this section.

Let us now consider the question of the arithmeticity of the graph of the Ackermann function, i.e., the question of the scope of Theorem 10.6. If we ignore the refined question of the Σ_1-nature of the definition of the graphs of primitive recursive functions, and only worry about the arithmetic definability of these graphs, then we can correctly state that Kurt Gödel first proved Theorem 10.6 at the end of 1930 and published a proof in his paper from which we have already extracted the formal definition of the class of primitive recursive functions, some details on these functions, and the β-function. We can also explain why, although Gödel knew about Ackermann's function, he did not bother to prove its arithmeticity: Gödel used the primitive recursive functions to develop a primitive recursive encoding of syntax enabling theories capable of reasoning about primitive recursive functions to reason about their syntax. Although this sounds like an odd thing to want, it led directly to his famous Incompleteness Theorems which we shall consider in various forms in Chapters III and IV, below. Showing that the primitive recursive functions could be defined in arithmetic would show that his Incompleteness Theorems held for theories of arithmetic. The arithmeticity of the graph of the Ackermann function simply wasn't relevant.

I should restate this last: in 1930, generalising Theorem 10.6 wasn't relevant; in 1934, Gödel mentioned the problem at the end of a series of lectures on his work given at Princeton University. He asserted one would need a generalisation of the β-function to handle the details as well as a general notion of recursiveness to to determine the generality of the result. Although he made a suggestion for the delineation of the general recursive functions, he gave no clue to the method of generalising the β-function.

Among those attending Gödel's lectures was Stephen Kleene, who, two years later, published a result which immediately entailed the appropriate generalisation of Theorem 10.6. Much of the next section is devoted to variations on Kleene's theme, so I won't discuss his work here. Instead, let me mention the work of Thoralf Skolem published the following year, 1937. Skolem's result was fairly general— not nearly as general as Kleene's, but general enough to exhibit the arithmetic nature of the graphs of all k-fold recursive functions.

Both Kleene and Skolem solved the problem of generalising the β-function primitive recursively. Appeal to Theorem 10.6 would then guarantee the arithmetic nature of the new β-function. As a preliminary illustration, let us consider a variant of Skolem's exhibition of the graph of the Ackermann function as arithmetic.

10.7. Example. *The graph,*

$$R(x, y, z):\ \ P(x, y) = z,$$

of the Ackermann function is a Σ_1*-relation.*

Proof. The idea is to use the primitive recursive encoding (7.8, 7.9) of finite sets to code up all the information needed to compute $P(x,y) = z$. This information is just the set of all triples $\langle s, t, w\rangle$ such that the computation of $P(s,t) = w$ is used in the computation of $P(x, y) = z$.

Suppose $P(x, y) = z$. If D_v is the set of triples occurring in the computation, which triples must D_v possess? Well, it must include $\langle x, y, z\rangle$:

i. $\langle x, y, z\rangle \in D_v$.

Further, by one of the recursion clauses, if $\langle s+1, t+1, P(s+1, t+1)\rangle \in D_v$, since $P(s+1, t+1) = P(s, P(s+1, t))$, D_v must also include the triples $\langle s, P(s+1, t), P(s+1, t+1)\rangle$ and $\langle s+1, t, P(s+1, t)\rangle$:

ii. $\forall syw \le v\,[\langle s+1, t+1, w\rangle \in D_v \Rightarrow \exists u \le v\,[\langle s, u, v\rangle \in D_v \wedge$

$$\wedge\ \langle s+1, t, u\rangle \in D_v]].$$

(The bounds $s, t, w, u \le v$ follow from the facts that a. v is greater than any element of D_v, and b. the code of any tuple is greater than any of its elements.) Next, the other recursion clause, $P(s+1, 0) = P(s, 1)$, tells us that, if $\langle s, 1, P(s+1, 0)\rangle \in D_v$, we will have to include $\langle s, 1, P(s, 1)\rangle$ in D_v:

iii. $\forall sw \le v\,[\langle s+1, 0, w\rangle \in D_v \Rightarrow \langle s, 1, w\rangle \in D_v]$.

Finally, a correctness condition which keeps us from putting extra false tuples into D_v is given by the basis, $P(0, t) = t + 1$, of the recursion:

iv. $\forall tw \le v\,[\langle 0, t, w\rangle \in D_v \Rightarrow w = t + 1]$.

Let $A(x, y, z, v)$ be the conjunction of clauses i - iv.

Claim. $P(x, y) = z$ *iff* $\exists vA(x, y, z, v)$.

The proof of the Claim is a good exercise in double induction, but it has no bearing on the matters being discussed. Thus, I relegate the proof of the Claim to the Exercises.

Granted the Claim, the proof is now complete. For, letting χ_A be the primitive recursive representing function for A, we have

$$A(x, y, z, v):\ \ \chi_A(x, y, z, v) = 0$$

exhibited as a Σ_1-relation by Theorem 10.6. But then $\exists vA$ is also Σ_1. QED

In the next section, we will look at Kleene's primitive recursive encoding of computations and see what it yields.

The last topic we have to discuss in this section is the simplification of the forms of Σ_1-definitions of Σ_1-relations. Stated differently, we shall look into the possibility of omitting closure conditions from Definition 10.5 without narrowing the class of relations obtainable. As already announced, in the next chapter we will see that all Σ_1-relations are Diophantine, whence all but one closure condition can be dropped. What we shall do here is merely illustrative.

We shall prove the equivalence between the notions of Σ_1-definability and *strict*-Σ_1-definability. To define this latter, we must first define Δ_0-definability.

10.8. Definition. A relation $R \subseteq \omega^n$ is a Δ_0-*relation* if it can be obtained by finitely many applications of the following rules:

i. if $S \subseteq \omega^n$ is a polynomial relation, then S and its complement $\omega^n - S$ are Δ_0-relations

ii. if S, T are Δ_0-relations and U is a conjunction of S, T or a disjunction of S, T, then U is a Δ_0-relation

iii. if $S \subseteq \omega^{n+1}$ is an $(n+1)$-ary Δ_0-relation and $0 \le i, j \le n$ with $i \ne j$, then

$$T(x_0, ..., x_{i-1}, x_{i+1}, ..., x_n):\ \exists x_i \le x_j\, S(x_0, ..., x_n)$$

$$U(x_0, ..., x_{i-1}, x_{i+1}, ..., x_n):\ \forall x_i \le x_j\, S(x_0, ..., x_n)$$

are Δ_0-relations.

This definition differs form those of Diophantine and Σ_1-relations in two important respects. First, it includes the complements of polynomial relations among the initial relations. This is not necessary for Diophantine or Σ_1-relations as

$$\begin{aligned} P(x_0, \ldots, x_{n-1}) \ne 0 \ &\text{iff}\ \ P < 0 \vee P > 0 \\ &\text{iff}\ \ \exists w\,[P + w + 1 = 0 \vee w + 1 = P]. \end{aligned}$$

Second, it does not include closure under existential quantification among its clauses. If we allow one instance of this, we obtain the class of strict-Σ_1-relations. The rationale for these choices is this: the initial relations— polynomial relations and their complements— are decidable, and the closure conditions chosen clearly preserve decidability (cf. 8.7 - 8.8, above). Unlike verifiability, which is preserved under existential quantification, decidability is not *clearly* preserved under such, and, in fact, it is not. But this is the story of the next two sections...

10.9. Definition. A relation $R \subseteq \omega^n$ is a *strict-Σ_1-relation* if there is a Δ_0-relation $S \subseteq \omega^{n+1}$ and an i, $0 \le i \le n$, such that, for all $x_0, ..., x_{n-1} \in \omega$,

$$R(x_0, ..., x_{i-1}, x_{i+1}, ..., x_n) \ \text{ iff } \ \exists x_i\, S(x_0, ..., x_n).$$

10.10. Theorem. *The classes of Σ_1- and strict-Σ_1-relations coïncide.*

Theorem 10.10 is not as impressive as the equivalence of the notions of Σ_1- and Diophantine definability, but it is something, as can be seen by the following corollary.

10.11. Corollary. *The Σ_1-relations $R \subseteq \omega^n$ are precisely those which can be represented,*

$$R(x_0, ..., x_{n-1}) \Leftrightarrow \exists x_n\, [F(x_0, ..., x_{n-1}, x_n) = 0], \qquad (*)$$

where F is primitive recursive.

Proof. If F is primitive recursive, the relation,

$$F(x_0, ..., x_n) = y,$$

is Σ_1 by Theorem 10.6, whence so is the relation,

$$F(x_0, ..., x_{n-1}, x_n) = 0,$$

and its existential quantification R.

Conversely, if R is a Σ_1-relation, it is also strictly-Σ_1 by Theorem 10.10,

$$R(x_0, ..., x_{n-1}) \Leftrightarrow \exists x_n\, S(x_0, ..., x_n), \qquad (**)$$

say, with S a Δ_0-relation. But Δ_0-relations are primitive recursive since they are generated from polynomial relations, which are evidently primitive recursive, by conjunction, disjunction, and bounded quantification— none of which operations leads us out of the class of primitive recursive relations. Thus, S has a primitive recursive representing function F and (*) follows then from (**). QED

Let us now prove Theorem 10.10.

Proof of Theorem 10.10. Since strict-Σ_1-relations are Σ_1-relations, we need only prove that every Σ_1-relation is strict-Σ_1. The proof is by induction on the number of steps used in generating a given Σ_1-relation R.

The basis of the induction is the case in which R is a polynomial relation: for all $x_0, ..., x_{n-1} \in \omega$,

$$R(x_0, ..., x_{n-1}):\ P(x_0, ..., x_{n-1}) = 0.$$

Then:

$$R(x_0, ..., x_{n-1}) \Leftrightarrow \exists x_n\,[P(x_0, ..., x_{n-1}) + 0{\cdot}x_n = 0],$$

and R is strict-Σ_1.

The induction step consists essentially of showing the class of strict-Σ_1-relations to be closed under the closure conditions defining the class of Σ_1-relations.

We start by considering closure under existential quantification. Let $R \subseteq \omega^{n+1}$ be a Σ_1-relation with a strict-Σ_1 representation: for all $x_0, ..., x_{n-1}, x_n \in \omega$,

$$R(x_0, ..., x_n) \Leftrightarrow \exists y\, R'(x_0, ..., x_n, y),$$

where R' is Δ_0. Observe,

$$\begin{aligned}\exists x_i\, R(x_0, ..., x_n) &\Leftrightarrow \exists x_i\, \exists y\, R' \\ &\Leftrightarrow \exists z\, \exists x_i \le z\, \exists y \le z\, R',\end{aligned}$$

(e.g., $z = \max\{x_i, y\}$) which last is strict-Σ_1.

For disjunction and conjunction, observe that if $R \subseteq \omega^n$, $S \subseteq \omega^m$ satisfy

$$\begin{aligned}R(x_{i_0}, ..., x_{i_{n-1}}) &\Leftrightarrow \exists y\, R'(x_{i_0}, ..., x_{i_{n-1}}, y) \\ S(x_{j_0}, ..., x_{j_{m-1}}) &\Leftrightarrow \exists z\, S'(x_{j_0}, ..., x_{j_{m-1}}, z),\end{aligned}$$

where R', S' are Δ_0, then, for $* \in \{\wedge, \vee\}$,

$$\begin{aligned}R * S &\Leftrightarrow \exists y\, R' * \exists z\, S' \\ &\Leftrightarrow \exists y\, \exists z\, (R' * S'),\end{aligned}$$

which is strict-Σ_1 by closure under existential quantification.

Bounded existential quantification offers no difficulty. Let R be a Σ_1-relation we have already shown to be is strict-Σ_1, say

$$R(x_0, ..., x_n) \Leftrightarrow \exists y\, R'(x_0, ..., x_n, y), \qquad (*)$$

where R' is Δ_0. Then

$$\begin{aligned}\exists x_i \le x_j\, R &\Leftrightarrow \exists x_i \le x_j\, \exists y\, R' \\ &\Leftrightarrow \exists y\, \exists x_i \le x_j\, R',\end{aligned}$$

whence $\exists x_i \le x_j\, R$ is also strict-Σ_1.

The heart of the proof is the case of bounded universal quantification. Let R, R' satisfy (*). To see that $\forall x_i \le x_j\, R$ is strict-Σ_1, use the β-function:

$$\begin{aligned}\forall x_i \le x_j\, R &\Leftrightarrow \forall x_i \le x_j\, \exists y\, R' \\ &\Leftrightarrow \exists cd\, \forall x_i \le x_j\, \exists y \le c\, [\beta(c, d, x_i) = y \wedge R']. \qquad (**)\end{aligned}$$

Given the closure conditions we have already proven for strict-Σ_1-relations, (**) would be strict-Σ_1 if the graph of the β-function were Δ_0. It almost is:

$$\begin{aligned}\beta(c,d,x_i) = y &\Leftrightarrow \mathrm{Rem}(c, 1 + (x_i + 1)d) = y \\ &\Leftrightarrow \exists z \le c\, [c = (1 + (x_i + 1)d)z + y \wedge y < 1 + (x_i + 1)d]\end{aligned}$$

$$\Leftrightarrow\ \exists z \leq c\, \exists w \leq 1 + (x_i + 1)d\, [c = (1 + (x_i + 1)d)z + y\ \wedge$$
$$\wedge\ y + w + 1 = 1 + (x_i + 1)d].$$

The problem here is that w is not bounded by a variable, but by a polynomial. This turns out not to be a problem for Σ_1-relations (cf. Exercise 2, below), but it is for Δ_0-relations. Either we have to broaden our definitions of Δ_0-relations or we have to move the w outside the bounded universal quantifier of (**). Fortunately, this latter task is easy: $1 + (x_j + 1)d$ is a uniform upper bound to all the $1 + (x_i + 1)d$'s bounding the w's. Thus, (**) becomes:

$$\forall x_i \leq x_j\, R \quad \Leftrightarrow\ \exists cdv\, \forall x_i \leq x_j\, \exists y {\leq} c [R'\ \wedge\ v = 1 + (x_j + 1)d\ \wedge$$
$$\wedge\ \exists w {\leq} v\, \exists z {\leq} c\, (c = (1 + (x_i + 1)d)z + y\ \wedge$$
$$\wedge\ y + w + 1 = 1 + (x_i + 1)d)].$$

We now have three existential quantifiers applied to a Δ_0-relation, which we can readily see to be strict-Σ_1 by the already used trick of bounding the variables of these quantifiers by their maximum.

QED

Let me make a couple of small remarks before we proceed to the exercises. The first of these concerns the substitutability of polynomials for variables in definitions of arithmetic, Σ_1-, Δ_0-, and strict-Σ_1-relations. We have substituted constants, alternate variables, and, in the proof of Theorem 10.6, even a polynomial of the form $z + 1$ for variables in Σ_1-definitions. The justification for this is given in Exercise 2, below: the classes of arithmetic, Σ_1-, and strict-Σ_1-relations are closed under polynomial substitutions. As for the class of Δ_0-relations, a similar closure property would hold if we were to allow polynomial upper bounds in the bounded quantifiers. The second point I wish to make is this: the great simplicity of the β-function has twice been used. First, we used its direct Σ_1-definition in establishing Theorem 10.6; second, we used its almost Δ_0-definition in proving Theorem 10.10. This is not characteristic of the β-function as a decoder for some sequence encoding scheme; the scheme to be discussed in Exercise 4, below, has these properties. What appears to be unique about the β-function— and here we are referring to its "direct" definition, not merely arbitrary definitions— is that it makes no use of the bounded universal quantifier. This fact will be of use in the next chapter.

Exercises.

1. i. Prove that the class of polynomial relations is closed under
   a. conjunction and disjunction
   b. universal quantification
   c. bounded universal quantification.

   ii. Prove that the class of Diophantine relations is closed under
   a. conjunction and disjunction
   b. bounded existential quantification.

   iii. Prove that the relation $\beta(c, d, i) = y$ is Diophantine.

   [Hints. i.a. cf. Lemma 8.7; i.b. what do you know about the coëfficients of a polynomial that vanishes everywhere?; and i.c. see section 2.]

2. i. Show, by induction on the number of steps in the generation of a Σ_1-relation, that, if $R \subseteq \omega^n$ is Σ_1 and $P_0, ..., P_{n-1}$ are m-ary polynomials, then $R(P_0(y_0, ..., y_{m-1}), ..., P_{n-1}(y_0, ..., y_{m-1}))$ is an m-ary Σ_1-relation.
ii. Do the same for arithmetic relations.
iii. Do the same for the class of relations obtained by allowing polynomial upper bounds to bounded quantifiers in Definition 10.8.

3. *De Morgan's Laws* for disjunction and conjunction are:

$$\neg(p \vee q) = \neg p \wedge \neg q; \quad \neg(p \wedge q) = \neg p \vee \neg q.$$

i. State and prove versions of de Morgan's Laws for bounded quantifiers:

$$\neg(\exists x \le y\, R) \Leftrightarrow \ ?$$
$$\neg(\forall x \le y\, R) \Leftrightarrow \ ?$$

ii. Show, by induction on the number of steps used in the generation of a Δ_0-relation that the class of Δ_0-relations is closed under complementation, i.e., if $R \subseteq \omega^n$ is a Δ_0-relation, then so is $\omega^n - R$.

4. Let $a_0, ..., a_n$ be a non-empty sequence of non-negative integers. A pair of natural numbers x, p can be viewed as a code for $(a_0, ..., a_n)$ if the following hold:
α. $x = \langle 0, a_0\rangle + \langle 1, a_1\rangle p + ... + \langle n, a_n\rangle p^n$
β. $p > \langle i, a_i\rangle$ for each i
γ. p is prime.

For a number of the more general form,

$$x = b_0 + b_1 p + ... + b_n p^n,$$

with each $b_i < p$, we call the terms $b_i p^i$, *terms of* x, p and the coëfficients b_i—*coëfficients of* x, p.

i. For any polynomial Q in several variables, show the following relations to be Δ_0:
a. $Q \mid v$ (Q divides v)
b. $Q < v$
c. $v < Q$
d. $Q \nmid v$ (Q does not divide v)
e. $\text{Rem}(x,Q) = y$ (y is the remainder of x on dividing by Q)

ii. Show that the following relations are Δ_0:
a. Prime(p) (p is prime)
b. Primpow(z, p) (z is a power of the prime p)

iii. For any polynomial Q, show the following relations to be Δ_0:
a. $Q \in \text{Term}(x,p)$ (Q is a term of x, p)
b. $Q \in \text{Coëf}(x,p)$ (Q is a coëfficient of x, p)

iv. Show the following to be Δ_0:
a. $x,p \in \text{Code}$ (x, p code a sequence)
b. $(x,p)_i = y$ ($y = a_i$ for x,p satisfying conditions α,β,γ, i.e., y is the $(i + 1)$-th element of the sequence coded by x, p).

[Hints. i.c. use $\forall x \le v$ instead of $\exists x \le Q$; i.e., consider the cases $Q > b$ and $Q \le b$.]

5. Prove the Claim,

$$P(x, y) = z \text{ iff } \exists v\, A(x, y, z, v),$$

of Example 10.7.

[Hint. $\Rightarrow$. Define properties

$Q(m, n)$: $\exists v\, A(m, n, P(m, n), v)$

$R(m)$: $\forall n\, Q(m, n)$.

Prove $R(m)$ by induction on m. Both the basis, $\forall n\, Q(0, n)$ and the inductive conclusion, $\forall n\, Q(m + 1, n)$, will require a subsidiary induction on n. Prove the converse by a similar double induction.]

11. Computability

The theory of computability, called Recursion Theory by logicians, largely derived from Gödel's work with primitive recursive functions. Prior to Gödel there was only one hint of a precursor. When Charles Babbage set about designing the Analytical Engine, an all-purpose mechanical computer, he naturally enough, gave some thought to the nature of computation and isolated such key ingredients as branching. However, he never developed a *theory*; he was primarily interested in constructing a machine. In the 1930s, when Stephen Kleene and Alan Turing were laying the foundations of recursion theory, the machines were not yet available. Turing invented a hypothetical machine on which to base his notion of computability, but this machine was based on paper and pencil computation; Kleene studied and proved the equivalence of three distinct notions of computation, none of which was mechanical in description, i.e., machine-based. Indeed, it was not until the 1960s that moderately realistic models of computing machines were introduced into recursion theory.

A Nobel Prize winning physicist, Eugene Wigner, wrote an article entitled, "The unreasonable effectiveness of mathematics in the natural sciences", an essay I strongly recommend to the reader. One of the points made by Wigner is that the mathematics appropriate to the physicist is "in many if not most cases" in existence before it is needed. He cites matrix algebra and its surprising usefulness in quantum mechanics as an instance. One can find better examples: one anecdote is that, early in the present century, Sir James Jeans, a noted physicist, and Oswald Veblen, a mathematician, were discussing the mathematics curriculum at Princeton and Jeans suggested dropping group theory as it would never be of use to physics. (Perhaps it was embarrassment over this suggestion that turned Jeans into a staunch supporter of pure mathematics.) The best example of all, however, is probably the calculus: this subject is so useful in physics and its textbooks are so full of physical applications that it is often incorrectly believed that the calculus owes a great deal of its development to physics and physical applications. The truth is a little less glorious: much of the calculus was developed because mathematicians wanted to draw curves.

I mention the facts of the previous paragraph, not so much because they are interesting in their own right, but for the light they shed on computability. Despite the modern identification of the the word "computability" with the phrase "electronic computability", our definition of "computability" will make no mention of electronic computers and, when we do introduce a model machine, we will not use the model to justify the definition of computability, but will use the definition to demonstrate the adequacy of the machine.

Computing has gone on for roughly at least five millennia. Throughout most of that time mechanical aids were inadequate. By the 1930s, paper and pencil computation was the most sophisticated (if not the speediest or most reliable) form of computation. The several nascent theories of computation were based on the idea of computing as rewriting according to rules. Alonzo Church's λ-calculus is essentially a substitute-and-simplify model of computation; Gödel suggested a calculus of equations, again allowing substitution and simplification; and Turing, in defining his Turing machine which erases and prints symbols on tape, made a careful analysis of what the human computer does when he sits down with paper and pencil. (And: Emil Post independently introduced a notion of computation remarkably similar to Turing's. Later, Post and A.A. Markov, Jr., offered additional formulations of computing as rewriting rules.)

Restricted to natural numbers, these notions of computability are equivalent to one another: Kleene proved the equivalence of λ-computability with Gödel's equational derivability, and, later, with the notion we will discuss below; Turing proved the equivalence of λ-computability with what is now termed Turing computability; and so on. It is often asserted that the equivalence of all such attempts to isolate the notion of computability is strong evidence for the success of these attempts. The equivalence is certainly a necessary condition for the correctness of all attempted explications of computability, but it is not a sufficient one: there is always the possibility of a common oversight. What I find most convincing is the nature of the proof of equivalence of any attempted definition with the one we will presently encounter. Some people are most convinced by Turing's analysis of what the human computer does.

How should we define computability? Well, we must first decide what we want to compute with. Modern programming languages like LISP are designed to handle many different data structures, almost lost among which are the integers. It should not come as any surprise to the reader to hear (I know: I should have written "read", but it looks so silly) that all the fancier data structures can be coded by integers with (loss of efficiency, but) no loss of generality. In this book we are, moreover, interested almost exclusively in the natural numbers. Therefore, we make the preliminary decision to consider only the computability of functions of natural numbers.

The last paragraph states the problem, but how do we solve it? Actually, one solution is quite easy and was eventually proposed by Kleene: regardless of how complex a computable function might be, its computation, when broken down into simple steps, ought to be primitive recursive. We saw this for the non-primitive recursive Ackermann function P in Example 10.7. If this holds in general, any computable function F satisfies, for all $x_0, ..., x_{n-1}, y \in \omega$,

$$F(x_0, ..., x_{n-1}) = y \quad \text{iff} \quad \exists z\, [G(x_0, ..., x_{n-1}, y, z) = 0],$$

for some primitive recursive function G. Thus, F can be computed by *searching* for z and reading the value of F off z, this latter because z is assumed to code the entire computation, which includes $x_0, ..., x_{n-1}$, and y along with other information. The search is most naturally handled by looking for the least z, whence we introduce the following unbounded analogue to the bounded μ-operator.

11.1. Definition. (Unbounded μ-Operator). Let $n > 0$. Let $G:\omega^{n+1} \to \omega$ be given and let $x_0, ..., x_{n-1} \in \omega$. By

$$\mu y\,[G(x_0, ..., x_{n-1}, y) = 0], \qquad (*)$$

we denote the least $y \in \omega$ such that $G(x_0, ..., x_{n-1}, y) = 0$; if no such y exists, we leave (*) undefined.

The fact that (*) must occasionally be left undefined is unpleasant, but it is a fact of life we must live with. The relations,

$$R(x_0, ..., x_{n-1}):\ \exists y\,[G(x_0, ..., x_{n-1}, y) = 0],$$

for G primitive recursive, are by Corollary 10.11 exactly the Σ_1-relations. They offer a tremendous class of examples where the search (*) is partially successful; to ignore them would be to ignore a lot. Moreover, in the theory of computability, it turns out that the problem of distinguishing those cases where success is always possible from the others is an unsolvable one. We will have to consider functions which are occasionally undefined, i.e., we will have to consider what are commonly called *partial* functions.

11.2. Definitions. Let n be given. A function $F:A \to \omega$, where $A \subseteq \omega^n$, is called a *partial* n-ary numerical function. If the domain A is all of ω^n, F is called a *total* function. If F, G are partial n-ary functions, and $x_0, ..., x_{n-1} \in \omega$, we write

$$F(x_0, ..., x_{n-1}) \simeq G(x_0, ..., x_{n-1})$$

to mean that either F and G are both undefined at $(x_0, ..., x_{n-1})$ or they are defined and equal.

Logicians like to use the symbol "$\simeq$" rather than "=" because equality is a *relation* that holds or doesn't hold between *existing* objects. The new symbol is more of a relation between notations which may or may not denote. In formal theories, terms are generally taken to denote existing objects; in computability theory, terms denote the outcome of a computation— which outcome can fail to exist if the computation (say, the search (*) of 11.1) never halts. Since it is convenient to use such terms, but easy to be misled by them, new notation is introduced to remind us of what we are dealing with. Other related popular notations are:

$$F(x_0, ..., x_{n-1})\downarrow:\quad \text{"}F(x_0, ..., x_{n-1}) \text{ is defined"}$$

$$F(x_0, ..., x_{n-1})\uparrow:\quad \text{"}F(x_0, ..., x_{n-1}) \text{ is undefined".}$$

These can be defined in terms of "$\simeq$", e.g.,

$$F(x_0, ..., x_{n-1})\downarrow:\quad \exists y\,[F(x_0, ..., x_{n-1}) \simeq y].$$

With Definitions 11.1 and 11.2, we can define what we mean by a computable partial function.

11.3. Definitions. A partial function is *partial recursive* if it can be generated after finitely many steps by the following rules:

F1. $Z(x) = 0$ (1,1)

F2. $S(x) = x + 1$ (2,1)

F3. $P^n_i(x_0, ..., x_{n-1}) = x_i \quad (0 \le i < n)$ $(3,n,i)$

F4. $F(\boldsymbol{x}) = G(H_0(\boldsymbol{x}), ..., H_{m-1}(\boldsymbol{x}))$ $(4,n,g,(h_0,...,h_{m-1}))$

F5. $$\begin{cases} F(0,\boldsymbol{x}) = G(\boldsymbol{x}) \\ F(x+1,\boldsymbol{x}) = H(F(x,\boldsymbol{x}),\boldsymbol{x}) \end{cases}$$ $(5,n+1,g,h)$

F6. $F(\boldsymbol{x}) \simeq \mu y\,[G(\boldsymbol{x}, y) = 0]$, $(6,n,g)$,

where in *F6* we assume that G is a total $(n+1)$-ary function and $\boldsymbol{x}$ denotes $x_0, ..., x_{n-1}$ in *F4* and *F6*, and $x_1, ..., x_n$ in *F5*. If a function F generated by the above rules is total, then F is called a *general recursive function*, or, more simply, a *recursive function.*

Terminology varies. The above definition is sometimes said to define the *μ-recursive functions*, the general recursive functions originally being taken to be those generated by Gödel's recursion equations alluded to above. If we do not take the purist route here, it is because i. the classes of μ-recursive and general recursive functions coïncide, ii. the term "general recursive" is in general use, and iii. Gödel's recursion equations are rarely mentioned nowadays, they being of merely historical interest.

The definition varies: we could demand that G, H be total in *F5* and that G be total in *F4*, or we could even weaken *F6* from demanding that G be total to merely requiring $G(x_0, ..., x_{n-1}, z)$ to be defined for all $z \le \mu y\,[G(\boldsymbol{x}, y) = 0]$. Moreover, the Normal Form Theorem will canonically associate every definition of a partial recursive function with a definition making only one use of *F6*.

The things about the definition most in need of explanation are the sequences to the right of the defining clauses. These are *codes* for the functions being defined— actually, codes for the *definitions* of the functions, but I shall eventually stop using the word "definition". The first entry in a sequence serving as a code for a (definition of a) function F tells which of *F1* - *F6* is the major (i.e., last) clause applied in the definition, and the second entry gives the arity of the function. Thus, (1,1) codes the unary Zero function and (2,1) the unary Successor function. The codes arising from applications of *F3* - *F6* carry extra information: the index i of P^n_i for *F3*; a code g and a sequence $(h_0, ..., h_{m-1})$ of codes for the function G and sequence $H_0, ..., H_{m-1}$ of functions, respectively, for *F4*; and, for *F5*, *F6*, codes g, h for the auxiliary functions G, H.

This explains what the codes *are*, but not what they are *for*. For our stated aim of exhibiting the graphs of partial recursive functions as Σ_1-relations by primitive recursively coding computations, they are hardly necessary. The proof of the Σ_1-nature of the graphs of primitive recursive functions can readily be extended to cover the new clause *F6* (Exercise 1, below). We code the computations anew for higher purposes. One of these is philosophico-æsthetic: a representation,

$$F(x_0, ..., x_{n-1}) = y \Leftrightarrow \exists z \text{ "z computes } F(x_0, ..., x_{n-1}) = y\text{"},$$

is obviously superior to a representation,

$$F(x_0, ..., x_{n-1}) = y \Leftrightarrow \exists z\, [G(x_0, ..., x_{n-1}, y, z) = 0],$$

in which G is merely some unspecified primitive recursive function. A second higher purpose is philosophico-pædagogical: the coding of computations for general recursive functions illustrates the coding power of the primitive recursive functions and the means by which the general recursive functions can simulate any computable numerical function. I.e., it illustrates how one can prove the general recursiveness of any function declared computable by one of the other formal definitions of computability. Finally, there is a mathematical higher purpose: uniformly coding computations via these function codes transforms the entire subject from one of considering individual functions to a study of these codes. This study will begin in the next section.

But enough of discussion; it is time for action. We must give a primitive recursive coding of computations of partial recursive functions. To do so, we must decide what a computation is. We can take the elements of a computation to be those triples, i.e., sequences of length three,

$$(e, x, y),$$

where i. e is the code of a partial recursive function F (which codes we shall term *indices*, in part not to use the word "code" for indices and computations),
ii. x is a sequence $(x_0, ..., x_{n-1})$ of length n, n being the arity of F, and
iii. $F(x_0, ..., x_{n-1}) = y$.
Before deciding how to put these elements together to obtain a computation, we should verify that conditions i and ii are primitive recursively decidable. (Condition iii is not generally primitive recursive— more anon.)

Recognition of indices is primitive recursive, provided one doesn't insist on actually checking if G is total in generating an index in clause *F6*, i.e., we can recognise primitive recursively the appropriate *candidates* for indices.

First, we need to recognise the initial indices:

$$Init(e):\ e = (1,1) \vee e = (2,1) \vee \exists ni \le e\, [e = (3,n,i) \wedge i < n].$$

Next, we need to check that a supposed index $e = (4,n,g,h)$ is of the correct *form* so that, if g is an index and h a sequence of indices, e will be an index. This means that the length of h must equal the arity of g, and the arities of the entries of h must equal that of e:

$$F4(e):\ \exists ngh \le e\, \{e = (4,n,g,h) \wedge n > 0 \wedge lh(h) = (g)_1 \wedge \forall i < lh(h)[((h)_i)_1 = n]\}.$$

Indices arising via primitive recursion are easier to check:

$$F5(e):\ \exists ngh \le e\, \{e = (5,n+1,g,h) \wedge n > 0 \wedge (g)_1 = n \wedge (h)_1 = n + 2].$$

Finally,

$$F6(e):\ \exists ng \le e\, [e = (6,n,g) \wedge n > 0 \wedge (g)_1 = n + 1].$$

With *Init*, *F4*, *F5*, and *F6*, we can define the representing function χ_I of the set of indices by a course-of-values recursion and definition-by-cases (the combination not leading out of the class of primitive recursive functions— cf. Exercise 2, below):

$$\chi_I(e) = \begin{cases} 0, & Init(e) \\ 0, & F4(e) \wedge \chi_I((e)_2) = 0 \wedge \\ & \quad \wedge \ \forall i < lh(\ (e)_3\ )[\chi_I(\ ((e)_3)_i\ ) = 0] \\ 0, & F5(e) \wedge \chi_I((e)_2) = 0 \wedge \chi_I((e)_3) = 0 \\ 0, & F6(e) \wedge \chi_I((e)_2) = 0 \\ 1, & \text{otherwise.} \end{cases}$$

We will write,

$$Index(e):\ \chi_I(e) = 0.$$

Condition ii on the triple (e, x, y) that x code a sequence of length equal to the arity of e is readily expressed:

$$h(x) = (e)_1.$$

Hence, we define

$$Accept(v):\ Index(\ (v)_0\ ) \wedge lh(\ (v)_1\ ) = ((v)_0)_1 \wedge lh(v) = 3,$$

asserting v to code an *acceptable* triple.

Condition iii on a triple (e, x, y) that the function with index e applied to the arguments coded by x yields the value y cannot be expressed primitive recursively. This correctness will have to be guaranteed by the structure of the computation, i.e., by how (e, x, y) fits with other elements of the computation. How should we put these elements together to create a whole? There are three natural ways to collect triples into a computation: we can declare a computation to be a set, a sequence, or a tree of acceptable triples related in the appropriate manners. We have already seen, with Skolem's Example 10.7, the use of sets to encode computations. We discuss here the use of sequences, that of trees will appear in the exercises.

Kleene, basing his coding of computations on Gödel's coding of formal derivations, used the sequential approach: a computation is a sequence of acceptable triples such that each triple is either immediately checkable by an initial function or follows from earlier triples in the list by one of the computation clauses *F4* -*F6* of 11.3. Thus, in this approach, we would define *Initcomp*(t):

$$\begin{aligned} \exists exy \le t\, \{t = (e,x,y) \wedge [(e = (1,1) \wedge lh(x) = 1 \wedge y = 0) \vee \\ \vee\ (e = (2,1) \wedge lh(x) = 1 \wedge y = (x)_0 + 1) \vee \\ \vee\ \exists ni \le e\,(e = (3,n,1) \wedge lh(x) = n \wedge i < n \wedge (x)_i = y)]\}, \end{aligned}$$

asserting that t is a triple correctly computing an initial function at some value. [The bounded existential quantifiers can be replaced by applications of the projections $(\cdot)_i$;

however, the resulting formulae would be less readable.] After defining *Initcomp*, one primitive recursively expresses that an acceptable triple follows from one or more triples in accordance with clauses *F4*, *F5*, or *F6* of Definition 11.3. *F5* is the easiest of these and we shall consider it first. There are two possibilities for a computation of $F(x,x_1,\ldots,x_n)$ defined by recursion from G, H as in *F5*— $x = 0$ and $x > 0$. For the former, we have $F5comp_G(t,u)$:

$$Accept(t) \wedge Accept(u) \wedge \exists nghxy \leq t\,\{(x)_0 = 0 \wedge$$
$$\wedge\ t = ((5,n+1,g,h),x,y) \wedge \exists z \leq x\,[x = (0)*z \wedge u = (g,z,y)]\}.$$

For the latter, we have $F5comp_H(t,u,v)$:

$$Accept(t) \wedge Accept(u) \wedge Accept(v) \wedge$$
$$\wedge\ \exists nghxy \leq t\,\{(x)_0 > 0 \wedge t = ((5,n+1,g,h),x,y) \wedge$$
$$\wedge\ \exists zw \leq x\,[x = ((x)_0)*z \wedge u = (h,(w,((x)_0 \dot{-} 1)*z,y) \wedge$$
$$\wedge\ v = ((5,n+1,g,h),((x)_0 \dot{-} 1)*z,w)]\}.$$

Both *F4* and *F6* will have the triple t following from a sequence of triples. First, we consider *F4*. We define $F4comp(t,u,v)$:

$$Accept(t) \wedge Accept(u) \wedge \forall i < lh(v)\, Accept(\,(v)_i\,) \wedge$$
$$\wedge\ \exists nghxy \leq t\,\{t = (4,n,g,h) \wedge \exists mzw \leq u\ [u = (g,w,y) \wedge$$
$$\wedge\ lh(w) = lh(v) \wedge \forall i < lh(w)\,[(v)_i = ((h))_i, x, (w)_i)]]\}.$$

Finally, we have $F6comp(t,u)$:

$$Accept(t) \wedge \forall i < lh(u)\, Accept(\,(u)_i\,) \wedge$$
$$\wedge\ \exists nghxy \leq t\,\{t = (6,n+1,g,h) \wedge lh(u) = y+1 \wedge$$
$$\wedge\ \forall i \leq y\, \exists vwz \leq u\ [(u)_i = (g,v,w) \wedge v = z*(i) \wedge$$
$$\wedge\ (i < y \wedge w > 0 \ \vee\ i = y \wedge w = 0)]\}.$$

None of these formulae is entirely readable, but a little study will either reveal that they merely assert the proper match-up between t and the other acceptable triples according to the rules *F4* - *F6*, or that they can be made to assert this after the reader makes some small corrections. (It is an established fact that one always makes minor mistakes in these matters.)

A *computation* consists of a sequence of acceptable triples, each of which is either an initially computable triple or follows from earlier triples in the sequence. A computation computes the last triple occurring in it. Modulo the determination of a couple of bounds, we can define $Comp(z,e,x,y)$ asserting z to code a computation of the triple (e,x,y) as follows. $Comp(z,e,x,y)$:

$$(z)_{lh(z)\dot{-}1} = (e, x, y) \wedge \forall i < lh(z) \{Initcomp((z)_i) \vee$$
$$\vee \exists j < i\, F5comp_G((z)_i, (z)_j) \vee \exists jk < i\, F5comp_H((z)_i, (z)_j, (z)_k) \vee$$
$$\vee \exists j < i \exists v \leq K(z)\, [F4comp((z)_i, (z)_j, v) \wedge \forall k < lh(v) \exists m < i\, [(v)_k = (z)_m]] \vee$$
$$\vee \exists u \leq K(z)\, [F6comp((z)_i, u) \wedge \forall k < lh(u) \exists m < i\, [(u)_k = (z)_m]]\}.$$

Again, thinking of the meanings of these clauses makes it clear that, once we've verified the primitive recursiveness of K, $Comp$ is a primitive recursive relation describing computations as desired. So, how do we determine K? The sequence u in the last disjunct is actually a subsequence of z and, for the encoding 7.8 which we have chosen as our standard, it is easily seen that the code of a subsequence is no larger than the code of the full sequence (for: one generally has fewer powers of 2, each with a smaller exponent, in the code of the subsequence). Thus, the second reference to K can be replaced by $\exists u \leq z$. With composition, however, some of the functions in the list $H_0, ..., H_{m-1}$ could be the same and the sequence y, although containing entries from z can have repetitions. But the length of v is bounded by an element of $(z)_i$, whence by z, and each entry is bounded by an element of z, whence by z. Thus, we can let $K(z)$ be the code of $(z, z, ..., z)$, the sequence having length z, i.e.,

$$K(z) = 2^z + 2^{z+z+1} + ... + 2^{(z+z+...+z)+(z-1)}$$
$$= \sum_{x=0}^{z-1} 2^{(x+1)z+x},$$

which is primitive recursive.

11.4. Definitions. Let $U(z) = ((z)_{lh(z)\dot{-}1})_2$ and, for $n > 0$, define T_n by

$$T_n(e, x_0, ..., x_{n-1}, z):\ \ Comp(z, e, (x_0, ..., x_{n-1}), U(z)).$$

We can now draw the main conclusion of this whole construction.

11.5. Theorem. (Normal Form Theorem). *Let* $n > 0$.

i. (Enumeration Theorem). *For every* Σ_1*-relation* $R \subseteq \omega^n$, *there is an* $e \in \omega$ *such that, for all* $x_0, ..., x_{n-1} \in \omega$,

$$R(x_0, ..., x_{n-1}) \Leftrightarrow \exists z\, T_n(e, x_0, ..., x_{n-1}, z). \tag{1}$$

Conversely, for every $e \in \omega$, *the relation* R *defined by* (1) *is a* Σ_1*-relation.*

ii. (Universal Partial Function). *For every partial recursive function* F *on* ω^n, *there is an* $e \in \omega$ *such that, for all* $x_0, ..., x_{n-1} \in \omega$,

$$F(x_0, ..., x_{n-1}) \simeq U(\mu z\, T_n(e, x_0, ..., x_{n-1}, z)), \tag{2}$$

where $\mu z\, T_n$ *indicates the least* z *satisfying* T_n, *i.e.,* $\mu z\, [\chi_{T_n} = 0]$. *Conversely, for every* $e \in \omega$, *the function* F *defined by* (2) *is partial recursive.*

Proof. i. The Σ_1-nature of any relation $\exists z\, T_n(e, x_0, ..., x_{n-1}, z)$ follows from Theorem 10.6 of the preceding section. Thus, we need only show that any Σ_1-relation has this form. This is fairly easy. By Corollary 10.11, there is a primitive recursive function F such that, for all $x_0, ..., x_{n-1} \in \omega$,

$$R(x_0, ..., x_{n-1}) \Leftrightarrow \exists y\, [F(x_0, ..., x_{n-1}, y) = 0].$$

Define G by

$$G(x_0, ..., x_{n-1}) \simeq \mu y\, [F(x_0, ..., x_{n-1}, y) = 0]$$

and choose e to be an index of G. Then,

$$\begin{aligned} R(x_0, ..., x_{n-1}) &\Leftrightarrow G(x_0, ..., x_{n-1})\downarrow \\ &\Leftrightarrow \exists z\, T_n(e, x_0, ..., x_{n-1}, z). \end{aligned}$$

ii. Again, the partial recursiveness of any function defined by (2) is obvious. But the direct assertion is also obvious: if F is partial recursive, choose e to be an index of F. QED

I am tempted to give a longer proof of the first part of the Theorem which avoids the quick appeal to Corollary 10.11. Such a proof would proceed by labelling the following corollary a "lemma" and proving the crucial implication by induction on the generation of Σ_1-relations.

11.6. Corollary. *Let $n > 0$ and $R \subseteq \omega^n$. R is a Σ_1-relation iff R is the domain of a partial recursive function.*

Proof. If R is a Σ_1-relation, there is an e such that formula (1) of 11.5 holds. For that e, R is clearly the domain of the function F given by (2). [Alternatively, R is the domain of the function G constructed in the proof of 11.5.]

Conversely, if R is the domain of some partial recursive function F, there is an index e for which (2) holds. Thus, for any $x_0, ..., x_{n-1} \in \omega$,

$$\begin{aligned} R(x_0, ..., x_{n-1}) &\Leftrightarrow U(\mu z\, T_n(e, x_0, ..., x_{n-1}, z))\downarrow \\ &\Leftrightarrow \mu z\, T_n(e, x_0, ..., x_{n-1}, z)\downarrow \\ &\Leftrightarrow \exists z\, T_n(e, x_0, ..., x_{n-1}, z), \end{aligned}$$

which reveals R to be Σ_1. [Alternatively, one could extend the proof of Theorem 10.6 to show the graph of any partial recursive function to be Σ_1.] QED

The reader can work out the alternative proof hinted at above for himself (Exercise 6, below). The reasons for carrying out this exercise are: i. the new proof shows more directly the semi-computable nature of Σ_1-relations, and ii. by not relying on Corollary 10.11, and proving in order 11.6, 11.5, and then 10.11, one sees these results established purely computationally— the syntactic component behind the proof of 10.11 is thus deëmphasised.

Two additional characterisations worth mentioning are the following.

11.7. Theorem. i. *Let F be a partial function on ω^n, for $n > 0$. Then: F is partial recursive iff its graph is Σ_1.*
ii. *Let $R \subseteq \omega$. Then: R is Σ_1 iff R is the range of a partial recursive function.*

The proof will be given in the Exercises (Exercise 7, below), along with some similar standard results.

Theorem 11.7.ii has one important terminological consequence.

11.8. Convention. Σ_1-relations are called *recursively enumerable*, or *r.e. relations*.

Occasionally, it is convenient to refer to these relations as r.e. relations and occasionally it is convenient to refer to them as Σ_1-relations. Thus, we shall use both terms (as well as the longer "recursively enumerable relations") interchangeably in the sequel.

It will not have escaped the reader's notice that we have not made full use of Theorem 11.5 in these applications— Corollary 10.11 would have sufficed. The full power of the Theorem will be felt in the next section, where the indices will completely transform the subject— or, as Goethe said, it will become "ganz etwas anderes".

What I wish to do now— still in this section— is to consider another definition of computability and prove its equivalence to the present one. This will afford another look at the primitive recursive simulation of computations. This model will appear so weak, however, that the primitive recursive simulation of the machine will not be as impressive as the machine's calculations of partial recursive functions.

The reader is advised that this material is a bit boring and that in the sequel it will only be used in the next chapter in giving an alternate proof of one result. Thus, the reader who is not interested in this material can easily skip ahead to the exercises and the next section.

The new definition of computability will be computability by a *register machine*. A register machine consists simply of a finite number of separately addressable registers R_0, ..., R_{n-1}. Each register can store a natural number. Modern electronic computers differ from register machines in that i. the size of the registers in computers is small— there is an upper bound on the size of the natural number that can be stored in a given register, and ii. the number of registers needed is quite large— 65000 for the small computers of a few years ago, 20,000,000 for the computer I'm typing this book on, and much more on the newer models.

Like modern computers, register machines can be programmed to do various things. A *program* consists of a list L_0, ..., L_{m-1} of instructions in a rather impoverished assembly language:

11.9. List of Commands. The register machine assembly language has the following commands:

i. INC R_j
ii. DEC R_j
iii. GO TO L_k
iv. IF $R_j = 0$ GO TO L_k
v. STOP.

The explanations of these commands are:

i. INC R_j: increment register R_j by 1, i.e., replace the number r_j in R_j by $r_j + 1$
ii. DEC R_j: decrement register R_j by 1, i.e., replace the number r_j in R_j by $r_j - 1$ (A convention will shortly guarantee that DEC R_j will only execute if $r_j \neq 0$; alternatively, we could use cut-off subtraction.)

iii. GO TO L_k: if L_i is GO TO L_k, control passes to instruction L_k (If there is no L_k, we imagine the machine hanging.)
iv. IF $R_j = 0$ GO TO L_k: if the content of R_j is 0, one next executes L_k and continues the program from that point; otherwise, one simply proceeds to instruction L_{i+1}
v. STOP: this halts the execution of the program and, presumably, tells the user that the execution is over.

A program is a consecutively indexed list $L_0, L_1, ..., L_{m-1}$ of commands. Execution of a program by a machine is sequential: $L_0, L_1, ..., L_{m-1}$ are executed in order except when a GO TO or a conditional GO TO says to do otherwise or when a STOP command halts the execution. Two obvious restrictions are:
i. the instruction L_k referred to in the two types of GO TO statements must exist or the program execution is interrupted; and
ii. any register R_j referred to in an instruction must exist or the program execution is interrupted.

A third restriction, although not necessary, is convenient:

11.10. Convention. A DEC command is only allowed in a program in the following context:

$$\begin{array}{ll} L_i & \text{IF } R_j = 0 \text{ GO TO } L_k \\ L_{i+1} & \text{DEC } R_j, \end{array}$$

where $k \neq i+1$. In words, one must check that R_j contains a non-zero number before decrementing; if R_j contains a 0, one must branch to some other command. Moreover, branching to a DEC command is not allowed.

Checking the Convention would be a nuisance in the primitive recursive simulation of register machine computations. Fortunately, we can ignore this during the simulation. It is in writing programs that the Convention is to be imposed in the name of "structured programming" or "programming style".

We know what it means to program a register machine, and how a program is to be executed. What does it mean to compute a partial function F by a register machine?

11.11. Definition. Let F be a partial n-ary numerical function. F is *register machine computable* (or: *r.m. computable*) if there is a machine containing at least the registers $R_0, ..., R_{n-1}$ and a program $L_0, ..., L_{m-1}$ such that, if we start with $x_0, ..., x_{n-1}$ in $R_0, ..., R_{n-1}$, respectively, and run the program, the execution will halt with a STOP command and $F(x_0, ..., x_{n-1})$ will be in R_0.

Notice that we allow extra registers for side computations. We make no assumptions on the final contents of the registers other than R_0; it is sometimes considered polite to empty them (i.e., put 0's in all of them).

11.12. Theorem. *Let F be a partial numerical function. Then: F is partial recursive iff F is computable by a register machine.*

Proof. One half of the proof, that showing partial recursive functions are r.m. computable, consists of choosing machines and writing programs for them. As this will help us to familiarise ourselves with register machine programming and computation, we shall present this half of the proof first.

We will show the r.m. computability of all partial recursive functions by induction on their generation. For the basis of the induction, we simply present machines and programs for computing the initial functions.

Zero. We need one register R_0 and the following program:

L_0	IF $R_0 = 0$ GO TO L_3
L_1	DEC R_0
L_2	GO TO L_0
L_3	STOP.

This program calculates the zero function by simply decrementing R_0 until R_0 contains 0 and then stopping.

Successor. Again we need one register R_0. The program is:

L_0	INC R_0
L_1	STOP.

Projection. Let $i < n$ be given. To calculate P^n_i, we need n registers $R_0, ..., R_{n-1}$ to hold the inputs $x_0, ..., x_{n-1}$. If $i = 0$, we already have $P^n_i(x_0, ..., x_{n-1}) = x_0$ in R_0 and can use the simple program:

L_0	STOP.

If $i \neq 0$, we must first reset R_0 to 0 and then copy R_i into R_0 by "simultaneously" incrementing R_0 and decrementing R_i until R_i has 0 in it. A program doing this is:

L_0	IF R_0= 0 GO TO L_3	
L_1	DEC R_0	Reset R_0 to 0
L_2	GO TO L_0	
L_3	IF $R_i = 0$ GO TO L_7	
L_4	DEC R_i	Copy R_i into R_0
L_5	INC R_0	
L_6	GO TO L_3	
L_7	STOP.	

The induction step of the proof consists of showing how we can take machines and programs for calculating some given functions and construct from them machines and programs for calculating functions arising from the given functions via *F4 - F6*. In this, I shall merely give an outline. When a portion of one program is copied into another one (as when all but the STOP command of the Zero program was used at the beginning of the

Projection program), both the lines of the program and their labels in GO TO statements may have to be renumbered. Moreover, we will sometimes need to apply such incorporated programs to registers other than $R_0, ..., R_{n-1}$. This requires relabelling of register numbers in command lines (e.g., in using the Zero program to reset a register other than R_0 to 0). Keeping track of the relabelling does not appeal to me and I forego the pleasure.

The programs to be described in the treatment of the induction step will require a lot of copying operations. The copying mechanism of the Projection program has the flaw of destroying the contents of R_i while copying them into R_0. To get around this, we have the copy program below, which copies R_i into R_j without losing what is in R_i. It uses an auxiliary register R_k. (It is assumed i, j, k are distinct.) The explanation of its working is written on the right:

L_0	IF $R_j = 0$ GO TO L_3	Reset R_j to 0
L_1	DEC R_j	
L_2	GO TO L_0	
L_3	IF $R_k = 0$ GO TO L_6	Reset R_k to 0
L_4	DEC R_k	
L_5	GO TO L_3	
L_6	IF $R_i = 0$ GO TO L_{11}	Copy R_i into R_j and R_k and Reset R_i to 0
L_7	DEC R_i	
L_8	INC R_j	
L_9	INC R_k	
L_{10}	GO TO L_6	
L_{11}	IF $R_k = 0$ GO TO L_{15}	Copy R_k into R_i
L_{12}	DEC R_k	
L_{13}	INC R_i	
L_{14}	GO TO L_{11}	
L_{15}	STOP.	

Composition. Suppose

$$F(x_0, ..., x_{n-1}) = G(H_0(x_0, ..., x_{n-1}), ..., H_{m-1}(x_0, ..., x_{n-1}))$$

and that we have register machines and programs for computing $G, H_0, ..., H_{m-1}$. An upper bound to the number of registers needed to compute F is 1 greater than the sum of the number of registers needed to compute $G, H_0, ..., H_{m-1}$. To compute $F(x_0, ..., x_{n-1})$, let $x_0, ..., x_{n-1}$ be put into registers $R_0, ..., R_{n-1}$, respectively. Begin the F-program by using a number of modified copies of the Zero program to set all registers other than $R_0, ..., R_{n-1}$ to 0. Then use enough copies of the copying program to copy R_0 into R_n, R_1 into R_{n+1}, etc. When this much of the program has been executed, one will have:

$$\begin{array}{cccccc} R_0 & \cdots & R_{n-1} & R_n & \cdots & R_{2n-1} \\ x_0 & \cdots & x_{n-1} & x_0 & \cdots & x_{n-1}. \end{array}$$

Now, run the H_0-program, modified so as to ignore $R_0, ..., R_{n-1}$ and leave its output in R_n:

$$\begin{array}{cccccc} R_0 & \cdots & R_{n-1} & R_n & \cdots & R_{2n-1} \\ x_0 & \cdots & x_{n-1} & H_0(\boldsymbol{x}) & \cdots & \end{array}$$

Now add to the end of the program so far constructed instructions to copy $R_0, ..., R_{n-1}$ into $R_{n+1}, ..., R_{2n}$, respectively. Letting $y_i = H_i(\boldsymbol{x})$, this yields:

$$\begin{array}{ccccccc} R_0 & \cdots & R_{n-1} & R_n & R_{n+1} & \cdots & R_{2n} \\ x_0 & \cdots & x_{n-1} & y_0 & x_0 & \cdots & x_{n-1}. \end{array}$$

Now add a modified H_1-program using $R_{n+1}, ..., R_{2n}$ and higher indexed registers to get

$$\begin{array}{cccccc} R_0 & \cdots & R_{n-1} & R_n & R_{n+1} & \cdots \\ x_0 & \cdots & x_{n-1} & y_0 & y_1 & \cdots, \end{array}$$

where, again $y_i = H_i(\boldsymbol{x})$. Repeat all of this $m-2$ more times, so that the execution will yield

$$\begin{array}{cccccc} R_0 & \cdots & R_{n-1} & R_n & \cdots & R_{n+m-1} \\ x_0 & \cdots & x_{n-1} & y_0 & \cdots & y_{m-1}. \end{array}$$

Now add to the end of the program enough copies of the Zero program to reset all registers from R_{n+m} onwards to 0 and a modification of the G-program using registers from R_n onwards. The result will be

$$\begin{array}{ccccc} R_0 & \cdots & R_{n-1} & R_n & \cdots \\ x_0 & \cdots & x_{n-1} & F(\boldsymbol{x}) & \cdots \end{array}$$

It remains only to incorporate the program for P_0^{n+1} to move $F(\boldsymbol{x})$ from R_n into R_0.

Once again I note that the actual construction of the program requires us to be careful about relabelling command indices and register indices. Writing a few sample programs based on the above will best illustrate the nuisance of it. Cf. Exercise 9, below.

Primitive recursion. I shall be briefer in describing this computation. Let

$$F(0, x_0, ..., x_{n-1}) \simeq G(x_0, ..., x_{n-1})$$
$$F(x+1, x_0, ..., x_{n-1}) \simeq H(F(x, \boldsymbol{x}), x, \boldsymbol{x}),$$

and suppose we can already compute G and H on register machines. Start with $x, x_0, ..., x_{n-1}$ in $R_0, ..., R_n$, respectively. Copy 0 into R_{n+1}, R_{n+2}, and R_{n+3} and copy $x_0, ..., x_{n-1}$ into $R_{n+4}, ..., R2_{n+3}$:

R_0	R_1	...	R_n	R_{n+1}	R_{n+2}	R_{n+3}	R_{n+4}	...	R_{2n+3}
x	x_0	...	x_{n-1}	0	0	0	x_0	...	x_{n-1}.

R_1 - R_n will store x_0, ..., x_{n-1}; R_{n+4} - R_{2n+3} and additional registers will be used to calculate G; R_{n+2} - R_{2n+3} and additional registers will be used to calculate H. R_0 and R_{n+1} will serve as counters: after each run of G or H, R_0 will decrement and R_{n+1} will increment. If R_0 is already 0, the latest value computed will go into R_0 and the computation will stop.

Run G to get

R_0	R_1	...	R_n	R_{n+1}	R_{n+2}	R_{n+3}	R_{n+4}	...
x	x_0	...	x_{n-1}	0	0	0	$G(\boldsymbol{x})$	...

If R_0 contains 0, copy R_{n+4} into R_0 and stop.

If $R_0 \neq 0$, decrement R_0, increment R_{n+1}, copy R_{n+4} into R_{n+2}, R_{n+1} into R_{n+3}, and R_1 - R_n into R_{n+4} - R_{2n+3}:

R_0	R_1	...	R_n	R_{n+1}	R_{n+2}	R_{n+3}	R_{n+4}	...	R_{2n+3}
$x-1$	x_0	...	x_{n-1}	1	y_0	1	x_0	...	x_{n-1},

where y_0 denotes $F(0,\boldsymbol{x})$. Run H to get

R_0	R_1	...	R_n	R_{n+1}	R_{n+2}	...
$x-1$	x_0	...	x_{n-1}	1	y_1	...,

where $y_1 = F(1,\boldsymbol{x})$. If R_0 is 0, copy R_{n+2} into R_0 and stop; otherwise, repeat the procedure of this paragraph (only, now leave R_{n+2} untouched).

Addition is rather special and a much simpler register machine program for calculating $x + y$ than one obtained by the above procedure can be given. Nevertheless, it would be instructive to the reader to develop a program for addition using the recursive definition.

Minimisation. Let

$$F(x_0, ..., x_{n-1}) \simeq \mu y\,[G(x_0, ..., x_{n-1}, y) = 0],$$

where G is total and we can calculate G by a register machine. To calculate F, we will need $n + 3$ additional registers beyond those needed to calculate G ($n + 2$ of these being visible and 1 for the copying mechanism). Let x_0, ..., x_{n-1} be put into R_0, ..., R_{n-1} in preparation for the calculation. Set register R_n to 0.

Copy R_0, ..., R_n into R_{n+1}, ..., R_{2n+1}:

R_0	...	R_{n-1}	R_n	R_{n+1}	...	R_{2n}	R_{2n+1}
x_0	...	x_{n-1}	0	x_0	...	x_{n-1}	0.

Ignoring registers R_0, ..., R_n, calculate G:

R_0	...	R_{n-1}	R_n	R_{n+1}	...
x_0	...	x_{n-1}	0	$G(\boldsymbol{x},0)$	...

If R_{n+1} is 0, copy R_n into R_0 and stop; otherwise, increment R_n and repeat the procedure of this paragraph.

With this last, we have finished the proof that every partial recursive function is r.m. computable. For the converse, we show how to code primitive recursively the computations of register machines. We can do this in one of two ways. We can, as in Theorem 11.5, create a uniform computation predicate, or we could simply handle each machine and program separately. It is not a great increase in effort to do the former, but there is no particular reason to do so and I shall simply do the latter. The energetic reader can carry out the uniform version on his own.

Let an n-register machine with program $L_0, ..., L_{m-1}$ be given. For convenience in coding the execution of the program, we assume an additional register K called the *counter*. K will contain the number i between 0 and $m-1$ of the next instruction to be executed. A *state description* of the machine is a sequence $(i, r_0, \ldots, r_{n-1})$, where i is the current instruction number in K and $r_0, ..., r_{n-1}$ are the numbers in $R_0, ..., R_{n-1}$, respectively. A *computation* will be a special sequence, $(s_0, \ldots, s_{k-1})$, of state descriptions. If the program and machine are to compute a function F at $x_0, ..., x_{n-1}$, the first state description must be $(0, x_0, ..., x_{n-1})$:

$$Initial(s, x_0, ..., x_{n-1}):\ s = (0, x_0, ..., x_{n-1}).$$

The final state must have the desired value y in R_0 and an index for the STOP instruction:

$$Result(s, y):\ y = (s)_1 \ \wedge\ \mathrm{STOP}(\ (s)_0\ ),$$

where

$$\mathrm{STOP}(i):\ i = i_0 \ \vee\ i = i_1 \ \vee\ \ldots\ \vee\ i = i_{h-1},$$

where $L_{i_0}, ..., L_{i_{h-1}}$ are the STOP instructions occurring in the program. (If there are no STOP instructions, then, by definition, the machine computes the empty function, which is trivially partial recursive, say,

$$\mathrm{Empty}(\mathbf{x}) = \mu y\ [\ \overline{sg}\ (Z(P_0^{n+1}(\mathbf{x}, y))) = 0].\ )$$

The transitions from one state description to the next depend on the instructions L_i. Each instruction shall have its own transition predicate depending on the type of instruction it is.

Increment. If L_i is INC R_j, define

$$L_i(s, t):\qquad (s)_0 = i \ \wedge\ (t)_0 = i+1 \ \wedge\ (t)_{j+1} = (s)_{j+1} + 1 \ \wedge$$
$$\wedge\ \forall h < n\ [h = 0 \ \vee\ h = j+1 \ \vee\ (s)_h = (t)_h].$$

Decrement. If L_i is DEC R_j, define

$$L_i(s, t):\qquad (s)_0 = i \ \wedge\ (t)_0 = i+1 \ \wedge\ (t)_{j+1} = (s)_{j+1} \dot{-} 1 \ \wedge$$
$$\wedge\ \forall h < n\ [h = 0 \ \vee\ h = j+1 \ \vee\ (s)_h = (t)_h].$$

GO TO. If L_i is GO TO L_k, define

$$L_i(s, t):\qquad (s)_0 = i \ \wedge\ (t)_0 = k \ \wedge\ \forall h < n\ [h = 0 \ \vee\ (s)_k = (t)_k].$$

Test. If L_i is IF $R_j = 0$ GO TO L_k, define

$$L_i(s,t): \quad (s)_0 = i \wedge \forall h < n\,[h = 0 \vee (s)_k = (t)_k] \wedge$$
$$\wedge\ [(s)_{j+1} = 0 \wedge (t)_0 = k \quad \vee \quad (s)_{j+1} \neq 0 \wedge (t)_0 = i+1].$$

With all of this, we can now define a primitive recursive predicate asserting "z is a computation of $F(x_0, ..., x_{n-1}) = y$":

$$Rmcomp_F(z, x_0, ..., x_{n-1}, y): \quad \forall k < lh(z)\,[lh((z)_k) = n+1] \wedge$$
$$\wedge\ Initial((z)_0, x_0, ..., x_{n-1}) \wedge Result((z)_{lh(z)\dot{-}1}, y) \wedge$$
$$\wedge\ \forall j < lh(z) \dot{-} 1\,[\bigvee_{i=0}^{m-1} [((z)_j)_0 = i \wedge L_i((z)_j, (z)_{j+1})]].$$

With this, one can define a result extracting function V to obtain y from z:

$$V(z) = \text{entry of } R_0 \text{ in the last state description in } z$$
$$= ((z)_{lh(z)\dot{-}1})_1.$$

Then:

$$F(x_0, ..., x_{n-1}) = V(\mu z\, Rmcomp_F(z, x_0, ..., x_{n-1}, V(z)))$$

is partial recursive. This completes the proof of Theorem 11.12. QED

(A small remark: Convention 11.10 is one on how we write programs. We *assume* L_0, ..., L_{m-1} satisfy it in the second half of the proof and thus we need not check for it in defining the predicates L_i in the case of a decrement instruction.)

Exercises.

1. Extend the proof of Theorem 10.6 to show that the graph of each partial recursive function is Σ_1. [Hint. If G is total, then $G(\mathbf{x}) \neq 0$ is equivalent to $\exists y\,[G(\mathbf{x}) = y+1]$.]
2. Show that the representing function χ_I for (candidates for) indices of partial recursive functions is primitive recursive. [Hint. First define H using the case definition for χ_I, but with new variables in all the positions where some value of χ_I occurs.]
3. Verify that (5, 2, (3,1,0), (4, 3, (2,1), ((3,3,0)))) is an index for addition.
   i. Find the set of acceptable triples occurring in a computation of 1+1=2.
   ii. What is an index for $D(x) = x + x$? List the triples occurring in a computation of $D(1) = 2$.

   [It is permissible to introduce abbreviations like $s = (2,1)$, $p_0^1 = (3,1,0)$, $a = (5, 2, p_0^1, (4,3,s, (p_0^3)))$.]
4. Our sequential coding of computations was not entirely true to Kleene in that we allowed sequence elements to be viewed as sequences. We can extend this to get a tree-like representation of computations. Instead of defining a computation of a triple to be a sequence of triples, we take it to be a sequence containing a triple and subcomputations. For example, a computation of

$$y = G(H_0(x), H_1(x))$$

could be a triple,

$$((e,(x),y),\ z,\ w), \qquad (*)$$

where e is an index of F, z is a computation of $y = G(v_0,v_1)$, and w is a pair (z_0, z_1) of computations of $v_0 = H_0(x)$ and $v_1 = H_1(x)$.

i. Give a complete coding of computations along these lines. Give a new definition of *Comp* based on this.

ii. Getting back to trees, (*) can be represented graphically by

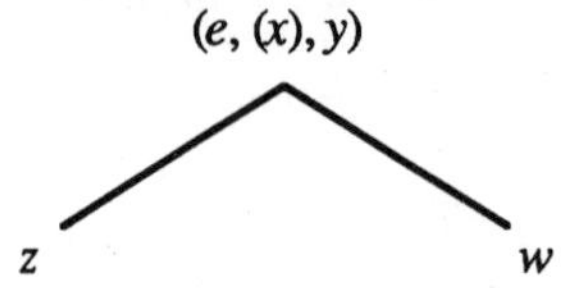

where each of z and w may also branch. Draw a picture of a tree representing the computation of $D(1) = 2$.

[Remarks. We use (x) instead of $x$in (*) because of the variable arities of functions. The tasks of parts i and ii can each be performed in a variety of ways.]

5. Give the Ackermann function P the code (7,2). Extend the computation predicate used in proving Theorem 11.5 to include computations of P and functions built up from it. Conclude that P is general recursive.

6. Give a direct proof by induction on the generation of Σ_1-relations that, if R is Σ_1, it is the domain of a partial recursive function.

[Do this by defining χ_R^+ so that

i. $\chi_R^+(x_0, ..., x_{n-1})\downarrow$ iff $R(x_0, ..., x_{n-1})$

ii. $R(x_0, ..., x_{n-1}) \Rightarrow \chi_R^+(x_0, ..., x_{n-1}) = 0$.]

7. i. Prove Theorem 11.7.i.

ii. Prove Theorem 11.7.ii.

iii. Prove that, if the set X is non-empty, then X is Σ_1 iff it is the range of a recursive function.

iv. Use i to show that, if $R, S \subseteq \omega^n$ are disjoint Σ_1-relations, and G, H are n-ary partial recursive functions, then the function

$$F(\boldsymbol{x}) \simeq \begin{cases} G(\boldsymbol{x}), & R(\boldsymbol{x}) \\ H(\boldsymbol{x}), & S(\boldsymbol{x}) \\ \text{undefined}, & \text{otherwise}, \end{cases}$$

is partial recursive.

[Hint. iii. Fix $x_0 \in X$ and use a primitive recursive definition-by-cases.]

8. Define a set $X \subseteq \omega$ to be recursive iff its representing function χ_X is recursive.

i. Show: X is recursive iff both X and $\omega - X$ are Σ_1.

ii. Show: X is recursive iff X is finite or there is a strictly increasing recursive function F such that X is the range of F.

9. i. Give a register machine program to compute addition. Assume the machine has only two registers.

ii. Give a register machine program to compute multiplication. Use as small a number of registers as possible.

iii. Use the program given in the text for Z to give a program to calculate $F(x) = Z(Z(x))$ by applying the method for handling compositions— even though there is an easier approach for this function.

iv. Find a register machine program to compute the function $D(x) = Add(P^1_0(x), P^1_1(x))$, using the technique to handle composition.

10. i. Give register machine programs for the functions

$$\chi_<(x, y) = \begin{cases} 0, & x \le y \\ 1, & x \ge y \end{cases}$$

$$\chi_=(x, y) = \begin{cases} 0, & x = y \\ 1, & x = y. \end{cases}$$

ii. Let $\mathcal{P}_1$ be the class of functions containing the initial functions $Z, S, P^n_i, \chi_<$, in addition to $A(x, y) = x + y$ and $M(x, y) = x{\cdot}y$, and closed under composition (*F4*) and minimisation (*F6*). Show: $\mathcal{P}_1$ coïncides with the class of partial recursive functions.

iii. Let $\mathcal{P}_2$ be defined like $\mathcal{P}_1$, but with $\chi_=$ in place of $\chi_<$. Show: $\mathcal{P}_2$ coïncides with the class of partial recursive functions.

iv. Show that Z, S can be dropped from the list of initial functions in the definition of $\mathcal{P}_1$.

v. Show that Z can be dropped from the list of initial functions in the definition of $\mathcal{P}_2$. Show that S cannot be so dropped.

[Hint. ii - iii. use Theorem 10.10.]

11. Invent codes for register machine commands, e.g., $(1, j)$ for INC R_j, $(2, j)$ for DEC R_j, etc. Give a uniform primitive recursive computation predicate for register machine computations. [For the sake of simplicity, you can limit the number of registers to some fixed number n (e.g., $n = 3$).]

12. Elementary Recursion Theory

The Normal Form Theorem, as already remarked, transforms the subject of Recursion Theory from the study of a special kind of definability into a study of indices and index constructions. In the present section we shall take a brief look at this new study, partially for its own interest, partially to establish results that will be applied in the next two chapters, and partially to introduce some techniques that will reäppear in a more complex setting in Chapter IV of volume II.

We begin by recalling Definition 11.4 and Theorem 11.5.

12.1. Definitions. i. $U(z) = ((z)_{lh(z)\dot{-}1})_2$.
ii. $T_n(e, x_0, \ldots, x_{n-1}, z)$: $Comp(z, e, (x_0, \ldots, x_{n-1}), U(z))$.

12.2. Theorem. (Normal Form Theorem). *Let $n > 0$.*
i. (Enumeration Theorem). *For every r.e. relation* (recall 11.8) $R \subseteq \omega^n$, *there is an* $e \in \omega$ *such that, for all* $x_0, ..., x_{n-1} \in \omega$,

$$R(x_0, ..., x_{n-1}) \Leftrightarrow \exists z\, T_n(e, x_0, ..., x_{n-1}, z). \tag{1}$$

Conversely, for every $e \in \omega$, *the relation R defined by* (1) *is an r.e. relation.*
ii. (Universal Partial Function). *For every partial recursive function F on* ω^n, *there is an* $e \in \omega$ *such that, for all* $x_0, ..., x_{n-1} \in \omega$,

$$F(x_0, ..., x_{n-1}) \simeq U(\mu z\, T_n(e, x_0, ..., x_{n-1}, z)). \tag{2}$$

Conversely, for every $e \in \omega$, *the function F defined by* (2) *is partial recursive.*

Ignoring the meanings of T_n, U, one can restate the two parts of this Theorem as follows:

Enumeration Theorem. *Let $n > 0$. There is an $(n + 1)$-ary r.e. relation W such that the n-ary r.e. relations are precisely those of the form,*

$$R(x_0, ..., x_{n-1}): \quad W(e, x_0, ..., x_{n-1}),$$

as e ranges over ω.

Universal Partial Function. *For each $n > 0$ there is an $(n + 1)$-ary partial recursive function* φ *such that the n-ary partial recursive functions are precisely those of the form*

$$F(x_0, ..., x_{n-1}) \simeq \varphi(e, x_0, ..., x_{n-1}).$$

Eventually we will be able to take this abstract view. For the moment, however, we will need to be aware of the exact forms of these universal functions and relations.

Theorem 12.2 readily suggests the following definitions.

12.3. Definitions. Let $n > 0$ and $e \in \omega$. The n-ary r.e. relation with *index e* is the relation

$$W_e^n = \{(x_0, ..., x_{n-1}): \exists z\, T_n(e, x_0, ..., x_{n-1}, z)\}.$$

[There is a slight ambiguity here as we use $(x_0, ..., x_{n-1})$ to denote a finite sequence, a code for the finite sequence, and an element of ω^n. We can identify all three and not worry about the ambiguity.] The n-ary partial recursive function with index e is

$$\varphi_e^n(x_0, ..., x_{n-1}) \simeq U(\mu z\, T_n(e, x_0, ..., x_{n-1}, z)).$$

If n is understood (and usually it is understood to be 1), we can omit the superscript "n" and write simply W_e or φ_e. Note that

$$W_e = \text{domain}(\varphi_e).$$

A more crucial observation is this: If e is an index generated by Definition 11.3 and is recognised as such by the predicate $Index(\cdot)$ constructed in the course of coding computations and $(e)_1 = n$, then φ_e^n is the function with index e as defined in 11.3. Thus, Definition 12.3 extends the assignment of functions to numbers from the indices of 11.3 to all numbers. What functions do these new indices index? Because of the insistence of the computation predicate *Comp* that all triples occurring in a computation z be acceptable, and acceptability entails the indices occurring in the triples be indices according to 11.3 (modulo the inability to express the totality of G in $F6$), no value will be given for the new indices. That is, if e is not an index according to $Index(\cdot)$ or if e is such an index but $n \neq (e)_1$, then φ_e^n is the empty function.

12.4. Lemma. i. *Let G be m-ary and $H_0, ..., H_{m-1}$ n-ary. If $g, h_0, ..., h_{m-1}$ are indices for $G, H_0, ..., H_{m-1}$, respectively, then $(4, n, g, (h_0,...,h_{m-1}))$ is an index for*

$$F(\boldsymbol{x}) \simeq G(H_0(\boldsymbol{x}), ..., H_{m-1}(\boldsymbol{x})).$$

ii. *If G is $(n + 1)$-ary with index g, then $(6,n,g)$ is an index for*

$$F(\boldsymbol{x}) \simeq \mu y\,[G(\boldsymbol{x}, y) = 0].$$

Proof. i. There actually is something to prove here because the indices $g, h_0, ..., h_{m-1}$ need not come from those generated by 11.3, and, if any of these indices is generated by 12.3 instead of 11.3, Definition 11.3 does not guarantee that $(4,n,g,(h_0,...,h_{m-1}))$ indexes the composition. However, the number does index the composition, for, if any of $g, h_0, ..., h_{m-1}$ is one of the new indices, the function it indexes is empty, i.e., never defined, whence the composition is empty. But, if one of $g, h_0, ..., h_{m-1}$ is not an index in the sense of 11.3, then $(4,n,g,(h_0,...,h_{m-1}))$ is not either, whence it indexes the empty function.

ii. The result is correct provided we carefully define $F(\boldsymbol{x}) \simeq \mu y\,[G(\boldsymbol{x},y) = 0]$ for partial functions G. For $(6,n,g)$ generated according to 11.3 and the primitive recursive predicate $Index(\cdot)$, which does not check the totality of G, the computation predicate will compute $F(\boldsymbol{x}) = z$ provided

i. $G(\boldsymbol{x}, z) = 0$

ii. for all $w < z$, $G(\boldsymbol{x}, w)$ is defined and different from 0;

in any other case, $F(\boldsymbol{x})$ is left undefined. If we take this as our definition of minimisation for partial functions, we see that the Lemma holds for those indices $(6,n,g)$ generated by 11.3. For the new indices, note that minimising the empty function yields the empty function.

QED

An alternative approach to obtaining 11.4 would be to drop the check on the legitimacy of indices in the definition of the computation predicate. Then $(4,n,g,(h_0,...,h_{m-1}))$ would automatically be calculated as a composition and $(6,n,g)$ would automatically be calculated as a minimisation.

The analogue to 12.4 for primitive recursion does not hold in general unless we change the computation predicate to accept triples with arbitrary numbers as indices. If we do that, it is no longer the case that the new indices all index the empty function.

To some extent we are trifling with curiosities. The important thing is that we can go primitive recursively from indices $g, h_0, ..., h_{m-1}$ for $G, H_0, ..., H_{m-1}$ to one for their composition. This is useful in displaying the effectiveness of certain operations and closure results.

12.5. Example. *Theorem* 11.7.ii *is effective in the sense that there are primitive recursive functions F, G such that (for n* $= 1$), *for all* $e \in \omega$,
i. $range(\varphi_{F(e)}) = domain(\varphi_e)$
ii. $domain(\varphi_{G(e)}) = range(\varphi_e)$.

I shall construct F and leave the construction of G to the reader. Given $W_e =$ domain(φ_e), we can find φ_f so that $W_e =$ range(φ_f) by choosing φ_f so that

$$\varphi_f(x) \simeq \begin{cases} x, & \varphi_e(x)\downarrow \\ \text{undefined}, & \text{otherwise.} \end{cases}$$

One possibility for φ_f is

$$\begin{aligned} \varphi_f(x) &\simeq x \cdot sg(\varphi_e(x) + 1) \\ &\simeq M(x, sg(S(\varphi_e(x)))), \end{aligned}$$

where M denotes multiplication. Observing that

$$\varphi_f(x) \simeq M(P^1_0(x), sg(S(\varphi_e(x)))),$$

we see that we can take

$$F(e) = (4, 1, m, H(e)),$$

where m is any index for multiplication and

$$H(e) = ((3,1,0), (4,1,s, (K(e))),$$

where s is any index for the signum function and

$$K(e) = (4, 1, (2,1), (e)).$$

The curious reader might enjoy calculating m and s and trying to write $F(e)$ out completely without making any mistakes with the parentheses. If he does this, he will be rewarded in seeing why I haven't done so ...

Effective constructions are most impressive against a background of non-effective results: if all were constructive, emphasising constructivity would merely be the belabouring of the obvious. But non-effectiveness abounds. As we saw in section 3, there are uncountably many functions from ω to ω, but there are only countably many recursive such functions because there are only countably many indices. The diagonalisation is, in fact, effective.

12.6. Theorem. *The partial recursive function*

$$F(x) \simeq \overline{sg}\,(\varphi_x(x))$$

has no total recursive extension.

Proof. First observe that

$$F(x) \simeq \overline{sg}\,(U(\mu z\, T_1(x, x, z)))$$

is indeed partial recursive.

Suppose φ_e is an extension of F. We show that $e \notin \text{domain}(\varphi_e)$. Suppose to the contrary that $\varphi_e(e)$ is defined. Then so is

$$F(e) = \overline{sg}\,(\varphi_e(e)).$$

But, if φ_e extends F, we also have

$$F(e) = \varphi_e(e),$$

whence $\varphi_e(e) = \overline{sg}\,(\varphi_e(e))$, which is impossible. QED

The function F of Theorem 12.6 assumes only the values 0 and 1. It is, thus, a partial representing function with no total extension. Recall from Exercise 8 of the preceding section the following definition.

12.7. Definition. A relation $R \subseteq \omega^n$ is ***recursive*** iff its representing function χ_R is recursive.

Theorem 12.6 has the following corollary.

12.8. Corollary. *There are r.e. non-recursive sets.*

Proof. Let $X = \{x\colon F(x) \simeq 0\}$, with F as in Theorem 12.6. X is r.e. but not recursive as its representing function extends F. QED

There are far more refined versions of the non-recursiveness of strange r.e. sets. The reader who has read over Exercise 8 of the immediately preceding section knows that a set X is recursive iff both it and its complement are r.e. For an r.e. set X, this means that X is recursive iff $\omega - X$ is r.e. Put differently, the complement of an r.e. non-recursive set is not r.e. One can ask that the complement be effectively non-r.e.

12.9. Definition. An r.e. set X is ***creative*** if there is a recursive function G such that, for any $e \in \omega$,

$$W_e \subseteq \omega - X \;\Rightarrow\; G(e) \in \omega - (X \cup W_e).$$

In words, X is creative if, given any attempt to enumerate its complement, one can effectively find an element of the complement that has been missed. A creative set is not recursive.

12.10. Corollary. *There is a creative set.*

Using the function F of 12.6, it is easy to define creative sets. Both sets,

$$A = \{x\colon F(x) \simeq 0\}, \qquad B = \{x\colon F(x) \simeq 1\},$$

are creative. These can be generalised: Define

$$K_n = \{x\colon \varphi_x(x) \simeq n\}.$$

Then, $B = K_0$ and $A = \bigcup_{n>0} K_n$ and each K_n, along with $K = \bigcup_{n\geq 0} K_n = A \cup B$, is creative. Of these, K is the easiest to be shown creative using the function $G(e) = e$. I leave the verification of the creativity of K as an exercise to the reader and discuss instead the creativity of A.

The idea behind the proof of the creativity of A is simple: if $W_e \subseteq \omega - A$, then the function

$$H(x) \simeq \begin{cases} 0, & x \in A \\ 1, & x \in B \cup W_e \\ \text{undefined}, & \text{otherwise}, \end{cases}$$

extends F. It can be shown (cf. Exercise 7.iii of section 11) that H is partial recursive. Hence, if $G(e)$ is an index of H, $\varphi_{G(e)}$ extends F and, by the proof of Theorem 12.6, $\varphi_{G(e)}(G(e))$ is undefined, i.e., $G(e) \in \omega - (A \cup B \cup W_e)$. Thus, once we show that G can be chosen recursively, we will have shown that G witnesses the creativity of A.

Before showing how to obtain G recursively, let us see how to obtain an even stronger result. Suppose W_{e_0} and W_{e_1} are disjoint r.e. sets such that $A \subseteq W_{e_0}$ and $B \subseteq W_{e_1}$. Then the function,

$$K(x) \simeq \begin{cases} 0, & x \in W_{e_0} \\ 1, & x \in W_{e_1}, \end{cases}$$

is a partial recursive extension of F. Hence, by the proof of Theorem 12.6, if K is $\varphi_{G(e_0,e_1)}$, then $G(e_0, e_1) \notin \text{domain}(\varphi_{G(e_0,e_1)}) = W_{e_0} \cup W_{e_1}$.

12.11. Definition. Let X, Y be disjoint sets. X, Y are *effectively inseparable* iff there is a recursive function G of two variables such that, for any $e_0, e_1 \in \omega$, if

i. $X \subseteq W_{e_0}$ and $Y \subseteq W_{e_1}$; and
ii. W_{e_0} and W_{e_1} are disjoint,

then $G(e_0, e_1) \notin W_{e_0} \cup W_{e_1}$.

Thus, once we show how to obtain a recursive G so that K is $\varphi_{G(e_0,e_1)}$, we will have shown the following.

12.12. Corollary. *There is a pair of effectively inseparable r.e. sets.*

Once again, there are pairs of sets which, if not as simply defined as A and B, are more easily proven effectively inseparable. The reader will find such an example in Exercise 5, below.

The concepts of creativity and effective inseparability, despite their possible appearance as oddities, are fundamental to recursion theory and its applications. Some of the more important of these will be given in Chapter III, below. For now we should prove the existence of such sets by effecting the index constructions needed.

The result needed to prove the creativity and effective inseparability of the given sets is the following.

12.13. Lemma. *There is a primitive recursive function F of four variables such that, if* $\varphi_{f_0}, \varphi_{f_1}$ *are partial functions and* W_{e_0}, W_{e_1} *are disjoint r.e. sets, then*

$$\varphi_{F(f_0,f_1,e_0,e_1)}(x) \simeq \begin{cases} \varphi_{f_0}(x), & x \in W_{e_0} \\ \varphi_{f_1}(x), & x \in W_{e_1}. \end{cases}$$

Proof. Two tricks are involved. First, we put off trying to define F by initially making f_0, f_1, e_0, e_1 arguments of the function we are trying to define; second, we initially make the computations of $\varphi_{f_i}(x)$ and $\varphi_{e_i}(x)$ arguments as well. Thus, we define the primitive recursive auxiliary functions:

$$J_1(f_0, f_1, e_0, e_1, x, z_0, z_1) = \begin{cases} 0, & T_1(f_0,x,z_0) \wedge T_1(e_0,x,z_1) \\ 0, & T_1(f_1,x,z_0) \wedge T_1(e_1,x,z_1) \\ 1, & \text{otherwise}, \end{cases}$$

$$J_2(f_0, f_1, e_0, e_1, x, z) = J_1(f_0, f_1, e_0, e_1, x, \pi_1^2(z), \pi_1^2(z)).$$

From these, we define the partial recursive

$$\varphi(f_0, f_1, e_0, e_1, x) \simeq U(\pi_1^2(\mu z\, [J_2(f_0, f_1, e_0, e_1, x, z) = 0])).$$

Assuming W_{e_0}, W_{e_1} to be disjoint, it is not too hard to see that

$$\varphi(f_0, f_1, e_0, e_1, x) \simeq \begin{cases} \varphi_{f_0}(x), & x \in W_{e_0} \\ \varphi_{f_1}(x), & x \in W_{e_1}. \end{cases} \qquad (*)$$

It now suffices to show that, given constants f_0, f_1, e_0, e_1, we can effectively find an index for the function obtained by fixing f_0, f_1, e_0, e_1, i.e., to show there is a recursive function F so that $\varphi_{F(f_0,f_1,e_0,e_1)}(x) \simeq \varphi(f_0, f_1, e_0, e_1, x)$.

12.14. Theorem. (Parameter Theorem). *For each m, n ≥ 1, there is an (m + 1)-ary primitive recursive function* s_n^m *such that, for any* $e, x_0, ..., x_{n-1}, y_0, ..., y_{m-1} \in \omega$,

$$\varphi^n_{s^m_n(e,y_0,...,y_{m-1})}(x_0, ..., x_{n-1}) \simeq \varphi^{m+n}_e(y_0, ..., y_{m-1}, x_0, ..., x_{n-1}).$$

A word or two before proving the Theorem. Usually one proves a lemma to prove a theorem, not conversely. The Parameter Theorem is, however, one of the cornerstones of Recursion Theory— as we shall shortly see. The lemmatic nature of the Lemma is this: Lemma 12.13 is a lemma to Corollaries 12.10 and 12.12, not to Theorem 12.14, which is a lemma to 12.13.

While I am digressing, I might also note that the Parameter Theorem is often called the *S-m-n Theorem* in honour of the notation Kleene originally used for the functions s^m_n. Kleene himself called it the *Iteration Theorem*, and Joseph Shoenfield introduced the name used here. I think Shoenfield's choice the most apt, but admit that "*S-m-n* Theorem" is the name most commonly used in the literature.

Proof of Theorem 12.14. Traditionally, one proves this by induction on m, noting that

$$s^{m+1}_n(e, y_0, ..., y_{m-1}, y_m) = s^1_n(s^m_{n+1}(e, y_0, ..., y_{m-1}), y_m).$$

For the indexing at hand this is a convenience rather than a necessity.

Thus, let $m = 1$. Observe:

$$\varphi_{s^1_n(e,y)}(\mathbf{x}) \simeq \varphi_e(y, \mathbf{x})$$

$$\simeq \varphi_e(K^n_y(\mathbf{x}), P^n_0(\mathbf{x}), ..., P^n_{n-1}(\mathbf{x})),$$

where K^n_y is the n-ary constant function with value y,

$$K^n_y(\mathbf{x}) = y.$$

Thus, we can take

$$s^1_n(e,y) = (4, n, e, (K^n(y), p^n_0, ..., p^n_{n-1})),$$

where $p^n_i = (3, n, i)$ is the index for P^n_i and K^n is the function coding indices for the K^n_y's:

$$K^n(0) = (4, n, (1,1), ((3,n,0))) \quad \text{(a code for } Z \circ P^n_0\text{)}$$

$$K^n(y + 1) = (4, n, (2,1), K^n(y)).$$

QED

Completion of the proof of Lemma 12.13. Let the function φ satisfying (*) have index e. Then we can take

$$F(f_0, f_1, e_0, e_1) = s^4_1(e, f_0, f_1, e_0, e_1).$$

QED

With Lemma 12.13, we can now complete the proofs of Corollaries 12.10 and 12.12. The easier of the two is 12.12 and it shall be treated here; treatment of Corollary 12.10 is reserved for the exercises (specifically, Exercise 6, below).

Proof of Corollary 12.12. Recall the function F of Theorem 12.5,

$$F(x) \simeq \overline{sg}\,(\varphi_x(x)).$$

F has the property that, for any index e, $F \subseteq \varphi_e$ (i.e., φ_e extends F) implies $e \notin \text{domain}(\varphi_e)$.

Let

$$A = \{x: F(x) \simeq 0\} = \{x: \exists y\,(y = \varphi_x(x) \wedge y \neq 0)\}$$

$$B = \{x: F(x) \simeq 1\} = \{x: \varphi_x(x) \simeq 0\}.$$

Let W_{e_0}, W_{e_1} be disjoint r.e. sets such that

$$A \subseteq W_{e_0} \text{ and } B \subseteq W_{e_1}.$$

Define

$$H(x) \simeq \begin{cases} 0, & x \in W_{e_0} \\ 1, & x \in W_{e_1}. \end{cases}$$

Choosing $f_0 = (1,1)$ and $f_1 = (4, 1, (2,1), ((1,1)))$ as codes for the constant functions 0 and 1, respectively, Lemma 12.13 yields

$$H(x) \simeq \varphi_{K(f_0,f_1,e_0,e_1)}(x),$$

for some primitive recursive function K. But f_0, f_1 are constants, whence

$$H(x) \simeq \varphi_{G(e_0,e_1)}(x)$$

for

$$G(x, y) = K(f_0, f_1, x, y).$$

The proof is completed, once again, by noting that H extends F, whence its index $G(e_0, e_1)$ is not in the domain $W_{e_0} \cup W_{e_1}$ of H. Thus, G witnesses the effective inseparability of A, B.

QED

If we view Corollaries 12.10 and 12.12 as *effective* non-recursiveness results and Lemma 12.13 as an *effective* closure result for the class of partial recursive functions, then we will recognise that the Parameter Theorem is a useful tool in establishing the recursive effectiveness of such results. We shall illustrate this with one additional example before turning to other uses of the Parameter Theorem.

We have already remarked that, contrary to the situation in the calculus, where the invariance of dimension makes single-variable calculus a different subject from multi-variable calculus, the existence of primitive recursive n-tupling functions and their inverses makes the theory of n-ary partial recursive functions identical to that of unary such functions. By the Parameter Theorem, the identification is effective: there are primitive recursive functions F, G such that

$$\varphi^1_{F(e)}(\langle x_0, ..., x_{n-1}\rangle_n) \simeq \varphi^n_e(x_0, ..., x_{n-1}) \tag{3}$$

$$\varphi^n_{G(e)}(x_0, ..., x_{n-1}) \simeq \varphi^1_e(\langle x_0, ..., x_{n-1}\rangle_n). \tag{4}$$

To find F, let φ^{n+1} by the $(n + 1)$-ary universal partial function for n-ary partial recursive functions, and observe that we want

$$\varphi_{F(e)}(x) \simeq \varphi^{n+1}(e, \pi^n_1(x), ..., \pi^n_n(x))$$

$$\simeq \varphi^2_{e_0}(e, x)$$

for some index e_0. Thus, we can take

$$F(e) = s^1_1(e_0, e).$$

Similarly, to define G, let φ^2 be the binary universal partial function for unary partial recursive functions and observe that we want

$$\begin{aligned}\varphi^n_{G(e)}(x_0, \ldots, x_{n-1}) &\simeq \varphi^2(e, \langle x_0, \ldots, x_{n-1}\rangle_n) \\ &\simeq \varphi^{n+1}_{e1}(e, x_0, \ldots, x_{n-1}),\end{aligned}$$

for some index e_1. Thus, we can take

$$G(e) = s^1_n(e_1, e).$$

As a quick application, let us prove the following binary variant of Lemma 12.13.

12.15. Application. *There is a recursive function K such that, for any f_0, f_1, e_0, e_1 such that $W^2_{e_0}$ and $W^2_{e_1}$ are disjoint,*

$$\varphi^2_{K(f_0,f_1,e_0,e_1)}(x, y) \simeq \begin{cases} \varphi^2_{f_0}(x,y), & (x,y) \in W^2_{e_0} \\ \varphi^2_{f_1}(x,y), & (x,y) \in W^2_{e_1}. \end{cases}$$

Proof. Let F, G be as just constructed. Observe,

$$\varphi^1_{F(f_i)}(z) \simeq \varphi^2_{f_i}(\pi^2_1(z), \pi^2_2(z))$$

and

$$z \in W^1_{F(e_i)} \Leftrightarrow (\pi^2_1(z), \pi^2_2(z)) \in W^2_{e_i}.$$

Thus, if we let J be the function constructed in Lemma 12.13, and define

$$K(f_0, f_1, e_0, e_1) = G(\, J(F(f_0), F(f_1), F(e_0), F(e_1))\,),$$

we have

$$\begin{aligned}\varphi^2_{K(f_0,f_1,e_0,e_1)}(x, y) &\simeq \varphi^1_{J(F(f_0),F(f_1),F(e_0),F(e_1))}(\langle x, y\rangle) \\ &\simeq \begin{cases} \varphi^1_{F(f_0)}(\langle x,y\rangle), & \langle x,y\rangle \in W^1_{F(e_0)} \\ \varphi^1_{F(f_1)}(\langle x,y\rangle), & \langle x,y\rangle \in W^1_{F(e_1)} \end{cases} \\ &\simeq \begin{cases} \varphi^2_{f_0}(x,y), & (x,y) \in W^2_{e_0} \\ \varphi^2_{f_1}(x,y), & (x,y) \in W^2_{e_1}. \end{cases}\end{aligned}$$

QED

Now for a confession: the construction of primitive recursive functions F and G satisfying (3) and (4) can be accomplished directly without appeal to the Parameter Theorem. For F, observe that we want

$$\varphi_{F(e)}(x) \simeq \varphi_e^n(\pi_1^n(x), \ldots, \pi_n^n(x)),$$

whence we can take

$$F(e) = (4, 1, e, (p_1, \ldots, p_n)),$$

where p_i is an index for π_i^n; G is similarly a composition. The reader might well wonder why, given the simplicity of the functions obtained by the direct approach, I bothered to give the construction using the s_n^m- functions. The reason, in one word, is: generality. The proof given applies to any indexing of the partial recursive functions for which a universal partial function exists and the Parameter Theorem holds.

What other indexings are there? There are the trivial variants, e.g., that discussed just after Lemma 12.4 or one in which one adds a few extra initial functions and indices for them. Similarly, one could add new closure conditions to the class of partial recursive functions and new indices for these conditions. For example, we could add:

$F7_k^n$. $F(x_0, \ldots, x_{n-1}) \simeq G(k, x_0, \ldots, x_{n-1}),$ $\qquad (7, n, k, g)$

where k is any constant. For such an indexing we clearly can define

$$s_n^1(e, x) = (7, n, x, e).$$

With a little explanation of what to do when W_{e_0} and W_{e_1} overlap, a new schema $F8$ can be added to simulate Lemma 12.13. (Cf. the remarks following Exercise 5, below.)

Another indexing can be had by using register machines. If one invents codes for register machines as in Exercise 11 of the previous section, the primitive recursive computation predicate so constructed will yield another indexing of the partial recursive functions— the indices being the codes of programs. The universal partial function is again quickly obtained from the computation predicate. How does s_n^1 work? Suppose we have a register machine program computing $G(x, x_0, \ldots, x_{n-1})$, and suppose k is given. A program computing $F(x_0, \ldots, x_{n-1}) \simeq G(k, x_0, \ldots, x_{n-1})$ is obtained as follows. We imagine $x_0, \ldots, x_{n-1}$ in $R_0, \ldots, R_{n-1}$:

R_0	$\ldots$	R_{n-1}
x_0	$\ldots$	x_{n-1}.

Copy the contents of R_{n-1} into R_n, R_{n-2} into R_{n-1}, etc., i.e., shift everything over one register. The last copying operation, of R_0 into R_1, can be assumed to leave R_0 empty. Thus, we have:

R_0	R_1	$\ldots$	R_{n-1}
0	x_0	$\ldots$	x_{n-1}.

Now apply k INC R_0 instructions to get

R_0	R_1	$\ldots$	R_{n-1}
k	x_0	$\ldots$	x_{n-1}.

Running the G-program now calculates F.

Thus, we see already a variety of indexings of the partial recursive functions. Using the more abstract proof of the existence of the effective constructions F and G underlying Application 12.15 will guarantee that they and the Application hold for any indexing which has a universal partial function and for which the Parameter Theorem holds. It turns out that these two results suffice for all applications we will give. That this must be the case was proven by Hartley Rogers, Jr.:

12.16. Definitions. Any assignment $e \mapsto \theta_e$ of unary partial recursive functions to natural numbers is called an *indexing* of the class of partial recursive functions. Such an indexing is called an *acceptable indexing* if
i. there is a universal partial function θ, i.e. a partial recursive function θ such that, for all $e, x \in \omega$,

$$\theta_e(x) \simeq \theta(e, x);$$

and ii. there is an s_1^1-function s, i.e., a recursive function s such that, for all $e, x, y \in \omega$,

$$\theta_{s(e,x)}(y) \simeq \theta_e(\langle x, y\rangle).$$

12.17. Theorem. *Let* $e \mapsto \theta_e$ *be an acceptable indexing. There is a recursive permutation* F *(i.e., a recursive one-one correspondence of* ω *with itself) such that, for all* $e \in \omega$, $\theta_e = \varphi_{F(e)}$.

Theorem 12.7 lends an invariant significance to the results we prove about indexings and operations on them: whatever we prove about one acceptable indexing holds, after translating, for any other acceptable indexing. Indeed, if our proofs rest only on the existence of a universal partial function and the existence of the s_1^1-function, we need not do any translating.

Theorem 12.17 is stated primarily for pædagogical reasons: to indicate the lack of dependence of the main results on accidents of our choice of indexing, and to emphasise the importance of the seemingly unimpressive Parameter Theorem. The importance of the Parameter Theorem is also displayed by its use in proving the forthcoming Recursion Theorem, but the Recursion Theorem is so impressive it readily eclipses the Parameter Theorem.

I shall not prove Theorem 12.17 here as it will play no rôle in the sequel. I note that its proof depends on the soon-to-be-discussed Recursion Theorem.

Before introducing the Recursion Theorem, let us pause and reconsider the function,

$$F(x) \simeq \overline{sg}\,(\varphi_x(x)).$$

The fact already noted, but easily obscured by the simple notation, is that this is the non-contradictory outcome of a diagonalisation argument: F is defined so as to disagree with the x-th function in the enumeration $\varphi_0, \varphi_1, \ldots$ of partial recursive functions in the x-th position. The partial recursiveness of the universal function $\varphi(e,x) \simeq \varphi_e(x)$ guarantees the partial recursiveness of F, whence the existence of an index e for F. For this index, we have

$$F(e) \simeq \overline{sg}\,(F(e)),$$

no contradiction since "$\simeq$" merely asserts equality if either side exists. We do not conclude that F is not partial recursive, but that it is *partial*: $F(e)$ is undefined. Attempting the same argument for total functions reveals something interesting:

12.18. Theorem. *The set* $\{e\colon \varphi_e$ *is total*$\}$ *is not r.e.*

Proof. It can be shown (Exercise 7.iii of section 11, above) that any non-empty r.e. set is the range of a general recursive function. Supposing $\{e\colon \varphi_e$ is total$\}$ to be r.e., pick G general recursive so that

$$\mathrm{range}(G) = \{e\colon \varphi_e \text{ is total}\}$$

and define

$$F(x) \simeq \varphi(G(x), x) + 1,$$

where φ is the universal partial function. F is partial recursive. Moreover,

$$\begin{aligned}\varphi_{G(x)}(x)\!\downarrow \;&\Rightarrow\; \varphi(G(x), x)\!\downarrow\\ &\Rightarrow\; F(x)\!\downarrow,\end{aligned}$$

whence F is total. But then $F = \varphi_{G(n)}$ for some $n \in \omega$. This quickly yields a contradiction:

$$\varphi_{G(n)}(n) = F(n) = \varphi(G(n), n) + 1 = \varphi_{G(n)}(n) + 1. \qquad \text{QED}$$

12.19. Corollary. *The set* $\{e\colon \varphi_e$ *is total*$\}$ *is not recursive.*

By the identification,

$$\text{recursive} = \text{computable} = \text{effective},$$

the non-recursiveness of the set $\{e\colon \varphi_e$ is total$\}$ means that there is no effective procedure which, given the index of a partial recursive function, will tell us whether or not the function is total. Since, for us, indices code definitions of partial recursive functions, this means that there is no effective way of telling from the definition of a partial recursive function whether or not the function defined by it is total. [*N.B.* The new indices of 12.3 form a primitive recursive, hence recursive, set of indices for the empty function and can be ignored.] Choosing indices to code register machine programs, this means that there is no effective way of telling from a program whether or not the function it defines is total.

This is an important matter. The abstract existence of sets, the membership problems of which are unsolvable, is interesting; the existence of non-recursive *r.e.* sets, i.e., simply definable such sets, is a bit more interesting; but the unsolvability of relatively natural problems, like that of deciding which programs compute total functions, is *very* interesting. This is a matter that deserves some fanfare, or, at least, *italics*. It is, however, a matter about which I have little to say, other than that we shall do much better very shortly, and that we shall present more interesting unsolvable problems in each of the next two chapters.

Getting back to the problem of diagonalisation, it is time to prove the Recursion Theorem, which is the result of attempting to diagonalise on the *indices* of the partial recursive functions. For the usual diagonalisation, one imagines an infinite array of (occasionally undefined) values:

$$\begin{array}{lll} \varphi_0(0) & \varphi_0(1) & \cdots \\ \varphi_1(0) & \varphi_1(1) & \\ \vdots & & \end{array}$$

One then attempts to change the diagonal; e.g., in defining F of Theorem 12.6, we looked at $\overline{sg}\ (\varphi_0(0))$, $\overline{sg}\ (\varphi_1(1))$, ... Instead of doing this, we now look at the array of functions:

$$\begin{array}{lll} \varphi_{\varphi_0(0)} & \varphi_{\varphi_0(1)} & \cdots \\ \varphi_{\varphi_1(0)} & \varphi_{\varphi_1(1)} & \\ \vdots & & \end{array}$$

Of course, some of these functions have undefined indices, in which cases the functions themselves are empty. The diagonal of this array is the sequence of partial functions,

$$\varphi_{\varphi_0(0)},\ \varphi_{\varphi_1(1)},\ \ldots$$

This sequence is itself a partial recursive binary function,

$$G(y, x) \simeq \varphi_{\varphi_y(y)}(x) \simeq \varphi(\varphi(y, y), x),$$

where φ is the universal partial function. By the Parameter Theorem, we can write

$$G(y, x) \simeq \varphi_{s(y)}\ (x),$$

for some primitive recursive function s. Now, let H be any total recursive function and see what happens when we try to use H to diagonalise on the indices of the diagonal: $\varphi_{H(\varphi_0(0))}$, $\varphi_{H(\varphi_1(1))}$, ..., i.e., consider $\varphi_{H(s(0))}$, $\varphi_{H(s(1))}$, ... The function $H \circ s$ is total and has an index, say e. Observe:

$$\varphi_{s(e)}(x) \simeq \varphi_{\varphi_e(e)}(x) \simeq \varphi_{H(s(e))}(x),$$

whence $\varphi_{s(e)} = \varphi_{H(s(e))}$. We have just proven Kleene's Recursion Theorem:

12.20. Theorem. (Recursion Theorem). *Let H be a recursive function. There is an n such that $\varphi_{H(n)} = \varphi_n$.*

Describing the Recursion Theorem in words is not quite easy. I would like to say that, if we think of H as inducing a map,

$$\hat{H}(\varphi_e) = \varphi_{H(e)},$$

then the Recursion Theorem asserts the existence of a fixed point φ_n to $\hat{H}$:

$$\varphi_n = \varphi_{H(n)} = \hat{H}(\varphi_n).$$

The problem with this formulation is that $\hat{H}$ need not be well-defined. The best I can offer is this: if we imagine H as a transformation of programs, there will be some program the transform of which calculates the same function as the original program.

The "fixed point" could be the empty function: if

$$\varphi_{H(e)}(x) \simeq \varphi_e(x) + 1,$$

then

$$\varphi_{H(n)} = \varphi_n \Rightarrow \varphi_n \text{ is empty.}$$

Despite such disappointing examples, the Recursion Theorem has some remarkable applications. For example, it can be used to give an alternate proof of the non-recursiveness of $\{e\colon \varphi_e \text{ is total}\}$.

Alternate proof of Corollary 12.19. Suppose $T = \{e\colon \varphi_e \text{ is total}\}$ were recursive. Choose e_0, e_1 such that $e_0 \in T$ and $e_1 \notin T$ (e.g., let e_0 be any index for Z and e_1 an index for $\mu y\, [S(P^2_0(x, y)) = 0]$). Define H recursive so that

$$\varphi_{H(x)}(y) \simeq \begin{cases} \varphi_{e_1}(y), & x \in T \\ \varphi_{e_0}(y), & x \notin T. \end{cases}$$

(Exercise: Prove that a recursive such H exists.) Apply the Recursion Theorem to find n so that $\varphi_{H(n)} = \varphi_n$, and observe that

$$n \in T \Rightarrow \varphi_n = \varphi_{H(n)} \Rightarrow \varphi_n = \varphi_{e_1} \Rightarrow \varphi_{e_1} \text{ is total,}$$

contrary to the choice of e_1, while

$$n \notin T \Rightarrow \varphi_n = \varphi_{H(n)} \Rightarrow \varphi_n = \varphi_{e_0} \Rightarrow \varphi_{e_0} \text{ is not total,}$$

again a contradiction.

QED

If we study this proof, we see that, aside from the supposed recursiveness of T, we only used two facts about T:

i. T is *non-trivial*: neither T nor $\omega - T$ is empty

ii. T is *extensional*: for any e, f,

$$\varphi_e = \varphi_f \Rightarrow [e \in T \Leftrightarrow f \in T].$$

Non-trivial extensional sets are often called *index sets* in the literature.

12.21. Theorem. (Rice's Theorem). *Every non-trivial extensional set is non-recursive.*

Proof. Let T be a recursive, non-trivial, extensional set. Let $e_0 \in T$, $e_1 \notin T$, and H be recursive such that, for all $x \in \omega$,

$$\varphi_{H(x)}(y) \simeq \begin{cases} \varphi_{e_1}(y), & x \in T \\ \varphi_{e_0}(y), & x \notin T. \end{cases}$$

Again, if $\varphi_{H(n)} = \varphi_n$, we have

$$n \in T \Rightarrow \varphi_n = \varphi_{H(n)} = \varphi_{e_1} \Rightarrow e_1 \in T, \text{ a contradiction}$$

$$n \notin T \Rightarrow \varphi_n = \varphi_{H(n)} = \varphi_{e_0} \Rightarrow e_0 \notin T, \text{ a contradiction.}$$

Hence, no set T can simultaneously be non-trivial, extensional, and recursive. QED

Rice's Theorem, proven by H. Gordan Rice, is a remarkable result. In terms of programs, it says: given any property holding of some computable functions and not holding of others, there is no effective procedure which, when applied to programs, will always tell whether or not the functions computed have the property. For example, one cannot effectively determine which programs are total, which programs are defined at 0, which programs are identically 0, which programs never assume the value 0, etc.

Oddly enough, it is properties of sets rather than of functions that will interest us most. Relative to sets, we say that a set X of numbers is *set-extensional* if, for all $e, f \in \omega$,

$$W_e = W_f \Rightarrow [e \in X \Leftrightarrow f \in X].$$

12.22. Corollary. (Rice's Theorem for R.e. Sets). *Every non-trivial, set-extensional set is non-recursive.*

Proof. Observe that the set-extensionality of X implies its extensionality:

$$\varphi_e = \varphi_f \Rightarrow W_e = W_f$$
$$\Rightarrow [e \in X \Leftrightarrow f \in X].$$ QED

Rice's Theorem is a generalisation of the non-recursiveness of the set of indices of total functions. By Theorem 12.18, however, this set is not even r.e. Rice conjectured a characterisation of non-trivial, extensional, r.e. sets. This conjecture was proven by various authors, presumably including Norman Shapiro as it is called the Rice-Shapiro Theorem. We shall finish this section with a discussion of this Theorem.

We shall restrict ourselves to set-extensional index sets as this is the case of the Theorem we will need in the next chapter. Let us think for a minute about how we could enumerate the entries of a set-extensional set. If we have enumerated e and if $W_e = W_f$, how will we know after *finitely* many steps to put f into our set. All we could hope to check is that W_e and W_f each contain some finitely many numbers. Moreover, if we specify some finitely many numbers, say $x_0, \ldots, x_{n-1}$, the set

$$X = \{e: x_0 \in W_e \wedge x_1 \in W_e \wedge \ldots \wedge x_{n-1} \in W_e\}$$

is clearly Σ_1-definable, whence r.e. More generally, if we have an enumeration $D_{F(0)}$, $D_{F(1)}$, ...of finite sets, the union of the sets,

$$\{e: D_{F(x)} \subseteq W_e\},$$

will be r.e.: e is in the union iff

$$\exists x \, \forall y \le F(x) \, [y \notin D_{F(x)} \vee y \in W_e], \tag{*}$$

a Σ_1-condition. This is as general as we can get.

12.23. Theorem. (Rice-Shapiro Theorem). *A non-empty, set-extensional set X is r.e. iff there is a recursive function F such that*

$$X = \{e: \exists x \, [D_{F(x)} \subseteq W_e]\}.$$

The proof will be outlined in the exercises. I make only two remarks here. First, note that the clause "$y \notin D_{F(x)}$" of (*) cannot be replaced by "$y \notin W_{F(x)}$" even if we knew the sets $W_{F(x)}$ all to be finite: how do we test membership in a finite set if all we know is its finiteness? If we knew *exactly* how many elements such a set had, we could list out the elements as we enumerated $W_{F(x)}$ and, when the list was complete, check y against this list. However, without knowing the *exact* number, we would have no way of knowing when the list is complete.

Second, the dependence of the membership of an index e in X on a finite amount of information is a sort of continuity condition. There is, in fact, a related result on the continuity of certain effective operators. I refer the interested reader to the text by Hartley Rogers, Jr., cited in the Reading List at the end of this chapter.

Finally, I ought to illustrate the application of the Rice-Shapiro Theorem. The following sets are r.e.:

$$I_n = \{e\colon n \in W_e\}$$
$$C_n = \{e\colon \text{card}(W_e) > n\}.$$

The following sets are not r.e.:

$$Tot = \{e\colon W_e = \omega\}$$
$$Inf = \{e\colon W_e \text{ is infinite}\}$$
$$Cof = \{e\colon W_e \text{ is cofinite}\} = \{e\colon \omega - W_e \text{ is finite}\}$$
$$Fin = \{e\colon W_e \text{ is finite}\}$$
$$Rec = \{e\colon W_e \text{ is recursive}\}.$$

Exercises.

1. If F is defined by

$$F(0, \mathbf{x}) \simeq G(\mathbf{x})$$
$$F(x+1, \mathbf{x}) \simeq H(F(x, \mathbf{x}), x, \mathbf{x}),$$

the index $(5,n,g,h)$ need not index F if one of g, h indexing G, H, respectively, is an index according to 12.3, but not according to 11.3.

i. If $(5,n,g,h)$ does not index F, which index g, h must be new?

ii. For each n, find an index e_n by 11.3 for the empty n-ary function. Find an index for F as above if one of g, h is not generated by 11.3.

iii. Using the primitive recursiveness of the predicate $Index(\cdot)$ recognising indices generated by 11.3, and the primitive recursiveness of the function $n \mapsto e_n$, construct a primitive recursive function $pr(g,h)$ which, given indices g, h for G, H, will produce an index f for F as defined above.

[Hint. iii. Use definition-by-cases.]

2. Construct a recursive G such that

$$\text{domain}(\varphi_{G(e)}) = \text{range}(\varphi_e),$$

as desired in 12.5.ii.

3. Show that $K = \{e\colon e \in W_e\}$ is creative.

4. Show that each $K_n = \{e\colon \varphi_e(e) = n\}$ is creative.

5. Define the sets

$$A = \{e: \exists y\, [T_1(\pi_2^2(e), e, y) \wedge \forall z \le y \neg T_1(\pi_1^2(e), e, z)]\}$$
$$B = \{e: \exists y\, [T_1(\pi_1^2(e), e, y) \wedge \forall z < y \neg T_1(\pi_2^2(e), e, z)]\}.$$

If we think heuristically of the computations as ordering the stages at which numbers get put into r.e. sets, we can read these definitions as follows. Put $e = \langle e_0, e_1\rangle$ into A if e is put into W_{e_1} before it can be put into W_{e_0}; and put $e = \langle e_0, e_1\rangle$ into B if e is put into W_{e_0} and it is not put into W_{e_1} any earlier.

i. Show: A, B are disjoint r.e. sets.

ii. Show: A, B are effectively inseparable via the function $G(e_0, e_1) = \langle e_0, e_1\rangle$.

[Remark. This trick of comparing witnesses to existential assertions is due to J. Barkley Rosser and will occur again in Chapter III. I also note that, natural as the effectively inseparable sets constructed in proving Corollary 12.12 are, the present pair was the historically first example of such a pair and most texts present them first. Moreover, the trick can be applied to modify Application 12.15 and yield a version of definition-by-r.e.-cases suitable for building a computation predicate on. The trick has, thus, greater depth than its initial appearance of artificiality suggests. Indeed, we will return to it in III.6.26 - 27.]

6. i. Construct recursive functions F, G such that

a. $W_{F(e,f)} = W_e \cap W_f$

b. $W_{G(e,f)} = W_e \cup W_f$.

ii. Complete the proof of Corollary 12.10: $A = \{x: \varphi_x(x) > 0\}$ is creative.

iii. Show: If X, Y are effectively inseparable r.e. sets, then each of X, Y is creative.

[Hint. ii - iii. use i.b.]

7. Let X be an r.e. non-recursive set.

i. Show: There is a one-one recursive function F such that range(F) = X.

ii. Let $D = \{x: \exists y > x\, [F(y) < F(x)]\}$, with F as in i. Show: D is r.e.

iii. Let D be as in ii. Show: $\omega - D$ is infinite.

iv. Show: $\omega - D$ contains no infinite r.e. set.

v. Show: D is a non-recursive r.e. set which is not creative.

[Hints. iii. if x_0 were an upper bound for $\omega - D$, one could find $x_1 < x_2 < \ldots$ such that $F(x_1) > F(x_2) > \ldots$; iv. if G enumerated an infinite subset of $\omega - D$, then $x \in X$ iff $\exists y \le G(z)[F(y) = x]$, where $x \le F(G(z))$; v. show that the complement of a creative set contains an infinite r.e. set. The construction of D is due to J.C.E. Dekker and simplifies an earlier construction due to Emil Post.]

8. Define the set of primitive recursive indices to be those generated according to *F1* - *F5* of Definition 11.3.

i. Show: The predicate *PRindex*(e) recognising primitive recursive indices is primitive recursive.

ii. Find a primitive recursive function F such that range(F) is the set of primitive recursive indices.

iii. Define

$$G(x) \simeq \varphi(F(x), x),$$

where φ is a universal partial function and F is as in ii. Show: G is a recursive, non-primitive recursive function.
[Hint: ii. define F by

$$F(0) = (1,1)$$
$$F(x+1) = \mu y\, [PRindex(y) \wedge y > F(x)],$$

and show that the y can be bounded primitive recursively.]

9. Without using the Rice-Shapiro Theorem, show that the set $\{e: W_e \text{ is infinite}\} = \{e: \varphi_e(x)\downarrow \text{ for infinitely many } x\}$ is not r.e.
[Hint. Given an enumeration F, define a recursive function G such that
i. for any $x \in \omega$, $G(x) \in W_{F(x)}$
ii. for all $x, y \in \omega$, $x < y \Rightarrow G(x) < G(y)$.]

10. i. Prove the following variant of Lemma 12.13: There is a recursive function F such that, for all $e_0, e_1, e_2, f_0, f_1, f_2$, if $W^2_{e_0}, W^2_{e_1}, W^2_{e_2}$ are pairwise disjoint, and if $\varphi^2_{f_0}, \varphi^2_{f_1}, \varphi^2_{f_2}$ are binary partial recursive functions, then

$$\varphi_{F(e_0,e_1,e_2,f_0,f_1,f_2)}(x,y) \simeq \begin{cases} \varphi^2_{f_0}(x,y), & (x,y) \in W^2_{e_0} \\ \varphi^2_{f_1}(x,y), & (x,y) \in W^2_{e_1} \\ \varphi^2_{f_2}(x,y), & (x,y) \in W^2_{e_2}. \end{cases}$$

ii. Show that there is a recursive F such that, for all $e \in \omega$,

$$\varphi_{F(e)}(x,y) \simeq \begin{cases} y+1, & x = 0 \\ \varphi_e(x-1,1), & x > 0 \wedge y = 0 \\ \varphi_e(x-1, \varphi_e(x, y-1)), & x > 0 \wedge y > 0. \end{cases}$$

iii. Show the partial recursiveness of the Ackermann function P. Can you readily conclude that P is total?

11. Let X be a non-trivial, set-extensional set.
i. Let Y be r.e. and G partial recursive. Show: There is a recursive function F such that

$$\varphi_{F(x)}(y) \simeq \begin{cases} G(y), & x \in Y \\ \text{und.}, & x \notin Y. \end{cases}$$

Conclude that

$$W_{F(x)} = \begin{cases} \text{domain}(G), & x \in Y \\ \varnothing, & x \notin Y. \end{cases}$$

ii. Let e_0 be an index for the empty set, let $e_1 \in X$, and assume $e_0 \notin X$. Choose $G = \varphi_{e_1}$ and $Y = K = \{e\colon e \in W_e\}$ in part i, and show that X is not recursive.

iii. Assume X is r.e., $e \in X$, and $W_e \subseteq W_f$. Find a recursive function F such that

$$W_{F(x)} = \begin{cases} W_f, & x \in X \\ W_e, & x \notin X. \end{cases}$$

Apply the Recursion Theorem to show that $f \in X$.

iv. Assume X is r.e., $e \in X$, and for no finite $W_{e_1} \subseteq W_e$ is $e_1 \in X$. Let $Y = W_f$ be a non-recursive r.e. set and define

$$\varphi_{F(x)}(y) \simeq \begin{cases} 0, & \exists z\, [T_1(e,y,z) \wedge \forall w \le z \neg T_1(f,x,w)] \\ \text{und}\,, & \text{otherwise.} \end{cases}$$

Show: a. $x \in Y \Rightarrow W_{F(x)} \subseteq W_e$ is finite
b. $x \in Y \Leftrightarrow F(x) \notin X$
c. $\omega - Y$ is r.e.

Conclude that, if X is r.e. and $e \in X$ then $e_1 \in X$ for some finite $W_{e_1} \subseteq W_e$.

v. Show that there is a recursive function F such that, for all $x \in \omega$, $W_{F(x)} = D_x$.

vi. Prove the Rice-Shapiro Theorem (Theorem 12.23).

[Hint. ii. Show $F(x) \in X$ iff $x \in K$. (Note that this gives a proof of Rice's Theorem that makes no appeal to the Recursion Theorem.)]

12. Let X, Y be disjoint non-empty, extensional sets (e.g., $\{e\colon \varphi_e(0) = 0\}$ and $\{e\colon \varphi_e(0) = 1\}$). Let $x_0 \in X$ and $y_0 \in Y$. Find F recursive so that

$$\varphi_{F(x,e,f)} = \begin{cases} \varphi_{y_0}, & x \in W_e \\ \varphi_{x_0}, & x \in W_f, \end{cases}$$

for disjoint W_e, W_f.

i. Modifying the proof of the Recursion Theorem, show that there is a recursive function H so that

$$\varphi_{F(H(e,f),e,f)} = \varphi_{H(e,f)}.$$

ii. Show that X, Y are effectively inseparable via the function

$$G(e,f) = F(H(e,f), e, f).$$

13. The Arithmetic Hierarchy

We have strayed a bit from our goal of studying arithmetic expressibility. This is particularly clear in the last section, where the emphasis was on manipulating indices and the recursive unsolvability of various problems. Moreover, we have primarily concerned ourselves with a narrow subclass of that of arithmetic relations, namely the class of Σ_1-relations. In this final working section of this chapter, we shall take a brief look at the class of all arithmetically definable sets and relations. It turns out that there is a natural hierarchy of the arithmetically definable relations into levels and each level behaves a bit like the class of Σ_1-relations.

The description of the levels is fairly easy, particularly if we take our cue from Theorem 10.10 or Corollary 10.11.

13.1. Definitions. Let $n > 0$. A relation $R \subseteq \omega^m$ is a Σ_n-*relation* if there is a primitive recursive $(m+n)$-ary relation P such that, for all $x_0, \ldots, x_{m-1} \in \omega$,

$$R(x_0, \ldots, x_{m-1}) \Leftrightarrow \exists y_0\, \forall y_1 \ldots Q_{n-1} y_{n-1}\, P(\mathbf{x}, \mathbf{y}),$$

where the quantifiers $\exists y_0 \forall y_1 \ldots Q_{n-1} y_{n-1}$ alternate in type. (Thus, Q_{n-1} is $\exists$ if n is odd and $\forall$ if n is even.) R is a Π_n-*relation* if $\omega^m - R$ is Σ_n. R is Δ_n if both R and $\omega^m - R$ are Σ_n.

For fixed n, the Σ_n-relations form a class that behaves very much like that of Σ_1-relations; the Δ_n-relations are the analogue, at that level, to the recursive relations. Indeed, for $n = 1$, the coïncidence of the classes of recursive and Δ_n-relations was proven by the reader in Exercise 8 of section 11, above.

These classes are readily seen to satisfy the following inclusions:

$$\begin{array}{ccccccccc}
 & & \Sigma_1 & & & & \Sigma_2 & & \\
 & \rotatebox{45}{$\subseteq$} & & \rotatebox{-45}{$\subseteq$} & & \rotatebox{45}{$\subseteq$} & & \rotatebox{-45}{$\subseteq$} & \\
\Delta_1 & & & & \Delta_2 & & & & \Delta_3 \subseteq \\
 & \rotatebox{-45}{$\subseteq$} & & \rotatebox{45}{$\subseteq$} & & \rotatebox{-45}{$\subseteq$} & & \rotatebox{45}{$\subseteq$} & \\
 & & \Pi_1 & & & & \Pi_2 & &
\end{array}$$

The Hierarchy Theorem asserts that there is no collapsing of this diagram, i.e., all inclusions are proper and, for each n, there are Σ_n-, non-Π_n-relations and Π_n-, non-Σ_n-relations. The Hierarchy Theorem is established by mimicking the construction of a non-recursive r.e. set. The first step in this direction is the proof of a generalisation of the Enumeration Theorem.

13.2. Theorem. (Σ_n-Enumeration Theorem). *Let $m, n > 0$. There is an $(m+1)$-ary Σ_n-relation $S_{m,n}$ such that, for any m-ary Σ_n-relation R, there is an $e \in \omega$ satisfying*

$$R(x_0, \ldots, x_{m-1}) \Leftrightarrow S_{m,n}(e, x_0, \ldots, x_{m-1})$$

for all $x_0, \ldots, x_{m-1} \in \omega$.

Proof. If n is odd, R has the form

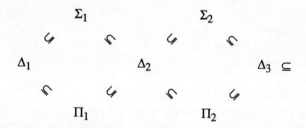

$$\exists y_0 \, \forall y_1 \ldots \exists y_{n-1} \, P(\mathbf{x}, \mathbf{y}),$$

which is equivalent to

$$\exists y_0 \, \forall y_1 \ldots \exists y_{n-1} \, T_{m+n-1}(e, \mathbf{x}, \mathbf{y}),$$

for some e, where T_k is as in Definition 11.4. If n is even, R has the form

$$\exists y_0 \, \forall y_1 \ldots \forall y_{n-1} \, P(\mathbf{x}, \mathbf{y}),$$

which is equivalent to

$$\exists y_0 \, \forall y_1 \ldots \exists y_{n-2} \, \neg \exists y_{n-1} \, T_{m+n-1}(e, \mathbf{x}, \mathbf{y}).$$

QED

Taking negations, the reader can easily derive the Π_n-Enumeration Theorem. From the Σ_n-Enumeration Theorem follows immediately the Hierarchy Theorem:

13.3. Theorem. (Hierarchy Theorem). *Let $n > 0$. There is a Σ_n-set X which is not a Π_n-set.*

Proof. Let $X = \{e\colon\ S_{1,n}(e, e)\}$. X is clearly Σ_n. If X were Π_n, its complement $\omega - X$ would have an index e_0. But then,

$$e_0 \in \omega - X \;\Leftrightarrow\; S_{1,n}(e_0, e_0)$$
$$\Leftrightarrow\; e_0 \in X,$$

a contradiction. QED

13.4. Corollary. *Let $n > 0$.*

i. *There is a Π_n-set which is not Σ_n.*
ii. *There is a Σ_n-set which is not Δ_n.*
iii. *There is a Σ_{n+1}-set which is not Σ_n.*

I leave these verifications to the reader.

The Hierarchy Theorem almost tells us that the class of arithmetic relations stratifies into levels and that each succeeding level contains something new. What we still need to complete this picture are i. a verification that these levels exhaust the class of arithmetic sets, and ii. the verification that each level behaves like the class of Σ_1-relations, i.e., a verification that each level is interesting.

Assuming for a moment the exhaustion of the class of arithmetic relations by the hierarchy of Σ_n-relations, we can easily conclude the following.

13.5. Corollary. (Tarski's Theorem). *There is no arithmetic enumeration of arithmetic sets, i.e., there is no arithmetic binary relation $A(\cdot,\cdot)$ such that the sets of the form*

$$A_e = \{x\colon\ A(e, x)\}$$

for $e \in \omega$, exhaust the class of arithmetic sets.

The reason is simple: Such a set A must be Σ_n, for some n, and each section A_e is also Σ_n. However, no set which is Σ_{n+1} but not Σ_n can occur among these sections.

The following lemma readily establishes the exhaustive nature of the Σ_n-hierarchy. It also offers a first indication that the class of Σ_n-relations is analogous to the class of Σ_1-relations.

13.6. Lemma. *Let* $n > 0$.
i. *The class of* Σ_n*-relations is closed under existential quantification.*
ii. *The class of* Π_n*-relations is closed under universal quantification.*
iii. *The classes of* Σ_n*- and* Π_n*-relations are each closed under conjunction, disjunction, bounded universal quantification, and bounded existential quantification.*

Proof. i and ii are easily established. For i, note simply that

$$\exists y\, \exists y_0 \ldots Q_{n-1} y_{n-1}\, P(\mathbf{x}, y, \mathbf{y})$$

is equivalent to

$$\exists z\, \forall y_1 \ldots Q_{n-1} y_{n-1}\, [\exists y y_0 \le z\, (z = \langle y, y_0 \rangle \ \wedge\ P(\mathbf{x}, y, \mathbf{y})].$$

The proof of ii is similar.

iii. By induction on n. In each case, the result for the class of Π_n-relations follows from that for Σ_n-relations by de Morgan's Laws:

$$\neg P \wedge \neg Q \Leftrightarrow \neg(P \vee Q)$$
$$\neg P \vee \neg Q \Leftrightarrow \neg(P \wedge Q)$$
$$\forall x \le y\, \neg P \Leftrightarrow \neg \exists x \le y\, P$$
$$\exists x \le y\, \neg P \Leftrightarrow \neg \forall x \le y\, P.$$

Moreover, the basis ($n = 1$) has already been established in section 10 (Corollary 10.11). The induction step is handled by familiar methods. E.g., to establish the closure of the class of Σ_{n+1}-relations under bounded universal quantification, observe that

$$\forall x_i \le x_j\, \exists y_0\, \forall y_1 \ldots Q_n y_n\, P(\mathbf{x}, y_0, \mathbf{y})$$

is equivalent to

$$\exists y\, \forall x_i \le x_j\, \forall y_1 \ldots Q_n y_n\, P(\mathbf{x}, (y)_{x_i}, \mathbf{y}),$$

where $\mathbf{y}$ abbreviates $y_1, \ldots, y_n$. An appeal to the closure result for the class of Π_n-relations yields the result.

I leave the rest of the proof to the reader. QED

As I said, Lemma 13.6 is only a first indication of how the Σ_n-relations constitute analogues to the Σ_1-relations. If we introduce *relativisations* of some basic notions, we can get a better view of the situation.

13.7. Definition. Let $\mathcal{P}$ be a class of sets. A relation $R \subseteq \omega^n$ is a $\Sigma_1(\mathcal{P})$*-relation* if R can be generated after finitely many steps by the following rules:
i. every polynomial relation is in $\Sigma_1(\mathcal{P})$
ii. if $S \subseteq \omega^m$ is in $\mathcal{P}$, then S and $\omega^m - S$ are $\Sigma_1(\mathcal{P})$
iii. if S, T are $\Sigma_1(\mathcal{P})$ and U is a conjunction or a disjunction of S, T, then U is $\Sigma_1(\mathcal{P})$
iv. if $S \subseteq \omega^{m+1}$ is $\Sigma_1(\mathcal{P})$ and $0 \le i, j \le n$ with $i \ne j$, then

$$T(x_0, ..., x_{i-1}, x_{i+1}, ..., x_n):\ \forall x_i \le x_j\, S$$
$$U(x_0, ..., x_{i-1}, x_{i+1}, ..., x_n):\ \exists x_i \le x_j\, S$$

are $\Sigma_1(\mathcal{P})$

v. if $S \subseteq \omega^{m+1}$ is $\Sigma_1(\mathcal{P})$ and $0 \le i \le n$, then

$$T(x_0, ..., x_{i-1}, x_{i+1}, ..., x_n):\ \exists x_i\, S$$

is $\Sigma_1(\mathcal{P})$.

13.8. Remarks. i. Dropping 13.7.ii gives us the definition of the class of Σ_1-relations. Hence, every Σ_1-relation is a $\Sigma_1(\mathcal{P})$-relation for any class $\mathcal{P}$.

ii. As with the class of Σ_1-relations, the class of $\Sigma_1(\mathcal{P})$-relations is closed under substitutions by recursive functions, even by substitution by functions with $\Sigma_1(\mathcal{P})$-graphs: if $S \subseteq \omega^m$ is $\Sigma_1(\mathcal{P})$ and $H_0, ..., H_{m-1}$ are n-ary with $\Sigma_1(\mathcal{P})$-graphs, then

$$R(x_0, ..., x_{n-1}):\ S(H_0(\boldsymbol{x}), ..., H_{m-1}(\boldsymbol{x}))$$

is $\Sigma_1(\mathcal{P})$. For, $R(\boldsymbol{x})$ is equivalent to

$$\exists y_0...y_{m-1}\, [S(y_0, ..., y_{m-1}) \wedge \bigwedge(y_i = H_i(\boldsymbol{x}))].$$

iii. (Most important). There is an asymmetry in the treatments in 13.7.i of polynomial relations and in 13.7.ii of relations from $\mathcal{P}$: we do not instantly accept the complement of a polynomial relation as $\Sigma_1(\mathcal{P})$, but we do accept such complements of elements of $\mathcal{P}$. Essentially, we are accepting polynomial relations via their representing functions, whence we get their complements by other closure conditions:

$$P(\boldsymbol{x}) \ne 0 \iff \exists y\, [y + 1 - P(\boldsymbol{x}) = 0 \vee P(\boldsymbol{x}) - (y+1) = 0];$$

allowing the complements of relations in $\mathcal{P}$ is a way of making the representing functions of sets in $\mathcal{P}$ $\Sigma_1(\mathcal{P})$. One can also introduce a stricter notion of a relation's being *positive* $\Sigma_1(\mathcal{P})$, say $\Sigma_1^+(\mathcal{P})$, by not allowing complements of elements of $\mathcal{P}$ in 13.7.ii. For our present purposes, the more liberal notion is more appropriate.

A couple of related definitions follow.

13.9. Definitions. A relation $R \subseteq \omega^n$ is a $\Pi_1(\mathcal{P})$-*relation* if $\omega^n - R$ is $\Sigma_1(\mathcal{P})$; R is $\Delta_1(\mathcal{P})$ if both R and $\omega^n - R$ are $\Sigma_1(\mathcal{P})$. If P is a single relation or set, we will write $\Sigma_1(P)$, etc., for $\Sigma_1(\{P\})$, etc.

If we let Σ_n, Π_n, etc., denote the classes of Σ_n-, Π_n-relations, etc., we can restate Lemma 13.6 as follows:

13.10. Theorem. *Let* $n > 0$.

i. $\Sigma_{n+1} = \Sigma_1(\Sigma_n) = \Sigma_1(\Pi_n) = \Sigma_1(S_{1,n})$

ii. $\Delta_{n+1} = \Delta_1(\Sigma_n) = \Delta_1(\Pi_n) = \Delta_1(S_{1,n})$

where $S_{1,n}$ *is the* Σ_n*-enumeration relation of Theorem* 13.2.

The proof is straightforward and I omit it.

A possibly more revealing characterisation is obtained by relativising the notion of recursiveness.

13.11. Definitions. Let $\mathcal{P}$ be a class of total numerical functions. A numerical partial function F is $\mathcal{P}$*-partial recursive* if F can be generated after finitely many steps by the following schemes:

$F0_n$. $F(x_0,...,x_{n-1}) = P(x_0,...,x_{n-1})$, some n-ary $P \in \mathcal{P}$

F1 - *F6*. as in Definition 11.3.

A total $\mathcal{P}$-partial recursive function is called $\mathcal{P}$*-recursive*. If $\mathcal{P} = \{P\}$, we speak of *P-recursive* and *P-partial recursive functions*.

13.12. Definitions. Let A be a set. A function F is *A-recursive* iff F is χ_A-recursive. Similarly, one can define *A-partial recursiveness* and $\mathcal{A}$*-recursiveness* and $\mathcal{A}$*-partial recursiveness* for a class $\mathcal{A}$ of sets.

13.13. Definitions. Let $\mathfrak{X}$ be a function, as set, or a class of functions or sets. A relation $R \subseteq \omega^n$ is $\mathfrak{X}$*-recursively enumerable*, or $\mathfrak{X}$*-r.e.*, if R is the domain of a $\mathfrak{X}$-partial recursive function. R is $\mathfrak{X}$*-recursive* if its representing function χ_R is $\mathfrak{X}$-recursive.

Via the Σ_n-Enumeration Theorem, it is easy to show that Σ_n-partial recursiveness and $S_{1,n}$-partial recursiveness are equivalent. When one has a single function P (such as $\chi_{S_{1,n}}$), one can add a code (such as (0,2)) for the new function and generalise the construction of the computability predicate *Comp* of section 11. This leads fairly directly to an equivalence proof between the relativisation of the concept of Σ_1-ness and that of recursive enumerability. In particular, for $P = \chi_{S_{1,n}}$, one can prove the following.

13.14. Theorem. (Post's Theorem). *Let* $m, n > 0$ *and* $R \subseteq \omega^n$.

i. *The following are equivalent*:

   a. *R is* Σ_{n+1}

   b. *R is* Σ_n*-r.e.*

   c. *R is* Π_n*-r.e.*

ii. *The following are equivalent*:

   a. *R is* Δ_{n+1}

   b. *R is* Σ_n*-recursive*

   c. *R is* Π_n*-recursive.*

In other words, $\Sigma_1(\Sigma_n) = \Sigma_n$-r.e. and $\Delta_1(\Sigma_n) = \Sigma_n$-recursive.

The proof is essentially the same as that showing the unrelativised results and offers nothing new; thus, I omit it.

With Post's Theorem we have completed our picture of arithmetic expressibility: via the coding tricks of the first half of this chapter, one can arithmetically express the computations of computable functions, whence anything computationally verifiable is arithmetically expressible. However, this is only the first level of arithmetic expressibility. There are computationally verifiable (i.e., Σ_1- or recursively enumerable) sets which are not computationally decidable (i.e., recursive). Relativising computations to the Σ_1-sets and relations, and repeating the process brings one to the second level— the class of Σ_2-relations. Iterating this brings one through infinitely many new levels and, ultimately, through the class of all arithmetic sets. Another way of putting this is: The class of arithmetic relations is the smallest class of relations containing the polynomial relations and closed under relative recursiveness— if R is arithmetic and S is R-recursive, then S is arithmetic.

The word "smallest" is important. By the Σ_n-Enumeration Theorem, each class Σ_n is countable. Thus, the class of arithmetic relations, being the union of a countable family of countable classes, is itself countable. There is, thus, a (necessarily not arithmetic) enumeration of arithmetic sets. Starting with such a relation and closing under relative recursiveness will generate a larger class of sets, the enumeration of which can be used again to generate an even larger class of sets... Done properly, this process eventually settles down with the class of *hyperarithmetic* sets, a very tiny subclass of the class of relations definable if one extends the arithmetic language to allow reference to arbitrary numerical functions or sets, or even arbitrary real numbers...

The goal of the present book, however, is not to study incredibly expansive hierarchies of definable relations, but to introduce the reader to the number theory of the logician. In the next chapter, we will look very closely at Σ_1-definability and show that all Σ_1-relations are Diophantine. This will demonstrate, among other things, the algorithmic unsolvability of the general Diophantine equation. In Chapter III, we will introduce formal logic and see what the existence of non-recursive r.e. sets and the arithmetic definability thereof say about attempts to formalise arithmetic, i.e., we will prove the necessary incompleteness of any reasonable formal theory of arithmetic.

14. Reading List

§1.

1. I.G. Bashmakova, *Diophant und diophantische Gleichungen*, UTB-Birkhäuser, Basel, 1974.
2. George Boole, *A Treatise on the Calculus of Finite Differences*, Dover, New York, 1960.
3. Howard Eves, *An Introduction to the History of Mathematics*, 6th ed., Saunders, Philadelphia, 1990.
4. Anthony Hyman, *Charles Babbage; Pioneer of the Computer*, Princeton, 1982.
5. Philip and Emily Morrison, *Charles Babbage and His Calculating Engines*, Dover, New York, 1961.

§2.

1. David Eugene Smith, *A Source Book in Mathematics, I*, Dover, New York, 1959. (Cf. in particular the entry, "Bernoulli on 'Bernoulli Numbers'".)
2. Dirk J. Struik, *A Source Book in Mathematics, 1200 - 1800*, Harvard, Cambridge (Mass.), 1969. (Cf. in particular the entry "Sequences and series" by Jakob Bernoulli.)

Another source book worth mentioning is

3. Ronald Calinger, *Classics of Mathematics*, Moore Pub. Co., Oak Park (Ill.), 1982.

Calinger's book repeats a lot of entries from other sources, including Smith and Struik. Struik has a number of errors and Calinger has copied these. Nonetheless, all three books are valuable. A more recent, and somewhat less technical, entry into the field is the following:

4. John Fauvel and Jeremy Gray, *The History of Mathematics; A Reader*, Macmillan Education, London, 1987.

§3.

1. Georg Cantor, *Contributions to the Founding of the Theory of Transfinite Numbers*, Dover, New York, 1955.

2. Richard Dedekind, "The nature and meaning of numbers", in: Richard Dedekind, *Essays in the Theory of Numbers*, Dover, New York, 1963.

3. Galileo Galilei, *Dialogues Concerning the Two New Sciences*, Dover, New York, 1954.

Some relevant papers can be found in the following source books.

4. Jean van Heijenoort, *From Frege to Gödel; A Source Book in Mathematical Logic, 1879 - 1931*, Harvard, Cambridge (Mass.), 1967.

5. Herbert Meschkowski, *Das Problem des Unendlichen*, Deutscher Taschenbuch Verlag, München, 1974.

§5.

1. Serge Lang, *Algebra*, Addison-Wesley, Reading (Mass.), 1965.

2. Serge Lang, *Complex Analysis*, Addison-Wesley, Reading (Mass.), 1977.

3. John S. Lew and Arnold L. Rosenberg, "Polynomial indexing of integer lattice points, I, II", *J. Number Theory* 10 (1978), 192 - 214, 215 - 243.

§6.

1. Leonard Eugene Dickson, *History of the Theory of Numbers, II*, Chelsea, New York, 1952.

2. Ho Peng-Yoke, "Ch'in Chiu-shao", *Dictionary of Scientific Biography*, Scribner's, New York, 1971.

Although it contains a few errors and is not meant to be read, Dickson's book definitely bears looking into. The *Dictionary of Scientific Biography* is an invaluable reference in the history of science.

§8.

1. Richard Dedekind, "The nature and meaning of numbers", in: Richard Dedekind, *Essays in the Theory of Numbers*, Dover, New York, 1963.

2. Kurt Gödel, "Über formal unentscheidbare Sätze der *Principia Mathematica* und verwandter Systeme, I", *Monatsheft f. Mathematik u. Physik* 38 (1931), 173 - 198; English translations in: Martin Davis, ed., *The Undecidable*, Raven Press, Hewlett (NY), 1965; Jean van Heijenoort, ed., *From Frege to Gödel; A Source Book in Mathematical Logic, 1879 - 1931*, Harvard, Cambridge (Mass.), 1967; and Solomon Feferman *et al.*, eds., *Kurt Gödel; Collected Works, I*, Oxford, 1986.

3. Thoralf Skolem, "Begründung der elementären Arithmetik durch die rekurrierende Denkweise ohne Anwendung scheinbarer Veränderlichen mit unendlichem Ausdehnungsbereich", most accessible in: Thoralf Skolem, *Selected Works in Logic*, Universitetsforlaget, Oslo, 1970; English translation in: Jean van Heijenoort, ed., *From Frege to Gödel; A Source Book in Mathematical Logic, 1879 - 1931*, Harvard, Cambridge (Mass.), 1967.

Two more substantial and more recent references are the following.

4. Rosza Péter, *Recursive Functions*, Academic Press, New York, 1967.

5. H.E. Rose, *Subrecursion; Functions and Hierarchies*, Oxford, 1984.

A good slim volume on number theory that the reader can consult for more on prime numbers is

6. K. Chandrasekharan, *Introduction to Analytic Number Theory*, Springer-Verlag, New York, 1968.

§9.

The pioneering papers are:

1. Wilhelm Ackermann, "Zum Hilbertschen Aufbau der reelen Zahlen", *Math. Annalen* 99 (1928), 118 - 133; English translation in: Jean van Heijenoort, ed., *From Frege to Gödel; A Source Book in Mathematical Logic, 1879 - 1931*, Harvard, Cambridge (Mass.), 1967.

2. David Hilbert, "Über das Unendliche", *Math. Annalen* 95 (1926), 161 - 190; English translation in: Jean van Heijenoort, ed., *From Frege to Gödel; A Source Book in Mathematical Logic, 1879 - 1931*, Harvard, Cambridge (Mass.), 1967.

3. Rosza Péter, "Konstruktion nichtrekursiver Funktionen", *Math. Annalen* 111 (1935), 42 - 60.

4. Gabriel Sudan, "Sur le nombre transfini ω^{ω}", *Bulletin mathematique de la Société roumaine des sciences* 30 (1927), 11 - 30; cf.: Christian Calude and Solomon Marcus, "The first example of a recursive function which is not primitive recursive", *Historia Mathematica* 6 (1979), 380 - 384.

The books by Péter and Rose cited for §8 above are also worth consulting for information on Ackermann's functions and their generalisations. Two good references on van der Waerden's Theorem are:

5. Ronald Graham, Bruce Rothschild, and Joel Spencer, *Ramsey Theory*, J. Wiley and Sons, New York, 1980.

6. A.Y. Khinchin, *Three Pearls of Number Theory*, Graylock Press, Baltimore, 1952.

Finally, an occurrence of the Ackermann function in analysis is reported in the following reference.

7. Steven F. Bellenot, "The Banach space *T* and the fast growing hierarchy from logic", to appear.

§10.

The first paper to cite is Gödel's fundamental one listed as reference 2 for §8, above. Complementary to this is the first paper cited below.

1. Kurt Gödel, "on undecidable propositions of formal mathematical systems", in: Martin Davis, ed., *The Undecidable*, Raven Press, Hewlett (NY), 1965; and Solomon Feferman, *et al.*, eds., *Kurt Gödel; Collected Works, I*, Oxford, 1986.

2. Thoralf Skolem, "Über die Zurückführbarkeit einiger durch Rekursionen definierter Relationen auf 'Arithmetische' ", most accessible in: Thoralf Skolem, *Selected Works in Logic*, Universitetsforlaget, Oslo, 1970.

3. Raymond Smullyan, *Theory of Formal Systems*, Princeton, 1961.

§11.

Wigner's paper cited in the text is

1. Eugene Wigner, "The unreasonable effectiveness of mathematics in the natural sciences", *Commun. in Pure and Applied Math.* 13 (1960), 1 - 14; reprinted in: Douglas Campbell and John Higgins, eds., *Mathematics; People, Problems, Results, III*, Wadsworth, Belmont (Cal.), 1984.

The following paper of Turing is recommended for its discussion of the nature of computability.

2. Alan M. Turing, "On computable numbers with an application to the Entscheidungsproblem", *Proc. London Math. Soc.*, ser. 2, vol. 42 (1936 - 37), 230 - 265; reprinted in: Martin Davis, ed., *The Undecidable*, Raven Press, Hewlett (NY), 1965.

Because of this paper, the identification of the classes of partial recursive and semi-computable functions is sometimes called *Turing's Hypothesis*. The more common designation of this identification is *Church's Thesis*, after Alonzo Church, whose less readable definition of computability in terms of λ-definability also appears in the Davis volume. Incidentally, the Davis volume contains other pioneering papers of recursion theory. The background to the founding of recursion theory from the perspective of one of its founders is given by:

3. Stephen Kleene, "Origins of recursive function theory", *Annals of the History of Computing* 3 (1981), 52 - 67.

Each of the following texts has a chapter on register machines. The reader who finds the present discussion inadequate is referred to them.

4. Heinz-Dieter Ebbinghaus, Jörg Flum, and Wolfgang Thomas, *Mathematical Logic*, Springer-Verlag, New York, 1984. (Chapter X)

5. John Bell and Moshe Machover, *A Course in Mathematical Logic*, North-Holland, Amsterdam, 1977. (Chapter 6)

§12.

This section contains results do to so many researchers that I have made no attempt to say anything about the history of the subject. The pioneering papers can be found in the book edited by Davis that has been cited several times (e.g., number 2 for §11, above). Although two decades old, the best book on recursion theory remains the following one by Rogers:

1. Hartley Rogers, Jr., *Theory of Recursive Functions and Effective Computability*, McGraw-Hill, New York, 1967.

This book is also a good source for §13. The reader who might find the huge code numbers unbelievable, might like to consult an introductory computer science text to see just how efficient the techniques of these last two sections can be if one does not restrict one's language to the arithmetic one:

2. Michael Burke and Ron Genise, *LOGO and Models of Computation*, Addison-Wesley, Menlo Park (Cal.), 1987.

Speaking of Computer Science, the reader might enjoy the following interpretation of Rice's Theorem:

3. William F. Dowling, "There are no safe virus tests", *Amer. Math. Monthly* 96 (1989), 835 - 836.

Chapter II. Diophantine Encoding

Denique fastuosum problema problematum ars Analytice, triplicem Zetetices, Poristices & Exegetices formam tandem induta, jure sibi adrogat, Quod est,
NULLUM NON PROBLEMA SOLVERE.

—François Viète

1. Diophantine Equations; Some Background

The main goal of the present chapter is to prove the effective unsolvability of the general problem of deciding which Diophantine equations have solutions. Once we've reached this goal, we will take a look at some applications and refinements. First, however, we must state clearly what the problem involves, i.e., what a Diophantine equation is. That, in part, is the purpose of the present section.

It can be said with confidence that Diophantus of Alexandria lived in Alexandria sometime between the middle of the second century B.C. (since Diophantus mentions a mathematician Hypsikles who lived then) and the middle of the fourth century A.D. (since Theon of Alexandria, who lived then, cites Diophantus). A letter from the eleventh century A.D. asserts that Anatolius of Alexandria (c. middle third century A.D.) had a friend named Diophantus. Moreover, Diophantus dedicated his *Arithmetica* to a "very respected Dionysius" and the best candidate for such an epithet was St. Dionysius, who lived in Alexandria in the third century A.D. Thus, scholars think that Diophantus lived c. 250 A.D.

We are more fortunate when it comes to the work of Diophantus: 6 of the 13 books of his *Arithmetica* have long been preserved and, relatively recently, more was found in Arabic translation. Thus, there is a good enough body of his work for scholars to assess. This assessment has become more positive over the years.

In rough terms, Diophantus posed and solved polynomial equations. As already mentioned in the introductory section of Chapter I, the available notation was rather limited. He had names for one variable and several of its powers. By asserting, for example, that some expression involving this variable and its powers is a square, he was able to express some equations in more than one unknown. Moreover, a felicitous substitution would occasionally reduce the number of variables of an equation to one, allowing Diophantus to handle such equations. For the most part, however, Diophantus treated equations in one variable and he looked for positive rational solutions, generally satisfying himself with producing one solution.

Here, I cannot resist quoting Eric Temple Bell:

Diophantus contented himself with special solutions of his problems; the majority of his numerous successors have done likewise, until diophantine analysis today is choked by a jungle of trivialities bearing no resemblance to cultivated mathematics. It is long past time that the standards of Diophantus be forgotten though he himself be remembered with becoming reverence.

Of course, Bell is generally quoted more for the passion of his prose than the soundness of his opinions; but his remark does illustrate an important point: a good method would produce *general* solutions, not merely *specific* solutions. Diophantus is often accused of lacking method. B.L. van der Waerden says simply that Diophantus' method varies from case to case. The historians J.E. Hofmann and O. Becker assert

Diophantus gives no general methods; rather it seems he uses a surprising new trick for each new problem.

A slightly more generous remark is to be found in the *Dictionary of Scientific Biography*:

In only a few cases can one recognize generally applicable methods of solutions in the computations that Diophantus presents, for he considers each case separately, often obtaining a solution by means of brilliant stratagems.

Against this negative tide, I.G. Bashmakova makes a strong case that Diophantus had general methods, later codified during the emergence of algebraic geometry, but that Diophantus simply did not have the notation necessary to make his methods apparent. Supporting this, André Weil says "there is much, in Diophantus and in Viète's *Zetetica*, which in our view pertains to algebraic geometry". [We shall have a look at such a geometric method in section 3, below, when we consider the solutions to the Pythagorean equation,

$$x^2 + y^2 = z^2,$$

and the Pell equation,

$$x^2 - Dy^2 = 1.]$$

For now, I suppose, the lesson to be learned from these remarks is that it requires great sophistication to make something of Diophantus' solutions. The history of things Diophantine prior to Pierre de Fermat as summed up by Fermat's son Samuel (quoted here from Weil's book) reflects this:

Bombelli, in his *Algebra*, was not acting as a translator for Diophantus, since he mixed his own problems with those of the Greek author; neither was Viète, who, as he was opening up new roads for algebra, was concerned with bringing his own inventions into the limelight rather than serving as a torch-bearer for those of Diophantus. Thus, it took Xylander's unremitting labors and Bachet's admirable acumen to supply us with the translation of Diophantus' great work.

Viète, once again, was the man who first introduced the modern variable. His five books of *Zetetica* (1593) are devoted to the algebraic solution of some 30 odd problems from Diophantus. That Viète's solutions are easier to wade through than Diophantus' justifies— for his and (probably) Diophantus' purposes— Viète's thrusting his own inventions into the limelight. Samuel's remark concerned his father's purpose.

Pierre de Fermat read Bachet's edition of Diophantus and discovered something new— number theory. This is not to say that there was a great deal of number theory in Diophantus, but that there was enough to inspire Fermat. The best known example of such inspiration is, of course, the Fermat Problem: in the margin of his copy of Bachet, next to a discussion of the Pythagorean equation,

$$x^2 + y^2 = z^2,$$

Fermat noted that he could prove that, for $n > 2$, the "Fermat equation",

$$x^n + y^n = z^n,$$

had only the trivial solutions consisting of 0's and 1's. The Fermat equation is homogeneous, whence its solvability in the rationals is equivalent to its solvability in the integers (cf. sections 3 and 7, below, for a fuller discussion). With Fermat comes the switch to the study of *integral* solutions to polynomial equations. The general problem of deciding which polynomials with rational coëfficients have rational zeroes does reduce to the general problem of deciding which such polynomials have integral zeroes (as we shall see in section 7, below). Thus, the switch from the search for rational zeroes to a search for integral zeroes is not so much a change in as an expansion of the general Diophantine problem. Moreover, working with integers gives one the advantage, as Fermat illustrated with the special case of his equation,

$$x^4 + y^4 = z^4,$$

of being able to use *induction* to prove there to be no integral solutions.

[Logically, Fermat's use of induction— called the *method of infinite descent*— is very interesting and affords a simple illustration of the rôle of transfinite ordinals in logical number theory. Unfortunately, this topic lies just beyond the scope of the present book.]

As I was saying, Fermat became interested in number theory— the theory of the properties of the integers. Many interesting number theoretic properties reduce to "Diophantine equations", perhaps better called "Fermatian equations" as the solutions now sought are to be integral. The problem, for example, of which numbers can be written as the sums of two squares is the Diophantine question: for which non-negative integers m is the equation,

$$x^2 + y^2 = m,$$

solvable in integers?

A Diophantine equation of particular interest to Fermat was the so-called Pell equation,

$$x^2 - Dy^2 = 1,$$

where D is not a perfect square. This equation has a long history and has played a central rôle in the development of the study of Diophantine equations. It also plays a central rôle in the proof of the algorithmic unsolvability of the general (Fermatian) Diophantine problem and I devote sections 3 and 4, below, to it. For the moment, let me simply note that the final solution to the Pell equation requires one to look at the more general equations,

$$x^2 - Dy^2 = m,$$

where D again is not a perfect square and m is a positive integer.

The mainstream of mathematics in Fermat's day was, however, pre-Newtonian and pre-Leibnitzian calculus. Everyone was graphing curves, constructing tangents, finding areas, and solving max-min problems. This work (to which Fermat contributed) was too interesting and Fermat could not seduce his contemporaries away from it. It wasn't until a couple of generations later that Leonhard Euler took up the subject and number theory got a new start. By the end of the 18th century, in the hands of Joseph Louis Lagrange and Adrien-Marie Legendre, number theory was going strong. This also held for Diophantine equations. Lagrange's contributions are particularly worth citing in this connexion:

a. in 1768, he proved the existence of solutions of the Pell equation;

b. inspired by this solution, he developed the theory of *binary quadratic forms*, i.e., the theory of equations,

$$ax^2 + bxy + cy^2 = m,$$

where a, b, c, and m are integral; and

c. in 1772, he published a proof that every non-negative integer is the sum of four squares, i.e., that the Diophantine equation,

$$x^2 + y^2 + z^2 + w^2 = m,$$

is solvable over the integers for any natural number m.

Thus, we see that by the end of the 18th century number theory and Diophantine equations were entering the mathematical mainstream. The next century, beginning in 1801 with the publication of Karl Friedrich Gauß' *Disquisitiones Arithmeticae*, deepened the involvement. Indeed, one major development in mathematics— the study of ideals in modern abstract algebra— began in the mid-1800s at least partly in attempts to solve the Fermat Problem cited above.

It should come as no surprise to the reader to read that, when David Hilbert decided to inaugurate the coming of the new century in his lecture at the Second International Congress of Mathematicians held in Paris in August of 1900 by proposing a list of 23 problems for future mathematicians, several of the problems he chose were number theoretical. Nor should we be surprised to find that one of the problems— the tenth— concerned Diophantine equations:

10. Determination of the Solvability of a Diophantine Equation

Given a diophantine equation with any number of unknown quantities and with rational integral numerical coëfficients: *To devise a process according to which it can be determined by a finite number of operations whether the equation is solvable in rational integers.*

Hilbert's tenth problem is unsolvable: there is no algorithm which, given a Diophantine equation, will decide in a finite number of steps whether or not the equation is solvable in integers.

Work towards the negative solution of Hilbert's tenth problem began around 1950. This is not as late as it may sound. The initial unsolvability results in logic and recursion theory were obtained in the mid-1930s. It wasn't until 1947 that an unsolvable problem of ordinary mathematics was exhibited and even then it— the word problem for semi-groups— was semi-logical in character. Thus, it was fairly early in the history of algorithmically unsolvable problems that Martin Davis and Julia Robinson independently set out to prove the unsolvability of Hilbert's tenth problem. Davis conjectured that every r.e. relation was Diophantine and proved every r.e. relation on ω^n could be written in the form,

$$\exists x\, \forall y \le x\, \exists y_0 \le x \ldots \exists y_{m-1} \le x\, [P(x, y, \mathbf{x}, \mathbf{y}) = 0], \qquad (*)$$

where P is a polynomial with integral coëfficients. Robinson began more directly by seeing what Diophantine relations she could find. Not finding many, she allowed exponentiation and looked to see what exponential Diophantine relations she could find. Having a little more success, she then proved that, in order to show exponentiation to have a Diophantine graph (hence, all exponential Diophantine relations to be Diophantine) it sufficed to show the Diophantine nature of any relation of roughly exponential growth. Neither Davis nor Robinson thought the other's approach to be too promising and not much was accomplished until around 1960.

In 1961, Davis, Hilary Putnam, and Robinson published a joint paper showing that every r.e. relation was an exponential Diophantine relation. This proof proceeds by using Robinson's exponential relations to eliminate the bounded universal quantifier from Davis' representation (*) of an r.e. relation. The proof is actually quite simple and is given in section 2, below.

The Davis-Putnam-Robinson Theorem was the first major breakthrough in the assault on Hilbert's tenth problem. It brought into the limelight Robinson's reduction of the desired theorem to the problem of showing any relation of roughly exponential growth to be Diophantine. There were a few interesting results established over the next decade, most notably Putnam's theorem (1960) that the problem of deciding which polynomials represent *all* natural numbers is algorithmically unsolvable, but, for the most part, the 1960s saw mainly a host of variant reductions, all relying on Robinson's original one.

In 1970, using such a function of roughly exponential growth, Juriĭ Matijasevič (or, using an alternate scheme of transliteration from the Russian, Yuri Matiyasevich) showed that all recursively enumerable relations are Diophantine. Where Robinson's reduction had used the Pell equation— or, rather, the sequence of solutions to the Pell equation— Matijasevič used the sequence $F_0, F_1, \ldots$ of Fibonacci numbers (Chapter I, section 8—just prior to 8.1 and again after 8.16). From a Diophantine definition of the relation,

$$y = F_{2x},$$

he was able to derive a direct Diophantine definition of exponentiation, thus bypassing Robinson's reduction formally, if not heuristically. Following Matijasevič's solution of Hilbert's tenth problem (if one may speak of the "solution" to an unsolvable problem), a number of researchers returned to the Pell equation and gave direct Diophantine definitions of the sequence of solutions thereto, these later definitions adapting Matijasevič's technique fairly directly. As the Pell equation is more central to matters Diophantine, I have decided to follow the post-Matijasevič crowd and work with the Pell equation. Sections 3 and 4, below, deal with the Pell equation; and section 5 presents the proof of Matijasevič's theorem on the unsolvability of Hilbert's tenth problem.

A few words are in order about sections 3 and 4. The establishment of an algorithmic unsolvability result, unlike the construction of an actual algorithm when such is possible, often does not require a really deep knowledge of the general subject in which the unsolvability result is to be proven. This is the case with Hilbert's tenth problem: the whole beautiful theory of binary quadratic forms can be ignored, and only the Pell equation need be looked at. In fact, only a very trivial special case of the Pell equation needs to be considered. Everything about this special case of the Pell equation and its sequence of solutions that one needs to know for the proof of Matijasevič's theorem can be found in section 4, and the reader can easily skip section 3. In section 3, the general Pell equation is considered and its solution given. This is done in part simply to give a fuller background against which to view the solution to Hilbert's tenth problem, and in part because, if we only need a special type of Pell equation in this book, this is not true in the literature.

Section 6 is devoted to more-or-less immediate applications of the algorithmic unsolvability of Hilbert's tenth problem— primarily further such unsolvability results. In section 7, the Diophantine problem over the rationals is related to that over the integers. It is still an open problem whether or not there is an algorithm to decide the general Diophantine problem over the rationals.

Sections 8, 9, and 10: Following the solution of Hilbert's tenth problem, both Matijasevič and Robinson took a look back, re-doing more elegantly some of the small steps in the path to the solution. In doing so, the binomial coëfficient received new attention and Matijasevič uncovered some semi-obscure results on binomial coëfficients by Ernst Kummer and Edouard Lucas which permitted new exponential Diophantine encoding techniques which bypassed the use of the bounded universal quantifier of (*) (recall the proof of I.10.6 where the bounded quantifier is introduced). Ultimately, this led to new proofs of the Davis-Putnam-Robinson Theorem. Section 8 discusses binomial coëfficients; section 9 presents a direct exponential Diophantine encoding of the running of programs on register machines; and section 10 presents another, slightly more difficult, application of the new coding technique.

The last few paragraphs above explain what all of the rest of the chapter other than the rest of this section will cover. In the remainder of this section, I shall state clearly what is meant by a "Diophantine equation".

We begin with the notion of a polynomial. But for a few notational conventions, this notion is clear enough and need not be explained.

1.1. Convention. Throughout the entire chapter, we will mean "polynomial with integral coëfficients" whenever we use the word "polynomial".

We could, with no greater generality, allow polynomials with rational coëfficients: if P has rational coëfficients, and d is any common denominator of these coëfficients, then the polynomial dP, which on simplifying has integral coëfficients, has exactly the same zeroes as P. Thus, we may feel free to violate Convention 1.1 and allow an occasional rational coëfficient.

1.2. Convention. To distinguish polynomials from their values $P(x_0,...,x_{n-1})$ and yet still exhibit the variables, we will use capital letters to denote these variables. Thus, $P(X_0, ..., X_{n-1})$ is a polynomial *qua* polynomial, and $P(x_0, ..., x_{n-1})$ is a value of the polynomial. The capitals P, Q will be reserved for polynomials.

Thus, in referring to the Pell equation above, I should have written,

$$X^2 - dY^2 = 1, \tag{1}$$

or,
$$X^2 - DY^2 = 1, \tag{2}$$
depending on whether D was considered to have a fixed (but unspecified) value d or to be another variable in the equation.

With the Pell equation, we will primarily be interested in the form (1) rather than (2), i.e., we will think of one or two fixed values d of the parameter D. For some equations, however, one is interested in the set of *parameters* for which the equation is solvable. For example, the representability problem for a quadratic form could be conceived of in terms of a polynomial,
$$aX^2 + bXY + cY^2 = M.$$
Here, a, b, and c are fixed (maybe 1, 0, and 1) and the problem is to decide for which values m of M the equation,
$$aX^2 + bXY + cY^2 = m,$$
is solvable in (say) integers x and y. [For $a = 1, b = 0$, and $c = 1$, the solution is: any prime divisor of m of the form $4k + 3$ must divide m to an even power.] Because of this latter sort of occurrence, we get a dichotomy of variables.

1.3. Convention. The variables of a polynomial fall into two classes— the *parameters* and the *solution-variables*. If $A_0, ..., A_{m-1}$ denote the parameters and $X_0, ..., X_{n-1}$ the solution-variables of a polynomial $P(A_0, \ldots, A_{m-1}, X_0, \ldots, X_{n-1})$, the *Diophantine relation* associated with P is the relation,
$$\{(a_0, \ldots, a_{m-1}) \colon \exists x_0 ... x_{n-1} \, [P(a_0, \ldots, a_{m-1}, x_0, \ldots, x_{n-1}) = 0]\}.$$

Notice that I did not say in Convention 1.3 where the values $a_0, ..., a_{m-1}$ of the parameters or the solutions $x_0, ..., x_{n-1}$ were supposed to come from. If we agree that $a_0, ..., a_{m-1}, x_0, ..., x_{n-1}$ are to be natural numbers, then the relation defined is a Diophantine relation in the sense of Chapter I. In this chapter, we will only concern ourselves with natural numerical parameters, whence the various types of Diophantine relations will be relations of natural numbers.

What remains to be determined are the various domains over which the solution-variables can range. Four natural candidates are:

i. $N = \omega =$ the set of natural numbers

ii. $N^+ = \omega^+ =$ the set of positive integers

iii. $Z =$ the set of integers

iv. $Q =$ the set of rational numbers.

[Of course, one can also consider the sets of positive or non-negative rational numbers, but the discussion becomes slightly more complicated and I wish to keep things simple in this introductory section. For more about the rational case, cf. section 7, below.]

1.4. Definitions. Let Γ be one of the domains just cited. By $DIOPH(\Gamma)$ is meant the problem of deciding which polynomial equations,

$$P(x_0, \ldots, x_{n-1}) = 0,$$

have solutions $x_0, \ldots, x_{n-1} \in \Gamma$. A relation $R \subseteq \omega^m$ is Γ-*Diophantine* if there is a polynomial $P(A_0, \ldots, A_{m-1}, X_0, \ldots, X_{n-1})$ such that

$$R = \{(a_0, \ldots, a_{m-1}) \in \omega^m: \exists x_0 \ldots x_{n-1} \in \Gamma\, [P(a_0, \ldots, a_{m-1}, x_0, \ldots, x_{n-1}) = 0]\}.$$

Thus, $DIOPH(Z)$ is Hilbert's tenth problem and the N-Diophantine relations are the Diophantine relations of Chapter I. The basic relations holding among the various Diophantine problems and classes of Diophantine relations are given by the following.

1.5. Facts. i. *The problems $DIOPH(N)$, $DIOPH(N^+)$, and $DIOPH(Z)$, are effectively inter-reducible.*
ii. *The problem $DIOPH(Q)$ effectively reduces to $DIOPH(Z)$.*
iii. *The classes of N-Diophantine, N^+-Diophantine, and Z-Diophantine relations coïncide and contain the class of Q-Diophantine relations.*

Proof of i. Let $P(X_0, \ldots, X_{n-1})$ be a polynomial with solution-variables as shown. Observe:
a. $P = 0$ is solvable in N iff $P(X_0 - 1, \ldots, X_{n-1} - 1) = 0$ is solvable in N^+.
b. $P = 0$ is solvable in N^+ iff $P(X_0 + 1, \ldots, X_{n-1} + 1) = 0$ is solvable in N.
c. $P = 0$ is solvable in N iff $P(X_{01}^2 + \ldots + X_{04}^2, \ldots, X_{n-1,1}^2 + \ldots + X_{n-1,4}^2) = 0$ is solvable in Z. [We are here using Lagrange's Theorem that every non-negative integer is the sum of four squares. A proof of Lagrange's Theorem will be given in section 3, below, as an adjunct to the discussion of the Pell equation.]
d. $P = 0$ is solvable in Z iff

$$\prod_e P(e_0 X_0, \ldots, e_{n-1} X_{n-1}) = 0$$

is solvable in N, where e ranges over all maps from $\{0, \ldots, n-1\}$ to $\{-1, 1\}$. QED

The proof of 1.5.ii, as well as that of the effective equivalence of $DIOPH(Q)$ with the *homogeneous* Diophantine problem over Z, is postponed to section 7, below. As for 1.5.iii, the coïncidence of the three classes of Diophantine relations over the three classes of integers is established by simply writing the parameters in P in the proof of 1.5.i— being careful not to tamper with the parameters while doing so with the solution-variables. The relation between the notions Q-Diophantine and Z-Diophantine is, again, deferred to section 7.

By Fact 1.5.i, in discussing Hilbert's tenth problem, it makes no difference if we discuss $DIOPH(N)$, $DIOPH(N^+)$, or $DIOPH(Z)$. In line with the emphasis of the rest of the book on N, we shall discuss $DIOPH(N)$.

1.6. Conventions. Except where otherwise noted, lower case roman letters will serve as variables for natural numbers. By a *Diophantine equation*, we will mean an equation from the problem $DIOPH(N)$.

The notion of an *exponential* Diophantine equation, already referred to in our historical sketch, has yet to be explained. As this is, largely, the topic of the next section, I defer the reader to that section.

2. Initial Results; The Davis-Putnam-Robinson Theorem

There were, as noted in the introduction to this chapter, two beginnings to the story that is to unfold. The simpler of the two was Julia Robinson's direct attempt to list interesting Diophantine relations.

In attempting to exhibit the Diophantine nature of a relation, it will often be convenient to use the following lemmas.

2.1. Lemma. *Let $P_0, \ldots, P_{n-1}$ be polynomials. For any values of the variables of the P_i's, we have the equivalences*:

i. $P_0 = 0 \wedge \ldots \wedge P_{n-1} = 0 \Leftrightarrow P_0^2 + \ldots + P_{n-1}^2 = 0$

ii. $P_0 = 0 \vee \ldots \vee P_{n-1} = 0 \Leftrightarrow P_0 \cdot P_1 \cdot \ldots \cdot P_{n-1} = 0.$

The proof of Lemma 2.1 is an easy exercise which I leave for the reader. I note that 2.1.ii was implicit in step d of the proof of Fact 1.5.i in the last section and that, moreover, the industrious reader has seen this before— in Exercise 1.i of Chapter I, section 10. Part ii of that exercise was almost the following corollary to Lemma 2.1.

2.2. Lemma. *The class of Diophantine relations is closed under conjunction, disjunction, and existential quantification.*

I leave the proof to the reader as another simple exercise.

2.3. Lemma. *If $R \subseteq \omega^{n+1}$ is Diophantine and F is an n-ary function with a Diophantine graph, then the relation $S(x_0, \ldots, x_{i-1}, x_{i+1}, \ldots, x_n)$ defined by*

$$S: \quad R(x_0, \ldots, x_{i-1}, F(x_0, \ldots, x_{i-1}, x_{i+1}, \ldots, x_n), x_{i+1}, \ldots, x_n)$$

is also Diophantine.

Proof. Observe that

$$S \Leftrightarrow \exists y\, [R(x_0, \ldots, x_{i-1}, y, x_{i+1}, \ldots, x_n) \wedge y = F(x_0, \ldots, x_{i-1}, x_{i+1}, \ldots, x_n)],$$

and apply Lemma 2.2. QED

Needless to say, application of Lemma 2.3 will not require *all* the variables $x_0, \ldots, x_{i-1}, x_{i+1}, \ldots, x_n$ to be arguments of F.

With these lemmas, we can list some Diophantine relations (Compare Examples I.8.9).

2.4. Examples. *The following relations are Diophantine*:

i. $x = y$: $x - y = 0$

ii. $x \mid y$: $\exists z(xz = y)$

iii. *x is even*: $2 \mid x$ (or: $\exists y(2y = x)$)

iv. *x is odd*: $2 \mid x+1$ (or: $\exists y(2y + 1 = x)$)

v. $x \le y$: $\exists z(x + z = y)$
   $x < y$: $\exists z(x + z + 1 = y)$

vi. $r = \mathrm{Rem}(x, y)$: $\exists q(x = qy + r \wedge r < y)$ (or: $y \mid x - r \wedge r < y$)

vii. $z = [x/y]$: $\exists r(x = zy + r \wedge r < y)$

viii. $z = \langle x, y\rangle$: $2z = (x+y)^2 + 3x + y$

ix. $x = \pi_1^2(z)$: $\exists y(z = \langle x, y\rangle)$
   $y = \pi_2^2(z)$: $\exists x(z = \langle x, y\rangle)$

x. $\beta(c, d, i) = a$: $\mathrm{Rem}(c, 1 + (i+1)d) = a$.

This list certainly does not exhaust the class of easily definable Diophantine relations, but it does seem to exhaust the class of such relations that have an established use in coding. The list seems very powerful. It does contain the β-function, which we were able to exploit back in the proof of Theorem I.10.6 in showing every primitive recursive function to have a Σ_1-definable graph. The graph of exponentiation to the base 2 would, for example, be given Σ_1-expression thus:

$$y = 2^x:\ \exists cd\, [\beta(c,d,0) = 1 \wedge \forall i \le x[\beta(c,d,i+1) = 2\cdot\beta(c,d,i)] \wedge \beta(c,d,x) = y].$$

This misses being Diophantine by the presence of the bounded universal quantifier, and, unfortunately, there isn't any obvious way of getting rid of this quantifier. (Using 2.4.v, one can eliminate bounded *existential* quantifiers, but that is not much help.)

At this point Robinson added exponentiation and extended the list 2.4 to include some further interesting *exponential* Diophantine relations. Martin Davis, on the other hand, sought, if not to eliminate the bounded universal quantifier, at least to limit it. Let us look first at Davis' result:

2.5. Theorem. (Davis Normal Form). *Let $R \subseteq \omega^m$ be a Σ_1-relation. There is a polynomial $P(A_0, \ldots, A_{m-1}, X, Y, X_0, \ldots, X_{n-1})$ such that, for all $a_0, \ldots, a_{m-1}$,*

$$R(\boldsymbol{a}) \Leftrightarrow \exists x\, \forall y \le x\, \exists x_0 \le x \ldots \exists x_{n-1} \le x\, [P(\boldsymbol{a}, x, y, \boldsymbol{x}) = 0]. \qquad (*)$$

By Davis' Theorem 2.5, which is a major refinement of Theorem I.10.10 on the strict Σ_1-definability of every r.e. relation, we need only eliminate one bounded universal quantifier in order to show a Σ_1-relation to be a Diophantine one.

Davis' Theorem is proven by showing the class of relations defined by an equivalence (*) to be closed under the closure properties defining the class of Σ_1-relations. An implicit induction on the number of steps in the generation of a Σ_1-relation then yields the Theorem. The proof, if not the result, really belongs in Chapter I, section 10, rather than here. Thus, I defer it to the end of this section, and continue with the more novel issues.

Returning to Robinson's approach, we would now like to see a nice stock of exponential Diophantine relations. However, we do not yet know what is meant by an "exponential Diophantine equation". Obviously, such an equation ought to be one involving "exponential polynomials". There are several candidates for the title "exponential polynomial".

2.6. Definition. The class of *iterated exponential polynomials* is generated inductively as follows:

i. variables and non-negative integers are iterated exponential polynomials,
ii. if P, Q are iterated exponential polynomials, then so are $P + Q$ and $P \cdot Q$,
iii. if P, Q are iterated exponential polynomials, then so is P^Q.

In other words, the iterated exponential polynomials are what we get from polynomials by allowing exponentiation and restricting all coëfficients to being non-negative. The restriction on coëfficients will guarantee that all values are integral, whence we can avoid things like

$$\sqrt{X} = X^{(2^{-1})}$$

or even

$$i = \sqrt{-1} = (-1)^{(2^{-1})}.$$

A narrower class of exponential polynomials is obtained by restricting 2.6.iii to allowing only constants and variables to occur in an exponentiation.

2.7. Definition. The class of *exponential polynomials* is generated inductively as follows:

i. variables and non-negative integers are exponential polynomials,
ii. if P, Q are exponential polynomials, then so are $P + Q$ and $P \cdot Q$,
iii. if n is a non-negative integer and X, Y are variables, then n^Y and X^Y are exponential polynomials.

It is not hard to see that every exponential polynomial can be obtained from an ordinary polynomial P with non-negative coëfficients by replacing some occurrences of variables Z in P by exponentials n^Y or X^Y. A very narrow class of exponential polynomials is obtained by restricting this still further and allowing only exponentiations of the form 2^Y.

2.8. Definition. A *dyadic exponential polynomial* is an expression of the form,

$$P(X_0, \ldots, X_{n-1}, 2^{X_{i_0}}, \ldots, 2^{X_{i_{k-1}}}),$$

where P is a polynomial with non-negative coëfficients.

Corresponding to each of these notions of "exponential polynomial" is its "exponential Diophantine problem", i.e., the problem of effectively deciding the solvability of "exponential Diophantine equations":

$$\exists \boldsymbol{x}\, [P(\boldsymbol{x}) = Q(\boldsymbol{x})],$$

where P, Q are "exponential polynomials". Now, the class of iterated exponential polynomials constitutes the most natural explication of the term "exponential polynomial", but the iterated exponential Diophantine problem is, at first sight, the most general. The dyadic exponential polynomials constitute the narrowest such explication and the algorithmic unsolvability of the dyadic exponential Diophantine problem is a nice refinement of the algorithmic unsolvability result we seek. It turns out, however, that all three problems are effectively equivalent: at the cost of a few variables, every exponential Diophantine equation reduces to a dyadic exponential equation, and every iterated exponential Diophantine equation reduces to an exponential Diophantine equation. [Cf. Exercises 2, 3, below.] Thus, except for the refined question of the number of variables needed for an unsolvability result, it does not much matter which of the three explications of "exponential polynomial" we choose to work with. The "exponential Diophantine" definitions we shall give will happen to be exponential Diophantine, so we shall choose the exponential Diophantine polynomials of Definition 2.7 as our standard.

In analogy to Convention 1.3, we can speak of exponential Diophantine relations. Analogues to Lemmas 2.1 to 2.3 hold for the class of exponential Diophantine relations. (For the one subtle point, cf. Exercise 1, below.) Moreover, since any polynomial equation,

$$P = 0,$$

can be rendered

$$P^+ = P^-,$$

where P^+ and P^- each has only non-negative coëfficients, it follows that all Diophantine relations are exponential Diophantine.

We escape from these trivialities with the following:

2.9. Examples. *The following relations are exponential Diophantine*:

i. $z = \binom{x}{y}$

ii. $y = x!$

iii. *p is prime*.

Proof. i. The binomial coëfficient $\binom{x}{y}$ is just the coëfficient of X^y in

$$(X + 1)^x = \sum_{i=0}^{x} \binom{x}{i} X^i.$$

If we replace X by 1, we see

$$\binom{x}{y} \le \sum_{i=0}^{x} \binom{x}{i} = (1 + 1)^x = 2^x,$$

whence $\binom{x}{y}$ is the digit corresponding to u^y in the u-adic expansion of

$$(u + 1)^x = \sum_{i=0}^{x} \binom{x}{i} u^i$$

for any $u > 2^x \geq \binom{x}{y}$. This yields

$$z = \binom{x}{y} \Leftrightarrow \exists uvw\,[u = 2x+1 \wedge (u+1)^x = vu^{y+1} + zu^y + w \wedge z < u \wedge w < u^y]$$

$$\Leftrightarrow \exists tuvw\,[u = 2x+1 \wedge t = u+1 \wedge z < u \wedge t^x = vu^{y+1} + zu^y + w \wedge w < u^y].$$

ii. If $z \geq x$, we can solve $\binom{z}{x} = \dfrac{z!}{x!(z-x)!}$ for $x!$ to get

$$\begin{aligned} x! &= \frac{z(z-1)\ldots(z-x+1)}{\binom{z}{x}} \\ &< \frac{z^x}{\binom{z}{x}} = \frac{z^x x!(z-x)!}{z!} \\ &< x!\frac{z}{z-1}\cdot\frac{z}{z-2}\cdots\frac{z}{z-x+1} \\ &< x!\frac{1}{1-1/z}\cdot\frac{1}{1-2/z}\cdots\frac{1}{1-(x-1)/z}. \end{aligned} \tag{1}$$

If we fix x and let $z \to +\infty$, the right-hand-side of (1) goes to $x!$. Thus, for very large z, we will have

$$x! = \left[\frac{z^x}{\binom{z}{x}}\right]. \tag{2}$$

If we can express "very large" in an exponential Diophantine way, we will have an exponential Diophantine definition of $y = x!$, namely:

$$y = x! \Leftrightarrow \exists z\,(z \text{ is very large } \wedge\ y = [z^x/\tbinom{z}{x}]).$$

For $x = 0$, any z is large enough. Thus, it suffices to define "z is very large" for $x > 0$.

The first thing to notice is that, other than $x!$, the factors on the right-hand-side of (1) can be uniformly replaced by a single repeated one:

$$\frac{1}{1-i/z} < \frac{1}{1-x/z}.$$

Thus, (1) yields

$$x! < z^x/\binom{z}{x} < x!\left(\frac{1}{1-x/z}\right)^x. \tag{3}$$

To handle (3), we need two elementary inequalities.

Claim. i. For $0 < \theta < 1/2$,

$$\frac{1}{1-\theta} < 1 + 2\theta \tag{*}$$

ii. For $0 < \theta < 1$ and $k \geq 1$,

$$(1+\theta)^k < 1 + 2^k\theta. \tag{**}$$

Proof of the Claim. i, Simple algebra.

ii. This is an application of the Binomial Theorem:

$$(1+\theta)^k = \sum_{i=0}^{k}\binom{k}{i}\theta^i = 1+\theta\cdot\sum_{i=1}^{k}\binom{k}{i}\theta^{i-1}$$

$$< 1+\theta\cdot\sum_{i=1}^{k}\binom{k}{i} < 1+2^k\theta.$$

QED

To continue the proof of 2.9.ii, assume $z > 2x$ so that $\theta = x/z < 1/2$. Then apply (*) to (3):

$$x! < z^x/\binom{z}{x} < x!(1+2(x/z))^x.$$

Applying (**) to this inequality, with $\theta = 2x/z$, yields

$$x! < z^x/\binom{z}{x} < x!(1+2^x(2x/z)) = x! + \frac{2^{x+1}x!x}{z} < x! + \frac{2^{x+1}x^{x+1}}{z}.$$

From this we see that we can take

$$z \text{ is very large:} \quad z > (2x)^{x+1}$$

in formula (2) to conclude the relation $Y = x!$ to be exponential Diophantine.

iii. Cf. Exercise 7.

QED

The exponential Diophantine nature of the set of primes is not particularly useful, but it does offer an interesting exponential Diophantine set which we would be hard put to express in any Diophantine way.

Before we prove anything, we need two additional examples, due to Davis and Putnam. These are forms of a generalised factorial,

$$\prod_{k=0}^{x}(a+bk).$$

Heuristically, the interest in the generalised factorial is this: following Gödel in his construction of the β-function, for fixed x we can find a, b so that the numbers $a+bk$ for $k = 1, ..., x$ are pairwise relatively prime. Thus, for any polynomial P and such a, b, x, one has

$$\prod_{k=0}^{x}(a+bk) \mid P \quad \text{iff} \quad \forall k \le x\,[a+bk \mid P].$$

In words, the product has simulated a bounded universal quantifier. Such a simulation in general is the goal of this section. We wish to eliminate the bounded universal quantifier from Davis' representation 2.5 of r.e. relations by exponential means.

2.10. Examples. *The following relations are exponential Diophantine*:

i. $y = \prod_{k=0}^{x}(a+bk)$

ii. $y = \prod_{k=0}^{x}(a-k)$.

Proof. i. (Robinson). This proof is somewhat inspired. Observe that

$$\prod_{k=0}^{x}(a+bk) = b^{x+1}\cdot\prod_{k=0}^{x}(k+a/b) = b^{x+1}(x+1)!\binom{x+a/b}{x+1}$$

where we define

$$\binom{\alpha}{n} = \frac{\alpha(\alpha-1)\ldots(\alpha-n+1)}{n!},$$

for rational values of α. If, as is usually the case, b does not divide a, the generalised binomial coëfficient $\binom{x+a/b}{x+1}$ is not a genuine binomial coëfficient and does not fall under the scope of Example 2.9.i. If, however, b does not divide a in real life, it does so modulo any prime p that doesn't divide a or b. Up to congruence modulo p, the generalised binomial coëfficient $\binom{x+a/b}{x+1}$ will be $\binom{x+z}{x+1}$ for some z.

Let us work out the details:

$$\begin{aligned}\binom{x+z}{x+1}(x+1)!b^{x+1} &= \frac{(z+x)(z+x-1)\ldots(z+1)z}{(x+1)!}(x+1)!b^{x+1} \\ &= (z+x)(z+x-1)\ldots(z+1)zb^{x+1} \\ &= (bz+xb)(bz+(x-1)b)\ldots(bz+b)bz. \qquad (*)\end{aligned}$$

Suppose now that p is a prime number greater than $(a+bx)^{x+1}$. p is relatively prime to b, so there are $s, t \in Z$ such that

$$sb + pt = 1,$$

whence

$$asb + apt = a,$$

i.e.,

$$asb \equiv a \pmod{p}.$$

Choose z between 0 and p such that $z \equiv as \pmod{p}$. Formula (*) then yields

$$\begin{aligned}\binom{x+z}{x+1}(x+1)!b^{x+1} &= \prod_{k=0}^{x}(bz+bk) \\ &\equiv \prod_{k=0}^{x}(a+bk) \pmod{p}.\end{aligned}$$

But $\prod_{k=0}^{x}(a+bk) \le (a+bx)^{x+1} < p$, whence

$$\text{Rem}(\binom{x+z}{x+1}(x+1)!b^{x+1}, p) = \prod_{k=0}^{x}(a+bk).$$

Conversely, suppose there are $p > (a+bx)^{x+1}$ and z such that

$$zb \equiv a \pmod{p}$$

and

$$\text{Rem}(\binom{x+z}{x+1}(x+1)!b^{x+1}, p) = y.$$

Once again, we can use the congruence and the size of p to conclude

$$\mathrm{Rem}\left(\binom{x+z}{x+1}(x+1)!b^{x+1}, p\right) = \prod_{k=0}^{x} (a+bk),$$

whence $y = \prod_{k=0}^{x} (a+bk)$.

This yields the exponential Diophantine definition:

$$y = \prod_{k=0}^{x} (a+bk) \Leftrightarrow \exists pz\, [p > (a+bx)^{x+1} \wedge p \mid bz - a \wedge$$
$$\wedge \mathrm{Rem}\left(\binom{x+z}{x+1}(x+1)!b^{x+1}, p\right) = y].$$

(To be perfectly correct, the right-hand-side ought to be cleaned up a bit using the exponential analogue to Lemma 2.3.)

ii. Observe

$$\prod_{k=0}^{x} (a-k) = a(a-1)\ldots(a-x) = \prod_{k=0}^{x} ((a-x)+k),$$

which is of the form of part i provided $a \ge x$. If $a < x$, however, one of the factors is 0 and the product is 0. Thus:

$$y = \prod_{k=0}^{x} (a-k) \Leftrightarrow (a < x \wedge y = 0) \vee (a \ge x \wedge y = \prod_{k=0}^{x} ((a-x)+k)),$$

$$\Leftrightarrow (a < x \wedge y = 0) \vee \exists z\, [z = a - x \wedge y = \prod_{k=0}^{x} (z+k)]. \qquad \text{QED}$$

Note that in 2.10.ii the coëfficient b was restricted to $b = 1$. The inequality $a < |b|x$ will not yield a 0 factor unless b divides a. Thus, there would be a lot of negative factors in $\prod_{k=0}^{x} ((a-x)+k))$. One could sort these out, but we will only need 2.10.ii in the form stated.

A second remark is that, for the use these products will be put to, more recent work has allowed the products to be bypassed in favour of the binomial coëfficients. However, I find the newer approach (cf. Exercise 9, below) less motivated.

This promised use of the generalised factorials is the following.

2.11. Bounded Quantifier Theorem. *Let $P(A_0, \ldots, A_{m-1}, X, Y, X_0, \ldots, X_{n-1})$ be a polynomial. There is a polynomial $Q(A_0, \ldots, A_{m-1}, X)$ such that, for any $a_0, \ldots, a_{m-1} \in \omega$,*

i. $\forall x\, [Q(a_0, \ldots, a_{m-1}, x) \ge x]$

ii. $\forall x\, \forall y x_0 \ldots x_{n-1} \le x\, [|P(\boldsymbol{a}, x, y, \boldsymbol{x})| \le Q(\boldsymbol{a}, x)]$

iii. *for any x, the following are equivalent:*

a. $\forall y \le x\, \exists x_0 \ldots x_{n-1} \le x\, [P(\boldsymbol{a}, x, y, \boldsymbol{x}) = 0]$

b. $\exists c t v_0 \ldots v_{n-1}\, [t = Q(\boldsymbol{a}, x)! \wedge 1 + (c+1)t = \prod_{m=0}^{x} (1 + (m+1)t) \wedge$

$$\wedge\ 1 + (c+1)t \mid \prod_{j=0}^{x} (v_0 - j) \wedge \ldots \wedge 1 + (c+1)t \mid \prod_{j=0}^{x} (v_{n-1} - j) \wedge$$

$$\wedge\ 1 + (c+1)t \mid P(a_0, \ldots, a_{m-1}, x, c, v_0, \ldots, v_{n-1})].$$

The construction of Q satisfying 2.11.i - ii is something of a red herring and can be isolated as a lemma: one simply replaces each coëfficient in P by its absolute value, substitutes X for each variable Y and X_i, and adds X as a summand to the result; if the final polynomial is called Q, conditions 2.11.i - ii are met.

The heart of the Theorem is the equivalence between the assertions iii.a and iii.b. The idea behind the construction of iii.b is fairly simple. Since $t = Q(\boldsymbol{a}, x)!$ and $Q(\boldsymbol{a}, x) \geq x$, the factors

$$1 + t,\ 1 + 2t,\ ...,\ 1 + (y + 1)t,\ ...,\ 1 + (x + 1)t$$

of $1 + (c + 1)t$ are pairwise relatively prime by Gödel's Lemma (I.7.1) on which the β-function was based. Thus, as remarked prior to proving Example 2.10, the divisibility statements $1 + (c + 1)t \mid R$ act like bounded universal quantifiers,

$$\forall y \leq x\ [1 + (y + 1)t \mid R].$$

Choosing, for each $y \leq x$ a prime divisor p_y of $1 + (y + 1)t$, the statements,

$$p_y \mid \prod_{j=0}^{x} (v_i - j),$$

act like bounded existential quantifiers,

$$\exists j \leq x\ [p_y \mid v_i - j], \quad \text{or: } \exists x_i \leq x\ [p_y \mid v_i - x_i].$$

Thus, iii.b begins to look like

$$\forall y \leq x\ \exists x_0 ... x_{n-1} \leq x\ [p_y \mid P(\boldsymbol{a}, x, c, \boldsymbol{x})].$$

The quantity c will turn out to be replaceable by y and p_y will be so large that $P \equiv 0 \pmod{p_y}$ will force $P = 0$.

Detailed proof of 2.11.iii. Notice first that the conditions

$$t = Q(\boldsymbol{a}, x)!, \qquad 1 + c + 1)t = \prod_{m=0}^{x} (1 + (m + 1)t)$$

determine t and c uniquely. Since $t = Q(\boldsymbol{a}, x)! \geq x!$, the numbers $1 + t, ..., 1 + (y + 1)t$ are pairwise relative prime. Further, for any $y \leq x$, $1 + (y + 1)t$ is relatively prime to $t = Q!$, whence to any number $\leq Q$. Thus,

$$p \text{ prime} \wedge p \mid 1 + (y + 1)t \ \Rightarrow\ p > x \tag{4}$$

and

$$p \text{ prime} \wedge p \mid 1 + (y + 1)t \ \Rightarrow\ p > |P(\boldsymbol{a}, x, y, \boldsymbol{x})|, \tag{5}$$

for all $y, x_0, ..., x_{n-1} \leq x$. (Here, of course, we are using 2.11.i and 2.11.ii, respectively.)

A final preparatory observation is that

$$c \equiv y \pmod{1 + (y + 1)t}. \tag{6}$$

To see this, note that

$$1 + (c + 1)t \equiv 1 + (y + 1)t \pmod{1 + (y + 1)t},$$

whence

$$(c + 1)t \equiv (y + 1)t \pmod{1 + (y + 1)t}.$$

But t and $1+(y+1)t$ are relatively prime, whence

$$ts \equiv 1 \pmod{1+(y+1)t}$$

for some s. Thus,

$$c+1 \equiv (c+1)ts \equiv (y+1)ts \equiv y+1 \pmod{1+(y+1)t},$$

from which (6) follows instantly.

Although neither implication of iii is trivial, the interesting one is iii.b $\Rightarrow$ iii.a, i.e., the verification that the bounded quantifier simulation described above really works. Thus, we begin with this implication.

Assume iii.b, i.e., let $v_0, \ldots, v_{n-1}$ exist satisfying all the necessary conditions. Let $y \leq x$ and let p be a prime divisor of $1+(y+1)t$. Choose

$$x_i = \mathrm{Rem}(v_i, p). \tag{7}$$

Since $p \mid 1+(y+1)t$ and $1+(y+1)t \mid 1+(c+1)t$, the assumptions $1+(c+1)t$ divide $\prod_{j=0}^{x}(v_i - j)$ and $P(\boldsymbol{a}, x, c, \boldsymbol{v})$ yield

$$p \mid \prod_{j=0}^{x}(v_i - j) \tag{8}$$

and
$$p \mid P(\boldsymbol{a}, x, c, \boldsymbol{v}). \tag{9}$$

Since p is prime, (8) yields $v_i \equiv j \pmod{p}$ for some $j \leq x$. With (7), this yields

$$x_i \leq x. \tag{10}$$

Using (6), (7), and (9), we have

$$P(\boldsymbol{a}, x, y, \boldsymbol{x}) \equiv P(\boldsymbol{a}, x, c, \boldsymbol{x}) \equiv P(\boldsymbol{a}, x, c, \boldsymbol{v}) \equiv 0 \pmod{p}.$$

But, by (5), $p > |P(\boldsymbol{a},x,y,\boldsymbol{x})|$, whence p can only divide $P(\boldsymbol{a},x,y,\boldsymbol{x})$ if $P(\boldsymbol{a},x,y,\boldsymbol{x}) = 0$. Thus, for $y \leq x$ we have found, by (10), $x_0, \ldots, x_{n-1} \leq x$ such that $P(\boldsymbol{a},x,y,\boldsymbol{x}) = 0$, i.e., we have derived iii.a.

For the converse, assume iii.a. For each $y \leq x$, choose $x_0(y), \ldots, x_{n-1}(y)$ such that

$$P(\boldsymbol{a}, x, y, x_0(y), \ldots, x_{n-1}(y)) = 0. \tag{11}$$

Since the numbers $1+(y+1)t$ are pairwise relatively prime for $y \leq x$, we can apply the Chinese Remainder Theorem to find $v_0, \ldots, v_{n-1}$ such that

$$v_i \equiv x_i(y) \pmod{1+(y+1)t}, \tag{12}$$

for all $y \leq x$. Using (6), (11), and (12), we get

$$\begin{aligned} P(\boldsymbol{a}, x, c, \boldsymbol{v}) &\equiv P(\boldsymbol{a}, x, y, x_0(y), \ldots, x_{n-1}(y)) \pmod{1+(y+1)t} \\ &\equiv 0 \pmod{1+(y+1)t}. \end{aligned}$$

The moduli being relatively prime, this last yields

$$\prod_{m=0}^{x}(1+(m+1)t) \mid P(\boldsymbol{a}, x, c, \boldsymbol{v}),$$

i.e.,

$$1 + (c + 1)t \mid P(\boldsymbol{a}, x, c, \boldsymbol{v}). \tag{13}$$

Similarly, (12) yields

$$1 + (y + 1)t \mid \prod_{j=0}^{x}(v_i - j),$$

whence again

$$1 + (c + 1)t \mid \prod_{j=0}^{x}(v_i - j). \tag{14}$$

Thus, we have the existence of $v_0, \ldots, v_{n-1}$ satisfying (13) and (14), i.e., we have iii.b. QED

Let $R \subseteq \omega^m$ be an r.e. relation. By Davis' Theorem 2.5, we can write

$$R(\boldsymbol{a}) \Leftrightarrow \exists x \,\forall y \le x \,\exists x_0 \ldots x_{n-1} \le x[P(\boldsymbol{a}, x, y, \boldsymbol{x}) = 0]$$

for some polynomial P. By the Bounded Quantifier Theorem, the right-hand-side of this is equivalent to an exponential Diophantine relation. Thus, we have the following.

2.12. Davis-Putnam-Robinson Theorem. *Every recursively enumerable relation is exponential Diophantine.*

One thing to notice about this proof is that the elimination of the bounded quantifier can*not* be iterated. For, the elimination introduces exponentiation which, unlike addition and multiplication, does not preserve congruence. For example, $5 \equiv 0 \pmod 5$, but

$$2^5 = 32 \equiv 2 \not\equiv 1 = 2^0 \pmod 5.$$

It is thus crucial to the proof of the Davis-Putnam-Robinson Theorem that Davis' normal form 2.5 of an r.e. relation has only one bounded universal quantifier.

Actually, in this last paragraph I should have applied the past tense. As we will see a few sections hence, Juriĭ Matijasevič has proven the Diophantine nature of exponentiation, whence one can obtain the Davis-Putnam-Robinson Theorem from representations of r.e. relations with several bounded universal quantifiers by alternately eliminating the quantifiers and eliminating the exponentiations. Moreover, new techniques allow one to bypass the Bounded Quantifier Theorem— cf. sections 8 and 9, below.

For now, however, the task is to complete the proof of Theorem 2.12 by proving Davis' Theorem 2.5. As I said earlier, this proof properly belongs in the last chapter and the reader might be tempted not to bother with it— especially after reading how more recent results render it ultimately unnecessary. Moreover, the most novel parts of the proof have been incorporated in the proof of Theorem 10.10 on the strict-Σ_1-nature of all Σ_1-relations. Nevertheless, going over the proof affords us a brief review, and another chance to manipulate quantifiers. The reader who is tempted to skip the proof is invited to treat Theorem 2.5 as an exercise.

Proof of Theorem 2.5. Let us first make a small reduction of the problem. Let a relation of the form

$$\exists x \,\forall y \le x \,\exists x_0 \ldots x_{n-1} \,[P(\boldsymbol{a}, x, y, \boldsymbol{x}) = 0] \tag{*}$$

be given, where P is again an ordinary polynomial. Whenever this relation holds, there is some x such that, for any $y \le x$, we can find $x_0(y), \ldots, x_{n-1}(y)$ making the polynomial vanish. Choose

$$z = \max \{x_i(y)\colon i \le n-1 \wedge y \le x\},$$

for some fixed choice of the $x_i(y)$'s. This new number z bounds the x_i's and allows us to transform (*) into

$$\exists x\, \exists z\, \forall y \le x\, \exists x_0 \ldots x_{n-1} \le z\, [P(\boldsymbol{a}, x, y, \boldsymbol{x}) = 0]. \tag{15}$$

Contracting x and z by using the pairing function transforms this into

$$\exists w\, \forall y \le w\, \exists x_0 \ldots x_{n-1} x z \le w\, [w = \langle x, z\rangle \wedge (y > x \vee P = 0)]. \tag{16}$$

There is still one hidden quantifier in this. Eliminating it yields

$$\exists w\, \forall y \le w\, \exists x_0 \ldots x_{n-1} x z r \le w\, [w = \langle x, z\rangle \wedge (y = x + r + 1 \vee P = 0)].$$

The closure of the class of polynomial relations under conjunction and disjunction transforms this last, whence (*), into the desired Davis Normal Form,

$$\exists w\, \forall y \le w\, \exists y_0 \ldots y_{k-1} \le w\, [Q(\boldsymbol{a}, w, y, \boldsymbol{y}) = 0].$$

Thus, it suffices to show that every recursively enumerable relation can be written in the form,

$$\exists x\, \forall y \le x\, D(\boldsymbol{a}, x, y), \tag{**}$$

where D is a Diophantine relation. We do this by showing the class of all relations of the form (**) to contain the polynomial relations and to be closed under conjunction, disjunction, existential quantification, and bounded universal quantification— the closure properties defining the class of Σ_1-relations (which class is coëxtensive with the class of r.e. relations).

A polynomial relation,

$$R(\boldsymbol{a})\colon\ P(\boldsymbol{a}) = 0,$$

can trivially be put into form (**) by padding the description with dummy variables. E.g., we can write

$$R(\boldsymbol{a})\colon\ \exists x\, \forall y \le x\, [P(\boldsymbol{a}) \cdot (x + y + 1) = 0].$$

Closure under existential quantification is established by the trick used in passing from (15) to (16) above:

$$\exists z\, \exists x\, \forall y \le x\, D \iff \exists w\, \forall y \le w\, \exists x z\, [(w = \langle x, z\rangle \wedge D) \vee y > x].$$

Closure under bounded universal quantification requires, as in the proof of I.10.10, the β-function. $\forall w \le z\, \exists x\, \forall y \le x\, D(\boldsymbol{a}, x, y, z, w)$ is equivalent to

$$\exists c d v\, \forall w \le z\, \forall y \le v\, \exists x \le v\, [(x = \beta(c, d, w) \wedge D(\boldsymbol{a}, x, y, z, w)) \vee y > x].$$

Since we can contract existential quantifiers, we see that we have reduced this case to the problem of showing every relation,

$$\exists x\, \forall y \le z\, \forall w \le v\, D(\boldsymbol{a}, x, y, z, w, v),$$

with D Diophantine, to be of the form (**). Here, one simply uses the pairing function and its inverses to obtain the equivalent:

$$\exists x\, \exists u\, \forall t \le u\, [u = \langle 0, z+v+1\rangle \wedge$$
$$\wedge\ (\pi_1^2(t) > z\ \vee\ \pi_2^2(t) > v\ \vee\ D(\boldsymbol{a}, x, \pi_1^2(t), z, \pi_2^2(t), v))],$$

which is put, by an existential quantifier contraction, into the desired form (**). To explain the choice of u, notice that, for $y \le z$ and $w \le v$, the pair (y, w) lies on the line $X + Y = y + w \le z + v < z + v + 1$. Thus, (y, w) gets enumerated by the inverse of the pairing function before any point on the line $X + Y = z + v + 1$, whence $\langle y, w\rangle < \langle 0, z+v+1\rangle$. [Cf. I.3.4.ff.]

Closure under conjunction and disjunction is now easy. Consider, e.g., conjunction. Observe,

$$\exists x\, \forall y \le x\, D_1 \wedge \exists u\, \forall v \le u\, D_2 \iff \exists x\, \exists u\, [\forall y \le x\, D_1 \wedge \forall v \le u\, D_2]$$
$$\iff \exists x\, \exists u\, \forall y \le x\, \forall v \le u\, [D_1 \wedge D_2].$$

The closure of the class of Diophantine relations under conjunction, along with the closure results already established for relations of the form (**), yield the result. QED

Exercises.

1. Let $P_0, Q_0, \ldots, P_{n-1}, Q_{n-1}$ be "exponential polynomials" (for your favourite interpretation of the phrase).
   i. Find P, Q also "exponential polynomials" such that
   $$P_0 = Q_0 \wedge \ldots \wedge P_{n-1} = Q_{n-1} \iff P = Q.$$
   ii. Do the same for disjunction.
2. Define the depth, $d(P)$, of an iterated exponential polynomial P by induction as follows:
   i. if P is a variable or a natural number, then $d(P) = 1$
   ii. $d(P + Q) = d(P \cdot Q) = \max\{d(P), d(Q)\}$
   iii. $d(P^Q) = d(P) + d(Q)$.
   Define the depth of an equation $P = Q$ to be the maximum of the depths of P and Q. Show by induction on this depth: Every iterated exponential Diophantine relation is an exponential Diophantine relation.
3. i. Show: For $y > 1$, $\mathrm{Rem}(2^{xy^2}, 2^{xy} - x) = x^y$.
   ii. Show: Every exponential Diophantine relation is a dyadic exponential Diophantine relation.
4. Show directly that the following relations are exponential Diophantine:
   i. $x \equiv y \pmod z$
   ii. $x \ne y$
   iii. $x \nmid y$ (x does not divide y)
   iv. $y \in D_x$
   v. $y \notin D_x$.
5. In the proof of 2.9.ii, we saw that one could define
   $$z \text{ is very large: } z > (2x)^{x+1}.$$

i. Show by induction on $n > 0$: Either $n + 1 > x$ or $(1 + \theta/2)^n < 1 + n\theta$, where $0 < \theta < 1/x$.

ii. Solve the inequality $1 + x\theta < 1 + 1/x^x$ for θ.

iii. Combining i and ii, show that we can define

$$z \text{ is very large: } z > 2x^{x+2}.$$

6. Show: If $z > (6x)^x$ and $y > z^x$, then

$$x! = \left[\frac{(z+1)^x y^x}{\mathrm{Rem}((y+1)^z, yy^x)}\right].$$

Check what kind of numbers arise for $x = 0, 1, 2$.
[Remark. The above definition, by avoiding the binomial coëfficients, yields a savings on variables in an exponential Diophantine (and, ultimately, a Diophantine) definition of the factorial.]

7. Let $p > 1$.

i. Show: p is prime iff p and $(p-1)!$ are relatively prime.

ii. Prove Wilson's Theorem: p is prime iff $(p-1)! \equiv -1 \pmod{p}$.

iii. (2.9.iii). Show twice that the set of primes is exponential Diophantine by using parts i and ii of this exercise. What is the difference between the two definitions?

8. Let a, b be positive integers. Show: The relation,

$$1 + y \mid \prod_{k=0}^{x} (a - bk)$$

is exponential Diophantine. [Hint. Replace a, if necessary, by something large enough to guarantee the product to have no negative factors.]

9. Let $P(A_0, ..., A_{m-1}, X, Y, Z, X_0, ..., X_{n-1})$ be a polynomial.

i. Construct a polynomial $Q(A_0, ..., A_{m-1}, X, Y)$ such that, for any $a_0, ..., a_{m-1}, x, y$,

$$\forall z \le x\, \forall x_0...x_{n-1} \le y\, [Q(\boldsymbol{a}, x, y) > |P(\boldsymbol{a}, x, y, z, \boldsymbol{x})| + 2x + y + 1]$$

ii. Let $t = Q(\boldsymbol{a}, x, y)!$ and observe that

$$\binom{t-1}{x+1} = (t-1)\left(\frac{t}{2} - 1\right)\cdots\left(\frac{t}{x+1} - 1\right).$$

Show that the factors exhibited are pairwise relatively prime integers.

iii. Prove the equivalence of the following:

a. $\forall z \le x\, \exists x_0...x_{n-1} \le y\, [P(\boldsymbol{a}, x, y, z, \boldsymbol{x}) = 0]$

b. $\exists b_0...b_{n-1}$

$$\binom{b_0}{y+1} \equiv \ldots \equiv \binom{b_{n-1}}{y+1} \equiv P(\boldsymbol{a}, x, y, t-1, \boldsymbol{b}) \equiv 0 \pmod{\binom{t-1}{y+1}},$$

where $t = Q(\boldsymbol{a}, x, y)!$.

[The proof of the Bounded Quantifier Theorem given by this Exercise avoids the generalised factorial and thereby reduces the overall number of variables needed for the algorithmic unsolvability of exponential Diophantine equations. The modest refinement in allowing different bounds on the two types of quantifiers can be obtained by the original method as well. Further refinement is possible: each existential quantifier can have its own upper bound.]

3. The Pell Equation, I

For any positive integer d that is not a perfect square, the equation

$$X^2 - dY^2 = 1, \qquad \text{(P)}$$

is called a *Pell equation* (after John Pell, an English mathematician who had nothing to do with the equation). The slightly more general equation,

$$X^2 - dY^2 = \pm k,$$

for $k \neq 0$, is also called a Pell equation and we will have to consider it at some point; but it is the more restricted form (P) that will really interest us. Looking at (P) and recognising that it defines an hyperbola when X and Y are treated as real variables may make such an interest seem unlikely; but, in the integers, (P) is a truly fascinating equation and it has had a long history.

One can begin the story of the Pell equation with Archimedes in the third century B.C., but Archimedes didn't really deal with the equation and the problem he posed, which leads to an instance of the Pell equation, really had no effect on later developments. This is as much as need be said about Archimedes and the Pell equation. Diophantus, some of whose problems gave rise to instances of the Pell equation, can similarly be lightly skipped over in this context. The contributions of the Hindu mathematicians cannot be ignored. Already in the 7th century A.D., Brahmagupta recognised that solutions (other than the trivial one $x = 1, y = 0$) to a Pell equation can be composed to yield new solutions; and by the 12th century, a general and quite effective method of finding solutions was known to Hindu mathematicians. What was lacking, however, was a proof that the method always provided a solution. The proof came somewhat later, in a Western mathematical culture that had to rediscover everything for itself and did not learn of the Hindu accomplishments until the early part of the 19th century.

I might mention quickly, before going on to discussing the Western development, that a moderately efficient solution to (P) yields a moderately efficient approximation x/y to $\sqrt{d}$, as

$$\frac{x^2}{y^2} - d = \frac{1}{y^2} \rightarrow \frac{x}{y} = \sqrt{d} + \frac{\theta}{y}, \text{ for some } 0 < \theta < 1.$$

Moreover, Brahmagupta's method of composing solutions allowed one to improve repeatedly on the rational approximation x/y to $\sqrt{d}$. The existence proof we will give begins a non-computational exploration of how good such an approximation can be.

The Western study of the Pell equation (P) begins with Pierre de Fermat. As mentioned earlier, Fermat tried unsuccessfully to interest his colleagues in number theory. One of his partial successes was his challenge to the English to solve (P). Unfortunately, the two main targets of Fermat's challenge— John Wallis and Lord Brouncker— got a miscopied statement of the challenge.

It must be remembered that we are discussing events of the middle 17th century. There was not a regular postal system, nor any typewriters with carbon paper, nor photo-copiers, nor electronic mail. Letters were laboriously written by hand and transmitted by intermediaries. Nonetheless, this was a century of great intellectual activity in Northern Europe and a lot of letters were written and slowly transmitted. Father Marin Mersenne, who organised the Paris Academy in 1635, kept up a vast correspondence with many scientists— including Fermat— during the second quarter of the century; and, later, when the English organised their Royal Society, its secretary Henry Oldenburg became the conduit for the transmission of information between England and the continent.

The letter of Fermat's challenge went by way of Sir Kenelm Digby, an incidental figure in the history of mathematics, whose secretary omitted the requirement that the solutions to (P) be integral from the copy of Fermat's letter he made to pass on to Wallis and Brouncker. When Brouncker determined all rational solutions, Fermat was unimpressed and declared that he had wanted integral solutions. The English declared Fermat had changed the problem, but solved it anyway— that is to say that Brouncker devised, like the Hindus some centuries earlier, a technique for finding the integral solutions, but he provided no proof that the technique worked.

Before finding integral solutions, however, it will be instructive to compare the Pell equation (P) and its rational solutions to the Pythagorean equation,

$$X^2 + Y^2 = Z^2, \tag{1}$$

and its rational solutions.

We begin with the Pythagorean equation. The solution $x = y = z = 0$ is trivial and can be ignored. That being the case, and our being momentarily interested in rational solutions, we can divide by the variable Z to get

$$(X/Z)^2 + (Y/Z)^2 = 1,$$

which, on relabelling, is the equation of the unit circle,

$$X^2 + Y^2 = 1. \tag{2}$$

Now, the rational points on the unit circle can be parametrised in various ways. One way is to begin with the point $(-1,0)$ on the circle and draw the chord from this point to some other point (x,y) on the circle as in Figure 3.1:

3.1. Figure.

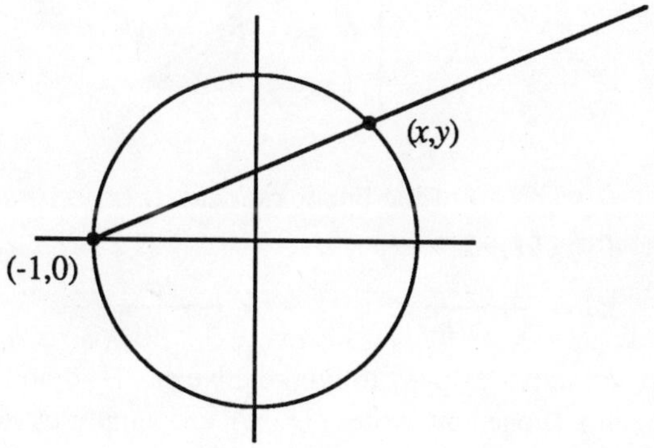

The equation of the line so drawn is $Y - 0 = t(X + 1)$, where t is the slope of the line, $t = y/(x+1)$. Obviously, if x, y are rational, then so is t. Less obvious, but also true, is the converse: if t is rational, so are x, y: a little algebra yields,

$$x = \frac{1 - t^2}{1 + t^2}, \qquad y = \frac{2t}{1 + t^2}. \tag{3}$$

In this way, the rational points (x, y) on the unit circle are rationally parametrised by (3) for $-\infty < t \leq +\infty$ ($t = +\infty$ being added to yield $(-1,0)$).

If one now sets $t = m/n$ in (3) and simplifies, one gets

$$x = \frac{n^2 - m^2}{n^2 + m^2}, \qquad y = \frac{2mn}{n^2 + m^2}. \tag{4}$$

Reïntroducing Z, we can let z be the common denominator in (4) and get

$$x = n^2 - m^2, \qquad y = 2mn, \qquad z = n^2 + m^2 \tag{5}$$

as solutions to the original Pythagorean equation (1). If m, n are relatively prime, these are called the *primitive* solutions to the Pythagorean equation and one can show that all positive integral solutions to (1) are of the form,

$$x = (n^2 - m^2)k, \quad y = 2mnk, \quad z = (n^2 + m^2)k,$$

where m, n, k are positive, $n > m$, and m, n are relatively prime.

With the Pell equation (P), we have, not a circle, but an hyperbola with asymptotes $Y = \pm(1/\sqrt{d})X$ and x-intercepts $(\pm 1, 0)$. We can parametrise the rational points on the right branch by drawing the line connecting $(-1, 0)$ on the left branch to an arbitrary point (x, y) on the right branch as in Figure 3.2.

3.2. Figure.

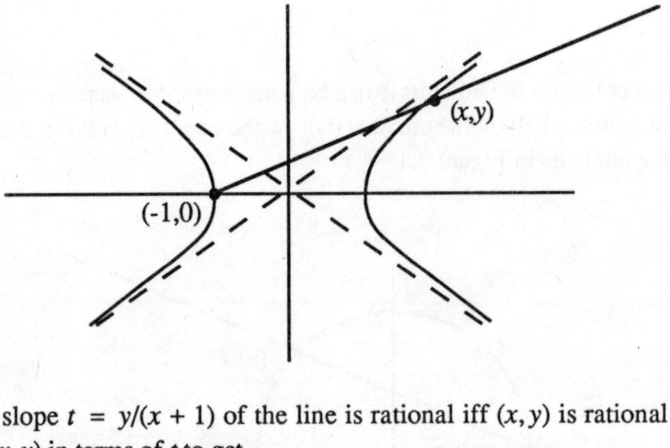

Again, the slope $t = y/(x+1)$ of the line is rational iff (x, y) is rational and once can solve (P) for (x, y) in terms of t to get

$$x = \frac{1 + dt^2}{1 - dt^2}, \qquad y = \frac{2t}{1 - dt^2}. \tag{6}$$

(For x to be positive, we must restrict t to lying between $-1/\sqrt{d}$ and $1/\sqrt{d}$. But let us ignore this consideration.) If one now writes $t = m/n$ and simplifies (6), one gets

$$x = \frac{n^2 + dm^2}{n^2 - dm^2}, \qquad y = \frac{2mn}{n^2 - dm^2}. \tag{7}$$

Unfortunately, we did not begin with a simple

$$X^2 - dY^2 = Z^2,$$

and have nothing to complete the proportion (4) : (5) : : (7) : ? Formula (7) is as far as the analogy will take us.

Formula (7) gives all rational solutions, and the integral ones must lie among them. So why wasn't Fermat satisfied with such a solution? The answer is close at hand. Suppose we want to use (7) to find integral solutions (x, y) other than the trivial $(1, 0)$. Without loss of generality, we can assume m, n are relatively prime. The denominator $n^2 - dm^2$ must divide the greatest common divisor of the numerators $2mn$ and $n^2 + dm^2$, which common divisor must divide 2 (exercise). Thus, we must have

$$n^2 - dm^2 = \pm 1 \quad \text{or} \quad n^2 - dm^2 = \pm 2.$$

(The choices -1 and -2 give us negative values of x and y, but taking absolute values will rectify this.) But all this means is that, in order to apply (7) to find all integral solutions to (P), we must find all such solutions to (P) *and* the extra equations

$$X^2 - dY^2 = 2,$$

etc. In short, for finding the integral solutions, Brouncker's method was useless.

Ultimately, as already noted, Brouncker devised a feasible method for finding non-trivial solutions to (P). However, he did not prove that his method always worked. One still needed a proof that, whenever d was not a square, the equation (P) had a natural numerical solution (x, y) other than the trivial $(1, 0)$. It was Joseph Louis Lagrange who first proved the general existence of solutions to (P). Lagrange's proof is quite constructive; it provides a feasible algorithm for finding the solutions and demonstrates that this algorithm works. One could scarcely ask for more and it may seem perverse to present a different proof. However, for the ostensibly negative purposes of the present chapter, we will only need to actually find solutions for a limited class of Pell equations, where d is of a special form; for the general equation, we will only need to know the existence of such solutions. The task of finding non-trivial solutions in the special case will be easy and is the starting point for section 4, below. (The curious reader can find several such easily solvable special Pell equations in the exercises to the present section. Indeed, he may want to solve Exercise 2 to acquire some concrete familiarity with the Pell equation before continuing on to the general case.) As for the task of proving existence, a shorter proof than Lagrange's exists. This proof, due to Gustav Peter Lejeune Dirichlet, is in part a combinatorial counting argument akin to, but more complicated than, that used in proving the Chinese Remainder Theorem with bounds (I.6.5), and it serves an an introduction to the combinatorial work in Chapter VI of volume II.

A first lemma towards Dirichlet's proof of the existence of non-trivial solutions to (P) is the following modern formulation of Brahmagupta's method of composition.

3.3. Lemma. *Let d be a positive integer that is not a perfect square. The set of real numbers of the form,*

$$x \pm y\sqrt{d},$$

where x, y form a non-negative integral solution (possibly trivial) to the Pell equation,

$$X^2 - dY^2 = 1,$$

form a subgroup of the multiplicative group of positive real numbers, i.e.,
i. *the product of two numbers in this set is in this set; and*
ii. *the multiplicative inverse of a number in this set is in the set.*

Proof. The factorisation,

$$1 + 0\sqrt{d} = 1 = x^2 - dy^2 = (x + y\sqrt{d})(x - \sqrt{d})$$

immediately yields ii. Assertion i follows from this factorisation and the identities,

$$(x_1 + y_1\sqrt{d})(x_2 + y_2\sqrt{d}) = (x_1x_2 + y_1y_2d) + (x_1y_2 + x_2y_1)\sqrt{d}$$
$$(x_1 - y_1\sqrt{d})(x_2 - y_2\sqrt{d}) = (x_1x_2 + y_1y_2d) - (x_1y_2 + x_2y_1)\sqrt{d}$$

I leave it to the reader to work out the details. QED

3.4. Corollary. *Let d be a positive integer that is not a perfect square. If the Pell equation,*

$$X^2 - dY^2 = 1, \qquad \text{(P)}$$

has a non-trivial solution, then it has infinitely many. In fact, there are x_1, y_1 *such that the non-negative solutions to this equation are exactly the pairs* (x_n, y_n) *given by*

$$x_n + y_n\sqrt{d} = (x_1 + y_1\sqrt{d}\,)^n, \qquad (*)$$

for $n \in \omega$.

Proof. Since $x^2 = 1 + dy^2$, for any solution (x, y) to (P), it follows that x grows with y. Thus, if there is any non-trivial solution at all, there is a minimum such given by the minimum $x_1 > 1, y_1 > 0$ satisfying (P). If we let $\delta = x_1 + y_1\sqrt{d}$, then

$$1 = \delta^0 < \delta < \delta^2 < \ldots$$

determines the infinite sequence (*) of distinct solutions to (P).

To see that these exhaust the non-negative integral solutions, assume that (x, y) is a solution of (P). By the Archimedean Property of the real line (cf. Exercise 8), there is a natural number n such that

$$\delta^n \le x + y\sqrt{d} < \delta^{n+1},$$

i.e.,

$$x_n + y_n\sqrt{d} \le x + y\sqrt{d} < x_{n+1} + y_{n+1}\sqrt{d}\,.$$

Multiplying by $x_n - y_n\sqrt{d}$ yields

$$1 \le (x + y\sqrt{d}\,)(x_n - y_n\sqrt{d}\,) < x_1 + y_1\sqrt{d}\,.$$

By the minimality of $x_1 + y_1\sqrt{d}$ and the fact that the product yields a solution to (P), it follows that

$$1 = (x + y\sqrt{d})(x_n - y_n\sqrt{d}).$$

Multiplying by $x_n + y_n\sqrt{d}$ yields

$$x + y\sqrt{d} = x_n + y_n\sqrt{d}.$$

QED

The next lemma might not appear to be related to the problem at hand. Its relevance is explained by recalling that, if (x, y) is a non-trivial solution to (P), then x/y is a good approximation to $\sqrt{d}$: for some $0 < \theta < 1$,

$$x/y = \sqrt{d} + \theta/y.$$

3.5. Lemma. (Rational Approximation to Irrationals). *Let $\xi > 0$ be an irrational number. There are infinitely many rational numbers x/y (with x, y relatively prime) such that*

$$0 < |\xi - x/y| < 1/y^2.$$

The proof is an application of Dirichlet's Pigeon-Hole Principle, or *Schubfachprinzip*, which is also used later on in the existence proof for the Pell equation. As, moreover, it will be generalised in Chapter VI of volume II, I shall make a formal declaration of it here even though the reader has probably seen it before.

3.6. Pigeon-Hole Principle. *Let X, Y be non-empty finite sets with cardinalities m, n, respectively. Suppose $m > n$ and F is a function mapping X to Y. Then there are $x_1, x_2 \in X$ with $x_1 \neq x_2$ and such that $F(x_1) = F(x_2)$.*

Proof. Is this a theorem to be proven or a Law of Thought? Most textbooks treat it like the latter and, like my using a counting argument to prove the Chinese Remainder Theorem (I.6.5), use the combinatorial result without supplying a proof. But for later chapters, I'd do the same here. (Have I used the Pigeon-Hole Principle without mention already?) However, I want to make (and stress) the point that the proof is an induction on n.

Let $A(n)$ be the assertion that no F mapping a set of cardinality greater than n to a set of cardinality n can be one-one. We show $\forall n > 0\, A(n)$ by induction on n.

Basis. $n = 1$. Any $F{:}X \to Y$ for a one-element set Y must be constant.

Induction step. Let $F{:}X \to Y$, card $(X) = m > n + 1 = \text{card}(Y)$. Choose $y_0 \in Y$. Either $F^{-1}(\{y_0\})$ has cardinality at least 2 and we are done, or $F^{-1}(\{y_0\}) = \{x_0\}$ for some x_0. In this case, the restriction of F to $X - \{x_0\}$ maps into $Y - \{y_0\}$ and the induction hypothesis yields the result.

QED

Note that the proof did not really require X to be finite.

Getting back on track, let us prove Lemma 3.5.

Proof of Lemma 3.5. We first prove that, for any integer $n > 1$, there are natural numbers x, y with

$$0 < y \leq n \tag{8}$$

such that

$$0 < |\xi - x/y| < 1/(ny). \tag{9}$$

Consider the $n+1$ numbers,

$$\{0\xi\},\ \{1\xi\},\ ...,\ \{n\xi\},$$

where $\{\alpha\} = \alpha - [\alpha]$ is the *fractional part* of α. Note first that these numbers are pairwise distinct. (If $\{i\xi\} = \{j\xi\}$, i.e., if $i\xi - [i\xi] = j\xi - [j\xi]$, then, unless $i = j$, we have $\xi = ([i\xi] - [j\xi])/(i-j)$, which is rational, contrary to the assumption on ξ.) Thus, if we partition the half-closed, half-open unit interval into n equal subintervals,

$$[0,\ 1/n),\quad [1/n,\ 2/n),\quad ...,\quad [(n\ -1)/n,\ 1),$$

and map $\{i\xi\}$ into the subinterval containing it, we see that two distinct fractional parts, say $\{i\xi\}$ and $\{j\xi\}$ with $i > j$, are mapped into the same subinterval:

$$0 < |\{i\xi\} - \{j\xi\}| < 1/n,$$

i.e.,

$$0 < |(i-j)\xi - ([i\xi] - [j\xi])| < 1/n.$$

Letting $y = i - j$ and $x = [i\xi] - [j\xi]$, we have

$$0 < |y\xi - x| < 1/n,$$

whence follows (9):

$$0 < |\xi - x/y| < 1/(ny).$$

Moreover, noting that $0 < y = i - j \leq i \leq n$, we see that (8) also holds.

To complete the proof, note that (8) and (9) immediately yield

$$0 < |\xi - x/y| < 1/(ny) < 1/y^2.$$

Reducing x/y to lowest terms, say x'/y', will only weaken the inequality:

$$0 < |\xi - x'/y'| = |\xi - x/y| < 1/y^2 \leq 1/(y')^2.$$

Thus, we have proven the existence of relatively prime natural numbers x, y satisfying the approximation required by the Lemma. To see that there are infinitely many such approximations, let $n \to +\infty$: for any given x, y, there are only finitely many n for which (9) is satisfied. QED

We now have all we need for proving the main result of this section.

3.7. Theorem. *Let d be a positive integer that is not a perfect square. The Pell equation,*

$$X^2 - dY^2 = 1,$$

has a non-trivial solution.

Proof. First, we apply Lemma 3.5 to show that, for some integer k with $|k| < 1 + 2\sqrt{d}$, the auxiliary equation,

$$X^2 - dY^2 = k, \tag{10}$$

has infinitely many solutions.

Let x, y satisfy the approximation,

$$|\sqrt{d} - x/y| < 1/y^2,$$

i.e.,

$$|y\sqrt{d}-x| < 1/y.$$

Note

$$\begin{aligned} |x^2-dy^2| &= |x-y\sqrt{d}|\cdot|x+y\sqrt{d}| \\ &= |x-y\sqrt{d}|\cdot|x-y\sqrt{d}+2y\sqrt{d}| \\ &< \frac{1}{y}\left(\frac{1}{y}+2y\sqrt{d}\right) \\ &< 1+2\sqrt{d}\,. \end{aligned}$$

Since this inequality holds for infinitely many x, y, some value k with $|k| < 1+2\sqrt{d}$ must be repeated inside the absolute value signs on the left infinitely often, i.e., equation (10) has infinitely many solutions. (Cf. Exercise 9, below.)

Lemma 3.3 now comes to mind. If multiplying solutions of (P) gives a solution of (P), then multiplying a solution of (P) by a solution of (10) ought to give a solution of (10). This is indeed the case (cf. Exercise 6, below). In the opposite direction, so to speak, this suggests that dividing two solutions to (10) will give us a solution to (P). It does, but—alas— the solution need not be integral. For example, the pairs $(83, 23)$ and $(5, 1)$ both satisfy

$$X^2-13Y^2 = 12.$$

But

$$\frac{83+23\sqrt{13}}{5+\sqrt{13}} = \frac{83+23\sqrt{13}}{5+\sqrt{13}}\cdot\frac{5-\sqrt{13}}{5-\sqrt{13}} = \frac{116+32\sqrt{13}}{12} = \frac{29}{3}+\frac{8}{3}\sqrt{13}$$

yields the rational, but non-integral, solution $(29/3,\ 8/3)$.

However, the situation is not hopeless. Let (x_1,y_1) and (x_2,y_2) be two solutions of the auxiliary equation (10) and consider

$$\begin{aligned} \frac{x_1+y_1\sqrt{d}}{x_2+y_2\sqrt{d}} &= \frac{x_1+y_1\sqrt{d}}{x_2+y_2\sqrt{d}}\cdot\frac{x_2-y_2\sqrt{d}}{x_2-y_2\sqrt{d}} \\ &= \frac{(x_1x_2-y_1y_2d)-(x_1y_2-x_2y_1)\sqrt{d}}{k} \\ &= \frac{x_1x_2-y_1y_2d}{k}-\frac{x_1y_2-x_2y_1}{k}\sqrt{d}\,. \end{aligned} \qquad (11)$$

Letting $x+y\sqrt{d}$ denote the right-hand-side of (11), we readily see that

$$\begin{aligned} k^2(x^2-dy^2) &= (kx)^2-d(ky)^2 \\ &= (x_1x_2-y_1y_2d)^2-d(x_1y_2-x_2y_1)^2 \\ &= x_1^2x_2^2+y_1^2y_2^2d^2-dx_1^2y_2^2-dx_2^2y_1^2 \\ &= x_1^2(x_2^2-dy_2^2)+dy_1^2(dy_2^2-x_2^2) \end{aligned}$$

$$= x_1^2k - dy_1^2k = k(x_1^2 - dy_1^2) = k^2,$$

whence

$$x^2 - dy^2 = 1,$$

and (x, y) is, as stated, a solution to (P). The problem is to show that (x_1, y_1) and (x_2, y_2) can be chosen so that

$$x = \frac{|x_1x_2 - y_1y_2d|}{|k|}, \qquad y = \frac{|x_1y_2 - x_2y_1|}{|k|} \tag{12}$$

are integral and $x \neq 1$ and $y \neq 0$. The Pigeon-Hole Principle will allow us to make such a choice.

Partition the solutions (x, y) to the auxiliary equation (10) by defining the equivalence relation,

$$(x_1, y_1) \sim (x_2, y_2) \text{ iff } x_1 \equiv x_2 \pmod{k} \wedge y_1 \equiv y_2 \pmod{k}.$$

By the infinitude of the set of such pairs and the Pigeon-Hole Principle, some class has at least two distinct members, say (x_1, y_1) and (x_2, y_2). For such solutions, we have

$$x_1x_2 - y_1y_2d \equiv x_1^2 - dy_2^2 \equiv 0 \pmod{k},$$

whence x is integral. Further,

$$x_1y_2 - x_2y_1 \equiv x_1y_1 - x_1y_1 \equiv 0 \pmod{k},$$

whence y is integral.

To see that (x, y) is not the trivial solution $(1, 0)$, suppose by way of contradiction that it were trivial: $x = 1, y = 0$. Then

$$x_1x_2 - y_1y_2d = \pm k, \qquad x_1y_2 - x_2y_1 = 0.$$

Thus,

$$\pm kx_1 = x_1^2x_2 - dy_1x_1y_2 = x_1^2x_2 - dx_2y_1^2 = x_2(x_1^2 - dy_2^2) = x_2k,$$

whence $x_1 = \pm x_2$. Since x_1, x_2 are both positive, this means $x_1 = x_2$, from which it follows that $y_1 = y_2$, contradicting the assumption that (x_1, y_1) and (x_2, y_2) were distinct solutions to (10). QED

Some proofs just cry out to be commented on. That of Theorem 3.7 is one of these. I have several comments to make concerning the proof.

The use of the Pigeon-Hole Principle was necessary for the proof of Theorem 3.7; isolating it as a lemma and supplying it with a proof of its own was not mathematically necessary at this level. Because Dirichlet first formulated the Principle (in 1842) in precisely this context, emphasising the Principle here would seem to be an historical necessity. In any event, the proof of 3.7 affords the most natural place in the book to insert the Pigeon-Hole Principle.

Dirichlet's lemma on the approximation of irrationals by rationals was the first of several results on the approximability or non-approximability of irrationals by rationals. Joseph Liouville's later (1844) negative result (cf. Exercise 11) begins the subject of transcendental number theory.

Getting down to Theorem 3.7 itself, it must be admitted that the proof given offers little computational help. For a feasible method, Lagrange's proof relating the Pell equation (P) to the continued fraction expansion of $\sqrt{d}$ is far better. The curious reader will find references in the reading list at the end of the chapter.

The composition of solutions (Lemma 3.3) can be de-mystified at the cost of some abstraction. For any non-square integer d (positive or negative), one can define

$$Q(\sqrt{d}) = \{\alpha + \beta\sqrt{d} : \alpha, \beta \in Q\}.$$

This is a perfectly good *number field*, much like the rationals or the reals. If d is negative, it is more like the field of complex numbers; if $d = -1$, it is the field of *complex rationals*. In the complex numbers, we have two important concepts— conjugation and absolute value:

$$\overline{\alpha + \beta\sqrt{-1}} = \alpha - \beta\sqrt{-1}$$

$$|\alpha + \beta\sqrt{-1}\,| = \sqrt{\alpha^2 + \beta^2}.$$

For complex numbers z_1, z_2, z, we have

$$\overline{z_1 z_2} = \overline{z_1} \cdot \overline{z_2}, \qquad |z| = \sqrt{z\,\overline{z}}, \qquad |z_1 z_2| = |z_1|\cdot|z_2|.$$

This can be generalised to $Q(\sqrt{d})$:

$$\overline{\alpha + \beta\sqrt{d}} = \alpha - \beta\sqrt{d}$$

$$N(\alpha + \beta\sqrt{d}) = (\alpha + \beta\sqrt{d})(\alpha - \beta\sqrt{d}) = \alpha^2 - d\beta^2.$$

If d is negative, we can take the square root of $N(\alpha + \beta\sqrt{d})$ and obtain a true absolute value. In the case we are interested in, however, $N(\alpha + \beta\sqrt{d})$ could be negative. It doesn't matter; the relevant thing is that conjugation and the signed *norm* $N(\cdot)$ commute with multiplication: Let $\theta_1, \theta_2 \in Q(\sqrt{d})$. Then,

$$\overline{\theta_1\theta_2} = \overline{\theta_1} \cdot \overline{\theta_2} \tag{13}$$

$$N(\theta_1\theta_2) = N(\theta_1)\cdot N(\theta_2). \tag{14}$$

Once it is proven, identity (14) yields Lemma 3.3. Moreover, from (14) also follows

$$N(\theta_1/\theta_2) = N(\theta_1)/N(\theta_2) \tag{15}$$

which was used in the proof of Theorem 3.7.

The abstract setting of the last paragraph is by no means the limit of abstraction relevant to Theorem 3.7. One can generalise the construction of $Q(\sqrt{d})$ to $Q(\theta)$ for any algebraic number θ, define a set of conjugates for any $\eta \in Q(\theta)$, and define $N(\eta)$ to be the product of η with its conjugates. Inside such a number field, one can define the appropriate set of *integers* and solve the *norm equation*, $N(\eta) = 1$ in "integers". Dirichlet's Unit Theorem describes completely the structure of the group of these solutions.

I cannot go into any detail at all on these matters. I mention them simply as a heuristic, or an indication that yet higher mathematics provides a heuristic that lends an air of naturalness to the proof given of the existence of solutions to Pell's equation (P).

What I wish to do for the remainder of this section is to present a proof of Lagrange's Theorem on the representability of every natural number as the sum of four squares. The result doesn't have much to do with the Pell equation itself, but the proof does superficially parallel that of Theorem 3.7.

3.8. Lagrange's Theorem. *Every natural number is the sum of four squares.*

The first lemma toward proving this theorem is an analogue to Lemma 3.3.

3.9. Euler's Identity.

$$\left(x_1^2 + x_2^2 + x_3^2 + x_4^2\right)\left(y_1^2 + y_2^2 + y_3^2 + y_4^2\right) =$$

$$= (x_1y_1 + x_2y_2 + x_3y_3 + x_4y_4)^2 + (x_1y_2 - x_2y_1 + x_3y_4 - x_4y_3)^2 +$$

$$+ (x_1y_3 - x_3y_1 + x_4y_2 - x_2y_4)^2 + (x_1y_4 - x_4y_1 + x_2y_3 - x_3y_2)^2.$$

There are some pretty symmetries in the four summands which could serve as aids to the memory, but one wonders how Euler could have dreamt such a thing up other than by crude analogy with the complex identity cited below. About a century after Euler proved this identity, in 1843 the Irish genius Sir William Rowan Hamilton invented his *quaternions* and quaternion algebra— a sort of four-dimensional analogue to complex algebra. The simplest way to describe quaternions is to think of the usual 3-dimensional vector space with its usual basis vectors $\boldsymbol{i}, \boldsymbol{j}, \boldsymbol{k}$, and to add a "temporal" or scalar dimension with basis vector 1. A quaternion is then any linear combination,

$$\alpha = a + b\boldsymbol{i} + c\boldsymbol{j} + d\boldsymbol{k}, \tag{16}$$

of these basic vectors (a, b, c, and d now being real numbers). The multiplication rules are:

i. 1 is a multiplicative identity
ii. $\boldsymbol{i}^2 = \boldsymbol{j}^2 = \boldsymbol{k}^2 = -1$
iii. $\boldsymbol{ij} = \boldsymbol{k}, \quad \boldsymbol{jk} = \boldsymbol{i}, \quad \boldsymbol{ki} = \boldsymbol{j}$
iv. $\boldsymbol{ji} = -\boldsymbol{ij}, \quad \boldsymbol{kj} = -\boldsymbol{jk}, \quad \boldsymbol{ik} = -\boldsymbol{ki}$.

Multiplications of sums are handled by the distributive law. If one defines the conjugate of (16) to be

$$\overline{\alpha} = a - b\boldsymbol{i} - c\boldsymbol{j} - d\boldsymbol{k},$$

then

$$\alpha\,\overline{\alpha} = a^2 + b^2 + c^2 + d^2,$$

and Euler's identity reduces to

$$(\alpha\,\overline{\alpha})(\beta\,\overline{\beta}) = (\alpha\beta)(\overline{\alpha\beta}),$$

analogous to the complex identity,

$$(x_1^2 + x_2^2)(y_1^2 + y_2^2) = (x_1y_2 - x_2y_1)^2 + (x_1y_2 + x_2y_1)^2,$$

or even to (14), above.

Euler didn't know about quaternions, so he had to verify his identity by multiplying both sides of his equation out, and, ultimately, Hamilton had to do the same. I suggest the reader do this too.

That Lemma 3.9 has to do with norms of quaternions is merely a heuristic remark tying the Lemma to the analogous parts of the proof of Theorem 3.7. Its relevance to Lagrange's Theorem is explained by saying that it proves the product of two numbers representable as sums of four squares also to be representable as sums of four squares. Since 0 and 1 are trivially so representable, the proof of Lagrange's Theorem reduces to showing that every prime is a sum of four squares. For $p = 2$, this is obvious.

The next lemma is broadly analogous to that part of the proof of Theorem 3.7 which produced infinitely many solutions to the auxiliary equation (10).

3.10. Lemma. *Let p be an odd prime. There are x, y such that*

$$x^2 + y^2 + 1 \equiv 0 \pmod{p}.$$

In fact, $x^2 + y^2 + 1 = mp$ for some $0 < m < p$.

Proof. Note that, for $0 \le i < j \le (p-1)/2$, we have

$$i^2 \not\equiv j^2 \pmod{p}.$$

For, otherwise we could factor $i^2 - j^2 \equiv 0 \pmod{p}$ and conclude

$$i \equiv \pm j \pmod{p},$$

which is impossible. It follows that, for such i, j,

$$-1 - i^2 \not\equiv -1 - j^2 \pmod{p}.$$

Now consider the $p + 1$ numbers,

$$0^2, 1^2, \ldots, \left(\frac{p-1}{2}\right)^2; -1, -1-1^2, \ldots, -1-\left(\frac{p-1}{2}\right)^2.$$

By the Pigeon-Hole Principle, two of them, say x^2 and $-1 - y^2$, are congruent modulo p. Thus,

$$x^2 + y^2 + 1 \equiv 0 \pmod{p}.$$

To obtain the bounds, write $x^2 + y^2 + 1 = mp$. Clearly $m > 0$. For the upper bound, observe

$$x^2 + y^2 + 1 < \frac{p^2}{4} + \frac{p^2}{4} + 1 = \frac{p^2}{2} + 1 < p^2,$$

whence

$$m = \frac{1}{p}(x^2 + y^2 + 1) < p. \qquad \text{QED}$$

Now, as in the proof of Theorem 3.7, where we discovered we couldn't always divide solutions of (10) to obtain integral solutions of (P), we have here a small problem. Observe, for instance, that

$$21 = 4^2 + 2^2 + 1^2 \text{ and } 3 = 1^2 + 1^2 + 1^2,$$

but $7 = 21/3$ is not a sum of three squares. We cannot necessarily factor a sum of k squares into a product of sums of k squares. Euler asserted that for $k = 4$ this could be done. His proof was not quite right and Lagrange eventually proved the result. Following Euler's line of reasoning, modern texts prove the following lemma.

3.11. Lemma. *Let p be an odd prime. If m is the least positive integer such that*

$$pm = x_1^2 + x_2^2 + x_3^2 + x_4^2, \qquad (*)$$

for some x_1, x_2, x_3, x_4, then $m = 1$.

Proof. By contradiction. Choose m minimal so that (*) holds and assume $m \neq 1$.

The number m cannot be even. For, otherwise, an even number of the x_i's are even. Sorting them so that $x_1 \equiv x_2 \pmod 2$ and $x_3 \equiv x_4 \pmod 2$, we would have

$$\frac{mp}{2} = \left(\frac{x_1 + x_2}{2}\right)^2 + \left(\frac{x_1 - x_2}{2}\right)^2 + \left(\frac{x_3 + x_4}{2}\right)^2 + \left(\frac{x_3 - x_4}{2}\right)^2,$$

contradicting the minimality of m.

Since m is odd, $m/2$ is not an integer and we can shift remainders on division by m to lying between $-m/2$ and $m/2$, i.e. we can write

$$x_i = mk_i + y_i \qquad (17)$$

with

$$-m/2 < y_i < m/2. \qquad (18)$$

Then,

$$y_1^2 + y_2^2 + y_3^2 + y_4^2 \equiv x_1^2 + x_2^2 + x_3^2 + x_4^2 \pmod m$$
$$\equiv 0 \pmod m.$$

Write

$$mm' = y_1^2 + y_2^2 + y_3^2 + y_4^2.$$

We claim that $0 < m' < m$. If $m' = 0$, i.e. if each $y_i = 0$, then

$$m^2 \mid x_1^2 + x_2^2 + x_3^2 + x_4^2 = mp,$$

and $m \mid p$. By Lemma 3.10, however, $0 < m < p$ and this is impossible. Thus, $0 < m'$.

To see that $m' < m$, note that by (18)

$$y_1^2 + y_2^2 + y_3^2 + y_4^2 < 4\left(\frac{m}{2}\right)^2 = m^2.$$

This means $mm' < m^2$, whence $m' < m$.

We shall now use m' to contradict the minimality of m. By Euler's Identity, $(mm')(mp)$ is a sum of four squares:

$$(mm')(mp) = \left(y_1^2 + y_2^2 + y_3^2 + y_4^2\right)\left(x_1^2 + x_2^2 + x_3^2 + x_4^2\right)$$
$$= z_1^2 + z_2^2 + z_3^2 + z_4^2,$$

with the z_i's given by 3.9. Looking at the forms of the z_i's, we see that m divides each one:

$$\begin{aligned} z_1 &= x_1y_1 + x_2y_2 + x_3y_3 + x_4y_4 \\ &\equiv x_1^2 + x_2^2 + x_3^2 + x_4^2 \pmod{m}, \text{ by (17)} \\ &\equiv 0 \pmod{m}, \text{ by choice of } m. \\ z_2 &= x_1y_2 - x_2y_1 + x_3y_4 - x_4y_3 \\ &\equiv x_1x_2 - x_2x_1 + x_3x_4 - x_4x_3 \pmod{m}, \text{ by (17)} \\ &\equiv 0 \pmod{m}. \end{aligned}$$

The calculations for z_3 and z_4 resemble that for z_2.

Now write $z_i = mu_i$ and observe

$$\begin{aligned} m^2m'p &= z_1^2 + z_2^2 + z_3^2 + z_4^2 \\ &= (mu_1)^2 + (mu_2)^2 + (mu_3)^2 + (mu_4)^2 \\ &= m^2(u_1^2 + u_2^2 + u_3^2 + u_4^2), \end{aligned}$$

whence

$$m'p = u_1^2 + u_2^2 + u_3^2 + u_4^2.$$

This contradicts the minimality of m and the Lemma is proved. QED

Theorem 3.8 now follows easily: By Lemma 3.10, some multiple mp of p is a sum of three— hence of four (add 0^2)— squares. By 3.11, the least such multiple is given by $m = 1$.

That is enough material for this section. The exercises that follow are something of a mixed bag, reflecting the nature of this section.

Exercises.

1. Consider the parametrisation,

$$x = \frac{1-t^2}{1+t^2}, \quad y = \frac{2t}{1+t^2}, \quad \text{for } t = y/(x+1),$$

of the unit circle.

i. Show: If we write $x = \cos\theta$, $y = \sin\theta$, then $t = \tan\theta/2$.

ii. Show: $d\theta = \dfrac{2}{1+t^2}dt$.

iii. Find $\int \sec\theta\, d\theta$.

[Most American calculus textbooks evaluate the integral of the secant by using some trick like

$$\int \sec\theta\, d\theta = \int \sec\theta \frac{\sec\theta + \tan\theta}{\sec\theta + \tan\theta} d\theta = \int \frac{\sec\theta\tan\theta + \sec^2\theta}{\sec\theta + \tan\theta} d\theta.$$

Later, they may introduce a tan $\theta/2$ substitution, but they never explain where it comes from. Worse still, they never mention that the integral of the secant was an historically important problem— cf. the paper by V. Frederick Rickey and Philip Tuchinsky cited in the references.]

2. There are some special classes of Pell equations for which solutions are readily found. For example, if $d = a^2 - 1$, then $(a, 1)$ is a non-trivial solution:

$$a^2 - (a^2 - 1)\cdot 1^2 = 1.$$

i. For each of the following values of d, find simple polynomials $P(X)$ and $Q(X)$ with integral coëfficients, P and Q not both constant, such that

$$P(a)^2 - dQ(a)^2 = 1. \qquad (*)$$

a. $d = a^2 + 1$

b. $d = a^2 - 2$

c. $d = a^2 + 2$.

ii. Verify that $P(a)$ and $Q(a)$ give the minimum non-trivial solutions to (*) in i.a - c and that $P(a) = a$, $Q(a) = 1$ give the minimum non-trivial solution for $d = a^2 - 1$.

iii. Find the minimum non-trivial solutions to

$$X^2 - dY^2 = 1$$

for $d = 2, 3, 5, 6, 7, 8, 10, 11, 14, 15, 17, 18$.

iv. Do the same as in iii for $d = 12$ and $d = 20$ by reducing these to the cases $d = 3$ and $d = 5$, respectively.

[Hint. In i, P is quadratic and Q is linear.]

3. i. Make a table with column I containing the numbers $n = 0, ..., 10$, column II containing the squares n^2, and column III containing the numbers $19n^2$. For each of these last entries, find $m^2 - 19n^2$, where m is the smallest number the square of which exceeds $19n^2$.

ii. Using two solutions to $X^2 - 19Y^2 = 9$, derive the solution $(170, 39)$ to $X^2 - 19Y^2 = 1$.

iii. Prove: This is the minimum non-trivial solution for $d = 19$.

4. Make a table similar to that in the previous exercise for $d = 13$. Find a solution to $X^2 - 13Y^2 = -1$. From this find the minimum non-trivial solution to $X^2 - 13Y^2 = 1$, and prove that it is minimum. [The solution is $(649, 180)$.]

5. Show: $X^2 - 3Y^2 = -1$ is not solvable in non-negative integers. Find a solution to $X^2 - 5Y^2 = -1$.

6. For

$$\overline{\alpha + \beta\sqrt{d}} = \alpha - \beta\sqrt{d} \quad \text{and} \quad N(\alpha + \beta\sqrt{d}) = (\alpha + \beta\sqrt{d})(\alpha - \beta\sqrt{d}),$$

prove formulae (13), (14), and (15) of the text.

7. (Raphael Robinson). Assume $P(X)$ and $Q(X)$ are polynomials with *complex* coëfficients, with not both P, Q constant.

i. Suppose

$$P(X)^2 - (X^2 - 1)Q(X)^2 = 1.$$

a. Show: The degree of P is one higher than that of Q, and, up to the sign, P and Q have the same leading coëfficient.

b. Replacing P by $-P$ if necessary, assume P and Q have the same leading coëfficient. Show: For some $n > 0$ and some choice of sign,

$$P(X) + Q(X)\sqrt{X^2 - 1} = \pm(X + \sqrt{X^2 - 1})^n.$$

ii. Let c be a non-zero integer (any complex number will work). Show: If

$$P(X)^2 - (X^2 - c)Q(X)^2 = 1,$$

then, for some $n > 0$ and some choice of sign,

$$P(X) + Q(X)\sqrt{X^2 - c} = \pm\left(\frac{X + \sqrt{X^2 - c}}{\sqrt{c}}\right)^n.$$

iii. Let c be a non-zero integer and suppose P, Q have *integral* coëfficients and satisfy

$$P(X)^2 - (X^2 - c)Q(X)^2 = 1.$$

a. Show: The leading coëfficient of $P(X)$ is

$$a_n = \pm(1/\sqrt{c})\left(\binom{n}{0} + \binom{n}{1} + \ldots + \binom{n}{2[n/2]}\right).$$

b. Show: $\binom{n}{0} + \binom{n}{1} + \ldots + \binom{n}{2[n/2]} = \frac{1}{2}\cdot 2^n$. Conclude that

$$a_n^2 = \frac{1}{4}\left(\frac{4}{c}\right)^n.$$

c. Show: $c = \pm 1, \pm 2$. Moreover, if $c = \pm 2$, then n must be even.

[Hints. i.b. use induction — multiplying $P(X) + Q(X)\sqrt{X^2 - 1}$ by $X - \sqrt{X^2 - 1}$ lowers the degrees; replace X by $\sqrt{c}X$ (*N.B.* This is where it is necessary to allow complex coëfficients in i.b.); iii.b. consider $(1 - 1)^n$ and $(1 + 1)^n$; iii.c. c must divide 4.]

8. The Archimedean Property of the real line, apparently due to Eudoxus, is usually stated additively:

AP: *For any* $\varepsilon > 0$ *and any* $x > 0$, *there is some positive integer* n *such that* $n\varepsilon > x$.

Derive from *AP* the multiplicative version:

For any $\delta > 1$ *and any* $x > 1$, *there is some positive integer* n *such that* $\delta^n > x$.

9. Prove the following variants of the Pigeon-Hole Principle:

i. Let X, Y be non-empty finite sets of cardinalities m, n, respectively, and let k be a positive integer. Suppose $m > kn$. For any function $F:X \to Y$ there is some $y \in Y$ such that $\mathrm{card}(F^{-1}(\{y\}))$ is greater than k.

ii. Let X be infinite and Y finite and non-empty. For any function $F:X \to Y$ there is some $y \in Y$ such that $F^{-1}(\{y\})$ is infinite.

10. (Multiplicative Pigeon-Hole Principle). Prove the following:

i. If $s \mid t_0 \ldots t_k$, then there is a p and a $j \le k$ such that $p \mid t_j$, $p \mid s$, and $p > s^{1/(k+1)}$, where $s, t_0, \ldots, t_k$ are all positive integers.

ii. Suppose $q > 0$ and we have $s \mid t_{i0} \ldots t_{ik_i}$, for $i = 0, \ldots, n-1$. Then there are p and $j_0, \ldots, j_{n-1}$ such that, for each i,

$$j_i \le k_i, \quad p \mid t_{ij_i}, \quad p \mid s, \quad \text{and } p > \left(\frac{s}{q}\right)^{1/(k_0+1)\ldots(k_{n-1}+1)},$$

where, again, s and all the t_{ij}'s are positive integers.

[The result of Exercise 10 was used by Juriĭ Matijasevič and Julia Robinson in generalising the proof via binomial coëfficients of the Bounded Quantifier Theorem given in Exercise 9 of section 2, above, to the case in which each bounded existential quantifier has its own bound. The interested reader will find this use in the paper by Matijasevič cited in the references for this section. Transcendental number theory relies on similar horrible-looking applications of the Pigeon-Hole Principle. I refer the reader to the first chapter of Stolarsky's book cited in the references to this section in the Reading List at the end of the chapter.]

11. (Liouville). Let α be an irrational algebraic number and let $P(X)$ be a polynomial with integral coëfficients of minimum degree d such that $P(\alpha) = 0$.
   i. Show: For any rational number p/q, $|P(p/q)| \geq 1/q^d$.
   ii. Show: For any rational number p/q, if $|p/q - \alpha| < 1$, then

$$|P(p/q)| \leq |p/q - \alpha| \cdot \left(\sum_{i=1}^{d} \left| \frac{d^i}{dx^i} P(\alpha) \right| \right).$$

   iii. Show: There is a constant $c > 0$ such that the inequality,

$$|p/q - \alpha| < c/q^d,$$

has only finitely many rational solutions p/q.
   iv. Show: The number $\theta = \sum_{n=0}^{\infty} 1/10^{n!}$ is transcendental.

12. The Three-Square Theorem asserts that a natural number n is a sum of three integral squares iff n is not of the form $4^m(8k+7)$, $m, k \in \omega$. Using this fact, show that every natural number is representable by the polynomial, $X^2 + Y^2 + Z^2 + Z$. Conclude that one can reduce the n-variable Diophantine problem over N to the $3n$-variable Diophantine problem over Z. (Cf. Fact 1.5.i.)

13. (Axel Thue). Let d be a positive integer that is not a perfect square, and suppose $(a,b) \neq (1,0)$ and (x,y) satisfy

$$a^2 - db^2 = 1, \quad x^2 - dy^2 = m, \quad \text{for some } m > 0.$$

   i. Show: $(ax - dby)^2 - d(ay - bx)^2 = m$. Where do $ax - dby$ and $ay - bx$ come from?
   ii. Assume $y > b\sqrt{m}$.
      a. Show: $ay - bx > 0$
      b. Show: $b^2 d > a^2 - 2a$
      c. Show: $b^2 x^2 > y^2(a^2 - 2a + 1)$
      d. Show: $y > ay - bx$.
   iii. Show: $X^2 - dY^2 = m$ has a solution with $y \leq b\sqrt{m}$.
   iv. Show: $X^2 - 6Y^2 = 2$ is not solvable in integers.

[Hints. ii.a. $y^2 = (a^2 - db^2)y^2$; ii.c. $b^2x^2 = b^2(m + dy^2)$.]

4. The Pell Equation, II

The goal of the present section is to establish a number of properties of the sequence of solutions of the Pell equation of the special form,

$$X^2 - (a^2 - 1)Y^2 = 1, \qquad (\mathrm{P}_a)$$

where $a \geq 2$. These properties fall into several categories: explicit definition, recursion equations, estimates of the growth of the sequence, and congruential and divisibility properties. In the next section, some of these properties will be used to give a Diophantine definition of the sequence of solutions to (P_a), and this will be exploited to give a Diophantine definition of exponentiation. To be precise, the Diophantine definition of the sequence of solutions to (P_a) will be uniform in a, i.e., a will be a parameter in the definition, not a fixed constant. In the present section, however, we can think of a as a fixed positive integer greater than 1.

For the benefit of the reader who has opted for the quickest approach to the algorithmic unsolvability of Hilbert's tenth problem (i.e., sections 1, 2, 4, and 5), the present section is completely self-contained. The reader who has read section 3 will find only one argument repeated.

We begin with a few elementary observations. The two most obvious are that $(1, 0)$ and $(a, 1)$ are solutions to (P_a). The first of these is called the *trivial* solution. The second is called the *minimum* solution, or minimum non-trivial solution. Justification for this latter terminology begins with the observation that, if (x, y) is a solution to $X^2 - (a^2 - 1)Y^2 = 1$, then

$$x^2 = (a^2 - 1)y^2 + 1,$$

and x grows with y. Thus, we can order the solutions by using the natural order on the x's or on the y's. Given this, the minimality of $(a, 1)$ is obvious: there is no y that can fit properly between its 1 and the 0 of the trivial solution.

As the name "minimum solution" suggests, there are a great many solutions to (P_a).

4.1. Lemma. *Let $a \geq 2$. The set of numbers of the form*

$$x \pm y\sqrt{a^2 - 1}, \qquad (*)$$

where (x, y) is a non-negative integral solution to (P_a), forms a subgroup of the multiplicative group of positive real numbers, i.e.,

i. *the product of two numbers in the set is again in the set, and*

ii. *the multiplicative inverse of a number in the set is in the set.*

Proof. Letting y range over negative integers as well, we can avoid the necessity of considering separately the cases in which a plus or a minus sign is used in (*).

It is best to begin with ii:

$$\left(x + y\sqrt{a^2 - 1}\right)\left(x - y\sqrt{a^2 - 1}\right) = x^2 - (a^2 - 1)y^2 = 1$$

$$= 1 + 0\sqrt{a^2 - 1}.$$

To establish i, let (x, y) and (s, t) be solutions to (P_a) and note that

$$\left(x + y\sqrt{a^2 - 1}\right)\left(s + t\sqrt{a^2 - 1}\right) = (xs + yt(a^2 - 1)) + (xt + ys)\sqrt{a^2 - 1}$$
$$\left(x - y\sqrt{a^2 - 1}\right)\left(s - t\sqrt{a^2 - 1}\right) = (xs + yt(a^2 - 1)) - (xt + ys)\sqrt{a^2 - 1}.$$

Let $u = xs + yt(a^2 - 1)$, $v = xt + ys$, and apply ii:

$$\begin{aligned} u^2 - (a^2 - 1)v^2 &= \left(u + v\sqrt{a^2 - 1}\right)\left(u - v\sqrt{a^2 - 1}\right) \\ &= \left(x + y\sqrt{a^2 - 1}\right)\left(s + t\sqrt{a^2 - 1}\right)\left(x - y\sqrt{a^2 - 1}\right)\left(s - t\sqrt{a^2 - 1}\right) \\ &= \left(x + y\sqrt{a^2 - 1}\right)\left(x - y\sqrt{a^2 - 1}\right)\left(s + t\sqrt{a^2 - 1}\right)\left(s - t\sqrt{a^2 - 1}\right) \\ &= 1 \cdot 1 = 1. \end{aligned}$$

Therefore, $(u, \pm v)$ are solutions and $u \pm v\sqrt{a^2 - 1}$ are in the set. QED

It turns out that the group of Lemma 4.1 is generated by the minimum solution.

4.2. Theorem. *Let $a \geq 2$. The non-negative integral solutions to*

$$X^2 - (a^2 - 1)Y^2 = 1 \qquad (P_a)$$

are given by $x = X_a(n)$, $y = Y_a(n)$ for $n \in \omega$, where $X_a(n)$ and $Y_a(n)$ are defined by

$$X_a(n) + Y_a(n)\sqrt{a^2 - 1} = (a + \sqrt{a^2 - 1})^n. \qquad (1)$$

Proof. (Cf. Exercise 1.) By Lemma 4.1, every $(X_a(n), Y_a(n))$ given by (1) is a solution to the given Pell equation. To see that (1) exhausts all solutions, let $(x, y) \neq (1, 0)$ satisfy (P_a) and consider $x + y\sqrt{a^2 - 1}$. Because $(a, 1)$ is the minimum solution, we have $a \leq x$ and $1 \leq y$, whence $a + \sqrt{a^2 - 1} \leq x + y\sqrt{a^2 - 1}$. However, $a + \sqrt{a^2 - 1} > 1$, so not every power of $a + \sqrt{a^2 - 1}$ lies below $x + y\sqrt{a^2 - 1}$. Choose n such that

$$(a + \sqrt{a^2 - 1})^n \leq x + y\sqrt{a^2 - 1} < (a + \sqrt{a^2 - 1})^{n+1}.$$

Multiply by $(a - \sqrt{a^2 - 1})^n$ to get

$$1 \leq (x + y\sqrt{a^2 - 1})(a - \sqrt{a^2 - 1})^n < (a + \sqrt{a^2 - 1}).$$

The minimality of $(a, 1)$ yields

$$1 = (x + y\sqrt{a^2 - 1})(a - \sqrt{a^2 - 1})^n.$$

Multiplying by $(a + \sqrt{a^2 - 1})^n$ finally yields

$$X_a(n) + Y_a(n)\sqrt{a^2 - 1} = (a + \sqrt{a^2 - 1})^n = x + y\sqrt{a^2 - 1}. \qquad \text{QED}$$

Formula (1) yields by itself some very interesting formulae.

4.3. Addition Rules. *Let $a \geq 2$. For any $m, n \in \omega$,*

i. $X_a(m \pm n) = X_a(m)X_a(n) \pm (a^2 - 1)Y_a(m)Y_a(n)$

ii. $Y_a(m \pm n) = \pm X_a(m)Y_a(n) + X_a(n)Y_a(m)$.

Proof. As in the proof of Lemma 4.1, we can use the fact that

$$\left(X_a(n) + Y_a(n)\sqrt{a^2-1}\,\right)^{-1} = X_a(n) - Y_a(n)\sqrt{a^2-1}$$

to extend (1) to negative n, where

$$X_a(-k) = X_a(k), \qquad Y_a(-k) = -Y_a(k),$$

and only work out the details for the plus sign:

$$\begin{aligned} X_a(m+n) &+ Y_a(m+n)\sqrt{a^2-1} \\ &= (a+\sqrt{a^2-1}\,)^{m+n} \\ &= (a+\sqrt{a^2-1}\,)^m(a+\sqrt{a^2-1}\,)^n \\ &= \left(X_a(m)+Y_a(m)\sqrt{a^2-1}\,\right)\left(X_a(n)+Y_a(n)\sqrt{a^2-1}\,\right) \\ &= \left(X_a(m)X_a(n) + (a^2-1)Y_a(m)Y_a(n)\right) \\ &\qquad + \left(X_a(m)Y_a(n) + X_a(n)Y_a(m)\right)\sqrt{a^2-1}, \end{aligned}$$

from which i and ii follow. QED

4.4. "Double-Angle Formula". *Let* $a \geq 2$.

i. $X_a(2n) = 2X_a(n)^2 - 1$

ii. $Y_a(2n) = 2X_a(n)Y_a(n)$.

Proof. ii follows immediately from 4.3.ii. For i, note that 4.3.i yields

$$\begin{aligned} X_a(2n) &= X_a(n)^2 + (a^2-1)Y_a(n)^2 \\ &= X_a(n)^2 + (X_a(n)^2 - 1), \text{ by } (P_a) \\ &= 2X_a(n)^2 - 1. \end{aligned}$$

QED

4.5. Simultaneous Recursion Equations. *Let* $a \geq 2$.

i. $X_a(0) = 1$

$X_a(n+1) = aX_a(n) + (a^2-1)Y_a(n)$

ii. $Y_a(0) = 0$

$Y_a(n+1) = X_a(n) + aY_a(n)$.

The proof is immediate. From 4.5, a little algebra yields the following.

4.6. Course-of-Values Recursion Equations. *Let* $a \geq 2$.

i. $X_a(0) = 1, \qquad X_a(1) = a$

$X_a(n+2) = 2aX_a(n+1) - X_a(n)$

ii. $Y_a(0) = 0, \qquad X_a(1) = 1$

$Y_a(n+2) = 2aY_a(n+1) - Y_a(n)$.

Proof. i. The initial values are obvious. For the recursion, note that 4.5 yields

$$\begin{aligned} X_a(n+2) &= aX_a(n+1) + (a^2-1)Y_a(n+1) \\ &= aX_a(n+1) + (a^2-1)(X_a(n) + aY_a(n)). \end{aligned} \tag{2}$$

But

$$X_a(n+1) = aX_a(n) + (a^2-1)Y_a(n),$$

whence

$$Y_a(n) = \frac{X_a(n+1) - aX_a(n)}{a^2-1},$$

whence (2) yields

$$\begin{aligned} X_a(n+2) &= aX_a(n+1) + (a^2-1)X_a(n) + a(X_a(n+1) - aX_a(n)) \\ &= 2aX_a(n+1) - X_a(n). \end{aligned}$$

ii. Similar. QED

Equations 4.6 are two-step linear recurrences. For one-step, non-linear recurrences, cf. Exercise 5, below.

Formula (1) tells us that $X_a(n) + Y_a(n)\sqrt{a^2-1}$ grows exponentially. In fact, each sequence X_a and Y_a grows exponentially. The following lemma gives fairly nice bounds on $Y_a(n)$.

4.7. Lemma on Bounds. *Let* $a \geq 2$.

i. *For* $n \geq 0$, $(2a-1)^n \leq Y_a(n+1) \leq (2a)^n$.

ii. *For* $n \geq 2$, $Y_a(n) \geq 2n$.

Proof. i. By induction on n. The basis is simple:

$$(2a-1)^0 = 1 = Y_a(1) = (2a)^0.$$

To establish the lower bound in the induction step, observe

$$\begin{aligned} (2a-1)^{n+1} &= (2a-1)^n(2a-1) \leq (2a-1)Y_a(n+1) \\ &\leq 2aY_a(n+1) - Y_a(n+1) < 2aY_a(n+1) - Y_a(n) \qquad (*) \\ &< Y_a(n+2), \end{aligned}$$

where (*) follows from the monotonicity of the sequence Y_a (by 4.5.ii).

For the upper bound, observe

$$\begin{aligned} Y_a(n+2) &= 2aY_a(n+1) - Y_a(n) \\ &\leq 2aY_a(n+1) \leq 2a(2a)^n = (2a)^{n+1}. \end{aligned}$$

[Assuming it settled that $Y_a(k) < X_a(k)$ for all k (Lemma 4.12, below), formula (1) almost yields the result directly:

$$\begin{aligned} 2Y_a(n+1) &\leq X_a(n+1) + Y_a(n+1)\sqrt{a^2-1} \\ &\leq (a + \sqrt{a^2-1})^{n+1} \\ &\leq (2a)^{n+1}, \end{aligned}$$

whence $Y_a(n+1) < a(2a)^n$.]

ii. The obvious thing to do is to apply part i:

$$\begin{aligned} Y_a(n) &\geq (2a-1)^{n-1} \geq 3^{n-1} = (1+2)^{n-1} \\ &\geq 1^{n-1} + \binom{n-1}{1} 2^{n-2} \\ &> 2^{n-2}(n-1) \\ &> 4(n-1), \text{ if } n \geq 4 \\ &> 2n. \end{aligned}$$

For $n = 2, 3$, one simply calculates $Y_a(n)$:

$$\begin{aligned} Y_a(2) &= X_a(1) + aY_a(1) = a + a\cdot 1 = 2a \geq 2\cdot 2 = 2n \\ Y_a(3) &= X_a(2) + aY_a(2) = 2a^2 - 1 + 2a^2 \\ &= 4a^2 - 1 \geq 16 - 1 = 15 > 2\cdot 3 = 2n. \end{aligned}$$

[Alternatively, and slightly more easily, one could use induction and the Simultaneous Recursion Equations.]

QED

There are also similar bounds for $X_a(n)$, but these won't be needed and are relegated to the exercises (Exercise 5). At about this point, the emphasis switches from the pair X_a, Y_a to the single sequence Y_a. This does not mean that X_a disappears from the exposition— indeed, it occurs in several additional lemmas— but that the sequence Y_a has such nice divisibility properties that it assumes a greater interest.

Before looking at the divisibility properties of the sequence Y_a, let us quickly note the following lemma relating sequences X_a, Y_a and X_b, Y_b for different Pell equations.

4.8. Congruence Rule. *Let $a, b \geq 2$, $a \neq b$. For any $n \geq 0$,*

i. $X_a(n) \equiv X_b(n) \pmod{b-a}$

$Y_a(n) \equiv Y_b(n) \pmod{b-a}$

ii. $Y_b(n) \equiv n \pmod{b-1}$.

Proof. If we define $Y_1(n) = n$, then we see that Y_1 satisfies the same course-of-values recursion as Y_a:

$$Y_1(0) = 0, \quad Y_1(1) = 1$$

$$Y_1(n+2) = 2Y_1(n+1) - Y_1(n).$$

Thus, for the Y_a's (the interesting case), we can treat the cases $a = 1$ and $a \geq 2$ simultaneously.

The proof is by induction on n. For $n = 0$ or 1, we have

$$Y_a(n) = n = Y_b(n),$$

whence

$$Y_a(n) \equiv Y_b(n) \pmod{b-a}.$$

For the induction step, observe

$$Y_a(n+2) = 2Y_a(n+1) - Y_a(n)$$

$$\equiv 2Y_b(n+1) - Y_b(n) \pmod{b-a}$$
$$\equiv Y_b(n+2) \pmod{b-a}.$$

I leave the congruence for the X's to the reader. QED

The chief divisibility property of the sequence Y_a, of which the other properties are primarily particularities, is stated in the following theorem.

4.9. Theorem. *Let $a \geq 2$, $m \geq 1$, $n \geq 0$, and $m \mid Y_a(n)$. The sequence of remainders of $Y_a(k)$ modulo m is periodic with a period dividing $2n$.*

Proof. Let $m \mid Y_a(n)$ and observe that

$$\begin{aligned} Y_a(k+2n) &= X_a(k)Y_a(2n) + X_a(2n)Y_a(k), \text{ by 4.3} \\ &= X_a(k)2X_a(n)Y_a(n) + (2X_a(n)^2 - 1)Y_a(k), \text{ by 4.4} \\ &\equiv (2X_a(n)^2 - 1)Y_a(k) \pmod{m}, \text{ since } m \mid Y_a(n) \\ &\equiv (2((a^2-1)Y_a(n)^2 + 1) - 1)Y_a(k) \pmod{m} \\ &\equiv Y_a(k) \pmod{m}. \end{aligned}$$

QED

It follows from the general existence theorem of the preceding section that $X^2 - dY^2 = 1$ has solutions so long as d is not a perfect square. Choosing $d = m^2(a^2-1)$, we see that $m \mid Y_a(n)$ for some $n > 0$. Hence Theorem 4.9 tells us that, for any $m \geq 1$, the sequence

$$\text{Rem}(Y_a(0), m), \text{Rem}(Y_a(1), m), \ldots$$

is periodic. We will not really need this result in such generality; for the values of m we will be interested in, it will be clear that an $n > 0$ such that $m \mid Y_a(n)$ exists. The next two lemmas give specific periodicity results and the exact periods we will need.

4.10. First Step-Down Lemma. *Let $a \geq 2$. For $m, n > 0$,*

i. $Y_a(m) \mid Y_a(n)$ iff $m \mid n$

ii. $Y_a(m)^2 \mid Y_a(n)$ iff $mY_a(m) \mid n$.

Proof. i. We begin with the right-to-left implication. Assume $n = mk$ and apply formula (1):

$$\begin{aligned} X_a(n) + Y_a(n)\sqrt{a^2-1} &= (a + \sqrt{a^2-1})^{mk} = ((a+\sqrt{a^2-1})^m)^k \\ &= (X_a(m) + Y_a(m)\sqrt{a^2-1})^k. \end{aligned}$$

Expanding the right-hand-side of this last by means of the Binomial Theorem and comparing coëfficients of $\sqrt{a^2-1}$, we see that

$$Y_a(n) = \binom{k}{1} Y_a(m) \cdot X_a(m)^{k-1} + Y_a(m)^3 \cdot \text{garbage}. \quad (3)$$

It follows immediately that $Y_a(m) \mid Y_a(n)$.

Conversely, suppose $Y_a(m) \mid Y_a(n)$. Assume, by way of contradiction, that $m \nmid n$. Then we can write $n = mq + r$ with $0 < r < m$. Apply the Addition Formula 4.3.ii to get

$$Y_a(n) = Y_a(mq+r) = X_a(mq)Y_a(r) + X_a(r)Y_a(mq). \tag{4}$$

We have already shown that

$$Y_a(m) \mid Y_a(mq) \tag{5}$$

and we assume $Y_a(m) \mid Y_a(n)$. Thus, (4) yields

$$Y_a(m) \mid X_a(mq)Y_a(r).$$

By (5) and the fact that $Y_a(mq)$ and $X_a(mq)$ are relatively prime (why?), it follows that $Y_a(m) \mid Y_a(r)$. But $r < m$ yields

$$0 < Y_a(r) < Y_a(m),$$

which contradicts $Y_a(m) \mid Y_a(r)$. Hence, it must be the case that $m \mid n$.

ii. Either condition implies $m \mid n$. Let $n = mk$ as before. By formula (3),

$$\begin{aligned} Y_a(m)^2 \mid Y_a(n) \quad &\text{iff} \quad Y_a(m)^2 \mid \binom{k}{1} Y_a(m)\cdot X_a(m)^{k-1} \\ &\text{iff} \quad Y_a(m) \mid k \qquad\qquad (*) \\ &\text{iff} \quad mY_a(m) \mid mk \\ &\text{iff} \quad mY_a(m) \mid n, \end{aligned}$$

where (*) follows from the relative primality of $X_a(m)$, $Y_a(m)$. QED

We shall shortly prove a very nice theorem which will shed some additional light on part i of this Lemma. Notice, by the way, that, if we agree that $0 \mid 0$ and $0 \nmid n$ for $n > 0$, then the Lemma holds for all $m, n \geq 0$.

4.11. Second Step-Down Lemma. *Let $a \geq 2$. For $n \geq 1$ and $i, j \in \omega$, if*

$$Y_a(i) \equiv Y_a(j) \pmod{X_a(n)},$$

then

$$j \equiv i \pmod{2n} \quad or \quad j \equiv -i \pmod{2n}.$$

This is an odd lemma indeed. The "Double-Angle Formula" tells us that $X_a(n)$ divides $Y_a(2n)$. This, with Theorem 4.9, tells us that the sequence of remainders of $Y_a(0)$, $Y_a(1)$, ... modulo $X_a(n)$ is periodic with period $4n$. The Second Step-Down Lemma is a sort of analysis of the period. With one exception ($a = 2$ and $n = 1$), the period is $4n$ and not $2n$. However, there are some curious inner congruences which complicate the picture.

Before launching into the proof of the Lemma, observe that, for $i \leq 2n$, the Addition Formula yields

$$\begin{aligned} Y_a(2n \pm i) &= \pm X_a(2n)Y_a(i) + X_a(i)Y_a(2n) \\ &\equiv \pm X_a(2n)Y_a(i) \pmod{X_a(n)} \\ &\equiv \pm(2X_a(n)^2 - 1)Y_a(i) \pmod{X_a(n)} \\ &\equiv \mp Y_a(i) \pmod{X_a(n)}. \end{aligned} \tag{6}$$

Letting $i = 1$, we see that

$$Y_a(1) \equiv Y_a(2n+1) \pmod{X_a(n)}$$

would yield via (6)

$$Y_a(1) \equiv -Y_a(1) \pmod{X_a(n)},$$

i.e., $X_a(n) \mid 2Y_a(1) = 2$. This can only happen for $a = 2$ and $n = 1$, in which case Theorem 4.13, below, will readily show the period to be exactly $2 = 2n$— in which case the Lemma holds.

Proof of Lemma 4.11. We can assume $a \neq 2$ or $n \neq 1$. Similarly to deriving (6), one can show

$$Y_a(4n \pm i) \equiv \pm Y_a(i) \pmod{X_a(n)},$$

for $i \leq n$. Together with (6), this yields the following four formulae for $0 \leq i \leq n$:

$$\left.\begin{aligned} Y_a(i) &\equiv Y_a(i) \pmod{X_a(n)} \\ Y_a(2n-i) &\equiv Y_a(i) \pmod{X_a(n)} \\ Y_a(2n+i) &\equiv -Y_a(i) \pmod{X_a(n)} \\ Y_a(4n-i) &\equiv -Y_a(i) \pmod{X_a(n)} \end{aligned}\right\}. \qquad (7)$$

The proof of the Lemma largely reduces to showing these to be the only congruences holding, i.e., to showing

$$Y_a(i) \not\equiv \pm Y_a(j) \pmod{X_a(n)} \qquad (8)$$

for $0 \leq i < j \leq n$, and

$$Y_a(i) \not\equiv -Y_a(i) \pmod{X_a(n)} \qquad (9)$$

for $0 < i \leq n$.

If $a > 2$, both (8) and (9) are simple matters. For,

$$X_a(n) = \sqrt{(a^2-1)Y_a(n)^2+1} > \sqrt{a^2-1}\,Y_a(n) > 2Y_a(n). \qquad (10)$$

Thus, (8) must be true because

$$X_a(n) > 2Y_a(n) > Y_a(j) \pm Y_a(i) > 0$$

for $n \geq j > i \geq 0$; and (9) must be true because

$$X_a(n) > 2Y_a(n) > 2Y_a(i) > 0$$

for $n \geq i > 0$.

If $a = 2$, the estimate (10) fails. If, however, one makes a small table of values of X_2 and Y_2, one readily conjectures, for $a = 2$, the following lemma. Working out the proof shows it to be valid for $a \geq 2$.

4.12. Lemma. *Let $a \geq 2$. For $n \geq 0$, $X_a(n) > \sum_{i=0}^{n} Y_a(i)$.*

Proof. By induction on n using the Simultaneous Recursion Equations.

For $n = 0$, $X_a(0) = 1 > 0 = Y_a(0)$.

For $n + 1$, observe

$$
\begin{aligned}
X_a(n+1) &= aX_a(n) + (a^2-1)Y_a(n), \text{ by 4.5} \\
&> X_a(n) + (a-1)X_a(n) + aY_a(n) \\
&> \sum_{i=0}^{n} Y_a(i) + (X_a(n) + aY_a(n)) \\
&> \sum_{i=0}^{n} Y_a(i) + Y_a(n+1) = \sum_{i=0}^{n+1} Y_a(i).
\end{aligned}
$$

QED

Continuation of the proof of Lemma 4.11. With Lemma 4.12, we can prove (8) and (9) uniformly for $a = 2$ and $a > 2$.

For (8), we cannot have $Y_a(i) \equiv \pm Y_a(j) \pmod{X_a(n)}$ for $0 \le i < j \le n$ because Lemma 4.12 tells us that

$$X_a(n) > \sum_{k=0}^{n} Y_a(k) \ge Y_a(j) \pm Y_a(i) > 0.$$

For (9), if $0 < i < n$, we have

$$X_a(n) > Y_a(n-1) + Y_a(n) > 2Y_a(i) > 0.$$

For $i = n$, note that

$$X_a(n) \mid 2Y_a(n) \Rightarrow X_a(n) \mid 2,$$

since $X_a(n)$ and $Y_a(n)$ are relatively prime. But, $X_a(n) \mid 2$ only in the case $a = 2$ and $n = 1$, which case was handled separately before beginning the proof of the Lemma.

This just about completes the proof of the Lemma. Let, for arbitrary $i, j \in \omega$,

$$Y_a(i) \equiv Y_a(j) \pmod{X_a(n)}.$$

Let i', j' be the remainders of i, j on dividing by $4n$. If $i' = j'$, then $i \equiv j \pmod{4n}$, whence $i \equiv j \pmod{2n}$. If $i' \ne j'$, then a glance at (7) shows $i' \equiv -j' \pmod{2n}$. QED

A couple of alternatives to the Second Step-Down Lemma appear in the exercises (Exercises 6 and 7, below). Before we use the Second Step-Down Lemma in the next section, I cannot offer any better explanation of the interest in such a lemma as this other than to say that such lemmas serve as the determination of the periods of Theorem 4.9 for specific divisors m.

The following theorem will not be used in the sequel, but it may shed a little light on Theorem 4.9 and Lemma 4.10.i. In any event, it is a lovely result.

4.13. Theorem. *Let $a \ge 2$ and $m, n \ge 0$. Then*

$$\gcd(Y_a(n), Y_a(m)) = Y_a(\gcd(m, n)),$$

where "gcd" abbreviates "greatest common divisor".

Proof. Without any loss of generality, we can assume $n > m > 0$. If $m \mid n$, then, by Lemma 4.10.i $Y_a(m) \mid Y_a(n)$ and

$$\gcd(Y_a(n), Y_a(m)) = Y_a(m) = Y_a(\gcd(m, n)).$$

If m does not divide n, apply the division algorithm to get

$$n = mq + r, \quad 0 < r < m,$$

and apply the Addition Formula to get

$$Y_a(n) = X_a(mq)Y_a(r) + Y_a(mq)X_a(r).$$

Since $Y_a(m) \mid Y_a(mq)$, any common divisor of $Y_a(n)$ and $Y_a(m)$ must divide $X_a(mq)Y_a(r)$. Since $Y_a(m) \mid Y_a(mq)$ and $Y_a(mq)$ is relatively prime to $X_a(mq)$, it follow that any common divisor of $Y_a(n)$ and $Y_a(m)$ must divide $Y_a(r)$. Conversely, it is clear that any common divisor of $Y_a(m)$ and $Y_a(r)$ must divide $Y_a(n)$. Thus,

$$\gcd(Y_a(n), Y_a(m)) = \gcd(Y_a(m), Y_a(r)). \tag{11}$$

But it is also true that

$$\gcd(n, m) = \gcd(m, r). \tag{12}$$

Since $m > r$, we can inductively assume

$$\gcd(Y_a(m), Y_a(r)) = Y_a(\gcd(m, r)),$$

and apply (11) and (12) to get

$$\gcd(Y_a(n), Y_a(m)) = Y_a(\gcd(n, m)).$$ QED

Noting that $Y_a(1) = 1$ and $Y_a(2) = 2a$, the Theorem immediately yields as a corollary the case $a = 2, n = 1$, of Lemma 4.11: $2 = X_2(1) \mid Y_2(i)$ iff $2 \mid i$.

With the exception of the Second Step-Down Lemma and the possible exception of the second half of the First Step-Down Lemma, the results of this section are fairly standard stuff of recreational mathematics. The Pell sequences X_a and Y_a, along with the famous Fibonacci sequence, are special cases of what are called *Lucas sequences*. Edouard Lucas was a 19th century number theorist and recreational mathematician. He is credited with having invented the Towers of Hanoi puzzle, the canonical example of recursion (in the computer scientist's sense of the word) in introductory programming courses. We will meet with one of Lucas' other mathematical contributions in sections 8 and 9, below.

Exercises.

1. In the text, it is implicitly assumed that, for d not a square,

$$x + y\sqrt{d} = a + b\sqrt{d} \Rightarrow x = a \wedge y = b.$$

   i. Show that this holds.
   ii. Show: If $a \geq 2$, then $a^2 - 1$ is not a square.

2. Prove Lemma 4.6.ii: $Y_a(n+2) = 2aY_a(n+1) - Y_a(n)$.
3. Let $a \geq 2$.
   i. Show: For $n \geq 0$,

$$n \leq X_a(n) < X_a(n+1) - X_a(n)$$
$$0 \leq Y_a(n) < Y_a(n+1) - Y_a(n)$$

   ii. Show: For $n \geq 0$, $X_a(n) \leq (2a)^n$
   iii. Show: For $n \geq 0$,

$$X_a(n) > \frac{(2a-1)^{n+1}}{2a}.$$

[Hint. iii. Apply 4.12.]

4. Let $a \geq 2$. For $n, k \geq 0$, show:

$$Y_a(n+k+1) = Y_a(n+1)Y_a(k+1) - Y_a(n)Y_a(k).$$

5. Let $a \geq 2$. Suppose $z > y$. Show: (z, y) is a solution to $Z^2 - 2aZY + Y^2 = 1$ iff, for some $n \in \omega$, $z = Y_a(n+1)$ and $y = Y_a(n)$.

6. (Alternative Second Step-Down Lemma). Let $a \geq 2$, $n \geq 0$.
   i. Show: $Y_a(2n+1) = (Y_a(n+1) + Y_a(n))(Y_a(n+1) - Y_a(n))$.
   ii. Show: $Y_a(n+1) + Y_a(n)$ and $Y_a(n+1) - Y_a(n)$ are relatively prime.
   iii. Show: For $0 \leq j \leq n+1$,

$$Y_a(n+j) \equiv -Y_a(n+1-j) \pmod{Y_a(n+1) + Y_a(n)}.$$

   iv. Show: The sequence of remainders of $Y_a(m)$ modulo $Y_a(n+1) + Y_a(n)$ is periodic with a period of $2n+1$.
   v. Show: For any i, j, if $Y_a(i) \equiv Y_a(j) \pmod{Y_a(n+1) + Y_a(n)}$, then $i \equiv j \pmod{2n+1}$.

7. (Another Alternative Second Step-Down Lemma). Let $a \geq 2$, $n \geq 0$.
   i. Show: For any i, $X_a(4n+i) \equiv X_a(i) \pmod{X_a(n)}$.
   ii. Show: For any i, j, if $X_a(i) \equiv X_a(j) \pmod{X_a(n)}$, then $i \equiv j \pmod{4n}$ or $i \equiv -j \pmod{4n}$.

5. The Diophantine Nature of R.E. Relations

The prize is now within reach. A Diophantine definition of the sequence Y_a, i.e., of the relation,

$$R(a, y, x):\ y = Y_a(x),$$

is an easy, if not immediate, application of various results of the preceding section. The exponential growth of Y_a then easily yields a Diophantine definition of exponentiation, and, therewith, the Davis-Putnam-Robinson Theorem yields the Diophantine nature of all r.e. relations and the algorithmic unsolvability of Hilbert's tenth problem.

The first order of business of the present section is the claiming of this prize, the proof that all r.e. relations are Diophantine and that, thus, Hilbert's tenth problem is algorithmically unsolvable. Other applications are deferred to the next section. Following the unsolvability of Hilbert's tenth problem, I shall finish the section with some remarks on the question of credit.

5.1. Theorem. *For $a, x, y \geq 2$, the following are equivalent:*

i. $y = Y_a(x)$

ii. *There are $w, u, v, s, t, b \in \omega$ such that*

a. $2x \leq y$	f. $y \mid b-1$
b. $w^2 - (a^2-1)y^2 = 1$	g. $u \mid t-y$
c. $u^2 - (a^2-1)v^2 = 1$	h. $y \mid t-x$
d. $s^2 - (b^2-1)t^2 = 1$	i. $y^2 \mid v$
e. $u \mid b-a$	j. $v > 0$.

Proof. What do these equations say? Equations ii.b - d identify several variables as solutions to the Pell equation $X^2 - (a^2 - 1)Y^2 = 1$ and an auxiliary Pell equation $X^2 - (b^2 - 1)Y^2 = 1$. The relevant ones are
ii.b. $y = Y_a(k)$, for some k
ii.c. $v = Y_a(m)$ and $u = X_a(m)$, for some m
ii.d. $t = Y_b(n)$, for some n.

Equation ii.j tells us that $m > 0$, whence the Second Step-Down Lemma will apply to congruences

$$Y_a(i) \equiv Y_a(j) \pmod{(X_a(m))}.$$

Such a congruence is supplied by equations ii.e and ii.g by the Congruence Rule:

$$Y_b(n) \equiv Y_a(n) \pmod{b-a}, \text{ by the Congruence Rule,}$$

whence

$$Y_b(n) \equiv Y_a(n) \pmod{(X_a(m))}, \text{ by ii.e.}$$

But

$$Y_b(n) \equiv Y_a(k) \pmod{(X_a(m))}, \text{ by ii.g.}$$

Thus,

$$Y_a(n) \equiv Y_a(k) \pmod{(X_a(m))}.$$

By the Second Step-Down Lemma, this yields

$$n \equiv k \pmod{2m} \quad \text{or} \quad n \equiv -k \pmod{2m}. \tag{1}$$

Equation ii.i,

$$Y_a(k)^2 \mid Y_a(m),$$

just begs for application of the second half of the First Step-Down Lemma:

$$kY_a(k) \mid m.$$

With (1), this yields

$$n \equiv k \pmod{y} \quad \text{or} \quad n \equiv -k \pmod{y}. \tag{2}$$

Equations ii.f and ii.h now come in: By the Congruence Rule, ii.f yields

$$Y_b(n) \equiv n \pmod{y},$$

while ii.h yields

$$Y_b(n) \equiv x \pmod{y}.$$

Thus, (2) yields

$$x \equiv k \pmod{y} \quad \text{or} \quad x \equiv -k \pmod{y}. \tag{3}$$

We can now bring in ii.a and the bound 4.7.ii:

$$2x \le y, \quad 2k \le y.$$

If $x \ne k$, then $x + k < \max\{2x, 2k\} \le y$ and we cannot have $x \equiv -k \pmod{y}$. By (3), we therefore have

$$x \equiv k \pmod{y}.$$

With the bounds, this means $k = x$, i.e., $y = Y_a(x)$.

We have just proven ii $\Rightarrow$ i. Before beginning the synthetic half of the proof, the reader ought to interpret this analytic half geometrically. Consider a table of numbers and their Y_a and Y_b values. We want to take $y = Y_a(k)$ and pin down k, i.e., to show it must be x. To do this, we find some $Y_b(n)$ and observe that we have various congruences in the table as shown in Figure 5.2.

5.2. Figure.

i	$Y_a(i)$	$Y_b(i)$
0	0	0
1	1	1
$\vdots$	$\vdots$	$\vdots$
k	$y = Y_a(k)$	$Y_b(k)$
$\vdots$	$\vdots$	$\vdots$
n	$Y_a(n)$	$t = Y_b(n)$

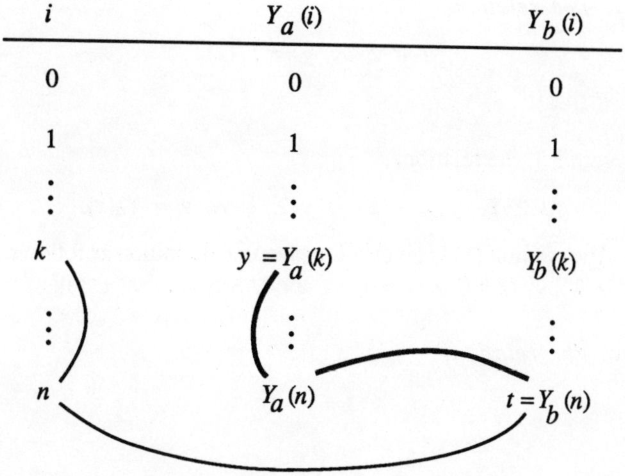

The fat lines indicate congruence modulo $X_a(m)$. The horizontal thin line indicates congruence modulo y and the vertical thin line indicates the $\pm$-congruence modulo y. Equations ii.b and ii.d postulate the existence of k and t. The horizontal fat line follows from ii.e and the Congruence Rule, the vertical from ii.g and the horizontal. Equation ii.f and the Congruence rule yield the horizontal thin line; and equations ii.c, ii.i, and ii.j with the vertical fat line yield the vertical thin line. We don't know what x is, but we do know that such congruences hold, and we want k to be x. So with ii.h we assert $x \equiv n$ $\pmod{y}$. With ii.a, we in fact get $x = \text{Rem}(n, y)$. The identical bound on k yields $k = \text{Rem}(n, y)$.

The real question emerging from the picture is: can we actually do this, i.e., can we actually find w, u, v, s, t, and b? The affirmative answer is the synthetic half of the proof.

First, let $y = Y_a(x)$. Equation ii.a is automatically satisfied, as is ii.b for $w = X_a(x)$. Let

$$m = xY_a(x).$$

By the First Step-Down Lemma, this guarantees

$$y^2 = Y_a(x)^2 \mid Y_a(m) > 0,$$

i.e., we can choose $u = X_a(m)$, $v = Y_a(m)$ and ii.c, ii.i, ii.j will all be satisfied.

$Y_a(m)$ and $X_a(m)$ are relatively prime and $y \mid Y_a(m)$. Thus, y and $X_a(m)$ are relatively prime and we can apply the Chinese Remainder Theorem to find b such that

$$b \equiv a \pmod{X_a(m)}$$
$$b \equiv 1 \pmod{y}.$$

This guarantees ii.e and ii.f.

The last step is to choose $t = Y_b(x)$ and $s = X_b(x)$. This guarantees ii.d, and the Congruence Rules yield ii.g and ii.h. QED

5.3. Corollary. *The relation,*

$$a \geq 2 \ \wedge \ y = Y_a(x), \tag{*}$$

is Diophantine.

Proof. By Theorem 5.1, the relation,

$$a \geq 2 \ \wedge \ x \geq 2 \ \wedge \ y \geq 2 \ \wedge \ y = Y_a(x), \tag{**}$$

is Diophantine. The relation (*) is given a Diophantine definition as follows:

$$(*): \quad a \geq 2 \ \wedge \ [x=0 \wedge y=0 \ \vee \ x=1 \wedge y=1 \ \vee \ (**)].$$ QED

5.4. Corollary. *The relation,*

$$z = x^y,$$

is Diophantine.

By the same sort of trick used in the proof of Corollary 5.3, it suffices to give a Diophantine definition that works for $x > 0$. There are several ways in which this can be done. The most straightforward approach is to note, by the Lemma on Bounds (4.7.i), that

$$Y_n(y) \sim (2n)^{y-1} \ \text{ as } n \to +\infty$$

and thus that

$$\frac{Y_{nx}(y+1)}{Y_n(y+1)} \sim x^y \text{ as } n \to +\infty.$$

One then finds a Diophantine bound $B(x,y)$ such that, if $n > B(x,y)$,

$$\left| \frac{Y_{nx}(y+1)}{Y_n(y+1)} - x^y \right| < 1/2,$$

and derives a Diophantine definition of exponentiation from this. Less intuitively, but more easily, we will derive the Corollary from the following.

5.5. Lemma. *Let* $a \geq 2, x > 0$.

i. $X_a(y) - Y_a(y)(a-x) \equiv x^y \pmod{2ax - x^2 - 1}$

ii. *For* $a = Y_{2x}(y+2)$, $2ax - x^2 - 1 > x^y$.

Proof. i. By induction on y. For $y = 0$, we have

$$X_a(0) - Y_a(0)(a-x) = 1 - 0(a-x) = 1 = x^0.$$

For the induction step, observe

$$\begin{aligned} X_a(y+1) - Y_a(y+1)(a-x) &= aX_a(y) + (a^2-1)Y_a(y) - (X_a(y) + aY_a(y))(a-x) \\ &= (a-a+x)X_a(y) + (a^2-1-a^2+ax)Y_a(y) \\ &= xX_a(y) + (ax-1)Y_a(y) \\ &\equiv xX_a(y) + (x^2-ax)Y_a(y) \pmod{2ax-x^2-1} \\ &\equiv x[X_a(y) - Y_a(y)(a-x)] \pmod{2ax-x^2-1} \\ &\equiv x \cdot x^y \pmod{2ax-x^2-1} \\ &\equiv x^{y+1} \pmod{2ax-x^2-1}. \end{aligned}$$

ii. By the Lemma on Bounds,

$$\begin{aligned} a = Y_{2x}(y+2) &\geq (4x-1)^{y+1} > (2x)^{y+1} \\ &> 2x^{y+1} = x^{y+1} + x^{y+1}. \end{aligned}$$

So

$$2ax > 2x^{y+2} + 2x^{y+2} \geq x^y + 2x^2 \geq x^y + x^2 + 1,$$

whence

$$2ax - x^2 - 1 > x^y.$$

QED

Proof of Corollary 5.4. Simply observe that

$$\begin{aligned} z = x^y \iff \exists a\, [a = Y_{2x}(y+2) \wedge 2ax - x^2 - 1 > z \wedge \\ \wedge X_a(y) - Y_a(y)(a-x) \equiv z \pmod{2ax-x^2-1}]. \end{aligned}$$

QED

From the Davis-Putnam-Robinson Theorem, Corollary 5.4 yields immediately the main result of this chapter.

5.6. Theorem. *Every recursively enumerable relation is Diophantine.*

By the existence of non-recursive r.e. sets, this has the immediate negative corollary:

5.7. Theorem. *There is no algorithm to decide which Diophantine equations have non-negative integral solutions, i.e., Hilbert's tenth problem is unsolvable.*

The next section is devoted to applications of these two fundamental theorems. I finish this section with a few remarks on crediting these results.

If we ignore such minor matters as how the proof given differs from the original and who is to be credited with making the adaptations, there is still a problem. Until now, I have only mentioned Juriĭ Matijasevič— not yet in this section, but in sections 1 and 2, above. Matijasevič, beyond all shadow of doubt, proved Theorem 5.6. Claims have been made on behalf of Gregory Chudnovsky for independent discovery.

At the time of the discovery, neither Matijasevič nor Chudnovsky was unknown to logicians; yet, despite the magnitude of the discovery, word did not get around too quickly. John McCarthy, the computer scientist who coined the phrase "artificial intelligence", was visiting the Soviet Union in early 1970 and attended a lecture given by a colleague of Matijasevič on the solution. McCarthy took notes and, when he returned to the United States, the whole world soon knew of Matijasevič's result and all logicians knew, up to some variation in spelling, Matijasevič's name. By the time the English translation of Matijasevič's paper appeared later in 1970, his proof of the Diophantine nature of the Fibonacci sequence had already been adapted to the Pell sequence by a number of researchers. Within a year, expositions in various languages were being written. Sometime after Matijasevič's fame had been established and the proof was well-known, rumours about Chudnovsky's independent discovery began to circulate.

Chudnovsky did publish an abstract in Russian in 1970 in which he claimed independent credit in the most neutral terms possible, stating simply his awareness of analogous work by Matijasevič. Unfortunately, the British translation of the journal, *Russian Math. Surveys*, did not then include the abstracts and most Western logicians never saw his abstract. A longer preprint in Russian, dated 1971, was circulated, but I do not know exactly when. I do know that it was no sooner than Autumn 1972 that Julia Robinson sent me a copy and I soon reciprocated with a very rough translation. The paper was a bit of a disappointment and quite a frustration. It was disappointing in that it did not give as general a construction as had been promised in the abstract; and it was frustrating first in that the detailed proofs of the first half of the paper gave nothing that wasn't already (i.e., by the time we were reading it) known and wasn't routine, and second in that the second half of the paper announced— without detail— two genuinely different approaches to proofs of Theorem 5.6.

There are two quite distinct issues here— credit for discovery, and contribution to the field. When I look at Chudnovsky's paper— it was eventually published in English translation in 1984 so anyone can now look at it— I get the impression it was primarily intended to serve as an historical document to establish Chudnovsky's share of credit for discovery. The most detailed material is that which does not significantly differ from Matijasevič's proof, which was already well-known by the time (mid-1970?) Chudnovsky was writing the paper. This material is accompanied by the closing remark that "the main results of §§1 - 5 were obtained by the author in 1969 and in the beginning of 1970... and were reported at the Steklov Institute of Mathematics (Moscow) and Kiev University". Moreover, when finally published in a volume of previously unpublished papers by Chudnovsky, this particular paper was given the least adequate description in the preface— the mere comment that it "is a translation of the author's paper of 1970 on diophantine sets, and represents an expanded version of his 1970 article", the 1970 article being the abstract referred to earlier. Of the new proofs (i.e., the two not already known) of Theorem 5.6, he offers only the barest of hints and says nothing about when they were discovered. (The natural conclusion is that they were obtained after the beginning of 1970, i.e., after Chudnovsky knew of Matijasevič's work.) Chudnovsky has not, at the time I'm writing this, published the details of these alternate proofs.

My own conclusion is that Chudnovsky was certainly capable of having proved Theorem 5.6 independently of Matijasevič. Whether he did so or not, is not, to me, the really important matter. Even if he deserves credit for co-discovery, he does not deserve credit for having made much of a contribution. His exposition of his version of the construction of Diophantine relations of exponential growth offers a minor contribution of perspective, but nothing of mathematics. The marvelous mathematical contribution he could have made, he withheld— namely, the two new proofs of Theorem 5.6. The latter of these proofs, showing the finiteness of the number of solutions to the equation,

$$9(U^2 + 7V^2)^2 - 7(W^2 + 7Z^2)^2 = 2,$$

(if that is indeed what he claims) would solve the major open problem in the area— Problem 9.11 of section 9, below.

It is not my task to criticise Chudnovsky, nor is it my responsibility (or even prerogative) to decide whether or not to credit him with independent co-discovery. The latter issue lacks proof in either direction, and the former— well, it is my responsibility as an expositor to tell the facts as I know them as clearly as I can and it is my further responsibility as a scholar and teacher not to skirt the issue or feign neutrality. Chudnovsky's situation, assuming he did make the discovery independently, can serve as an object lesson to the future professional mathematician.

A slim volume by Raymond Wilder, *Mathematics as a Cultural System*, can easily be read in one or two sittings and I recommend it heartily to the prospective mathematician. Among other things, he explains in the book not only that, but also why, many discoveries are multiply made, often simultaneously. A beautiful example is the independent discovery of the geometric representation of complex numbers by Jean Robert Argand, Abbé Buée, Karl Friedrich Gauss, C.V. Mourey, John Warren, and Caspar Wessel. The co-discovery of one of the non-Euclidean geometries by Gauss and János Bolyai is a better known example, and, of course, the most famous example in mathematics is the independent discovery by Isaac Newton and Gottfried Wilhelm Leibnitz of the collection of algorithms we call the calculus. In matters Diophantine, the adaptation of Matijasevič's technique to the Pell equation was carried out independently by Chudnovsky, Martin Davis, Simon Kochen and Julia Robinson, N.K. Kossovsky, and Kurt Schütte (with the word "adaptation" being perhaps not the correct one in Chudnovsky's case).

It has happened to me at least 5 or 6 times that a result of "mine" was also proven by someone else. The prospective mathematician might as well accept the fact now that most probably he will occasionally not be the sole discoverer of his results, possibly even his favourite results. The mathematician with a keen interest in mathematics and the requisite social skills can find a partner in his co-discoverer; the mathematician with an eye to glory can only create for himself two enemies— his co-discoverer and himself, the latter generally being the harmful one.

The most famous priority dispute in the history of mathematics, and the one with the most profound consequences, was that initially between the supporters of Newton and those of Leibnitz. The fight left Newton an embittered old man, it hurt Leibnitz, and it isolated British mathematics from the more vigorous continental development for over a century. Twentieth century scholarship finally answered the underlying question: who invented the calculus? Newton did this first in the mid-1660s, but kept his invention a secret. A decade later, Leibnitz independently re-discovered it and, after almost another decade, started publishing. Further development in the hands of Leibnitz, the Bernoullis, Euler, and the rest of the continental mathematicians followed in Leibnitz's, not Newton's, footsteps. We should stand in awe of the intellect of a man like Newton, who could invent the calculus and do such pioneering work in optics and mechanics as he did; but we should also be aware of the damage he did, of the dishonour he brought upon himself and the Royal Society in manipulating the latter in its pretended investigation into the matter of priority. Even a Newton-worshipper like Augustus de Morgan (who actually went so far as to write a monograph proving Newton's niece to have been secretly married to Newton's benefactor and that Newton had therefore not, as Voltaire said, prostituted her for his own advancement) declared Leibnitz had not been justly treated by the Royal Society. Mathematics does not have Newton to thank for the differential and integral calculus; it owes that debt to Leibnitz.

The dispute between Newton and Leibnitz was inevitable. Each man believed he had priority and, despite the vast number of priority disputes of the 17th century, the phenomenon of independent discovery was not yet recognised. Was it recognised in the 19th century when Gauss wrote to Farkas Bolyai to announce his pleasure that the latter's son had discovered exactly the same non-Euclidean geometry which Gauss had, but had not dared to publish for fear of the "clamour of the Boetians"? Whether yes or no, the young generally do not learn from history and János Bolyai bitterly assumed the great Gauss was trying to steal his invention.

The future mathematicians among the readers should also be aware that equal credit does not accrue to equal discovery. There are certain rules governing this. One is the so-called *Matthew effect*, popularly put "them that's got shall get; them that's not shall lose". All things being equal, the more famous of the co-discoverers will receive greater acknowledgement. In the case of the geometric representation of the complex numbers cited above, the French refer to the "Argand plane" and almost everyone else refers to Gauss. Wessel is occasionally cited as the first to publish, but who ever mentions Buée, Mourey, or Warren? This inequality is not really unfair. The greater the fame of a discoverer, the more the prestige that rubs off on the discovery and the greater the likelihood of the discovery entering into an active mathematical development: there are numerous cases where a discovery lay dormant for years and only its re-discovery by a more famous mathematician brought it to everybody's attention. When the earlier discovery is unearthed, the right words are said, but the greater credit— that for the ensuing development— still accrues to the famous re-discoverer.

Sometimes things can be unfair. One friend of mine sat through a lecture given by the advisor of someone whose work duplicated that of my friend. After hearing his own work negatively compared to that of the lecturer's student for an hour, he approached the lecturer, expressed great interest, and said he would like to know more. In the end, my friend spent a year at the lecturer's university and made a number of contacts as well as friends. Was he merely playing some sort of game? Perhaps, but in retrospect it seems more constructive than brooding over being treated unfairly.

But enough! My head is swimming with anecdotes that I'd love to tell, but I shall probably get into enough hot water over this last one when the parties involved recognise themselves. The main point has been made. The reader who goes on to an academic career in mathematics will find on occasion that his work, however original, is not unique, and that he may not be getting all the recognition he feels he deserves. When this happens, he can join the "opposition", turning competition into coöperation; he can leave the area and work on something else; or, he can impose a heavy emotional penalty on himself. The choice will really be up to him; he need not suffer if he doesn't want to.

Exercises.

1. Let $a \geq 2$.
   i. Give a Diophantine definition of the relation $z = X_a(x)$.
   ii. Let $x > 1, a > x^y$. Show:
   $$x^y = \left[\frac{X_{ax}(y)}{X_a(y)}\right].$$
   iii. Give a Diophantine definition of exponentiation based on ii.
2. Find a Diophantine bound $B(x, y)$ such that
   $$\left|\frac{Y_{nx}(y+1)}{Y_n(y+1)} - x^y\right| < 1/2$$
   for $n > B(x,y)$ and give a Diophantine definition of exponentiation based on this.
3. Give a Diophantine definition of the relation $y = Y_a(x)$ based on one of the Alternative Second Step-Down Lemmas of Exercises 6 and 7 of section 4, above.
4. Consider the equation $X^2 - 7Y^2 = 2$.
   i. Apply Thue's result (Exercise 13 of section 3) to show: If x, y are natural numbers satisfying this equation, then, for some $n \in \omega$,
   $$x + y\sqrt{7} = (3+\sqrt{7})(8+3\sqrt{7})^n.$$
   ii. Show: If x, y satisfy this equation, then $3 \mid x$
   iii. Show: The equation,
   $$9(U^2+7V^2)^2 - 7(W^2+7Z^2)^2 = 2,$$
   has only finitely many solutions.

6. Applications

In the present section we shall consider some applications of Theorem 5.6 (every r.e. relation is Diophantine) and 5.7 (Hilbert's tenth problem is unsolvable). Most of these are unsolvability results and depend only on Theorem 5.7. A few applications require the more fundamental Theorem 5.6. As one might expect from its title, the present section is a miscellany with no discernible structure to it. The minimal structure it has is this: it starts with simple applications of Theorem 5.7 to questions of unsolvability, and switches to more recursion theoretic applications of Theorem 5.6 following 6.8.

For the sake of the record, Fact 1.5.i on the equivalence of $DIOPH(N)$, $DIOPH(N^+)$, and $DIOPH(Z)$ ought to be invoked to yield our first application.

6.1. Application. *The Diophantine problems over the integers and the positive integers are algorithmically unsolvable.*

As stated in section 1, the Diophantine problem over the rationals is yet open. The available partial results will be discussed in the next section.

In section 10, below, we will discuss the problem of reducing the number of variables needed for an unsolvable Diophantine problem over the natural numbers. At the cost of adding variables, the degree is easily lowered to 4.

6.2. Application. *The general problem of deciding which systems of quadratic equations are solvable in natural numbers is algorithmically unsolvable. The general Diophantine problem of degree 4 is algorithmically unsolvable.*

Proof. The idea of the proof is very simple. One keeps introducing new variables to lower the degree. Suppose we have a polynomial,

$$P:\ \Sigma c_{i_0 \ldots i_{n-1}} X_0^{i_0} \ldots X_{n-1}^{i_{n-1}}. \tag{1}$$

Introduce new variables $Y_{i_0 \ldots i_{n-1}}$ and equations

$$Y_{i_0 \ldots i_{n-1}} = X_0^{i_0} \ldots X_{n-1}^{i_{n-1}} \tag{2}$$

$$\Sigma c_{i_0 \ldots i_{n-1}} Y_{i_0 \ldots i_{n-1}} = 0. \tag{3}$$

Solvability of (1) in the X's is equivalent to the solvability of the system (2) and (3) in the X's and Y's. The degree of (3) is 1, whence (3) need no longer be tampered with. An instance of (2) of degree 1 or 2 need not be tampered with. Any other instance can be replaced by a system of lower degree. For example, if $i_0 > 1$ and $i_1 + \ldots + i_{n-1} > 0$, one can replace the corresponding instance of (2) by the pair

$$Y_{i_0 \ldots i_{n-1}} = Z X_1^{i_1} \ldots X_{n-1}^{i_{n-1}} \tag{4}$$

$$Z = X_0^{i_0}, \tag{5}$$

with each equation of degree less than $i_0 + \ldots + i_{n-1}$.

This method can be used in a rigorous proof by induction on degree by choosing the correct induction hypothesis. The correct choice of assertion to induct on is that every *system* of equations of degree $d \geq 2$ is equivalent to a system of degree 2. I leave the details to the reader.

As to the degree 4 problem, let P be an arbitrary polynomial. Find a system $P_0, \ldots, P_{m-1}$ of quadratic polynomials, the solvability problem of which is equivalent to that of P. Then observe:

$$\exists x\, [P(x) = 0] \text{ iff } \exists y\, [\bigwedge_{i=0}^{m-1} P_i(y) = 0]$$

$$\text{iff } \exists y\, [\sum_{i=0}^{m-1} P_i(y)^2 = 0],$$

and ΣP_i^2 has degree at most 4. QED

Linear algebraic techniques can be applied to the question of which systems of linear equations have solutions in natural numbers. The solvability of this problem will be established along logical lines in Chapter III, below. The solvability of the quadratic Diophantine problem is a deep result due to Carl Ludwig Siegel around 1972. The problem of providing a general method or showing there is none is open for the cubic case.

Incidentally, the reduction used in the proof of Application 6.2 has been known for at least half a century. It seems first to have appeared in print in Thoralf Skolem's book on Diophantine equations published in 1938. It is a curious thing that Skolem, who worked in both Diophantine equations and recursion theory, never sought to prove the unsolvability of Hilbert's tenth problem. His only recorded comment on it appears to be a remark from the end of a paper he published in 1962, "It has been asked whether the recursively enumerable sets are all of them Diophantine sets... I regret not having had the opportunity to study this question seriously."

Another easy, but by no means trivial, application is the following.

6.3. Application. (Putnam). *The problem of determining which polynomials represent all natural numbers (all integers) is unsolvable.*

Proof. (Davis, Putnam, Robinson). Putnam's original proof of 1960 pre-dated the solution of Hilbert's tenth problem and wasn't this easy. Let $P(A, X_0, ..., X_{n-1})$ define a non-recursive r.e. set S,

$$a \in S \Leftrightarrow \exists x\, [P(a, x) = 0].$$

Define the polynomials

$$Q(A, X_0, ..., X_{n-1}, Y, Z) = (1 - 2\cdot P(A, X_0, ..., X_{n-1})^2)Y - Z$$

and $$Q_a(X_0, ..., X_{n-1}, Y, Z) = Q(a, X_0, ..., X_{n-1}, Y, Z).$$

Observe that

$$a \in S \Rightarrow \text{range}(Q_a) = Z$$

$$a \notin S \Rightarrow \text{range}(Q_a) \subseteq Z - N = \{z \in Z\colon z \text{ is non-positive}\}.$$ QED

We shall shortly generalise the trick used. First, however, I want to present the following complementary result due to Dorothy Bollman and Miguel Laplaza.

6.4. Application. *The problem of deciding which polynomials $Q{:}N^n \rightarrow N$ are one-one is unsolvable.*

Proof. (Smoryński). We need an auxiliary polynomial $P{:}N^2 \rightarrow N$ with the following properties:

i. for all $x, y \in \omega$, $P(x, 0) = P(0, y) = 0$
ii. for all $x, y > 0$, $P(x, y) > 0$
iii. restricted to pairs of positive integers, P is one-one.

Assuming for the moment we have such a P, let R be any n-ary polynomial and define

$$Q_R(X_0, ..., X_{n-1}) = P(R(X_0, ..., X_{n-1})^2, \langle X_0, ..., X_{n-1}\rangle_n),$$

where $\langle\cdot,...,\cdot\rangle$ is any polynomial n-tupling function. Observe that

$$Q_R \text{ is one-one} \quad \text{iff} \quad R \text{ has no solution other than } \langle 0, ..., 0\rangle_n \text{ in } \omega^n.$$

A solution to the left-hand problem would provide a solution to the right-hand one, which, however, is (essentially) Hilbert's tenth problem stated negatively.

And how do we find P? Lou van den Dries and Albert Visser came up with the ultra-simple

$$P(X, Y) = \langle X^2Y, XY^2\rangle.$$

Cf. Exercise 1, below. QED

A couple of remarks about this proof: First, if we don't mind adding a variable, we could take Q_R to be

$$Q_R(X_0, ..., X_{n-1}, Y) = P(R(X_0, ..., X_{n-1})^2, Y + 1).$$

Second, note that the polynomial Q_R assumes no negative values. Thus, the problem of distinguishing those polynomials which assume only non-negative values (unsolvable— as seen by inserting minus signs in the right places in the proof of Application 6.3) does not enter into the unsolvability of the problem at hand. In fact, one can obtain double surety in the matter by squaring: the problem of deciding for which polynomials Q the square $Q^2:N^n \rightarrow N$ is one-one is unsolvable.

The combined problem of surjectivity and injectivity, i.e., that of deciding which polynomials $Q:N^n \rightarrow N$ are one-one correspondences, is an open problem. Conjecture I.4.4 of Chapter I expresses the hope that the n-tupling polynomials can be explicitly described; Bollman and Laplaza have shown, however, that the corresponding problem for *algebraic functions* (functions the graphs of which are polynomial relations) is unsolvable.

A last simple result to look at before considering more recursion theoretic matters is a nice result due to Martin Davis. It requires a little notation.

6.5. Notations. For any polynomial $P(X_0, ..., X_{n-1})$, let

$$\#(P) = \text{card}(\{(x_0, ..., x_{n-1}): P(x_0, ..., x_{n-1}) = 0\}),$$

be the number of zeroes of P. This cardinal is either a natural number or it is the cardinality of the set of natural numbers. (Cf. I.3.2 and its ensuing discussion.) Since the only infinite cardinal we will be interested in in this book is that of the natural numbers, we can simply write "∞" for the infinite cardinal. We write

$$C = \{0, 1, ..., \infty\} = \omega \cup \{\infty\}$$

for the set of possible values of $\#(P)$. Finally, if $A \subseteq C$ we set

$$A^* = \{P: \#(P) \in A].$$

Davis' Theorem is the following analogue to Rice's Theorem (I.12.21).

6.6. Application. (Davis' Theorem). *Let $A \subseteq C$ with $A \neq \emptyset$ and $A \neq C$. Then: A^* is not recursive.*

The proof requires a simple lemma.

6.7. Lemma. i. *Let $m \in \omega$. There is a recursive operation T^m on polynomials such that, for all P,*

$$\#(T^m(P)) = \#(P) + m.$$

ii. *There is a recursive operation* T^∞ *on polynomials such that, for all* P,

$$\#(T^\infty(P)) = \begin{cases} 0, & \#(P) = 0 \\ \infty, & \#(P) > 0. \end{cases}$$

Proof. If $m = 0$, let $T^m(P) = P$. Otherwise, let $P(X_0, ..., X_{n-1})$ be given, let Y be a new variable, and define

$$T^m(P):\ (P(X_0, ..., X_{n-1})^2 + Y^2)\cdot\prod_{i=1}^{m} (X_0^2 + ... + X_{n-1}^2 + (y-i)^2).$$

The zeroes of $T^m(P)$ are just the tuples $(x_0, ..., x_{n-1}, 0)$, where $P(x_0, ..., x_{n-1}) = 0$ and $(0, ..., 0, i)$, for $i = 1, ..., m$. Thus, $\#(T^m(P)) = \#(P) + m$.

ii. Given $P(X_0, ..., X_{n-1})$, let Y be a new variable and define

$$T^\infty(P):\ P(X_0, ..., X_{n-1})\cdot(Y+1).$$

The zeroes of $T^\infty(P)$ are exactly the tuples $(x_0, ..., x_{n-1}, y)$, where $P(x_0, ..., x_{n-1}) = 0$ and $y \in \omega$. QED

Proof of Application 6.6. Let $\varnothing \subset A \subset C$ be given (inclusions proper). There are two cases.

Case 1. $\infty \in A$. Choose $m \in \omega$ such that $m \notin A$ and observe, for any P,

$$\#(P) = 0 \text{ iff } \#(T^m T^\infty(P)) \notin A,$$

i.e.,

$$\#(P) > 0 \text{ iff } T^m T^\infty(P) \in A^*.$$

If A^* were recursive, we could decide $\#(P) > 0$, i.e., Hilbert's tenth problem would be decidable.

Case 2. $\infty \notin A$. Then $\infty \in C - A$ and $(C-A)^* = C^* - A^*$ is not recursive (by Case 1), whence A^* is not recursive. QED

The meaning of Application 6.6 is clear enough, but it might be worthwhile to note two simple consequences.

6.8. Corollary. *There is no effective procedure to solve either of the following two problems*:

i. *which polynomials have unique solutions*

ii. *which polynomials have infinitely many solutions.*

Simple as it is, Application 6.6 demands comment. The simplest point to dispose of concerns the extra variables in T^m and T^∞. Using the facts that a polynomial on ω^n which is 0 for all but finitely many n-tuples is identically 0 (Exercise 2, below) and that we can effectively decide if a polynomial is identically 0 (by putting it into a normal form $\Sigma c_{i_0...i_{n-1}} X_0^{i_0} ... X_{n-1}^{i_{n-1}}$ and checking that all the coefficients $c_{i_0...i_{n-1}}$ are 0), we can easily find a recursive version of T^m which does not require a new variable. Given P, decide if P is identically 0 or not. If it is, set $T^m(P) = P$. Otherwise, let $(a_{00}, ..., a_{0,n-1}), ...,$

$(a_{m-1,0}, ..., a_{m-1,n-1})$ be the first m n-tuples in some natural ordering which are not zeroes of P, and define

$$T^m(P) = P(X_0, ..., X_{n-1})\cdot \prod_{i=0}^{m-1}(\sum_{j=0}^{n-1}(X_j - a_{ij})^2).$$

There seems to be no similar natural definition of T^∞ which does not require an additional variable. That there is such a recursive T^∞ which does not raise the number of variables (provided P has enough variables to begin with) can be established recursion theoretically.

Another quick remark: The extension of Davis' Theorem to an analogue to the Rice-Shapiro Theorem can be made in the exponential Diophantine case (cf. section 9, below), but is yet an open problem in the Diophantine case.

The mention of recursion theory becomes necessary with Davis' Theorem. For the previous applications, even for Theorem 5.7 on the unsolvability of Hilbert's tenth problem, it was not necessary to be very formal about effectiveness in a non-numerical context. Once we have formally isolated the notion of recursiveness of sets and functions of natural numbers, have proven the existence of a non-recursive r.e. set, and have had a little experience with recursive functions, we can generally forego the formality and deal with intuitively effective operations. Hilbert's tenth problem is not unsolvable because we have recursively reduced a recursively unsolvable problem to it— the notion of recursiveness does not yet apply outside a strictly numerical context— but because the reduction was effective and an effective solution to the tenth problem would yield an effective solution to a recursively unsolvable problem about natural numbers, where "effective" does mean "recursive". So, why did I say "recursive" in Application 6.6 and Lemma 6.7, where "effective" was the appropriate adjective? The reason is that we shall now take a more formal look at the notion of effectiveness when applied to polynomials.

Before attempting to define a formal notion of effectiveness or recursiveness of functions of polynomials, we must decide what a polynomial is, how we are to grab hold of one to manipulate it. It won't do to think of a polynomial as a function with infinite domain; that is too much to handle effectively. We must think of a polynomial as a formal expression of some sort— a string of symbols. One possibility is to deal with the normal forms,

$$\Sigma c_{i_0 \ldots i_{n-1}} X_0^{i_0} \ldots X_{n-1}^{i_{n-1}},$$

of polynomials. This is a definite, but inconvenient, possibility. Instead, we shall think of polynomials as terms built up from variables and constants by additions and multiplications, with one final subtraction. Constants will be natural numbers and variables will be from the fixed list $X_0, X_1, ...$— only the letter X subscripted by a natural number. We shall also demand that all parentheses be inserted so that no ambiguity is possible.

6.9. Definition. The *pre-polynomial expressions* (*ppe*'s) are defined inductively by:

i. each natural number n is a ppe
ii. each variable X_n is a ppe
iii. if P, Q are ppe's, then so are $(P + Q)$ and $(P\cdot Q)$.

6.10. Definition. A *polynomial expression* is an expression $(P - Q)$, where P, Q are pre-polynomial expressions.

In these definitions, +, ·, and – are symbols, not operations. Moreover, the parentheses are also symbols and form integral parts of the strings. Because of the parentheses, there is a unique readability theorem asserting that a polynomial expression can be read in one and only one way as a polynomial. In introductory logic texts one often finds proofs of such results. I think I will skip this.

Another thing I will skip is a proof that every polynomial has at least one polynomial expression representing it and that, if we are given a decent representation of the polynomial, we can effectively find a polynomial expression for it. The process is simple enough. Begin with the polynomial, e.g.,

$$X_3 + (X_2 - 3X_1)X_1 + 7X_0 - 5X_1,$$

and expand it:

$$X_3 + X_2X_1 - 3X_1^2 + 7X_0 - 5X_1.$$

Separate the positive and negative parts,

$$(X_3 + X_2X_1 + 7X_0) - (3X_1^2 + 5X_1),$$

and then insert parentheses, say by associating additions and multiplications to the left:

$$(((X_3 + (X_2 \cdot X_1)) + (7 \cdot X_0)) - (((3 \cdot X_1) \cdot X_1) + (5 \cdot X_1))).$$

If we wish to compute with polynomial expressions, we must either extend our formal definition of recursiveness to the language of polynomial expressions, a straightforward but boring task, or we must code the expressions as numbers and use the notion of recursiveness for functions of natural numbers to explain what we mean by a recursive operation on polynomial expressions. More specifically, the second approach assigns to polynomial expressions P numerical codes $\ulcorner P \urcorner$ and declares an operation T on polynomial expressions to be *recursive* if the induced function,

$$F_T(\ulcorner P \urcorner) = \ulcorner T(P) \urcorner,$$

is recursive. This latter approach is not totally unfamiliar to us— recall the recursive simulation of finite sequences— and is a simpler task than the former of working out a general recursion theory. We shall thus follow this simpler approach.

6.11. Numerical Codes for Polynomial Expressions. We begin with a simple one-one correspondence between pre-polynomial expressions and natural numbers:

i. the number n has code $\ulcorner n \urcorner = 4n$

ii. the variable X_n has code $\ulcorner X_n \urcorner = 4n + 1$

iii. the expression $(P + Q)$ has code $\ulcorner (P + Q) \urcorner = 4\langle \ulcorner P \urcorner, \ulcorner Q \urcorner \rangle + 2$

iv. the expression $(P \cdot Q)$ has code $\ulcorner (P \cdot Q) \urcorner = 4\langle \ulcorner P \urcorner, \ulcorner Q \urcorner \rangle + 3$.

From this, one gets a one-one correspondence between polynomial expressions and natural numbers by simply taking

$$\ulcorner P - Q \urcorner = \langle \ulcorner P \urcorner, \ulcorner Q \urcorner \rangle.$$

For each natural number e, we let P_e denote the unique polynomial expression with code e: $\ulcorner P_e \urcorner = e$.

That 6.11 does indeed provide one-one correspondences is an easy exercise I leave to the reader. (Cf. Exercise 4, below.) This particular coding is due to Julia Robinson, who used it to give the following proof of the Normal Form Theorem for r.e. relations.

6.12. Application. (Normal Form Theorem). *For each $m > 0$, there is an r.e. relation T_m such that, for all $a_0, ..., a_{m-1} \in \omega$,*

$$\exists x_m ... x_{n-1} \, [P_e(\boldsymbol{a}, \boldsymbol{x}) = 0] \quad \text{iff} \quad T_m(e, a_0, ..., a_{m-1}),$$

where $n > m$ is large enough so that the variables of P_e are from the list $X_0, ..., X_{m-1}, X_m, ..., X_{n-1}$.

Proof. Observe that $\exists x_m ... x_{n-1} \, [P_e(\boldsymbol{a}, \boldsymbol{x}) = 0]$ is equivalent to

$$\exists st \, \{\forall y \le e \, [(t)_{4y} = y \;\wedge\; (t)_{4y+1} = (s)_y \;\wedge\; (t)_{4y+2} = (t)_{\pi_1^2(y)} + (t)_{\pi_2^2(y)} \;\wedge$$

$$\wedge\; (t)_{4y+3} = (t)_{\pi_1^2(y)} \cdot (t)_{\pi_2^2(y)}] \;\wedge\; \bigwedge_{i=0}^{m-1} [(s)_i = a_i] \;\wedge\; (t)_{\pi_1^2(e)} = (t)_{\pi_2^2(e)}\}.$$

This latter relation is clearly Σ_1 and can be put into the desired form. QED

6.13. Corollary. (Universal Diophantine Relations). *For each $m > 0$, there is an $(m+1)$-ary Diophantine relation D_m universal for m-ary Diophantine relations, i.e., for every m-ary Diophantine relation D there is an e such that, for all $a_0, ..., a_{m-1} \in \omega$,*

$$D(a_0, ..., a_{m-1}) \;\Leftrightarrow\; D_m(e, a_0, ..., a_{m-1}).$$

The proof is simple. One applies Theorem 5.7 to the r.e. relation T_m of Application 6.12. Notice, by the way, that, by appeal to D_1, there is a fixed number n of variables needed to give a Diophantine definition of any r.e. set. By the methods given so far, n is very large. It can, however, be improved and the currently best known value of n is $n = 9$. Section 10, below, will give a similar small n for (iterated) exponential Diophantine relations. Incidentally, the same number n applies to all Diophantine relations as one can use the polynomial m-tupling functions to reduce m-ary Diophantine relations to unary ones.

The immediate value of Application 6.12 is as a novelty. We can, however, pair it with the Parameter Theorem and begin to apply some recursion theory.

6.14. Application. (Parameter Theorem). *For each $m > 0$, there is a primitive recursive function S such that, for any $e, a, a_0, ..., a_{m-1} \in \omega$,*

$$T_m(S(e,a), a_0, ..., a_{m-1}) \;\Leftrightarrow\; T_{m+1}(e, a, a_0, ..., a_{m-1}).$$

Proof. Fix e. Consider the relation

$$R(a, a_0, ..., a_{m-1}): \quad T_{m+1}(e, a, a_0, ..., a_{m-1})$$

i.e., $R(a, a_0, ..., a_{m-1})$: $\exists y\, [P_e(a, a_0, ..., a_{m-1}, y) = 0]$.

We want to fix the parameter a and replace $P_e(X_0, X_1, ..., X_m, X_{m+1}, ..., X_{n-1})$ by $P_e(a, X_0, ..., X_{m-1}, X_{m+1}, ..., X_{n-1})$, and find the code of this new polynomial recursively from the code e of the old one and the parameter a.

We first construct an auxiliary substitution function for pre-polynomial expressions by a course-of-values recursion by cases. Define S_1 by

$$S_1(4y, x) = 4y$$

$$S_1(4y+1, x) = \begin{cases} 4x, & y = 0 \\ 4y - 3, & 0 < y \le m \\ 4y + 1, & m < y \end{cases}$$

$$S_1(4y+2, x) = 4\langle S_1(\pi_1^2(y), x), S_1(\pi_2^2(y), x)\rangle + 2$$

$$S_1(4y+3, x) = 4\langle S_1(\pi_1^2(y), x), S_1(\pi_2^2(y), x)\rangle + 3.$$

The function S on codes of polynomial expressions is then

$$S(y, x) = \langle S_1(\pi_1^2(y), x), S_1(\pi_2^2(y), x)\rangle.$$ QED

In this last proof, we have implicitly invoked our convention of considering an operation on polynomials recursive if the corresponding operation on the codes of polynomial expressions is recursive. We shall, for the most part, continue to keep the convention implicit and merely manipulate the codes.

In section 12 of Chapter I, we defined an indexing of partial recursive functions to be *acceptable* if the indexing had a universal partial function and satisfied the Parameter Theorem. Adapting the notion to indexings of r.e. relations, so that the universal function becomes the universal relation, we see that Applications 6.13 and 6.14 yield the acceptability of the indexing 6.12 of r.e. relations by codes for polynomial expressions. This is a useful observation. For one thing, if we apply the relational analogue to Roger's Theorem (I.12.17) on the recursive isomorphism of any two acceptable indexings of partial recursive functions, we can conclude the isomorphism of our polynomial indexing of r.e. relations with the indexing given in Chapter I. An application is this: if the reader has lost track of the effectiveness of the proof of Theorem 5.6 that every r.e. relation is Diophantine, it doesn't matter; Rogers tells us that there is a recursive F_m such that, for any e,

$$W_e^m = \{(a_0, ..., a_{m-1}): \exists y\, [P_{F_m(e)}(a, y) = 0]\}.$$

(But cf. Exercise 5.)

The thing to notice about this latest application is not the application itself— the proof of Theorem 5.6 given is constructive and a recursive F_m can be obtained from it by inspection; the thing to notice is that it opens the door to recursion theoretic techniques. To begin with, since our new indexing is acceptable, let us use the notation "W_e^m" for it rather than for the old indexing:

$$W_e^m = \{(a_0, ..., a_{m-1}): T_m(e, a_0, ..., a_{m-1})\}$$

$$= \{(a_0, ..., a_{m-1}): \exists y\,[P_e(a, y) = 0]\}.$$

One might even write W_P^m for $W_{\ulcorner P \urcorner}^m$.

Returning to Application 6.6, we can take a recursion theoretic look at Lemma 6.7.ii:

6.15. Lemma 6.7.ii Revisited. *There is a recursive operation T^∞ on polynomials such that, for any P,*

$$\#(T^\infty(P)) = \begin{cases} 0, & \#(P) = 0 \\ \infty, & \#(P) > 0. \end{cases}$$

Proof. The binary r.e. relation,

$$R(e, x): \exists y\,[y \in W_e \wedge x = x],$$

has an index e_0 relative to the Normal Form Theorem 6.12. Thus,

$$R(e, x): T_2(e_0, e, x).$$

By the Parameter Theorem, there is a recursive $G(e) = S(e_0, e)$ such that, for any e,

$$R(e, x): T_1(G(e), x),$$

i.e.,

$$W_{G(x)} = \{x: R(e, x)\} = \begin{cases} \emptyset, & W_e = \emptyset \\ \omega, & W_e \neq \emptyset. \end{cases}$$

If we define $T(P) = P_{G(\ulcorner P \urcorner)}$, we quickly see that T is the desired T^∞:

$$\#(P_e) = 0 \Rightarrow W_e = \emptyset \Rightarrow W_{G(e)} = \emptyset \Rightarrow \#(T(P_e)) = 0$$

$$\#(P_e) > 0 \Rightarrow W_e \neq \emptyset \Rightarrow W_{G(e)} = \omega \Rightarrow \#(T(P_e)) = \infty.$$ QED

The function T^∞ constructed in this proof could produce polynomials with arbitrarily many variables. Writing

$$D_1(e, x): \exists y_0 ... y_{n-1}\,[P(e, x, y_0, ..., y_{n-1}) = 0],$$

the polynomial

$$Q_e(X_0, ..., X_n) = P(G(e), X_0, ..., X_n)$$

has exactly $n + 1$ variables, regardless of the number of variables P_e has. Thus, setting $T^\infty(P_e) = Q_e$, we have the result promised earlier that, if P has enough variables to begin with, $T^\infty(P)$ need not have any more variables than P. [Unfortunately, this economy starts one step too late! The number n can be taken to be 9. The Diophantine representation of a non-recursive r.e. set uses a polynomial in 10 variables— the 9 existentially quantified ones and the parametric one. The undecidability result holds in 9 variables as the parametric variable is successively replaced by the constants 0, 1, ... The proof of Application 6.6 adds one variable to the polynomials as (as we have already seen) T^m need not require any new variable; similarly, the present T^∞, even restricted to 9-variable polynomials, produces a 10-variable polynomial. Thus either construction of T^∞ yields the same number of variables for Davis' Theorem.]

The proof of Lemma 6.15 is just an indication. Using polynomial expressions as indices of r.e. sets, Rice's Theorem and the Rice-Shapiro Theorem yield the following two results.

6.16. Application. *No non-trivial property of Diophantine sets can be effectively determined form the polynomial expressions indexing the sets.*

6.17. Application. *Suppose we can recursively enumerate all polynomial expressions indexing Diophantine sets with a given non-trivial property. There is a recursive function F such that a polynomial expression P is in the enumeration iff, for some n,*

$$D_{F(n)} \subseteq W_{\ulcorner P \urcorner}.$$

We needn't give the proofs— those given in Chapter I are valid for any acceptable indexing.

Applications 6.16 and 6.17 are not particularly interesting in that they offer nothing really new. Generalising the construction in the proof of Putnam's Theorem (Application 6.3, above), we can get something a bit more interesting.

6.18. Lemma. T*here is a recursive function G which, given any polynomial expression P, produces a polynomial expression Q such that the non-negative range of (the polynomial defined by) Q is the Diophantine set defined by P, i.e., every r.e. set is effectively the non-negative range of a polynomial.*

Proof. Let $P(X_0, ..., X_n)$ be given, X_0 serving as the parametric variable, and define Q_1 by

$$Q_1(X_0, ..., X_n) = (1 - P(X_0, ..., X_n)^2)X_0.$$

Q_1 is not quite what we want: its positive range is just the set of positive integers in $W_{\ulcorner P \urcorner}$, but 0 is trivially in its non-negative range. To avoid adding 0 if it doesn't belong, modify Q_1 to get

$$Q(X_0, ..., X_n) = (1 - P(X_0 - 1, ..., X_n)^2)X_0 - 1.$$

I leave to the reader the verification that Q works. QED

The effectiveness of the passage from P to Q is obvious. Because we only allowed one subtraction in a polynomial expression, the formal recursiveness of the passage from P to a polynomial expression for Q may not be so clear. Recursion theory provides an easy, albeit ridiculous, proof that we can recursively find such a Q. The first step is to find Q_1 recursively. Note that Q_1 is not formally a polynomial expression, but it is easily given such an expression. Write $P = (P_1 - P_2)$. Then

$$\begin{aligned} Q_1 &= (1 - (P_1 - P_2)^2)X_0 \\ &= (1 - (P_1^2 - 2P_1P_2 + P_2^2))X_0 \end{aligned}$$

$$= (1 + 2P_1P_2)X_0 - (P_1^2 + P_2^2)X_0$$

$$= (((1 + ((2{\cdot}P_1){\cdot}P_2)){\cdot}X_0) - (((P_1{\cdot}P_1) + (P_2{\cdot}P_2)){\cdot}X_0)),$$

this last (assuming I got all the parentheses right) being a polynomial expression. For $\ulcorner P \urcorner = e = \langle \ulcorner P_1 \urcorner, \ulcorner P_2 \urcorner \rangle$, we see that we can define

$$G_1(e) = \langle \mathit{Firsthalfq}(e), \mathit{2ndhalfq}(e) \rangle,$$

where

$$\mathit{Firsthalfq}(e) = \mathit{Xmult}(\ \mathit{Crossmult}(\pi_1^2(e), \pi_2^2(e))\ )$$

$$\mathit{2ndhalfq}(e) = \mathit{Xmult}(\ \mathit{Sumsq}(\pi_1^2(e), \pi_2^2(e))\ ),$$

where

$$\mathit{Xmult}(a) = 4\langle a, 1\rangle + 3$$

$$\mathit{Crossmult}(p_1, p_2) = 4\langle 4,\ 4\langle 4\langle 8, p_1\rangle + 3, p_2\rangle + 3\rangle + 2$$

and

$$\mathit{Sumsq}(p_1, p_2) = 4\langle 4\langle p_1, p_1\rangle + 3,\ 4\langle p_2, p_2\rangle + 3\rangle + 2.$$

A little work will show $G_1(\ulcorner P \urcorner) = \ulcorner Q_1 \urcorner$.

Now, as in the proof of Lemma 6.15, find H recursive such that

$$W_{H(e)} = \{x + 1\colon x \in W_e\}$$

and observe that we want

$$Q = P_{G_1H(e)}(X_0, ..., X_n) - 1.$$

This gives

$$G(e) = \langle \pi_1^2(G_1H(e)), 4\langle \pi_2^2(G_1H(e)), 4\rangle + 2\rangle. \qquad \text{QED}$$

I have the feeling I have been a bit excessive in actually giving the details and promise that this is the last index construction we will meet in the present volume.

Lemma 6.18 was proven so that it could be applied to yield the following.

6.19. Application. (Rice's Theorem for Ranges of Polynomials). *Let $\mathfrak{P}$ be a property holding of some r.e. sets and not holding of others. The set of polynomial expressions whose non-negative ranges have property $\mathfrak{P}$ is not recursive.*

Proof. Note that $D = \{\ulcorner P \urcorner\colon W_{\ulcorner P \urcorner} \text{ has property } \mathfrak{P}\}$ is not recursive by application 6.16. If

$$R = \{\ulcorner Q \urcorner\colon \text{range}(Q) \cap \omega \text{ has property } \mathfrak{P}\}$$

were recursive, then D would be recursive as we would have

$$\ulcorner P \urcorner \in D \Leftrightarrow G(\ulcorner P \urcorner) \in R,$$

where G is as in Lemma 6.18. QED

In a similar manner, one can prove an analogue to the Rice-Shapiro Theorem for the non-negative ranges of polynomials. Two special instances of such applications are

i. we cannot recursively enumerate all polynomials which map onto all non-negative integers, and

ii. we cannot recursively enumerate all "prime-enumerating polynomials", i.e., all polynomials which have the set of prime numbers as non-negative range.
The first of these statements is a refinement of Putnam's Application 6.3. The second is a curiosity. Formulae for prime numbers have been sought for ages. It has long been known that no non-constant polynomial can have only prime numbers in its range (cf. Exercise 8, below). The fact that all composite numbers can be pushed into the negative portion of the range is surprising and popular accounts of the unsolvability of Hilbert's tenth problem stress the existence of such prime-enumerating polynomials. What never gets stressed is the sheer number and complexity of such prime-enumerating polynomials: we cannot necessarily recognise one when we have one, and we cannot even enumerate them all.

This is a good place to stop— at least in the direction we've been going. That is, at least for a while— at the end of section 9 we will return briefly to such recursion theoretic considerations.

Before returning to strictly Diophantine matters, however, I would like quickly to mention that, although Theorem 5.6 can be viewed dually as a theorem about Diophantine equations and as a theorem about r.e. relations, its application is not limited to Diophantine matters and recursion theory proper. There are some quite impressive applications to the calculus. For the class of *elementary expressions* naming the elementary functions of the calculus, one cannot effectively decide

i. when two elementary expressions define the same function

ii. which elementary expressions define elementary functions possessing elementary antiderivatives

iii. which definite integrals are convergent.

The starting point to all of this is the simple observation of the equivalence, for any polynomial $P(X_0, ..., X_{n-1})$, of the following two assertions

i. P has a non-negative integral zero

ii. the function

$$P(X_0, ..., X_{n-1})^2 + \sum_{i=0}^{n-1} \sin^2(\pi X_i)$$

has a non-negative real zero.

The number π can be existentially quantified away— it is the unique z satisfying

$$\sin z = 0 \ \wedge \ 3 \leq z \ \wedge \ z \leq 4.$$

The sines cannot be got rid of: the Diophantine problem over the reals (as well as over the complex numbers) is effectively solvable. This has to do with the fact that in the field of real numbers (and even more so in the field of complex numbers) it is easy for a polynomial to have a root. If I may quickly digress: Jan Denef has begun the study of the Diophantine problem over rings of integers in "small" fields intermediate between the rationals and the reals and has obtained a number of algorithmic unsolvability results. Also, it might be mentioned that Davis and Putnam proved, back in 1963, the unsolvability of the Diophantine problem over the ring $Z[X]$ of polynomials in one variable. (Cf. Exercise 11, below, for a modern proof.)

Getting back to the real case, I'd like to remark that an extra coding trick reduces to one the number of variables needed. It is possible to construct continuous elementary functions $F_0, ..., F_{n-1}$ of a real variable such that, for all real $x_0, ..., x_{n-1}$ and all real $\delta > 0$, there is a real x satisfying

$$|F_i(x) - x_i| < \delta.$$

The proofs of this last remark and the more interesting unsolvability results from calculus would take up too much space for me to present them here. Besides, it gives the reader the opportunity to read through a couple of well-written research papers. These are the papers of Daniel Richardson and Paul Wang cited in the Reading List at the end of this chapter. Richardson's paper was written before Matijasevič proved Theorem 5.6 and unnecessarily assumes P to be of the form $Q(X_0, ..., X_{n-1}, 2^{X_0}, ..., 2^{X_{n-1}})$ for some polynomial Q. The reader may thus set himself the little exercise of seeing what refinements of Richardson's results are yielded by Theorem 5.6. After completing this exercise, he should look at Wang's paper written after Theorem 5.6 had been proved.

Exercises.

1. Show that the pseudo-pairing functions P of i and ii, below, satisfy the conditions
   a. $P(x, 0) = P(0, y) = 0$, for all $x, y \in \omega$
   b. $P(x, y) > 0$ for all positive $x, y \in \omega$
   c. P is one-one when restricted to pairs of positive integers.
   i. (Lou van den Dries— Albert Visser). $P(X, Y) = \langle X^2Y, XY^2\rangle$.
   ii. (Murray Koppel). $P(X, Y) = XY[(XY)^3 + \langle X, Y\rangle]$.
2. Let $P(X_0, ..., X_{n-1})$ be a polynomial of n variables. Suppose P is equal to 0 for all but finitely many n-tuples $x_0, ..., x_{n-1} \in \omega$. Show: P is identically 0.
   [Hint. Induction on n.]
3. Show: $\{P\colon \#(P) \text{ is finite}\}$ is not r.e.
   [Hint. $\{P\colon \#(P) = 0\}$ is not r.e.]
4. Prove that the codings of 6.11 do indeed give one-one correspondences between the set of pre-polynomial expressions and the set of natural numbers and between the set of polynomial expressions and the set of natural numbers.
5. Invoking the isomorphism of acceptable indexings of r.e. relations is not necessary for the simple purpose of proving the effectiveness of Theorem 5.6. Letting W_e denote the r.e. set with index e under the indexing of Chapter I, and W_e^P the r.e. set with index e under the new polynomial indexing, find a recursive G such that, for all e, $W_e = W^P_{G(e)}$.
   [Hint. Look at the proof of Theorem 6.15.]
6. Let S be an r.e. set and $P(X_0,...,X_n)$ a polynomial such that, for all $a \in \omega$,
   $$a \in S \Leftrightarrow \exists \mathbf{x}\, [P(a, \mathbf{x}) = 0].$$
   Modify the construction of Q in the proof of Lemma 6.18 so that
   $$\text{range}(Q) = S \cup \{-n\colon n > 0\}.$$
   [Hint. Look at the construction in the proof of Application 6.3.]

7. (Effectiveness of Lemma 6.18). i. Note that, if P, Q are polynomial expressions, $(P + Q)$ is not a polynomial expression because it has the form $((P_1 - P_2) + (Q_1 - Q_2))$. The obvious polynomial expression defining the function intended by $(P + Q)$ is $((P_1 + Q_1) - (P_2 + Q_2))$. Find recursive functions *Plus*, *Times*, and *Minus* which take pairs of (codes for) polynomial expressions as arguments and produce (codes for) polynomial expressions representing the sum, product, and difference of the polynomials represented by the inputs.

ii. Find a recursive substitution function *SUB* on codes of polynomial expressions such that, for any $P(X_0, X_1, ..., X_n)$ and $Q(X_0)$, $SUB(\ulcorner P \urcorner, \ulcorner Q \urcorner)$ produces a code for a polynomial expression defining $P(Q(X_0), X_1, ..., X_n)$.

iii. Find a recursive G such that

$$G(P) = (1 - P(X_0 - 1, X_1, ..., X_n)^2)X_0 - 1,$$

i.e., $G(\ulcorner P \urcorner)$ is a code for a polynomial expression for the polynomial exhibited.

8. (Prime-Enumerating Polynomials, I). i. Consider the polynomials

a. $P(X) = X^2 + X + 17$

b. $P(X) = 2X^2 + 29$

c. $P(X) = X^2 + X + 41$.

Evaluate each of these for $x = 0, 1, ...$ until a non-prime appears.

ii. Prove: Let $P(X)$ be a non-constant polynomial. For some $x \in \omega$, $P(x)$ is composite.

iii. Prove: Let $P(X_0, ..., X_{n-1})$ be a non-constant polynomial. For some $x_0, ..., x_{n-1} \in \omega$, $P(x_0, ..., x_n)$ is composite.

[Hint. ii. Let $a_0 = P(0)$ and consider $P(ka_0)$ for $k = 0, 1, ...$]

9. (Prime-Enumerating Polynomials, II). Using the techniques introduced so far in the chapter, write down explicitly a prime-enumerating polynomial. If you have faithfully done all the exercises and are a little careful, you should be able to find one with fewer than 30 variables.

10. Let $m < n$. Show that there are *not* polynomials $P_0, ..., P_{n-1}$ in the variables $X_0, ..., X_{m-1}$ satisfying

$$\forall y_0 ... y_{n-1} \exists x_0 ... x_{m-1} \left[\bigwedge_{i=0}^{n-1} P_i(\mathbf{x}) = y_i \right].$$

[Hint. How many m-tuples are there modulo 2? How many n-tuples?]

11. Let $Z[X]$ be the set of all polynomials in one variable X and with integral coëfficients. The Diophantine problem over $Z[X]$ is the following: given a polynomial $P(X_0, ..., X_{n-1})$ in the given variables and with integral coëfficients, do there exist $P_0, ..., P_{n-1} \in Z[X]$ such that $P(P_0, ..., P_{n-1})$ is identically 0?

i. Let $P, Q \in Z[X]$ satisfy the identity

$$P^2 - 3Q^2 = 1.$$

Show: P, Q are constant.

ii. Identifying Z with the set of constant polynomials, show that Z is a Diophantine set in $Z[X]$.

iii. Show: *DIOPH*($Z[X]$) is unsolvable.

[Remark-Hints: The 3 in part i can be replaced by any number d which is not a square; 3 was chosen as the simplest number of the form $a^2 - 1$. An obvious proof of part i refers to the slow growth of $Q(0)$, $Q(1)$, ... There is a more elementary approach— compare the leading coëfficients of P, Q. For ii, to test if $R \in Z[X]$ is constant, either replace 3 by $3R^2$ and appeal to the solution in section 3 of the general Pell equation, or replace Q by $R^2 + S^2 + T^2 + U^2 + V^2$.]

7. Forms

In this section we shall finally consider the Diophantine problem over the rationals. Although it easily reduces to the Diophantine problem over the integers— which latter problem is unsolvable— the converse reduction is by no means obvious and, if it exists, it has yet to be found. The rational Diophantine problem is, however, equivalent to a narrower subproblem— the *homogeneous* Diophantine problem over either the integers or the rationals. The proofs of these equivalences are the main goals of this section. Before getting to them, we will have to acquaint ourselves with homogeneous polynomials— their definition and a characterisation; after proving the equivalences, we will look into a couple of lesser, but relevant matters.

A few words before beginning. In past sections we have concerned ourselves primarily with non-negative solutions. The reason for this restriction was simply that we already had the Σ_1-form of r.e. relations for non-negative integers. In the rational case, there is no particular reason to stick to non-negative solutions. For historical reasons— i.e., for a superficial faithfulness to Diophantus— one might want to impose a restriction to positive solutions. It doesn't matter— the Diophantine problems in the rational numbers, non-negative rationals, and positive rationals are all effectively inter-reducible just as the corresponding problems in integers are inter-reducible— cf. Exercise 1, below. For convenience, we shall consider arbitrary rational solutions.

7.1. Definition. Let $P(X_0, \ldots, X_{n-1})$ be a polynomial written in its normal form,

$$P: \quad \Sigma c_{i_0 \ldots i_{n-1}} X_0^{i_0} \ldots X_{n-1}^{i_{n-1}}.$$

P is a *homogeneous polynomial*, or a *form*, if the degrees $i_0 + \ldots + i_{n-1}$ of the monomials are all the same. This constant, say d, is called the *degree* of the form.

For example,

$$X^2 + Y^2 - Z^2, \quad X - Y + Z, \quad X^2 - 3YZ$$

are forms, but

$$X^2 + Y^2 + Z^2 + Z, \quad X^2 + 2X + 1$$

are not. If one refers to an equation $P = Q$ as homogeneous when $P - Q$ is homogeneous, then the equation

$$X^2 + Y^2 = Z^2$$

is homogeneous, but

$$X^2 + Y^2 = 1, \quad X^2 - dY^2 = 1$$

are not.

Because all monomials in a form have the same degree, forms have the curious homogeneity property,

$$P(TX_0, \ldots, TX_{n-1}) = T^d P(X_0, \ldots, X_{n-1}),$$

where d is the degree of P. This property is characteristic.

7.2. Theorem. *Let $P(X_0, \ldots, X_{n-1})$ be a polynomial of degree d. Then: P is homogeneous iff $P(TX_0, \ldots, TX_{n-1}) = T^d P(X_0, \ldots, X_{n-1})$, where T is a new variable.*

Proof. Half of this is trivial. Write

$$P = \Sigma c_{i_0 \ldots i_{n-1}} X_0^{i_0} \ldots X_{n-1}^{i_{n-1}}, \tag{1}$$

with all $i_0 + \ldots + i_{n-1} = d$ and observe

$$\begin{aligned} P(TX_0, \ldots, TX_{n-1}) &= \Sigma c_{i_0 \ldots i_{n-1}} (TX_0)^{i_0} \ldots (TX_{n-1})^{i_{n-1}} \\ &= \Sigma c_{i_0 \ldots i_{n-1}} T^{i_0 + \ldots + i_{n-1}} X_0^{i_0} \ldots X_{n-1}^{i_{n-1}} \\ &= \Sigma c_{i_0 \ldots i_{n-1}} T^d X_0^{i_0} \ldots X_{n-1}^{i_{n-1}} \\ &= T^d P(X_0, \ldots, X_{n-1}). \end{aligned}$$

For the converse, we can assume without loss of generality that P has degree $d > 0$. Suppose

$$P(TX_0, \ldots, TX_{n-1}) = T^d P(X_0, \ldots, X_{n-1}). \tag{2}$$

Write

$$P(X_0, \ldots, X_{n-1}) = \sum_{i=0}^{d} P_i(X_0, \ldots, X_{n-1}),$$

where each P_i is homogeneous of degree i (i.e., collect all the monomials of degree i and call the result P_i). Note that

$$P(TX_0, \ldots, TX_{n-1}) = \sum_{i=0}^{d} T^i P_i(X_0, \ldots, X_{n-1}) \tag{3}$$

by what we have already proven. By the assumption (2) we also have

$$P(TX_0, \ldots, TX_{n-1}) = \sum_{i=0}^{d} T^d P_i(X_0, \ldots, X_{n-1}).$$

Subtracting (3) from this yields

$$0 = \sum_{i=0}^{d-1} (T^d - T^i) P_i(X_0, \ldots, X_{n-1}).$$

Collecting like terms, one easily sees that the coëfficient of $T^d X_0^{i_0} \ldots X_{n-1}^{i_{n-1}}$ for $i_0 + \ldots + i_{n-1}$

$< d$ is just $c_{i_0 \ldots i_{n-1}}$ (as in (1)). Since a polynomial is identically 0 iff each of its coëfficients is 0, we see that $c_{i_0 \ldots i_{n-1}} = 0$ for $i_0 + \ldots + i_{n-1} < d$. Thus, every term of (1) has degree d, whence P is homogeneous of degree d. QED

The proof yields something just a bit stronger— namely, that the homogeneity property,

$$P(TX_0, \ldots, TX_{n-1}) = T^d P(X_0, \ldots, X_{n-1}),$$

forces P to have degree d. For, notation aside, the proof of the relevant implication did not require one to assume that P had degree d; it showed that any other monomial in P of degree other than d had coëfficient 0. For us, however, the crucial thing about Theorem 7.2 is not so much the nice syntax-free characterisation of homogeneity it offers, but merely the fact that homogeneous polynomials have the property given by the characterisation. One quick application is this: if P has degree $d > 0$, then

$$P(0, \ldots, 0) = P(0^d \cdot 0, \ldots, 0^d \cdot 0) = 0^d \cdot P(0, \ldots, 0) = 0.$$

In other words, every non-constant form has a zero, but a trivial one. The homogeneous Diophantine problem is to determine the existence of non-trivial zeroes. More formally:

7.3. Definition-Convention. Let Γ be one of N, Z, or Q. The *homogeneous Diophantine problem over* Γ is the problem of determining which homogeneous Diophantine equations $P = 0$ have non-trivial solutions in Γ, i.e., which forms $P(X_0, \ldots, X_{n-1})$ have zeroes in Γ other than $(0, \ldots, 0)$.

Given this convention, we can easily apply Theorem 7.2 to prove the following result connecting the homogeneous Diophantine problem over Q with that over Z.

7.4. Theorem. *The homogeneous Diophantine problems over N, Z, and Q are effectively equivalent, i.e., they reduce effectively to one another.*

Proof. The proof in section 1 (1.5.i) of the equivalence of the Diophantine problems over N and Z carries over unchanged to the homogeneous case. It thus suffices to prove the equivalence of the homogeneous Diophantine problems over the integers and the rationals, respectively.

Let $P(X_0, \ldots, X_{n-1})$ be a non-constant form of degree d. Note that any non-trivial zero $x_0, \ldots, x_{n-1} \in Z$ is a rational zero as $Z \subseteq Q$. Suppose $x_0, \ldots, x_{n-1} \in Q$ is a rational zero. Letting z be the common denominator of the x's, we can write $x_i = y_i/z$, with y_i integral. Thus,

$$0 = P(x_0, \ldots, x_{n-1}) = P(y_0/z, \ldots, y_{n-1}/z) = (1/z)^d P(y_0, \ldots, y_{n-1}).$$

Since $(1/z)^d \neq 0$, we have $P(y_0, \ldots, y_{n-1}) = 0$ for $y_0, \ldots, y_{n-1} \in Z$. Moreover, the solution is non-trivial as $y_i = 0$ iff $x_i = 0$, and some x_i was non-zero. Thus, we have shown that a form has a non-trivial integral zero iff it has a non-trivial rational zero, and the two homogeneous problems are therefore equivalent. QED

We have just achieved our first major goal of the section. The second is to prove the following Theorem.

7.5. Theorem. *The Diophantine problem over Q is effectively equivalent to the homogeneous Diophantine problem over Z.*

Proof. The key idea to one direction in the proof, the reduction of the homogeneous to the non-homogeneous problem, we have seen in section 3, above, when we replaced the equation,

$$X^2 + Y^2 = Z^2, \tag{4}$$

by

$$(X/Z)^2 + (Y/Z)^2 = 1,$$

and then by

$$X^2 + Y^2 = 1. \tag{5}$$

The point was that, in a non-trivial solution to (4), the value of Z could not be 0 and we could divide by Z^2 to obtain a rational solution to (5). Conversely, our rational solution to (5) gave us an integral (whence, rational) solution to (4). With one minor snag, this generalises.

Let $P(X_0,...,X_{n-1})$ be a form. We want to replace the homogeneous equation,

$$P(X_0, ..., X_{n-1}) = 0, \tag{6}$$

over the rationals by an equivalent inhomogeneous equation,

$$Q(X_0, ..., X_{n-1}) = 0, \tag{7}$$

over the rationals. In a non-trivial solution to (6), we will have $x_0, ..., x_{n-1}$ integral with some $x_i \neq 0$. By Theorem 7.2 we have

$$0 = P(x_0, ..., x_{n-1}) = x_i^d \cdot P(x_0/x_i, ..., x_{i-1}/x_i, 1, x_{i+1}/x_i, ..., x_{n-1}/x_i),$$

which implies

$$P(x_0/x_i, ..., x_{i-1}/x_i, 1, x_{i+1}/x_i, ..., x_{n-1}/x_i) = 0.$$

We are thus tempted to obtain (7) by taking

$$Q(X_0, ..., X_{i-1}, X_{i+1}, ..., X_{n-1}) = P(X_0, ..., X_{i-1}, 1, X_{i+1}, ..., X_{n-1}).$$

The problem, of course, is that we do not know which variable X_i will have a non-zero entry in the non-trivial solution. We have to consider all the possibilities and take

$$Q(X_0, ..., X_{n-1}) = \prod_{i=0}^{n-1} P(X_0, \ldots, X_{i-1}, 1, X_{i+1}, \ldots, X_{n-1}).$$

The reader can easily verify that, with this choice of Q, the existence of a non-trivial solution to (6) is equivalent to the existence of a rational solution to (7). Thus, we have reduced the homogeneous Diophantine problem over the integers to the ordinary Diophantine problem over the rationals.

Before proceeding to the converse reduction, note that, if P has degree d, Q could have degree as high as d^n. At the cost of 4 additional variables, we can provide a reduction with a linear growth in the degree. Let $Q(X_0, ..., X_{n-1}, Z_0, ..., Z_3)$ be

$$P(X_0, ..., X_{n-1})^2 + \left[\sum_{i=0}^{n-1} X_i^2 - (Z_0^2 + ... + Z_3^2 + 1)\right]^2,$$

of degree max $\{2d, 4\}$. The trick is that, by Lagrange's Theorem (3.8), an integer like Σx_i^2 is positive iff it is of the form $z_0^2 + ... + z_3^2 + 1$. Thus, if $x_0, ..., x_{n-1}, z_0, ..., z_3$ is a rational zero of Q, then Σx_i^2 is positive, whence $(x_0, ..., x_{n-1}) \neq (0, ..., 0)$, and $P(X_0, ..., X_{n-1}) = 0$ has a non-trivial rational solution, and, by the proof of Theorem 7.4, it has a non-trivial integral solution. The converse also holds: if $x_0, ..., x_{n-1}$ is a non-trivial integral zero of P, then Σx_i^2 is a positive integer and we can find $z_0, ..., z_3$ such that $\Sigma x_i^2 = z_0^2 + ... + z_3^2 + 1$. This gives $x_0, ..., x_{n-1}, z_0, ..., z_3$ as an integral— whence rational— zero of Q. (But, cf. Remark 7.6, below.)

The heart of the proof, however, is the converse reduction, i.e., the reduction of the Diophantine problem over the rationals to the homogeneous Diophantine problem over the integers. The obvious thing to do is to attempt to reverse the process used above in going from (4) to (5). Let $P(X_0, ..., X_{n-1})$ be an arbitrary polynomial of degree $d > 0$. A rational zero could be written $x_0/z, ..., x_{n-1}/z$, where z is the common denominator. This gives the intermediate equation,

$$P(X_0/Z, ..., X_{n-1}/Z) = 0,$$

which becomes a polynomial on multiplying by Z^d and simplifying:

$$Q(X_0, ..., X_{n-1}, Z) = Z^d P(X_0/Z, ..., X_{n-1}/Z).$$

The polynomial Q happens to be homogeneous of degree d. (Why?)

The form Q is the one we want to use to test the solvability of P with. It is obvious that a rational zero $x_0/z, ..., x_{n-1}/z$ of P yields the non-trivial integral zero $x_0, ..., x_{n-1}, z$ of Q (non-trivial since $z \neq 0$), and that a zero of Q, say, $x_0, ..., x_{n-1}, z$ with $z \neq 0$ will yield a zero $x_0/z, ..., x_{n-1}/z$ of P. What is not obvious is that in a non-trivial zero of Q, the value z will be non-zero. Indeed, for

$$P(X_0, X_1) = X_0^2 \cdot X_1^2 + 1,$$

the corresponding form

$$Q(X_0, X_1, Z) = X_0^2 \cdot X_1^2 + Z^4$$

has the non-trivial zero $x_0 = 1, x_1 = 0, z = 0$, and no non-trivial zero with $z \neq 0$. The paradigmatic polynomial $X^2 + Y^2 - 1$ of equation (5) was a bit too special.

If Q does not provide us with a solution, it does at least provide us with the beginning of one. If we can express in a homogeneous way the two conditions

$$Q(X_0, ..., X_{n-1}, Z) = 0 \tag{8}$$

and

$$Z \neq 0, \tag{9}$$

then we will be finished. Condition (8), itself, is practically all we will need to express (8). The Pell equation will allow us to express condition (9). Let $R(X_0, ..., X_{n-1}, Z, Y_0, ..., Y_3, V_0, ..., V_3)$ be the polynomial

$$(X_0^2 + \ldots + X_{n-1}^2 + Y_0^2 + \ldots + Y_3^2)^2 - 3(V_0^2 + \ldots + V_3^2)^2 - Z^4. \tag{10}$$

The equation $R = 0$ expresses condition (9), whence (8) and (9) are together expressed by the single homogeneous equation,

$$Q^8 + R^{2d} = 0. \tag{11}$$

Let us prove the equivalence of the pair of conditions (8) and (9) with the single condition (11). First, assume integers $x_0, \ldots, x_{n-1}, z$, not all 0, exist satisfying (8) and (9). Let $u \geq x_0^2 + \ldots + x_{n-1}^2$ and v be integral solutions to the Pell equation

$$X^2 - 3Y^2 = 1.$$

Choose, by Lagrange's Theorem, $y_0, \ldots, y_3$ and $v_0, \ldots, v_3$ integral such that

$$z^2u = \Sigma x_i^2 + y_0^2 + \ldots + y_3^2, \quad z^2v = v_0^2 + \ldots + v_3^2.$$

Then

$$(z^2u)^2 - 3(z^2v)^2 = z^4(u^2 - 3v^2) = z^4,$$

whence

$$(\Sigma x_i^2 + y_0^2 + \ldots + y_3^2)^2 - 3(v_0^2 + \ldots + v_3^2)^2 - z^4 = 0.$$

By (10), this equation just asserts $R(\boldsymbol{x}, z, \boldsymbol{y}, \boldsymbol{v}) = 0$. Together with the assumption $Q(\boldsymbol{x}, z) = 0$, we have

$$Q(\boldsymbol{x}, z)^8 + R(\boldsymbol{x}, z, \boldsymbol{y}, \boldsymbol{v})^{2d} = 0,$$

i.e., $x_0, \ldots, x_{n-1}, z, y_0, \ldots, y_3, v_0, \ldots, v_3$ is a non-trivial solution to (11).

Conversely, suppose $x_0, \ldots, x_{n-1}, z, y_0, \ldots, y_3, v_0, \ldots, v_3$ is a non-trivial solution to (11). Obviously, $x_0, \ldots, x_{n-1}, z$ is a solution to (8). We must see that $z \neq 0$. Suppose, by way of contradiction, that $z = 0$. From (11), it follows that $R(\boldsymbol{x}, z, \boldsymbol{y}, \boldsymbol{v}) = 0$, i.e.,

$$(\Sigma x_i^2 + y_0^2 + \ldots + y_3^2)^2 = 3(v_0^2 + \ldots + v_3^2)^2.$$

Since $\sqrt{3}$ is irrational, this means that all of the x's, y's, and v's are 0, whence the integral solution was trivial after all.

But now we are done. We have seen that the existence of a rational zero of P entails the existence of a zero of Q with $z \neq 0$, which entails the existence of a non-trivial integral solution to the homogeneous equation (11). We have also seen, conversely, that the existence of a non-trivial integral solution of (11) yields an integral solution to (8) and (9), i.e., a non-trivial zero of Q with $z \neq 0$, and that this implies the existence of a rational zero of P.

QED

Although Theorem 7.5 is an oft-cited result, its proof appears to be little-known. The above proof, i.e., the invention of R, is due to Raphael Robinson, who produced it when Juriĭ Matijasevič wrote to Julia Robinson to announce that someone had asked him in a seminar how the proof went and he suddenly realised that he didn't know. Perhaps there is somewhere in an old German treatise on algebraic geometry a better proof buried, i.e., one that doesn't require an 8-fold increase in degree and 9 extra variables. I note that there is a reduction of the quadratic Diophantine problem over the rationals to the quadratic homogeneous Diophantine problem over the rationals (hence over the integers) which only adds one new variable. Through various substitutions (arising, e.g., by completing the square), every quadratic polynomial transforms into one of the following three types of polynomials:

i. $\sum_{i=0}^{m-1} a_i Y_i^2$

ii. $\sum_{i=0}^{m-1} a_i Y_i^2 + \sum_{i=m}^{n-1} a_i Y_i$

iii. $\sum_{i=0}^{m-1} a_i Y_i^2 + c,\ c \neq 0,$

where no a_i is 0. Type i is a form and, if that case arises, the reduction is complete. Type ii has a solution and, if that is the case that arises, there is no need to go further. In the remaining type iii case, the homogenising trick works. For

$$Q(Y_0, \ldots, Y_{m-1}, Z) = \sum a_i Y_i^2 + cZ^2,$$

it can be shown that the existence of a non-trivial rational zero entails the existence of another rational zero with $z = 1$, whence an integral zero with $z \neq 0$. The proof, taken from the paper of D.J. Lewis cited in the references, is outlined in Exercises 3 and 4, below.

The following remark is perhaps less important, but it is formally labelled because it is referred to in the proof of Theorem 7.5.

7.6. Remark. The easy reduction, that from the homogeneous to the non-homogeneous case, need not increase the degree or the number of variables. Given $P(X_0, \ldots, X_{n-1})$ homogeneous, we could, instead of looking for a *single* polynomial Q such that $P = 0$ has a non-trivial integral solution iff $Q = 0$ has a rational solution, produce the n polynomials $Q_0, \ldots, Q_{n-1}$ in $n-1$ variables given by

$$Q_i(X_0, \ldots, X_{i-1}, X_{i+1}, \ldots, X_{n-1}) = P(X_0, \ldots, X_{i-1}, 1, X_{i+1}, \ldots, X_{n-1})$$

and observe that P has a non-trivial integral zero iff at least one of these n polynomials has a rational zero. We thus reduce the n-variable homogeneous problem of degree d over the integers (or rationals) to the $(n-1)$-variable inhomogeneous problem of degree d over the rationals as follows. Given P, construct $Q_0, \ldots, Q_{n-1}$. Then ask successively if $Q_0, \ldots, Q_{n-1}$ has a rational zero until either an affirmative answer has been found or all n answers have been negative. Such a reduction is not as elegant as one of the form,

$$P \in X \text{ iff } Q = F(P) \in Y,$$

for some effectively computable F, but it is effective and, in the present case, it provides a better result than the more straightforward reductions offered in the proof of Theorem 7.5.

At the moment, however, Remark 7.6 has only one application. The single-variable Diophantine problem over the rationals is easily solved (cf. Exercise 6, below), whence the two-variable homogeneous Diophantine problem over the integers is solvable. All else remains open— except, in the linear case, which we need not discuss, and in the quadratic case, where the relevant reduction goes the other way: the quadratic Diophantine problem over the rationals reduces to the homogeneous quadratic Diophantine problem over the rationals. This latter problem is equivalent to a special case of the quadratic Diophantine problem over the integers, which general problem Carl Ludwig Siegel solved in 1972. This is, however, probably not the most accessible or even interesting approach. Besides the field of real numbers, the field of rational numbers has a great many metric completions— one for each positive prime integer. A deep and beautiful result of the theory of such completions asserts that a quadratic form has a non-trivial rational zero iff it has a non-trivial zero in each metric completion of the rationals. The Diophantine problems over these metric completions turns out to be more amenable to study than that over the rationals. A curious result is this: a quadratic form in 5 or more variables has a non-trivial zero in every metric completion with the possible exception of the field of real numbers itself. Thus, a quadratic form in 5 or more variables has a non-trivial rational zero iff it has a non-trivial real zero. As the general Diophantine problem over the field of real numbers is effectively solvable, there is thus a decision method for determining the rational solvabiilty of homogeneous quadratic equations in 5 or more variables. The cases of 2, 3, and 4 variables require a bit more work. The theory behind all of this is rather deep and I cannot begin to discuss it here. The reader with a background in algebra, topology, and number theory will find D.J. Lewis' survey article cited in the Reading List to offer a good, if concise, introduction and orientation.

The solvability of the quadratic Diophantine problem over the rationals is as much of a general positive nature about the problem that is known and can be stated without introducing major new concepts. Theorems 7.4 and 7.5 are the chief reductive results. There is one more reduction worth citing before discussing what little that is negative is known. This is the reduction of an arbitrary homogeneous equation to a system of quadratic homogeneous equations. This result is due to Andrew Adler.

7.7. Theorem. *The homogeneous Diophantine problem over the integers is equivalent to the problem of deciding which systems of homogeneous quadratic equations are non-trivially solvable in integers.*

Proof. Adler's original proof translated a homogeneous equation directly into a system of homogeneous quadratic equations. Here, we shall give a "simpler" proof by appealing to powerful tools.

Let Q be a form. Via the first reduction in the proof of Theorem 7.5, find a polynomial P such that P has a rational zero iff Q has a non-trivial integral zero.

Apply Skolem's reduction (6.2) of $P = 0$ to a system, say,

$$P_0 = 0, ..., P_{m-1} = 0,$$

of quadratic equations. (Note that the proof of 6.2 carries over to the rational case.) The direct quadratic-to-quadratic reduction of the Diophantine problem over the rationals to the homogeneous Diophantine problem, if it extends to systems, does not obviously do so. However, Raphael Robinson's reduction of the general Diophantine problem over the rationals to the homogeneous one over the integers is readily modified to reduce quadratic systems to quadratic systems.

Letting $X_0, ..., X_{n-1}$ denote the variables of $P_0, ..., P_{m-1}$, define $Q_0, ..., Q_{m-1}$ by

$$Q_i(X_0, ..., X_{n-1}, Z) = Z^2 P_i(X_0/Z, ..., X_{n-1}/Z).$$

Note that we use the same Z as the homogenising variable for all the P_i's— it is, after all, supposed to be the common denominator of all of the variables in all of the equations. These Q_i's are quadratic forms and give us m equations of the desired system, namely the equations

$$Q_i(X_0, ..., X_{n-1}, Z) = 0. \tag{12}$$

Coupled with (12) should be the equation

$$R(X_0, ..., X_{n-1}, Z, Y_0, ..., Y_3, V_0, ..., V_3) = 0,$$

with R as in (10). However, R has degree 4, not 2. This equation can, however, be replaced by three quadratic equations:

$$X_0^2 + ... + X_{n-1}^2 + Y_0^2 + ... + Y_3^2 = W_0 Z \tag{13}$$

$$V_0^2 + ... + V_3^2 = W_1 Z \tag{14}$$

$$W_0^2 - 3W_1^2 = Z^2, \tag{15}$$

with the extra new variables W_0, W_1.

I claim that $P = 0$ is solvable in the rationals iff the quadratic system given by (12) - (15) has a non-trivial solution in integers. I leave the details as an exercise to the reader. (Exercise 6, below.) QED

7.8. Corollary. *The homogeneous Diophantine problem over the integers is effectively equivalent to the homogeneous Diophantine problem of degree* 4 *over the integers.*

With Corollary 7.8, we have exhausted the major reductions. It is now time to discuss negative results. The following is due, again, to Adler.

7.9. Theorem. *The problem of deciding which forms represent* 1 *over the integers is algorithmically unsolvable.*

Proof. Let $P(X_0, ..., X_{n-1})$ be a polynomial of degree d and define

$$Q(X_0, ..., X_{n-1}, Z) = 2(Z^d P(X_0/Z, ..., X_{n-1}/Z))^2 + Z^{2d}.$$

Observe that

$$\exists x\, [P(x) = 0] \Leftrightarrow \exists x z\, [Q(x, z) = 1],$$

quantification being over the integers. QED

7.10. Corollary. *The problem of deciding which forms represent all natural numbers over the integers is algorithmically unsolvable.*

Proof. Multiply Q in the proof of Theorem 7.9 by $Y_0^2 + ... + Y_3^2$. QED

Note that, although homogeneous problems,

$$Q = 0,$$

are solvable in the rationals iff they are solvable in the integers, the same does not generally hold for inhomogeneous equations,

$$Q = 1,$$

for forms Q. Thus, Theorem 7.9 and Corollary 7.10 are unsolvable problems about forms over the integers and, apparently, tell us nothing about the rationals.

In the next chapter, we shall introduce formal languages, formal logic, and the notion of a theory. Julia Robinson has shown that the theory of the field of rational numbers is undecidable, i.e., the problem of deciding which sentences in the language are true and which are false is an unsolvable problem. Thus, there are unsolvable problems about the rationals and we have some reason to believe the Diophantine problem over Q is unsolvable. Once we have the notion of a formal theory, we will, however, find ourselves heading in a different direction. Nonetheless, we will pause briefly in the next chapter to consider the theory of the rational number field.

Exercises.

1. i. Show that every non-negative rational number can be written in the form,

$$\frac{x^2 + y^2 + z^2 + w^2}{q^2 + r^2 + s^2 + t^2 + 1},$$

where $x, y, ..., s, t$ are rational (or even integral).

ii. Find a similar representation for positive rational numbers.

iii. Prove the effective equivalence of the general Diophantine problems over the rationals, the non-negative rationals, and the positive rationals.

2. Let $P(X_0, \ldots, X_{n-1})$ be a polynomial of degree d. Show that $Q(X_0, \ldots, X_{n-1}, Z) = Z^d P(X_0/Z, ..., X_{n-1}/Z)$ is homogeneous of degree d.
[Hint. For a quick proof (assuming Q a polynomial), apply Theorem 7.2.]

3. Let $P(X_0, ..., X_{n-1})$ be a polynomial. A polynomial $Q(Y_0, ..., Y_{n-1})$ arising from P by a change of variables,

$$X_i = \Sigma b_{ij} Y_j, \text{ all } b_{ij} \text{ rational}, \qquad (*)$$

is said to be *equivalent* to P if the system (*) is solvable for the Y_j's in terms of the X_i's (i.e., if the matrix (b_{ij}) is invertible).

i. Show: If P and Q are equivalent, then P has a rational zero iff Q has a rational zero.

ii. Let P be a quadratic polynomial. Show that, by completing the square, if P has non-zero terms $aX^2 + bXY$ a change of variables can eliminate the cross term. Do the same to eliminate the linear term if $aX^2 + bX$ occurs.

iii. Let P be a quadratic polynomial. If aXY occurs with no corresponding X^2 or Y^2 term, show that the change of variables,

$$X = \frac{U+V}{2}, \qquad Y = \frac{U-V}{2},$$

will eliminate the XY-term.

iv. Let P be a quadratic polynomial. Show that, by a succession of changes of variables, P can be made equivalent to a quadratic polynomial of one of the following three types:

i. $\sum_{i=0}^{m-1} a_i Y_i^2$

ii. $\sum_{i=0}^{m-1} a_i Y_i^2 + \sum_{i=m}^{n-1} a_i Y_i$

iii. $\sum_{i=0}^{m-1} a_i Y_i^2 + c$,

where c and all the a_i's are not equal to 0.

4. Let $P(X_0, ..., X_{n-1}) = \sum_{i=0}^{n-1} a_i X_i^2 + c$ with rational non-zero $a_0, ..., a_{n-1}, c$. Define $Q(X_0, ..., X_{n-1}, Z)$ to be $\sum_{i=0}^{n-1} a_i X_i^2 + cZ^2$.

   i. Suppose $Q(b_0, ..., b_{n-1}, 0) = 0$ with, say, $b_{n-1} \neq 0$. Define

   $$R(Y_0, ..., Y_{n-1}, Z) = Q(b_0 Y_0, b_1 Y_0 + Y_1, ..., b_{n-1} Y_0 + Y_{n-1}, Z).$$

   Expand R and find $c_0, ..., c_{n-1}$ such that $R(c_0, ..., c_{n-1}, 1) = 0$, i.e., $Q(b_0 c_0, b_1 c_0 + c_1, \ldots, b_{n-1} c_0 + c_{n-1}, 1) = 0$.

   ii. Prove that the quadratic Diophantine problem over the rationals reduces to the homogeneous quadratic Diophantine problem over the rationals.

   [Hint. i. R assumes the form $cZ^2 + Y_0 L(Y_1, ..., Y_{n-1}) + F(Y_0, ..., Y_{n-1})$, where F is homogeneous quadratic and L is linear and not identically 0.]

5. Show that the one-variable Diophantine problem over the rationals is effectively solvable. [Hint. What must the numerator and denominator of a zero of $P(X) = \Sigma a_i X^i$ divide?]

6. Complete the proof of Theorem 7.7.

7. Prove the effective equivalence of the Diophantine problem over the positive rationals with the homogeneous Diophantine problem over the positive integers. [*N.B.* The word "positive" refers to the solutions, not the coëfficients, which are arbitrary integers.]

*8. Binomial Coëfficients

Following Matijasevič's solution of Hilbert's tenth problem, researchers in the area retraced the path to the solution to see what improvements could be made. New exponential Diophantine definitions using fewer variables were found and the products used in the original proof of the Davis-Putnam-Robinson Theorem were replaced by a clever application of the binomial coëfficients. (Some of these improvements have been incorporated into the text and some can be found hidden in the exercises of the preceding sections.) This look back was partly due to the search for a more elegant and more intelligible presentation of the proof, and partly to conserve variables to show that Hilbert's tenth problem was unsolvable for a small number of variables. (It is now known to be unsolvable for 9 variables; cf. section 10 for references.) One of the useful developments here was the Relation Combining Lemma, which allowed one to express one inequality, one divisibility statement, and one assertion that a number is a square, using only one extra variable instead of the three needed when expressing the three assertions separately.

These are solid developments, but they are very technical and are mainly suited for the specialist. Of somewhat broader interest, and more appropriate for inclusion here, was the shift in general coding strategy. In any number of papers following his solution to Hilbert's tenth problem, Matijasevič used b-adic expansions, where b was a large power of 2. Ultimately, he found a reference to a curious result of Ernst Kummer which allowed him to bypass the use of the bounded universal quantifier and establish the Davis-Putnam-Robinson Theorem directly. This construction was simplified and made very intelligible in joint work of Matijasevič and James P. Jones.

In the next section we will encounter the Jones-Matijasevič proof of the Davis-Putnam-Robinson Theorem, and in section 10 we will see Matijasevič's proof, via Kummer, of the algorithmical unsolvability of the general iterated exponential Diophantine problem in three variables. The present section concerns itself with Kummer's result on binomial coëfficients— its derivation and a few hints to its use.

Kummer's result is a bit obscure. It is not unknown to number theorists, but it has been forgotten and rediscovered and it has never made it into the elementary number theory texts. This is a bit odd as the result is a corollary to a result that appears in all but the most elementary introductory textbooks on number theory— namely, the following result of Legendre.

8.1. Theorem. *Let p be a prime and m a positive integer. If p^μ is the exact power of p that divides $m!$ (i.e., $p^\mu | m!$ and $p^{\mu+1} \nmid m!$), then*

$$\mu = \left[\frac{m}{p}\right] + \left[\frac{m}{p^2}\right] + \ldots = \sum_{k=1}^{\infty} \left[\frac{m}{p^k}\right]. \tag{1}$$

Moreover, if we write m in base p,

$$m = a_0 + a_1 p + \ldots + a_n p^n, \quad 0 \le a_i < p,$$

and set

$$s = a_0 + a_1 + \ldots + a_n,$$

then

$$\mu = \frac{m - s}{p - 1}. \tag{2}$$

Two quick comments before proving the Theorem: First, the infinite sum in (1) is actually finite as, from some point on, p^k is greater than m and the greatest integers $[m/p^k]$ are then all 0. Second, I must be honest and state that only (1) appears universally in textbooks on number theory, where it is a key lemma in the proofs of Chebyshev's Theorem and Bertrand's Postulate:

Chebyshev. For $x > 1$, let $\pi(x)$ denote the number of primes $< x$. There are positive constants a, A such that, for all sufficiently large x,

$$a\frac{x}{\log x} < \pi(x) < A\frac{x}{\log x}.$$

Bertrand. For all positive integers n, there is a prime p such that $n < p \le 2n$.

For our purposes, (1) is merely a lemma to be used to establish (2). The reader who would like to see proofs of Chebyshev's Theorem or Bertrand's Postulate from (1) is referred to Chandrasekharan's book cited in the Reading List at the end of the chapter.

Proof of Theorem 8.1. (1). Observe that, for any k,

$$[m/p^k] = \text{the number of integers from } 1, 2, ..., m \text{ divisible by } p^k.$$

Thus,

$$[m/p] - [m/p^2] = \text{number of integers from } 1, 2, ..., m \text{ divisible by } p, \text{ but not by } p^2$$

$$[m/p] - [m/p^2] = \text{number of integers from } 1, 2, ..., m \text{ divisible by } p^2, \text{ but not by } p^3,$$

etc. Therefore,

$$\begin{aligned}\mu &= \sum_{k=1}^{\infty} k([m/p^k] - [m/p^{k+1}]) \\ &= [m/p] - [m/p^2] + 2([m/p^2] - [m/p^3]) + 3([m/p^3] - [m/p^4]) + ... \\ &= [m/p] + [m/p^2] + [m/p^3] + ...\end{aligned}$$

To prove (2), observe that, if $m = a_0 + a_1 p + ... + a_n p^n$, with $0 \le a_i < p$, then

$$\begin{aligned}\left[\frac{m}{p^k}\right] &= \left[\frac{a_0 + a_1 p + ... + a_k p^k + ... + a_n p^n}{p^k}\right] \\ &= a_k + a_{k+1} p + ... + a_n p^{n-k}.\end{aligned}$$

So

$$\begin{aligned}\mu &= \sum_{k=1}^{\infty}[m/p^k] = \sum_{k=1}^{n}[m/p^k] \\ &= (a_1 + a_2 p + ... + a_n p^{n-1}) + (a_2 + a_3 p + ... + a_n p^{n-2}) + ... + (a_n) \\ &= a_1 + a_2(p+1) + a_3(p^2 + p + 1) + ... + a_n(p^{n-1} + ... + 1) \\ &= a_1\frac{p-1}{p-1} + a_2\frac{p^2-1}{p-1} + a_3\frac{p^3-1}{p-1} + ... + a_n\frac{p^n-1}{p-1} \\ &= \frac{1}{p-1}\left[a_1(p-1) + a_2(p^2-1) + ... + a_n(p^n-1)\right] \\ &= \frac{1}{p-1}[a_0 + a_1 p + a_2 p^2 + ... + a_n p^n - a_0 - a_1 - a_2 ... - a_n] \\ &= \frac{m-s}{p-1}.\end{aligned}$$

QED

We can now state Kummer's Theorem.

8.2. Kummer's Theorem. *Let a, b be positive integers, p a prime, and suppose p^μ is the highest power of p dividing $\binom{a+b}{a}$. Then: μ is the number of carrying operations that have to be performed when adding a and b in base p.*

Before launching into the proof of this, let us prove a little lemma.

8.3. Lemma. *Let p be any base (prime or otherwise). When adding two numbers in base p notation, one only needs to carry* 0's *and* 1's.

Proof. Let a and b be the numbers in question. By padding one or the other number's p-adic expansion with 0's, we can assume the two numbers to have the same number of "digits" in base p, say:

$$a = a_n p^n + \dots + a_0, \quad 0 \le a_i < p \tag{3}$$

$$b = b_n p^n + \dots + b_0, \quad 0 \le b_i < p. \tag{4}$$

The proof of the lemma is by induction on the position of the digits being added, starting at the right.

The basis is easy: $a_0 + b_0$ can be $2p - 2$ at most, whence either 0 (if $a_0 + b_0 < p$) or 1 (if $p \le a_0 + b_0 \le 2p - 2$) must be carried.

In adding $a_{k+1} + b_{k+1}$, once again one gets a sum of at most $2p - 2$. By induction hypothesis, one has carried at most a 1, yielding an overall sum of at most $2p - 1$, whence again only a 0 or a 1 can be carried. QED

Proof of Theorem 8.2. Write a, b as in (3) and (4) of the proof of the Lemma. Define $\varepsilon_0, \dots, \varepsilon_n$ to be the carries, i.e., ε_i is 1 or 0 according as there is or is not a carry in adding the i-th digits:

$$\left.\begin{aligned} a_0 + b_0 &= \varepsilon_0 p + c_0 \\ \varepsilon_0 + a_1 + b_1 &= \varepsilon_1 p + c_1 \\ &\vdots \\ \varepsilon_{i-1} + a_i + b_i &= \varepsilon_i p + c_i \\ &\vdots \end{aligned}\right\} \tag{5}$$

with $0 \le c_i < p$ for $i = 0, \dots, n$. Thus,

$$a + b = c_0 + c_1 p + \dots + c_n p^n + \varepsilon_n p^{n+1}.$$

We wish to apply Legendre's Theorem to each of the factorials of

$$\binom{a+b}{a} = \frac{(a+b)!}{a!b!}.$$

To this end, define

$$\alpha = \sum_{i=0}^{n} a_i, \quad \beta = \sum_{i=0}^{n} b_i, \quad \gamma = \sum_{i=0}^{n} c_i.$$

With these, Legendre's Theorem yields

$$\mu = \frac{a+b-\gamma-\varepsilon_n}{p-1} - \frac{a-\alpha}{p-1} - \frac{b-\beta}{p-1} = \frac{\alpha+\beta-\gamma-\varepsilon_n}{p-1}. \tag{6}$$

To calculate $\alpha+\beta-\gamma-\varepsilon_n$, add the equations (5):

$$\Big(\sum_{i=0}^{n-1}\varepsilon_i\Big) + \alpha + \beta = \Big(\sum_{i=0}^{n}\varepsilon_i p\Big) + \gamma,$$

whcncc

$$\alpha+\beta-\gamma = \sum_{i=0}^{n}\varepsilon_i p - \sum_{i=0}^{n-1}\varepsilon_i,$$

whence

$$\alpha+\beta-\gamma-\varepsilon_n = \sum_{i=0}^{n}\varepsilon_i p - \sum_{i=0}^{n}\varepsilon_i = (p-1)\sum_{i=0}^{n}\varepsilon_i\,,$$

and (6) yields

$$\mu = \frac{a+b-\gamma-\varepsilon_n}{p-1} = \sum_{i=0}^{n}\varepsilon_i\,,$$

which, by Lemma 8.3, is the desired result: μ is the number of carrying operations performed in adding a and b in base p. QED

8.4. Corollary. *Let n be a positive integer, and suppose 2^μ is the exact power of 2 that divides $\binom{2n}{n}$. Then: μ is the number of 1's in the dyadic expansion of n.*

The idea of the proof is simple. When adding n to itself, a carry can occur only when a 1 pops up in the dyadic expansion of n. A carried 1 can cause no new carries, for it can only hit either two 0's and there will be no carry or two 1's and there is already a carry to be performed allowing the carried 1 to stay.

Corollary 8.4 can be restated as:

8.4'. Corollary. *Let n be a positive integer, and suppose 2^μ is the exact power of 2 that divides $\binom{2n}{n}$. Then: $\mu = \mathrm{card}(D_n)$.*

[Recall Definition I.7.6 of the canonical indexing of finite sets.]

Because the relation "2^μ is the exact power of 2 that divides $\binom{2n}{n}$" is exponential Diophantine (it just says $2^\mu \mid \binom{2n}{n} \wedge 2^{\mu+1} \nmid \binom{2n}{n}$ and divisibility, non-divisibility, and the graph of the binomial function are all exponential Diophantine), it follows that the relation,

$$y = \mathrm{card}(D_x),$$

is exponential Diophantine. Relative to this coding of finite sets, membership and non-membership are obviously exponential Diophantine:

$$y \in D_x \Leftrightarrow [x/2^y] \text{ is odd} \quad \text{(cf. I.8.9)}$$

$$y \notin D_x \Leftrightarrow [x/2^y] \text{ is even.}$$

Equality of sets is even more trivially exponentially Diophantine:

$$D_x = D_y \Leftrightarrow x = y.$$

But, short of appeal to the Davis-Putnam-Robinson Theorem, it has not until now been clear that the cardinality of a finite set can be expressed in terms of its index in an exponential Diophantine way. Can more properties of finite sets D_x and D_y be expressed exponentially Diophantinely in x and y? The answer is "yes" and Kummer's Theorem 8.2 shows this.

The following result (originally stated in different terms) is due to Edouard Lucas.

8.5. Corollary. (Lucas' Theorem). *Let x, y be natural numbers. Then:*

$$D_x \subseteq D_y \text{ iff } \binom{y}{x} \text{ is odd.}$$

Proof. Note that either condition implies that $y \geq x$. For $x = 0$ or $x = y$, the equivalence is trivial. For $x \neq y$ positive, we will apply Kummer's Theorem in the case $p = 2$. Observe

$D_x \subseteq D_y \Leftrightarrow$ When one lines up the dyadic expansions of x and y, there is a 1 in the expansion of y wherever there is a 1 in that of x.

$\Leftrightarrow$ The 1's in the dyadic expansion of $y - x$ correspond exactly to the 1's in the expansion of y which are not in that of x.

$\Leftrightarrow$ There are no carries when adding x to $y - x$ in base 2.

$\Leftrightarrow$ 2^0 is the exact power of 2 dividing $\binom{x + y - x}{x} = \binom{y}{x}$.

$\Leftrightarrow$ $\binom{y}{x}$ is odd. QED

Lucas' Theorem is the chief tool we will use in the next section to give a direct proof of the Davis-Putnam-Robinson Theorem without using the bounded-universal-quantifier-elimination as in section 2, above. Heuristically, we can say that the bounded quantifier is implicit in the inclusion assertion and is eliminated by saying "$\binom{y}{x}$ is odd". The more mysterious Corollary 8.4 is the key to Matijasevič's earlier avoidance of the bounded universal quantifier and it will be used in section 10, below, to prove the unsolvability of iterated exponential Diophantine equations in three variables.

Corollaries 8.4' and 8.5 do not exhaust all the applications of Kummer's Theorem to the exponential Diophantine expressibility of relations among finitely coded sets. The following corollary collects a few more interesting equivalences.

8.6. Corollary. *Let x, y, z be natural numbers.*

i. $D_x \cap D_y = \emptyset \Leftrightarrow \binom{x+y}{x}$ *is odd*

ii. $D_x \cap D_y = D_z \Leftrightarrow \binom{x}{z}\binom{x+y-z}{x}$ *is odd*

iii. $D_x \cup D_y = D_z \Leftrightarrow \binom{z}{x}\binom{x}{z-y}$ *is odd*

iv. $D_x - D_y = D_z \Leftrightarrow \binom{x}{z}\binom{y+z}{x}$ *is odd.*

The proof of Corollary 8.6 is left to the exercises immediately following.

Exercises.

1. Prove Corollary 8.6. [Hint. ii. $\binom{x}{z}\binom{x+y-z}{x} = \binom{y}{z}\binom{x+y-z}{y}$.]
2. (Alternate proof of Lucas' Theorem). Write

$$s = \sum_{i=0}^{n} s_i 2^i, \qquad r = \sum_{i=0}^{n} r_i 2^i, \qquad 0 \le r_i, s_i < 2$$

and suppose $r < s$.

i. Show: $(1+X)^{2^i} \equiv 1 + X^{2^i} \pmod 2$.

ii. Show: $(1+X)^s \equiv (1+X)^{s_0}(1+X^2)^{s_1}...(1+X^{2^n})^{s_n} \pmod 2$.

iii. Show: The coëfficient of X^r in

$$\prod_{i=0}^{n}(1+X^{2^i})^{s_i} \text{ is } \binom{s_n}{r_n}\binom{s_{n-1}}{r_{n-1}}...\binom{s_0}{r_0}.$$

iv. Let $P(X) = a_0 + a_1X + ... + a_mX^m$. Show: If $P(X) \equiv 0 \pmod 2$, then $a_i \equiv 0 \pmod 2$ for each a_i.

v. Show: $\binom{s}{r} \equiv \binom{s_n}{r_n}\binom{s_{n-1}}{r_{n-1}}...\binom{s_0}{r_0} \pmod 2$.

vi. Show: $D_r \subseteq D_s$ iff $\binom{s}{r}$ is odd.

[Remark. In i, ii, and iv, we talk of congruence of polynomials modulo 2. One definition of $P(X) \equiv Q(X) \pmod 2$ is that $P(n) \equiv Q(n) \pmod 2$ for all integral n. Use this.] [Hints. i. induction on i; iii. $X^r = X^{r_n 2^n} \cdot X^{r_{n-1}2^{n-1}} \cdot ... \cdot X^{r_0}$; iv. induction on m. Parts i - v of this exercise generalise to congruences modulo any prime p.]

Yet another proof of Lucas' Theorem can be found in Hausner's paper cited in the references. Hausner's proof is a simple counting argument. Both the proof given in Exercise 2 and Hausner's proof avoid Kummer's and Legendre's Theorems. The approach via Legendre and Kummer strikes me, however, as less serendipitous and, in any event, although we will only need Lucas' Theorem in the next section, we will require Kummer's 8.4 in the section following that.

*9. A Direct Proof of the Davis-Putnam-Robinson Theorem

In this section, we shall construct a direct exponential Diophantine encoding of the running of a register machine. This will immediately yield anew the algorithmic unsolvability of the general exponential Diophantine problem.

This proof is, in fact, a bit simpler than that given in section 2 and it yields some extra information. The advantage of the earlier proof is in the later task of formalising the proof of the equivalence of the notions Σ_1 and Diophantine (which task will not be carried out in this book).

As mentioned in the last section, the key to the proof is Lucas' Theorem, which we shall restate with some more convenient notation.

9.1. Definition. Let x, y be non-negative integers and suppose

$$x = \sum_{i=0}^{n} x_i 2^i, \quad y = \sum_{i=0}^{n} y_i 2^i$$

with $0 \le x_i, y_i < 2$. We write $x \preccurlyeq y$ if $x_i \le y_i$ for $i = 0, 1, ..., n$, i.e., $x \preccurlyeq y$ iff $D_x \subseteq D_y$.

The relation $\preccurlyeq$ has been called *masking* in the literature. For lack of better terminology, we shall occasionally use this term in the sequel.

9.2. Lucas' Theorem. *For $x, y \in \omega$, $x \preccurlyeq y$ iff $\binom{y}{x}$ is odd.*

This was proven as Corollary 8.5 in the last section.

If one multiplies x and y by any power of 2, the masking relation will remain unchanged:

$$x \preccurlyeq y \Leftrightarrow x \cdot 2^m \preccurlyeq y \cdot 2^m,$$

for

$$x \cdot 2^m = \sum_{i=0}^{n} x_i 2^{m+i}, \quad \text{and} \quad y \cdot 2^m = \sum_{i=0}^{n} y_i 2^{m+i}$$

have exactly the same coëfficients as x and y, albeit translated m spaces. In terms of sets, if one defines

$$X + m = \{x + m\colon x \in X\},$$

one sees that

$$D_{x2^m} = D_x + m \text{ and } D_{y2^m} = D_y + m,$$

and multiplication by 2^m merely translates the sets being indexed by m. Such translation can be applied to yield a very useful result:

9.3. Lemma. *Let $u_0, ..., u_{n-1}, v_0, ..., v_{n-1}$ be given and suppose q is a power of 2 greater than each u_i and v_i. Then:*

$$\sum_{i=0}^{n} u_i q^i \preccurlyeq \sum_{i=0}^{n} v_i q^i \text{ iff } \forall i \le n{-}1\, [u_i \preccurlyeq v_i].$$

Proof. In terms of sets, if $q = 2^m$, $\Sigma u_i q^i$ is the index of the set

$$X = D_{u_0} \cup (D_{u_1} + m) \cup (D_{u_2} + 2m) \cup ... \cup (D_{u_{n-1}} + (n-1)m)$$

and $\Sigma v_i q^i$ is the index of

$$Y = D_{v_0} \cup (D_{v_1} + m) \cup (D_{v_2} + 2m) \cup ... \cup (D_{v_{n-1}} + (n-1)m).$$

The number m is so large that the components $D_{u_i} + mi$ and $D_{v_i} + mi$ lie entirely in the interval $[mi, m(i+1)) = \{x \in \omega\colon mi \le x < m(i+1)\}$. Thus,

$$\begin{aligned}
\sum_{i=0}^{n} u_i q^i \preccurlyeq \sum_{i=0}^{n} v_i q^i \quad &\Leftrightarrow \quad X \subseteq Y \\
&\Leftrightarrow \quad \forall i \le n-1\, [(D_{u_i} + mi) \subseteq (D_{v_i} + mi)] \\
&\Leftrightarrow \quad \forall i \le n-1\, [D_{u_i} \subseteq D_{v_i}] \\
&\Leftrightarrow \quad \forall i \le n-1\, [u_i \preccurlyeq v_i].
\end{aligned}$$

QED

The dyadic expansion of a number can be thought of not only as a coding of a set, but also as a number written in binary. When this is done, the number $q = 2^m$ given above acts simply to chop a large binary number up into smaller *blocks* of m binary digits apiece. Lemma 9.3 asserts that, if q is big enough, we can store $u_0, ..., u_{n-1}$ and $v_0, ..., v_{n-1}$ in corresponding blocks of two large binary numbers without losing any masking information: the fact that u_i, v_i begin their binary digits at the same powers of 2 in the two large numbers means that the corresponding binary digits of the block holding u_i are less than or equal to those of the block holding v_i iff $u_i \preccurlyeq v_i$. It will later be convenient to think solely in terms of strings of 0's and 1's partitioned into blocks forming q-adic digits and I recommend to the reader the simple exercise of reworking the proof of Lemma 9.3 in such terms.

But for their variable size, we could compare these blocks with the *bytes* of computer scientists. The *b*inary dig*its* are exactly their *bits*. Given a number x with its dyadic and q-adic expansions,

$$x = \Sigma x_{2,i} 2^i = \Sigma x_{q,i} q^i, \text{ with } q = 2^m,$$

we shall follow the computer scientists and call the digits in base 2 the *bits* of the number x. The *i-th bit* (for $i = 0, 1, ...$) of x is the bit $x_{2,i}$ associated with the power 2^i. The digits in base q we shall simply refer to as the *digits* of x. The *i-th digit* is the digit

$$x_{q,i} = \sum_{j=0}^{m-1} x_{2,mi+j} 2^j$$

associated with q^i. Since we shall not use base 10, this use of the word "digit" will not cause any confusion— except for the reader who skips this paragraph.

Let us now recall from chapter I, section 11, a few facts about register machines. In doing so, it will now be convenient to violate some conventions announced in section 1 about the use of upper and lower case letters. Throughout the remainder of this section, we will use such letters as most convenient.

To begin with, a register machine has a finite set of registers $R_0, R_1, ..., R_{r-1}$ ($r > 0$). A program for a register machine is a finite list of instructions, $L_0, L_1, ..., L_{l-1}$ ($l > 0$) of the forms (I.11.9):

i. INC R_j

ii. DEC R_j

iii. GO TO L_k

iv. IF $R_j = 0$ GO TO L_k

v. STOP.

Two restrictions (I.11.10) were imposed on the use of these commands. First, one could not GO TO a DEC command. Second, a DEC command could only occur in a context of the form,

$$L_i: \quad \text{IF } R_j = 0 \text{ GO TO } L_k$$

$$L_{i+1}: \quad \text{DEC } R_j,$$

where $k \neq i+1$. A third restriction will be introduced now: the STOP command can only be the last instruction L_{l-1} of the program; moreover, the last instruction must be the STOP command. This new restriction is not essential: given any program, we can obtain an equivalent program satisfying this restriction by i. appending a STOP instruction to the end of the given program, and ii. replacing all other STOP instructions occurring in the program by GO TO statements transferring control to the new STOP command.

Theorem I.11.12 asserted that every partial recursive function is computable by a register machine. We can normalise such a computation so that it always ends in a STOP command with 0's in all the registers other than R_0. If we consider the computations of functions,

$$F(x) \simeq \begin{cases} 0, & x \in S \\ \text{und.}, & \text{otherwise,} \end{cases}$$

for r.e. sets S, we can state the following form of Theorem I.11.12.

9.4. Theorem. *Let S be an r.e. set. There is a register machine and a program such that, for any x, $x \in S$ iff when one starts the machine running the given program with x in register R_0 and 0 in all other registers, the machine will halt with a* STOP *command and* 0*'s in all registers. Moreover, if $x \notin S$, we can assume the computation never halts.*

The new insistence that the registers other than R_0 start and end with 0's in them, as well as the new restriction on the STOP command, is primarily an aesthetic consideration: the exponential Diophantine definition of the set S obtained by coding the running of the program on the machine will be simpler than it would be without the insistence.

As mentioned in the last section, the method of exponentially coding the running of a register machine is due to James Jones and Juriĭ Matijasevič. The exposition below follows a lecture by Martin Davis.

Suppose now that S is any r.e. set and that we have a register machine with registers $R_0, ..., R_{r-1}$ and program $L_0, ..., L_{l-1}$ which semi-computes S as in Theorem 9.4. Suppose $x \in S$. Then there is a computation on the machine which follows the program, executing instructions at times $t = 0, 1, ..., s$, with the machine executing a stop command at time s and all registers having 0's at that time. For each register R_j and each time t, let

$$r_{jt} = \text{contents of register } R_j \text{ at time } t. \tag{1}$$

For each instruction L_i and each time t, let

$$l_{it} = \begin{cases} 1, & L_i \text{ is executed at time } t \\ 0, & \text{otherwise.} \end{cases} \qquad (2)$$

For a very large power q of 2, we can code the histories of the individual registers and instructions as follows:

$$\hat{R}_j = \sum_{t=0}^{s} r_{jt} q^t \qquad (3)$$

$$\hat{L}_i = \sum_{t=0}^{s} l_{it} q^t . \qquad (4)$$

Obviously, we want $q > r_{jt}$ and $q > l_{it}$ for all i, j, t. It will be convenient to have, in addition, $q > 2r_{jt}$ and $q > l+1$. These inequalities are achieved by choosing

$$q = 2^{x+s+l+1}. \qquad (5)$$

Before explaining why this is large enough, let us quickly define one last quantity:

$$I = \sum_{t=0}^{s} q^t . \qquad (6)$$

About (5): Recall that R_0 starts with x in it, that all other registers begin with 0's in them, that the greatest increase that can occur in a register in one unit of time is for its contents to be incremented by 1, that the computation runs only $s+1$ steps, and that the last instruction executed is merely a STOP. Thus,

$$r_{jt} \leq x + s$$

for all j, t. Hence,

$$r_{jt} \leq x + s < 2^{x+s} < q/2.$$

(Note that this gives $r_{jt} < q/2 - 1$.) Similarly,

$$l + 1 < 2^{l+1} \leq q.$$

I cannot readily explain the purpose of the I introduced by (6). Suffice it to say that the introduction of this term will become transparent on application.

The problem before us is to express in an exponential Diophantine way that given numbers $x, q, I, \hat{R}_0, ..., \hat{R}_{r-1}, \hat{L}_0, ..., \hat{L}_{l-1}$ are indeed of this form, i.e., that they satisfy (3) - (6), and, moreover, that the $\hat{R}_j$'s and $\hat{L}_i$'s do indeed code the histories of the registers $R_0, ..., R_{r-1}$ and instruction lines $L_0, ..., L_{l-1}$ during the execution of the program by the machine (so that (1) and (2) will also hold).

The number q is readily given by (5):

$$q = 2^{x+s+l+1}. \qquad \text{(I)}$$

The number I of (6) is uniquely determined by q and s and can be defined by summing the geometric progression:

$$I = \frac{q^{s+1} - 1}{q - 1},$$

i.e.,

$$1 + Iq = q^{s+1} + I. \tag{II}$$

The number I can be used to assert that the $\hat{L}_i$'s are of the form (4):

$$\hat{L}_i \preccurlyeq I. \tag{III}$$

The fact that the greatest power of q entering into I is q^s implies, by (III) and Lemma 9.3, the same to hold for $\hat{L}_i$. Thus, (4) holds for some digits l_{it}, $t = 0, 1, ..., s$. But by Lemma 9.3 we have $l_{it} \le 1$, i.e., each l_{it} is either a 0 or a 1 as desired. Another equation,

$$\hat{L}_0 + \hat{L}_1 + ... + \hat{L}_{l-1} = I, \tag{IV}$$

yields more: comparing digits in (IV), we see that, for each t,

$$l_{0t} + l_{1t} + ... + l_{l-1,t} + \text{carry} = 1 + \text{carry}\cdot q.$$

However, the left-hand-side is at most l + carry (since the l_{it}'s are ≤ 1), and $l < q-1$ (by (I)). A simple induction on t (the position of the digits being compared in the q-adic expansions) shows there will be no carries. Thus,

$$l_{0t} + l_{1t} + ... + l_{l-1,t} = 1,$$

and we see that, for each t, exactly one l_{it} is 1. Thus, equations (III) and (IV) express a. that each $\hat{L}_i$ is of the form (4) with the digits being 0 or 1, and b. that, for each t, exactly one instruction is executed.

Two additional equations,

$$1 \preccurlyeq \hat{L}_0 \tag{V}$$

$$\hat{L}_{l-1} = q^s, \tag{VI}$$

assert that the execution of the program begins with instruction L_0 and ends with the STOP instruction L_{l-1},

Stating that the transfer of command is correct, i.e., that if L_i is followed in execution by L_k, then $l_{it} = 1$ implies $l_{k,t+1} = 1$, is a bit more involved. For each instruction L_i, depending on the type of command L_i is, there will be one or more masking assertions. For L_i an INC or DEC command, one takes

$$q\hat{L}_i \preccurlyeq \hat{L}_{i+1}, \tag{VII}$$

and, for every GO TO L_k command, one takes

$$q\hat{L}_i \preccurlyeq \hat{L}_k. \tag{VIII}$$

In (VII), for example, observe that

$$l_{it} \preccurlyeq l_{i+1,t+1},$$

whence

$$l_{it} = 1 \Rightarrow l_{i+1,t+1} = 1.$$

Thus, (VII) says exactly what it should.

The remaining command— namely, the branching command— depends on the contents of the registers. Moreover, we are not finished with the DEC and INC commands until we state exponentially how they affect the registers. Thus, we consider next the registers.

A first condition on the registers is:

$$\hat{R}_j \preccurlyeq (q/2-1)I. \tag{IX}$$

(Note. Since negative coëfficients are not allowed in our exponential polynomials, we should more properly write

$$\exists z\,[2z+2 = q \wedge \hat{R}_j \preccurlyeq zI]$$

in place of (IX). However, such might be carrying pedantry too far and I write (IX) in the form given. Similar abbreviations will appear in equations to come.) As with (III), (IX) tells us that $\hat{R}_j$ can be written in the form (3), but with each

$$r_{jt} \preccurlyeq q/2-1 \tag{*}$$

(by 9.3). Now, $q/2-1$ is just a string of 1's when written in binary, whence the masking assertions (*) reduce to the simple inequalities

$$r_{jt} \le q/2-1.$$

It turns out that, for each register R_j, the history $\hat{R}_j$ can be adequately described by a single equation, which we call its *register equation*. First, define

$$+(j) = \{i\colon\ L_i \text{ is INC } R_j\}$$
$$-(j) = \{i\colon\ L_i \text{ is DEC } R_j\}.$$

The register equations are

$$\hat{R}_0 = q\hat{R}_0 + \sum_{i\in+(0)} q\hat{L}_i - \sum_{i\in-(0)} q\hat{L}_i + x \tag{X$_0$}$$

for $j = 0$, and

$$\hat{R}_j = q\hat{R}_j + \sum_{i\in+(j)} q\hat{L}_i - \sum_{i\in-(j)} q\hat{L}_i, \tag{X$_j$}$$

for $j > 0$.

Notice that these equations concern the digits rather than the bits: First, look at the 0-th digits. For $j = 0$, (X)$_0$ gives

$$r_{00} = x,$$

and for $j > 0$, (X)$_j$ gives

$$r_{j0} = 0,$$

both equations being what one wants. Next look at the $(t+1)$-th digits on either side of the register equations. On the left one has $r_{j,t+1}$— the contents of register R_j at time $t+1$. On the right, one has

r_{jt},	if neither an INC R_j nor a DEC R_j is executed at time t
$r_{jt}+1$,	if INC R_j is executed at time t
$r_{jt}-1$,	if DEC R_j is executed at time t,

i.e., the digit is exactly the contents of R_j at time $t+1$— provided $t+1 \le s$. For $t = s$, the $(t+1)$-th digit on the left of (X) is 0 and the corresponding digit on the right is r_{js}— showing each register to end with a 0.

Finally, we must express the correctness of the transfer of control corresponding to a branching command. Let L_i be IF $R_j = 0$ GO TO L_k. If $k = i+1$, simply use (VII). Otherwise, use the two masking assertions,

$$q\hat{L}_i \preccurlyeq \hat{L}_{i+1} + \hat{L}_k \tag{XI}$$

$$q\hat{L}_i \preccurlyeq \hat{L}_{i+1} + (qI - 2\hat{R}_j). \tag{XII}$$

As with (VII), condition (XI) tells us that, for any time t,

$$l_{it} = 1 \;\Rightarrow\; l_{i+1,t+1} = 1 \vee l_{k,t+1} = 1.$$

Thus, by (XI), if L_i is executed at time t, one of L_{i+1} and L_k is executed at time $t+1$. We know by (IV) that they cannot both be executed then. Condition (XII) will tell us which of the two is executed at time $t+1$.

Let us look at (XII). By condition (IX), $qI > 2\hat{R}_j$, so the subtraction is possible. Multiplying $\hat{R}_j$ by 2 simply shifts each *bit* of $\hat{R}_j$ one position to the left. The numbers $2r_{jt}$ are less than q, so individual changes do not carry over to the next digits. Multiplying I by q shifts the digits (all 1's) of I one digital position to the left. Graphically, when we align qI and $2\hat{R}_j$ for subtraction, the numbers look like this:

bit position:	...	$2m+1$	$2m$	$2m-1$	...	$m+1$	m	$m-1$	...	1	0
qI:	...	0	1	0	...	0	1	0	...	0	0
$2\hat{R}_j$:	...	0	0	bits of r_{ij}		0	0	bits of r_{j0}			0

The thing to observe is that the subtractions can be isolated in blocks— not the digits themselves, but blocks starting one bit to the left of the starts of the digits.

If $r_{jt} = 0$, the subtraction looks like

$m(t+1)$		...		$mt+1$
1	0	...	0	0
0	0	...	0	0
1	0	...	0	0

yielding a 1 in bit $m(t+1)$. But the bit in that position in $q\hat{L}_i$ is just $l_{it} = 1$. By (XII), $l_{i+1,t+1}$ cannot also be 1 or the two 1's in that bit in $\hat{L}_{i+1}$ and $qI - 2\hat{R}_j$ would cancel and carry. Thus, if $r_{jt} = 0$, we have $l_{i+1,t+1} = 0$ and, by (XI), $l_{k,t+1} = 1$.

If, on the other hand, $r_{jt} \neq 0$, the subtraction looks like

$m(t+1)$		...		$mt+1$
1	0	...	0	0
0	$\mid \leftarrow$	bits	$\rightarrow \mid$	0
0	$\mid \leftarrow$	bits	$\rightarrow \mid$	0

i.e., the 1 in bit $m(t+1)$ of qI had to be borrowed to perform the subtraction. This leaves a 0 in that bit in $qI - 2\hat{R}_j$, whence the $m(t+1)$-bit of $\hat{L}_{i+1} + qI - 2\hat{R}_j$ is that of $\hat{L}_{i+1}$. Again, the corresponding bit of $q\hat{L}_i$ is assumed to be 1, whence (XII) implies that $l_{i+1,t+1} = 1$.

With this last, we have completed our task of writing exponential Diophantine conditions expressing that the $\hat{R}_j$'s and $\hat{L}_i$'s code the running of the program on the machine and that the membership of x in S has been verified. This immediately yields:

9.5. Theorem. *Let S be any r.e. set. There are exponential polynomials P, Q such that, for any $x \in \omega$,*

$$x \in S \Leftrightarrow \exists sqI\hat{R}_0 \dots \hat{R}_{r-1}\hat{L}_0 \dots \hat{L}_{l-1}[P(s,q,I,\hat{R}_0,\dots,\hat{L}_{l-1}) = Q(s,q,I,\hat{R}_0,\dots,\hat{L}_{l-1})].$$

9.6. Corollary. (Davis-Putnam-Robinson). *Every r.e. relation is exponential Diophantine.*

To prove this, use a polynomial n-tupling function.

9.7. Corollary. *The general exponential Diophantine problem is not algorithmically solvable.*

As in section 2, one derives this by simply choosing S non-recursive in Theorem 9.5.

The Jones-Matijasevič proof of the Davis-Putnam-Robinson Theorem yields a nice refinement that was first proven by Matijasevič a few years after he had proven the Diophantine representability of r.e. relations. This result, to which we now turn, can be proven via the Davis-Putnam-Robinson method (cf. Exercise 5, below), and this was how Matijasevič first proved it. However, it almost suggests itself in the Jones-Matijasevič proof of Theorem 9.5. The thing to notice is this: if $x \in S$, the numbers $s, q, I, \hat{R}_0, \dots, \hat{R}_{r-1}, \hat{L}_0, \dots, \hat{L}_{l-1}$ solving the "equations" (I) - (XII) are unique, i.e., if the system (I) - (XII) has any solution at all, it has a unique one. Such unicity suggests the following definition.

9.8. Definition. Let $R \subseteq \omega^n$ be given. A representation

$$R(x_0, \dots, x_{n-1}): \exists y_0 \dots y_{m-1}\,[P(\mathbf{x}, \mathbf{y}) = Q(\mathbf{x}, \mathbf{y})],$$

with P, Q exponential polynomials, is a *singlefold exponential Diophantine representation* of R if P, Q also satisfy

$$\forall \mathbf{xyz}\,[P(\mathbf{x}, \mathbf{y}) = Q(\mathbf{x}, \mathbf{y}) \wedge P(\mathbf{x}, \mathbf{z}) = Q(\mathbf{x}, \mathbf{z}) \Rightarrow \bigwedge_{i=0}^{m-1}(y_i = z_i)],$$

i.e., if whenever $P = Q$ is solvable for some given values $x_0, ..., x_{n-1}$ of the parameters, the solution is unique.

9.9. Theorem. *Every r.e. relation has a singlefold exponential Diophantine representation.*

Obviously, it suffices to prove the Theorem for unary relations, i.e., sets. The uniqueness of s, q, I, etc. satisfying (I) - (XII) does not quite yield the Theorem because (I) - (XII) are not exponential Diophantine *equations*, but exponential Diophantine *relations* with hidden quantifiers of three sorts:

i. quantifiers needed to eliminate negative coëfficients,

ii. quantifiers needed to define the binomial coëfficients used in expressing masking assertions:

$$u \preccurlyeq v \Leftrightarrow \binom{v}{u} \text{ is odd; and}$$

iii. quantifiers needed to express the oddness of these binomial coëfficients.

The first type of quantifier offers no trouble. If we rewrite, e.g., (XII) to eliminate the negative coëfficients, we get

$$\exists z[q\hat{L}_i \preccurlyeq \hat{L}_{i+1} + z \wedge z + 2\hat{R}_j = qI].$$

Clearly the choice of z is unique: $z = qI - 2\hat{R}_j$. Similarly, quantifiers of the third type are no trouble:

$$v \text{ is odd} \Leftrightarrow \exists z\,(v = 2z + 1)$$

$$\Leftrightarrow \exists! z\,(v = 2z + 1),$$

where $\exists! z\ \varphi$ abbreviates

$$\exists z\ \varphi z \wedge \forall wz(\varphi w \wedge \varphi z \rightarrow w = z).$$

The crucial quantifier is that used in defining the binomial coëfficients:

9.10. Lemma. *The relation* $z = \binom{x}{y}$ *has a singlefold exponential Diophantine representation.*

The exponential Diophantine definition of Example 2.9.i,

$$z = \binom{x}{y} \Leftrightarrow \exists tuvw\,[u = 2^x + 1 \wedge t = u + 1 \wedge z < u \wedge$$

$$\wedge\ t^x = vu^{y+1} + zu^y + w \wedge w < u^y],$$

is, in fact, singlefold.

From 9.10 and the above comments, the reader should be able to piece together a decent proof of Theorem 9.9.

Before discussing a couple of applications of Theorem 9.9, let me note quickly that one can analogously define the notion of a *singlefold Diophantine representation* of an r.e. relation and ask if every r.e. relation has such a representation.

9.11. Open Problem. Does every r.e. relation have a singlefold Diophantine representation?

Problem 9.11 is one of the three main open problems in the logical study of Diophantine equations, the other two problems being the Diophantine problem over the rationals and the problem of the exact number of variables needed for an unsolvable Diophantine problem.

Part of the interest in Problem 9.11 stems from the history of the study of Diophantine equations in the present century. In 1909, Axel Thue proved the first powerful result asserting certain equations had only finitely many solutions. Among the various extensions and applications of his result, is L.J. Mordell's result of 1922 that what is now called the Mordell equation,

$$Y^2 = X^3 + K, \tag{M}$$

has, for each non-zero integral value of the parameter K, at most finitely many integral solutions. For over forty years, number theorists only knew that similar equations had only finitely many solutions. They had effective bounds on the *numbers* of solutions, but not on the *sizes* of such. Without bounds on the sizes of the solutions, they could not effectively decide the solvability or unsolvability of such equations. In 1968, the number theorist Alan Baker was able to provide such effective bounds for many of these equations. For the Mordell equation (M), for example, he showed that, for any parameter k, any integral solutions x, y had to satisfy

$$\max\{|x|, |y|\} \le e^{10^{10}|k|^{10^4}}.$$

The situation that almost occurred, where there are effective bounds on the number of solutions but no such bounds are possible on the sizes of solutions, does occur in the exponential Diophantine case.

9.12. Application. *There are exponential polynomials P, Q with parameter A and variables $X_0, ..., X_{n-1}$ such that the number of solutions to the equations,*

$$P(a, X_0, ..., X_{n-1}) = Q(a, X_0, ..., X_{n-1}), \tag{*}$$

is uniformly bounded, but there is no recursive bound on the size of the possible solutions $x_0, ..., x_{n-1}$, i.e., there is no recursive F such that, for any a, (*) is solvable in ω iff there are $x_0, ..., x_{n-1} \le F(a)$ which satisfy (*).

The proof is, of course, simple: let P, Q yield a singlefold exponential Diophantine representation of a non-recursive r.e. set S. For any $a \in \omega$, there is at most one solution to the equation (*); a recursive bound F on the sizes would, however, make S recursive:

$$a \in S \Leftrightarrow \exists x_0...x_{n-1} \le F(a)\ [P(a, \mathbf{x}) = Q(a, \mathbf{x})].$$

It is an open problem whether or not such behaviour actually occurs among ordinary (i.e., non-exponential) Diophantine equations. The best that can be said is that, for a while, one had the one type of bound but did not *know* the other.

A second application of the existence of singlefold exponential Diophantine representations of r.e. relations is my lovely refinement of Martin Davis' Diophantine analogue of Rice's Theorem for the cardinalities of the sets of solutions to Diophantine equations (Theorem 6.6). I shall prove the corresponding exponential Diophantine analogue to the Rice-Shapiro Theorem.

We begin by recalling a few notions and notations from section 6, above. The main notation we want is $\#(P)$— the number of zeroes of a polynomial. With exponential polynomials, we are dealing with equations of the form $P = Q$ instead of $P = 0$ and we must either write $\#(P, Q)$ or expand our concept of exponential polynomial so that we can treat $P - Q$ as one entity P' and write $\#(P')$.

9.13. Notational Conventions. In the following application (and corresponding exercise) we will term an *exponential polynomial* any function (or: expression) $P - Q$, where P, Q are exponential polynomials as in Definition 2.7. We will denote such an exponential polynomial with some negative coëfficients by P or Q or so, just as we denoted exponential polynomials with only non-negative coëfficients.

With this, several definitions form section 6, above, extend to the exponential case.

9.14. Definitions. i. $C = \omega \cup \{\infty\}$

ii. Let $P(X_0, ..., X_{n-1})$ be an exponential polynomial. Then:

$$\#(P) = \mathrm{card}(\{(x_0, ..., x_{n-1}) \colon P(x_0, ..., x_{n-1}) = 0\}).$$

iii. Let $A \subseteq C$. Then:

$$A^* = \{P \colon P \text{ is an exponential polynomial} \wedge \#(P) \in A\}.$$

iv. $A \subseteq C$ is *non-trivial* if $A \neq \emptyset$ and $A \neq C$.

Theorem 6.6 asserted for polynomials that, if $A \subseteq C$ is non-trivial, then A^* is not recursive. The proof given carries over immediately to the exponential case. In the exponential case, however, a stronger result holds:

9.15. Theorem. *Let $A \subseteq C$ be non-empty. Then: A^* is r.e. iff, for some $m \in \omega$, $A = \{c \colon c \geq m\}$.*

The r.e. nature of the sets A^* for A of the given form is obvious. We thus need only to prove the converse. As in section 6, we will use a pair of effective operations.

9.16. Definitions. Let $P(X_0, ..., X_{n-1})$ be an exponential polynomial. Define

i. $T^{\infty}(P) = (Y + 1) \cdot P$

ii. for $m > 0$,

$$T^m(P) = (P^2 + Y^2) \cdot \prod_{i=1}^{m} \left(\sum_{j=0}^{n-1} (X_j^2 + (Y - i)^2) \right).$$

The following are easily seen and will be used with little or no mention:

$$\#(T^{\infty}(P)) = \begin{cases} \infty, & \#(P) > 0 \\ 0, & \#(P) = 0 \end{cases}$$

$$\#(T^m(P)) = \#(P) + m.$$

Using these, we shall prove Theorem 9.15 through a series of simple lemmas.

9.17. Lemma. *Let $A \neq \emptyset$ and suppose A^* is r.e. Then: $\infty \in A$.*

Proof. By contradiction. Assume $\infty \notin A$. Let m be the least element of A. Observe that

$$\#(P) = 0 \Leftrightarrow \#(T^m T^{\infty}(P)) \in A$$
$$\Leftrightarrow T^m T^{\infty}(P) \in A^*.$$

Thus, if A^* were r.e.,

$$\{0\}^* = \{P\colon \#(P) = 0\}$$

would also be r.e. As $\{P\colon \#(P) \neq 0\}$ is clearly r.e., this would, however, make the question

$$\#(P) = 0?$$

recursive— contrary to the Davis-Putnam-Robinson result. QED

Note that the proof of this lemma does not use the singlefold nature of the representation of r.e. sets, whence it carries over *mutatis mutandis* to the ordinary Diophantine case. In particular, we can conclude the non-recursive-enumerability of the set of polynomials with only finitely many non-negative integral zeroes.

The next lemma also holds for polynomials. It is a reduction and is of no particular interest in its own right.

9.18. Lemma. *Let A^* be r.e. and suppose, for some $m \in \omega$, that $m \in A$, but $m+1 \notin A$. Then: $\{c \in C\colon c \neq 1\}$ is r.e.*

Proof. Suppose $m \in A$, $m+1 \notin A$. Let

$$X = \{P\colon T^m(P) \in A^*\} \cup \{c \in C\colon c \geq 2\}^*.$$

The set X is clearly r.e. Moreover, it is readily seen that

$$X = \{c \in C\colon c \neq 1\}^*.$$

QED

9.19. Lemma. $\{c \in C\colon c \neq 1\}^*$ *is not r.e.*

Proof. Let S be a non-recursive r.e. set and choose $P(A, X_0, \ldots, X_{n-1})$ so that the representation

$$a \in S \Leftrightarrow \exists x_0 \ldots x_{n-1}\, [P(a, x_0, \ldots, x_{n-1}) = 0]$$

is singlefold. Let $P_a(X_0, \ldots, X_{n-1})$ be $P(a, X_0, \ldots, X_{n-1})$ and observe that

$$a \notin S \Leftrightarrow \#(P_a) \neq 1.$$

Thus, the recursive enumerability of $\{c \in C\colon c \neq 1\}^*$ would yield the recursive enumerability of $\omega - S$. This, with the recursive enumerability of S would yield the recursiveness of S, a contradiction. QED

By Lemmas 9.17 - 9.19, if $A \subseteq C$ is non-empty and A^* is r.e., then A is upward closed. Hence, if $m = \min(A)$, we have $A = \{c \in C\colon c \geq m\}$. To complete the proof of the Theorem, we need only show that this minimum cannot be ∞.

9.20. Lemma. *$\{P\colon \#(P) = \infty\}$ is not r.e.*

Proof. Let $P(X, Y, Z_0, \ldots, Z_{n-1})$ give a singlefold representation of the universal binary relation $y \in W_x$:

$$y \in W_x \iff \exists z\, [P(x, y, z) = 0].$$

For each $x \in \omega$, let $P_x(Y, Z_0, \ldots, Z_{n-1})$ be $P(x, Y, Z_0, \ldots, Z_{n-1})$ and observe:

$$W_x \text{ is infinite} \iff \#(P_x) = \infty.$$

A recursive enumeration of $\{\infty\}^*$ would yield a recursive enumeration of $\{x\colon W_x \text{ is infinite}\}$, contrary to the Rice-Shapiro Theorem (I.12.23). QED

With the proof of Lemma 9.20 we have completed the proof of Theorem 9.15.

Lemma 9.20 is really a result of recursion theory, unlike the other lemmas used in proving Theorem 9.15. Nevertheless, it seems almost scandalous to have to appeal to such a powerful result as the Rice-Shapiro Theorem (even if 9.15 is an analogue thereof) to prove the non-recursive-enumerability of so simple a set as $\{x\colon W_x \text{ is infinite}\}$. This non-enumerability can be established by a mere diagonal argument, as the reader will prove in Exercise 6, below. For those who like to apply high-powered tools, the Rice-Shapiro Theorem is applied in Exercise 7 to give a more direct proof of 9.15. It is also applied in Exercise 8 to characterise all recursive operations T such that $\#(T(P)) = F(\#(P))$ for some (not necessarily recursive) function F. This yields a third proof of Theorem 9.15.

As for ordinary polynomials, I note that Lemmas 9.16 and 9.17 reduce the analogue to Theorem 9.15 to the problem of showing the non-recursive-enumerability of the sets,

$$\{P\colon P \text{ is a polynomial} \wedge \#(P) \neq 1\}$$
$$\{P\colon P \text{ is a polynomial} \wedge \#(P) = \infty\}.$$

An affirmative solution to Open Problem 9.11 on the existence of singlefold Diophantine representations of r.e. relations would yield these non-enumerabilities.

Exercises.

1. Let q be a positive integer. Show: q is a power of 2 iff $q \preccurlyeq 2q - 1$.
2. Suppose that, instead of computing the partial representing function of a set S, the register machine of this section was to compute a partial recursive function F leaving the value $y = F(x)$ in R_0 at the end of the computation. Show that the proof of Theorem 9.5 carries over with a small change in (I) - (XII). What is the change?
3. Suppose we extend our assembly language to include commands of the form IF $R_i \leq R_j$ GO TO L_k. Write conditions analogous to (I) - (XII) so that the proof of Theorem 9.5 still goes through.

4. Complete the proof of Theorem 9.9 as outlined in the text.

5. (Alternate proof of Theorem 9.9). Let S be a given r.e. set and suppose $P(A, X_0, ..., X_{n-1})$ is an ordinary polynomial such that

$$a \in S \Leftrightarrow \exists x_0...x_{n-1}\,[P(a, \mathbf{x}) = 0].$$

i. Construct a polynomial $Q(A, X_0, ..., X_{n-1}, X)$ such that

a. $a \in S \Leftrightarrow \exists x_0...x_{n-1}x\,[Q(a, \mathbf{x}, x) = 0]$

b. for any $a, \mathbf{x}, \mathbf{y}$,

$$Q(a, \mathbf{x}, x) = 0 \wedge Q(a, \mathbf{y}, x) = 0 \Rightarrow \bigwedge_{i=0}^{n-1} x_i = y_i$$

c. $Q(a, \mathbf{x}, x) = 0 \Rightarrow \bigwedge_{i=0}^{n-1} x_i \le x.$

ii. Show that there is a polynomial R such that

a. $a \in S \Leftrightarrow \exists x\, \forall y \le x\, \exists x_0...x_{n-1} \le x\,[R(a, x, y, \mathbf{x}) = 0]$

b. the x in a is unique and, for such x and any $y \le x$, the numbers $x_0, ..., x_{n-1}$ are unique.

iii. Prove Theorem 9.9 by applying the Bounded Quantifier Theorem (Theorem 2.11) to ii.a.

[Hints. i. use $\langle \cdot, ..., \cdot \rangle_n$; ii. minimise the x of Q; iii. use the exponential bounds available for the Chinese Remainder Theorem (I.6.5).]

6. Show by direct diagonalisation that $\{x\colon W_x \text{ is infinite}\}$ is not r.e.: Let $e_0, e_1, ...$ be a recursive enumeration of indices of infinite r.e. sets. Find an infinite recursive set X such that $X \ne W_{e_i}$ for any e_i. [Hint. Define the complement of X to be the range of a strictly increasing sequence $x_0, x_1, ...$ with $x_i \in W_{e_i}$.]

7. i. Use the Rice-Shapiro Theorem to show: If $A \subseteq C$ is non-empty and $\{x \in \omega\colon \mathrm{card}(W_x) \in A\}$ is r.e, then, for some $m \in \omega$,

$$A = \{c \in C\colon c \ge m\}.$$

ii. Assuming the effective passage between r.e. sets and their singlefold exponential Diophantine representations, derive Theorem 9.15 from part i of this exercise.

8. Consider the following conditions on a function $F\colon C \to C$, C as in Definition 9.14.i:

a. (monotonicity). $x \le y \Rightarrow F(x) \le F(y)$

b. (continuity). $F(\infty) = \sup\{F(n)\colon n \in \omega\}$

c. (lower semi-recursiveness). the restriction to ω^2 of the relation $F(x) \ge y$ is r.e.

i. Suppose F satisfies conditions a - c. Show: There is a recursive function G such that, for all $x \in \omega$,

$$\mathrm{card}(W_{G(x)}) = F(\mathrm{card}(W_x)). \qquad (*)$$

ii. Show: There is a recursive function H such that, for all $x \in \omega$, $\mathrm{card}(W_{H(x)}) = x$.

iii. Let $F\colon C \to C$ and suppose G is a recursive function satisfying (*).

a. Show: F is monotone

b. Show: F is continuous

c. Show: F is lower semi-recursive.

iv. For each exponential polynomial $P(X)$ find an exponential polynomial $Q(A, Y)$ such that

$$\#(P) = \#(Q) = \mathrm{card}(\{a\colon \exists y\, [Q(a, y) = 0]\}).$$

v. Prove: For any $F\colon C \to C$, the following are equivalent:

a. There is a recursive operation T on exponential polynomials such that, for any P,

$$\#(T(P)) = F(\#(P))$$

b. F is monotone, continuous, and lower semi-recursive.

vi. Use iv and v to give another proof of Theorem 9.15.

[Hints. i. either you see it or you don't— let

$$W_{G(x)} = \{y\colon \exists n\, [F(n) \geq y + 1 \wedge \mathrm{card}(W_x) \geq n]\};$$

iii.a - b. apply the Rice-Shapiro Theorem; iii.c. apply part ii; v. imagine the polynomial Q of part iv as the index of the set $W_Q = \{a\colon \exists y\, [Q(a, y) = 0]\}$ and apply i and iii; and vi. for A^* r.e., find a recursive T such that

$$W_{T(P)} = \begin{cases} \omega, & P \in A^* \\ \varnothing, & P \notin A^*. \end{cases}]$$

9. Following section 6 (6.11 *ff.*), give a formal treatment of recursiveness of operations on exponential polynomials. Apply this to Exercises 7.ii and 8.iv.

*10. The 3-Variable Exponential Diophantine Result

W have seen in section 6, above, that there is some positive integer m such that the m-variable Diophantine problem over the natural numbers is unsolvable. Juriĭ Matijasevič and Julia Robinson gave the first respectable upper bound on the least such m. Using the unsolvability of Hilbert's tenth problem as a starting point, they gave a relatively economical exponential Diophantine encoding of Diophantine equations— using modified b-adic expansions for large b's— and then eliminated the exponentiations by using a super-economically defined relation of exponential growth and various variable-saving devices. In all, they used 13 existentially quantified variables to give a Diophantine representation of an arbitrary r.e. relation. Later, Matijasevič learned of Kummer's Corollary 8.4 and replaced the modified b-adic expansions by a new encoding exploiting Kummer's result. In doing so, he reduced the number of variables needed to give a Diophantine representation of any r.e. relation from 13 to 9. This is currently the best-known bound on the number of variables needed to obtain an unsolvable result. The gap between the solvable and the unsolvable is rather wide: the single variable Diophantine problem over the natural numbers is solvable, and the 2-variable problem is suspected to be solvable.

Both the 13-variable and the 9-variable results are too technical to be presented here. Fortunately, there is a nice limited-number-of-variables-result that can appear here— namely, Matijasevič's 3-variable iterated exponential Diophantine result. The proof is not only more accessible than those of the 9- or 13-variable results, but, for the interested reader, it serves as an introduction to the proof of the 9-variable result, which the reader can find in the paper of James P. Jones cited in the Reading List.

There are two small preparatory steps that must be taken before beginning the proof. First, let $P(X_1, ..., X_n)$ be a polynomial with the variables $X_1, ..., X_n$ and possibly some parametric variables, mention of which we will suppress throughout this section. Replace P by

$$Q: \quad P(X_1, ..., X_n)^2 + (Z - \langle X_1, ..., X_n \rangle_n - 1)^2,$$

or, more appropriately by d^2Q (which we shall still call Q), where d is the common denominator of the rational coëfficients of the n-tupling polynomial $\langle \cdot, \ldots, \cdot \rangle_n$. Note the following.

10.1. Lemma. *For a given polynomial P, and Q as just defined from P,*
i. *Q has a non-negative integral zero iff P has one*
ii. *for any zero $z, x_1, ..., x_n$ of Q, z uniquely determines $x_1, ..., x_n$*
iii. *for any zero $z, x_1, ..., x_n$ of Q, and any $1 \le i \le n$, $z > x_i$.*

Henceforth we shall work with an arbitrary polynomial Q satisfying 10.1.ii - iii, ignoring where it came from. To state exactly what we will prove, we need first to introduce a little notation, and to recall Kummer's 8.4.

10.2. Kummer's Corollary. *Let $\kappa(t)$ denote the exact power of 2 in the prime factorisation of t, and $\sigma(t)$ the number of 1's occurring in the dyadic expansion of t. Then:*

$$\kappa\left(\binom{2t}{t}\right) = \sigma(t).$$

The "κ" notation is of fleeting importance. We see it once and it is gone. The "σ" will be handy below.

Matijasevič's proof of the unsolvability of the 3-variable iterated exponential Diophantine problem proceeds as follows: given $Q(Z, X_1, \ldots, X_n)$ satisfying 10.1.ii -iii, two nearly iterated exponential polynomials $M(z)$ and $A(z)$ are constructed such that

$$z \text{ codes the first zero of } Q \quad \text{iff} \quad \sigma(A(z)) \ge M(z).$$

Here, we say that z codes the first zero of Q if z is the smallest number for which there are $x_1, ..., x_n$ such that $Q(z, x) = 0$. The word "nearly" in describing the iterated exponential polynomiality of M and A will be explained in Lemma 10.12, below. (Actually, M is a polynomial and only A is only "nearly" an iterated exponential polynomial.) The elimination of the "nearness" (one division and two subtractions) will be attended to at the end of this part of the proof.

Once the Diophantine problem,

$$\exists z\boldsymbol{x}\, [Q(z, \boldsymbol{x}) = 0],$$

has been reduced to

$$\exists z[\sigma(A(z)) \ge M(z)], \tag{1}$$

Kummer's Corollary transforms this to

$$\exists z\,[2^{M(z)} \mid \binom{2A(z)}{A(z)}]. \qquad (2)$$

Expressing this exponentially will introduce two additional existential quantifiers, thus yielding:

10.3. Theorem. *Every r.e. relation R has a singlefold iterated exponential Diophantine representation of the form,*

$$R(\boldsymbol{a}):\ \exists xyz\,[E(\boldsymbol{a},x,y,z) = F(\boldsymbol{a},x,y,z)].$$

A bit more will hold: E and F will be *dyadic* iterated exponential polynomials, i.e., any exponentiation with a non-constant exponent will have 2 as its base.

Let me make a couple of quick comments before beginning the proof. The super-quick comment is the reminder that I will suppress mention of the parameters in Q in the proof of Theorem 10.3. The quick, but not super-quick comment concerns the singlefold nature of the representation in 10.3: the uniqueness of z in (1) follows from the fact that z codes the (unique) *least* zero of q. That the binomial coëfficient can be eliminated in a singlefold manner then yields the singlefold nature of the representation. The first part of this— the uniqueness of z— does not come for free, but has complicated slightly the construction of A and M. Following Jones' translation of Matijasevič's paper, I will note parenthetically the simplifications possible if one drops the requirement that the final representation be singlefold.

Let us now begin the proof of Theorem 10.3. We start with two lemmas isolating a couple of obvious properties of σ.

10.4. Lemma. i. *Let $t_0 < s$, and let $t_0, t_1, s \in \omega$. Then,*

$$\sigma(t_1 \cdot 2^s + t_0) = \sigma(t_1) + \sigma(t_0).$$

ii. *Let $t_0, ..., t_n$ and $s_0 \leq ... \leq s_n$ be given with $t_i < 2^{s_{i+1}-s_i}$. Then,*

$$\sigma\Big(\sum_{i=0}^{n} t_i 2^{s_i}\Big) = \sum_{i=0}^{n} \sigma(t_i).$$

Proof. i. Multiplying t_1 by 2^s shifts all the bits of t_1 to the left by s places. Since $t_0 < 2^s$, t_0 has fewer than s bits when written in binary. Thus, when adding $t_1 2^s$ and t_0, there will be no overlap in 1's, i.e., no carrying operations— the 1's of $t_1 \cdot 2^s + t_0$ will occur exactly where the 1's occur in $t_1 2^s$ and in t_0.

ii. By induction on n— I leave the details to the reader. QED

10.5. Lemma. *Let q be an integer such that $|q| < 2^s \leq 2^r$, with s, r natural numbers. Then:*

$$\sigma(2^r + q) \in \begin{cases} [1, s+1], & \text{for } q \geq 0 \\ [r-s+1, r], & \text{for } q < 0. \end{cases}$$

Here, the interval notation is applied to natural numbers:

$$[x, y] = \{z \in \omega: x \le z \le y\}.$$

Proof of 10.5. If $q \ge 0$, Lemma 10.4.i applies:

$$\sigma(2^r + q) = \sigma(1) + \sigma(q) = 1 + \sigma(q).$$

That $1 + \sigma(q) \ge 1$ is obvious. On the other hand, since $q < 2^s$, we have $1 + \sigma(q) \le 1 + s$.

If $q < 0$, since $|q| < 2^s \le 2^r$, the number $2^r + q = 2^r - |q|$ lies strictly between 0 and 2^r, whence it has at most r bits, i.e.,

$$\sigma(2^r + q) \le r.$$

Moreover, since $q \ne 0$ and 2^r is a 1 followed by 0's when written in binary, some borrowing must occur in subtracting $|q|$ from 2^r. To see how much borrowing is needed, imagine 2^r and $|q|$ lined up for the subtraction:

bit #:	r	$r-1$	...	s	$s-1$	...	0
2^r:	1	0	...	0	0	...	0
$\lvert q\rvert$:					q_{s-1}	...	q_0
$2^r + q$:		1	...	1	a_{s-1}	...	a_0

The block from bit s to bit $r-1$ in $2^r + q$ contains $r - s$ 1's. Moreover, if q_i is the first bit of $|q|$ which is a 1, then the corresponding a_i is 1. Thus,

$$\sigma(2^r + q) \ge r - s + 1.$$

QED

10.6. Definition-Lemma. *For natural numbers a and integers b, define*

$$F^-(a,b) = (2^a + b)2^{a+1} + (2^a - b)$$

$$F^+(a,b) = (2^a + b - 1)2^{a+1} + (2^a - b - 1).$$

Let q be an integer such that $|q| < 2^s \le 2^r$ *and let* $s \ge 2$. *Then:*

$$\sigma(F^*(r,q)) \in \begin{cases} [r - s + 2 * (r - s),\ r + s + 2 * (r - s)], & q = 0 \\ [r - s + 2,\ r + s + 2], & q \ne 0, \end{cases}$$

*where * denotes either + or –.*

A couple of words about F^- and F^+ before beginning the proof: Lemma 10.4.i will tell us that, for $|q| < 2^s \le 2^r$,

$$\sigma(F^-(r,q)) = \sigma(2^r + q) + \sigma(2^r - q)$$

$$\sigma(F^+(r,q)) = \sigma(2^r + q - 1) + \sigma(2^r - q - 1).$$

These expressions are *even* functions of q, i.e.,

$$\sigma(F^*(r,-q)) = \sigma(F^*(r,q)).$$

Moreover, $\sigma(F^+(r,q))$ is maximised for $q = 0$ and $\sigma(F^-(r,q))$ is minimised for $q = 0$. The functions F^* are, thus, the initial building blocks from which we will construct $A(z)$ for which $\sigma(A(z))$ will be relatively large only for the first z to code a zero of Q. [If we were not interested in a singlefold representation, we would only need the zero-maximising F^+.]

Proof of Lemma 10.6. By cases according as * is + or –, and subcases as q is or is not 0.

We begin with the – case. Observe that

$$F^-(r,q) = (2^r+q)2^{r+1}+2^r-q,$$

with $2^r-q < 2^{r+1}$, whence 10.4 yields

$$\sigma(F^-(r,q)) = \sigma(2^r+q)+\sigma(2^r-q).$$

If $q = 0$, we quickly see that

$$\sigma(F^-(r,0)) = 2 \in [2,2s+2] = [r-s+2-(r-s), r+s+2-(r-s)].$$

If $q \neq 0$, 10.5 yields

$$\sigma(2^r+|q|) \in [1,s+1]$$
$$\sigma(2^r-|q|) \in [r-s+1,r],$$

whence

$$\sigma(2^r+q)+\sigma(2^r-q) \in [r-s+2,r+s+1] \subseteq [r-s+2,r+s+2].$$

In the + case we again begin with the observation that

$$F^+(r,q) = (2^r+q-1)2^{r+1}+2^r-q-1,$$

with $2^r-q-1 < 2^{r+1}$, whence

$$\sigma(F^+(r,q)) = \sigma(2^r+q-1)+\sigma(2^r-q-1).$$

For $q = 0$,

$$\sigma(2^r-1) = r,$$

whence

$$\sigma(F^+(r,0)) = 2r \in [2r-2s+2,2r+2] \subseteq [r-s+2+(r-s), r+s+2+(r-s)].$$

For $q \neq 0$, there are two subcases. If neither $q+1$ nor $-q+1$ is 2^s, then, as in the – case, we apply 10.5 to get

$$\sigma(2^r+|q|-1) \in [1,s+1]$$
$$\sigma(2^r-|q|-1) \in [r-s+1,r],$$

whence

$$\sigma(F^+(r,q)) \in [r-s+2,r+s+1] \subseteq [r-s+2,r+s+2].$$

In the other subcase, $|q| = 2^s-1$ and

$$\sigma(2^r+|q|-1) = \sigma(2^r+2^s-2) = s,$$

as the reader can readily verify. But,

$$\sigma(2^r - |q| - 1) = \sigma(2^r - 2^s) = r - s,$$

as the reader can also verify. Thus,

$$\sigma(F^+(r,q)) = r - s + s = r \in [r - s + 2, r + s + 2],$$

since $s \geq 2$. QED

We are going to use the F^*'s for $* = +$ or $-$ to construct the function $A(z)$. For q we will want to use values $Q(w, x_1, \ldots, x_n)$ for $w \leq z$ and $x_1, \ldots, x_n < z$. For this, we will need $R(z)$ and $S(z)$ such that

$$2 \leq S(z)$$

and

$$|Q(w, x_1, \ldots, x_n)| < 2^{S(z)} \leq 2^{R(z)}$$

hold for any $z \in \omega$ and any $w, x_1, \ldots, x_n \leq z$. To this end, we can choose

$$S(Z) = 2(\Sigma(c_{i_0 \ldots i_n})^2)(Z + 1)^d, \tag{3}$$

where

$$Q(Z, X_1, \ldots, X_n) = \Sigma c_{i_0 \ldots i_n} Z^{i_0} X_1^{i_1} \ldots X_n^{i_{n-1}}$$

and d is the degree of Q. [*N.B.* i. The reader who wishes to keep track of the parameters in Q will first have to replace each *polynomial* $c_{i_0 \ldots i_n}$ in the parametric variables by the polynomials obtained by replacing each coëfficient of $c_{i_0 \ldots i_n}$ by its absolute value; ii. the introduction of the variable w is forced on us by the desire for a singlefold representation.]

$S(Z)$ is a polynomial in Z with no negative coëfficients. In particular, it is a dyadic exponential polynomial.

The definition, if not the choice, of $R(Z)$ is fairly simple:

$$R(Z) = (4Z^{n+1} + 2)S(Z) \tag{4}$$

where $n + 1$ is the number of variables of Q. $R(Z)$ is a polynomial in Z with only non-negative coëfficients, and clearly satisfies the desired inequality,

$$R(z) \geq S(z), \text{ for all } z \in \omega.$$

The necessity for making R so much larger than S as is done in (4) will emerge in the proof of Lemma 10.10, below.

Finally, define

$$T(Z) = R(Z) - S(Z) + 2. \tag{5}$$

T is also a polynomial in Z and, using (4), T can be rewritten in a form with only non-negative coëfficients:

$$T(Z) = (4Z^{n+1} + 1)S(Z) + 2.$$

Using Q, R, S, T, Lemma 10.6 yields the following simple lemma.

10.7. Lemma. *Let $z, w, x_1, \ldots, x_n \in \omega$, with $z \geq w, x_1, \ldots, x_n$. Let $*$ be $+$ or $-$, and let $\sigma = \sigma(F^*(R(z), Q(w, \boldsymbol{x})))$. Then:*

$$\sigma \in \begin{cases} [T(z) * (R(z) - S(z)), T(z) + 2S(z) * (R(z) - S(z))], & \text{if } Q(w, \boldsymbol{x}) = 0 \\ [T(z), T(z) + 2S(z)], & \text{otherwise.} \end{cases}$$

The proof is simply a matter of writing $r = R(z)$, $s = S(z)$, $r - s + 2 = T(z)$ and comparing the given mess with that in Lemma 10.6.

The next lemma has some content.

10.8. Definition-Lemma. *For* $*$ *either* $+$ *or* $-$, *define* $B^*(z,w)$ *to be the function*

$$\sum_{x_1=0}^{z-1} \cdots \sum_{x_n=0}^{z-1} F^*(R(z), Q(w,\boldsymbol{x}))\cdot 2^{(2R(z)+3)(x_1+(z+1)x_2+\ldots+(z+1)^{n+1}x_n)}.$$

Then:

$$\sigma(B^*(z,w)) = \sum_{x_1=0}^{z-1} \cdots \sum_{x_n=0}^{z-1} \sigma(F^*(R(z), Q(w,\boldsymbol{x}))).$$

The proof of Lemma 10.8 amounts to noticing that, for $z > 0$, the numbers $x_1 + (z+1)x_2 + \ldots + (z+1)^{n-1}x_n$ for $x_1, \ldots, x_n < z$ are distinct numbers written in base $z+1$. (Base z would also yield this, but Lemma 10.12, below, requires $z+1$.) Thus, B^* assumes the form

$$\sum_{i=0}^{z^n-1} t_i 2^{(R(z)+3)a_i}, \quad 0 < a_0 < a_1 < \ldots < a_{z^n-1}$$

and Lemma 10.4.ii will apply once we've seen

$$0 < t_i < 2^{2R(z)+3}.$$

But the t_i's are of the form $F^*(R(z), q)$ with $|q| < 2^{S(z)} \leq 2^{R(z)}$, and

$$\begin{aligned} F^*(R(z),q) &\leq F^-(R(z),q) \\ &\leq (2^{R(z)} + q)2^{R(z)+1} + 2^{R(z)} - q \\ &< 2^{R(z)+1}2^{R(z)+1} + 2^{R(z)+1} \\ &< 2^{R(z)+1}(2^{R(z)+1} + 1) \\ &< 2^{R(z)+1}2^{R(z)+2} = 2^{2R(z)+3}. \end{aligned}$$

I leave the details of the proof of Lemma 10.8 as an exercise to the reader (Exercise 2, below). [For the non-singlefold representation, one can drop B^-.]

Combining Lemmas 10.7 and 10.8, we get the following:

10.9. Lemma. *Let* $*$ *be* $+$ *or* $-$. *For* $w \leq z \in \omega$, *let* $\sigma = \sigma(B^*(z,w))$. *We have*

$$\sigma \in \begin{cases} [z^nT(z)*(R(z)-S(z)), z^n(T(z)+2S(z))*(R(z)-S(z))], & \exists \boldsymbol{x}\,[Q(w,\boldsymbol{x}) = 0] \\ [z^nT(z), z^n(T(z)+2S(z))], & \textit{otherwise.} \end{cases}$$

Proof. By Lemma 10.8,

$$\sigma(B^*(z,w)) = \Sigma \ldots \Sigma\, \sigma(F^*(R(z), Q(w,\boldsymbol{x}))),$$

where there are z^n summands. If there is no zero corresponding to w, then each summand lies, by Lemma 10.7, between

$$T(z) \text{ and } T(z) + 2S(z).$$

In this case, the sum lies between z^n copies of $T(z)$ and z^n copies of $T(z) + 2S(z)$, i.e.,

$$z^n T(z) \leq \sigma(B^*(z,w)) \leq z^n(T(z) + 2S(z)).$$

If $Q(w, \boldsymbol{x}) = 0$ for some $x_1, ..., x_n$, then for that unique list $x_1, ..., x_n < w \leq z$, $\sigma(F^*(R(z), Q(w,\boldsymbol{x})))$ lies between

$$T(z) * (R(z) - S(z)) \text{ and } T(z) + 2S(z) * (R(z) - S(z)).$$

(Uniqueness follows from the fact 10.1.ii that w determines $x_1, ..., x_n$.) All other summands lie between

$$T(z) \text{ and } T(z) + 2S(z).$$

Thus, there are $z^n - 1$ lower bounds of $T(z)$ and one of $T(z) * (R(z) - S(z))$, yielding

$$\begin{aligned} \sigma(B^*(z,w)) &\geq (z^n - 1)T(z) + T(z) * (R(z) - S(z)) \\ &\geq z^n T(z) * (R(z) - S(z)). \end{aligned}$$

Similarly,

$$\sigma(B^*(z,w)) \leq z^n(T(z) + 2S(z)) * (R(z) - S(z)). \qquad \text{QED}$$

We are now ready to define $A(z)$:

$$A(z) = B^+(z,z) + \sum_{w=0}^{z-1} B^-(z,w) \cdot 2^{(2R(z)+3)(w+1)}. \tag{6}$$

[If a singlefold representation is not required, one can take $B^+(z,z)$ for $A(z)$.]

The idea behind the definition is simple. We want $\sigma(A(z))$ to be large only when z codes the least zero of Q. The summands $B^-(z,w)$ reduce σ whenever $w < z$ codes a zero of Q and $B^+(z,z)$ increases σ whenever z codes a zero of Q. Two things remain to be seen. The first is that $\sigma(A(z))$ is relatively large only when z codes the least zero of Q, and the second is that A is a dyadic exponential polynomial. The first of these is a mere computation and will be handled in the next lemma. The second is not clear— note the variable upper bounds in the sums of B^+, B^-, and A— and, in fact, is only almost true: $A(z)$ is actually of the form,

$$\frac{E_1(z) - E_2(z)}{E_3(z) - E_4(z)},$$

for iterated exponential polynomials $E_1, ..., E_4$. This is the content of Lemma 10.12, below. Lemmas 10.10 and 10.12 complete the first, more difficult half of the proof of the three-variable result.

To prove that, for appropriate values of z, $A(z)$ satisfies inequality (1),

$$\sigma(A(z)) \geq M(z),$$

we first have to define M:

$$M(Z) = (Z^n + Z^{n+1})T(Z) + R(Z) - S(Z). \tag{7}$$

Note that $M(Z)$ is a polynomial in Z and it can be rewritten so that all coëfficients are non-negative. [In the non-singlefold case, take $M(Z) = Z^nT(Z) + R(Z) - S(Z)$.]

10.10. Lemma. *For any* $z \in \omega$ *we have*

$$\sigma(A(z)) \geq M(z)$$

iff Z *is the smallest number coding a zero* $z, x_1, ..., x_n$ *of* Q.

Proof. Note first that $A(0)$ is an empty sum, whence $A(0) = 0$ and $\sigma(A(0)) = 0$. Moreover, $M(0) = R(0) - S(0) = S(0) > 0$, whence $\sigma(A(0)) < M(0)$. But $Q(0,x_1,\ldots,x_n) \neq 0$ by Lemma 10.1.iii.

Now for the real proof! For $z > 0$, there are three cases— z is the least number coding a zero of Q, there are no zeroes of Q coded by any $w \leq z$, and there is a number $w < z$ coding a zero of Q. We consider these cases in this order.

Case 1. If z is the least code of a zero of Q, then Lemma 10.9 yields

$$\sigma(B^+(z,z)) \geq z^nT(z) + R(z) - S(z)$$

$$\sigma(B^-(z,w)) \geq z^nT(z), \text{ for } w < z,$$

whence (assuming the analogue for $A(z)$ to Lemma 10.8 holds— cf. Exercise 2, below),

$$\begin{aligned}\sigma(A(z)) &= \sigma(B^+(z,z)) + \sum_{w=0}^{z-1}\sigma(B^-(z,w)) \\ &\geq z^nT(z) + R(z) - S(z) + \sum_{w=0}^{z-1} z^nT(z) \\ &\geq z^nT(z) + R(z) - S(z) + z \cdot z^nT(z) \\ &\geq (z^n + z^{n+1})T(z) + R(z) - S(z) = M(z), \text{ by (7).}\end{aligned}$$

Case 2. There are no zeroes of Q for $w = 0, ..., z$. Then 10.9 yields

$$\begin{aligned}\sigma(A(z)) &= \sigma(B^+(z,z)) + \sum_{w=0}^{z-1}\sigma(B^-(z,w)) \\ &\leq z^n(T(z) + 2S(z)) + \sum_{w=0}^{z-1} z^n(T(z) + 2S(z)) \\ &\leq z^n(T(z) + 2S(z)) + z \cdot z^n(T(z) + 2S(z)) \\ &\leq (z^n + z^{n+1})T(z) + 2(z^n + z^{n+1})S(z).\end{aligned}$$

It thus suffices to show that

$$2(z^n + z^{n+1})S(z) < R(z) - S(z).$$

But,

$$\begin{aligned}R(z) - S(z) &= (4z^{n+1} + 2)S(z) - S(z) = 4z^{n+1}S(z) + S(z) \\ &> 4z^{n+1}S(z) > 2(z^n + z^{n+1})S(z).\end{aligned}$$

Case 3. For some $w < z$, and some $x_1, ..., x_n < w$, $Q(w, \mathbf{x}) = 0$. The largest possible value of $\sigma(A(z))$ would occur if z codes a zero of Q and there is only one $w < z$ which also codes a zero of Q. Thus, consider

$$\begin{aligned}\sigma(A(z)) &= \sigma(B^+(z,z)) + \sum_{w=0}^{z-1} \sigma(B^-(z,w)) \\ &\le z^n(T(z) + 2S(z)) + R(z) - S(z) + \sum_{w=0}^{z-1} z^n(T(z) + 2S(z)) \;-\; (R(z) - S(z)) \\ &\le (z^n + z^{n+1})(T(z) + 2S(z)) \\ &\le (z^n + z^{n+1})T(z) \;+\; 2(z^n + z^{n+1})S(z)),\end{aligned}$$

which, as we saw in Case 2, is $< M(z)$. QED

We are now ready to prove that $A(z)$ is "nearly" an iterated dyadic exponential polynomial. It is, in fact, nearly a dyadic exponential polynomial and we can get a more refined statement of what sort of function A is if we liberalise Definitions 2.7 and 2.8 of "exponential polynomial" and "dyadic exponential polynomial".

10.11. Definitions. The class of *exponential polynomials* is inductively generated as follows:

i. variables and non-negative integers are exponential polynomials
ii. if P, Q are exponential polynomials, then so are $P + Q$ and $P \cdot Q$
iii. if n is a non-negative integer, X is a variable, and P is a polynomial with non-negative coëfficients, then n^P and X^P are exponential polynomials.

The class of *dyadic exponential polynomials* is inductively generated as above, but with the third clause replaced by,

iii'. if P is a polynomial with non-negative coëfficients, then 2^P is a dyadic exponential polynomial.

The non-dyadic part of this definition could be made yet more liberal by allowing P^Q to be an exponential polynomial when P, Q are polynomials with non-negative coëfficients. What we intend here is that exponentiation not be iterated.

The final crucial lemma needed to reach our first major goal is the following.

10.12. Lemma. *There are dyadic exponential polynomials E_1, E_2, E_3, E_4 such that*

$$A(z) = \frac{E_1(z) - E_2(z)}{E_3(z) - E_4(z)}.$$

Moreover, E_1, E_2, E_3, E_4 can be chosen so that

$$E_1(z) \ge E_2(z) \;\; and \;\; E_3(z) > E_4(z)$$

for all $z \in \omega$.

Proof sketch. Recall (6),

$$A(z) = B^+(z,z) + \sum_{w=0}^{z-1} B^-(z,w)\cdot 2^{2(R(z)+3)(w+1)}.$$

Let us look at the second sum:

$$\sum_{w=0}^{z-1}\sum_{x_1=0}^{z-1}\cdots\sum_{x_n=0}^{z-1} F^-(R(z),Q(w,\boldsymbol{x}))\cdot 2^{sum}, \tag{8}$$

where

$$sum = (2R(z)+3)((w+1)+\sum_{i=1}^{n}(x_i(z+1)^{i-1})).$$

The exponent *sum* can be compactly written as $x_0Q_0(z) + \ldots + x_nQ_n(z)$ for polynomials $Q_i(z)$ and $x_0 = w + 1$. $F^-(R(z), Q(w,\boldsymbol{x}))$ can be written as

$$(2^{2R(z)+1} + 2^{R(z)}) + (2^{R(z)+1} - 1)Q(w,\boldsymbol{x}).$$

Distributing this out of (8) yields the sums

$$(2^{2R(z)+1} + 2^{R(z)})\Sigma\Sigma\ldots\Sigma 2^{x_0Q_0(z)+\ldots+x_nQ_n(z)} \tag{9}$$

and

$$(2^{R(z)+1} - 1)\Sigma\Sigma\ldots\Sigma Q(w,\boldsymbol{x})2^{x_0Q_0(z)+\ldots+x_nQ_n(z)}. \tag{10}$$

Q itself can be written as a sum, say,

$$\sum_{i_0+\ldots+i_n\le d} c_{i_0\ldots i_n} w^{i_0}x_1^{i_1}\ldots x_n^{i_n}.$$

We can distribute this last out of (10) to get

$$\sum_{i_0+\ldots+i_n\le d} (2^{R(z)+1} - 1)c_{i_0\ldots i_n}\Sigma\Sigma\ldots\Sigma w^{i_0}x_1^{i_1}\ldots x_n^{i_n}2^{x_0Q_0(z)+\ldots+x_nQ_n(z)}. \tag{11}$$

Thus, looking at (9) and (11), we see that the second sum (8) occurring in $A(z)$ can be written as a linear combination of sums

$$\sum_{w=0}^{z-1}\sum_{x_1=0}^{z-1}\cdots\sum_{x_n=0}^{z-1} w^{i_0}x_1^{i_1}\ldots x_n^{i_n}2^{x_0Q_0(z)+\ldots+x_nQ_n(z)}, \tag{12}$$

where the coëfficients of the linear combinations are differences of dyadic exponential polynomials in z.

The first sum in $A(z)$, namely $B^+(z,z)$, can also be written in this form, but without the w and the outer sum of (12). Thus, $A(z)$ is a linear combination of sums of the form (12) and similar sums without the variable w, where the coëfficients of the combination are differences of dyadic exponential polynomials in z.

We now set about simplifying (12). The innermost sum can be rewritten,

$$w^{i_0}x_1^{i_1}\ldots x_{n-1}^{i_{n-1}}2^{x_0Q_0(z)+\ldots+x_{n-1}Q_{n-1}(z)}\sum_{x_n=0}^{z-1} x_n^{i_n}(2^{Q_n(z)})^{x_n}. \tag{13}$$

We can simplify this last if we can simplify sums of the form,

$$\sum_{x=0}^{z-1} x^i p^x, \tag{14}$$

where $p \neq 1$. [For $p = 2^{Q_n(z)}$, we have $p \neq 1$ since $q_n(z) \neq 0$ for $z \in \omega$.] It is at this point that the word "sketch" in the description of this proof as a "proof sketch" becomes operative. I simply assert by fiat that (14) has the form,

$$\frac{G_i(z,p)p^z + H_i(z,p)}{(p-1)^{i+1}}, \tag{15}$$

where G_i and H_i are polynomials with integral coëfficients. (Cf. Exercise 4, below, if the fiat is unacceptable.)

One can now replace (14) (for $x = x_n$) by (15) in (13) and reduce (12) to something like

$$\sum_{w=0}^{z-1} \sum_{x_1=0}^{z-1} \cdots \sum_{x_{n-1}=0}^{z-1} w^{i_0} x_1^{i_1} \ldots x_{n-1}^{i_{n-1}} 2^{x_0 Q_0(z) + \ldots + x_{n-1} Q_{n-1}(z)} F_n(z),$$

the innermost sum of which can be rewritten as

$$w^{i_0} x_1^{i_1} \ldots x_{n-2}^{i_{n-2}} 2^{x_0 Q_0(z) + \ldots + x_{n-2} Q_{n-2}(z)} F_n(z) \sum_{x_{n-1}=0}^{z-1} x_{n-1}^{i_{n-1}} (2^{Q_{n-1}(z)})^{x_{n-1}}.$$

The inner sum of this can now be simplified and the entire process iterated until (12) looks like

$$\left(\sum_{w=0}^{z-1} w^{i_0} 2^{Q_0(z) w}\right) 2^{Q_0(z)} F_1(z) \ldots F_n(z) = 2^{Q_0(z)} F_0(z) \ldots F_n(z).$$

It follows that $A(z)$ is a linear combination, with coëfficients being differences of dyadic exponential polynomials, of functions of the form

$$2^{Q_0(z)} F_0(z) \ldots F_n(z)$$

and (for B^+)

$$F_1(z) \ldots F_n(z),$$

where each $F_i(z)$ has a form (15) with G_i, H_i polynomials. If one puts everything over a common denominator, multiplies everything out, and separates terms with positive and negative coëfficients, one readily sees that $A(z)$ can be written in the form,

$$A(z) = \frac{E_1(z) - E_2(z)}{E_3(z) - E_4(z)},$$

where the E_i's are dyadic exponential polynomials. Moreover, $E_3(z) - E_4(z)$ is the product of things of the form

$$(2^{Q_j(z)} - 1)^{i+1},$$

where $Q_j(z) > 0$ for all $z \in \omega$. Thus,

$$E_3(z) - E_4(z) > 0 \text{ for all } z \in \omega.$$

Since $A(z) \geq 0$ for all $z \in \omega$, we also have

$$E_1(z) - E_2(z) = A(z)(E_3(z) - E_4(z)) \geq 0$$

for all such z. QED

With Lemma 10.12 we have proven the following.

10.13. Theorem. *Every r.e. relation can be represented in the form,*

$$R(\boldsymbol{a}):\ \exists z\,[\sigma(A(\boldsymbol{a}, z)) \geq M(\boldsymbol{a}, z)],$$

where

i. $M(A_0, ..., A_{m-1}, Z)$ *is a polynomial with non-negative coëfficients; and*

ii. $A(A_0, ..., A_{m-1}, Z)$ *is of the form*

$$\frac{E_1(A_0,\ldots,A_{m-1},Z) - E_2(A_0,\ldots,A_{m-1},Z)}{E_3(A_0,\ldots,A_{m-1},Z) - E_4(A_0,\ldots,A_{m-1},Z)},$$

with E_1, E_2, E_3, E_4 *dyadic exponential polynomials and, for all* $\boldsymbol{a}, z \in \omega$,

$$E_3(\boldsymbol{a}, z) > E_4(\boldsymbol{a}, z) \ \ and \ \ E_1(\boldsymbol{a}, z) \geq E_2(\boldsymbol{a}, z).$$

Moreover, z, if it exists, is unique.

We have attained our first goal and are half-way to proving Theorem 10.3 on the 3-variable iterated exponential Diophantine representability of r.e. relations. The second half is, of course, the transformation of the expression $\exists z\,[\sigma(A(\boldsymbol{a}, z)) \geq M(\boldsymbol{a}, z)]$, i.e.,

$$\exists z\left[2^{M(z)} \mid \binom{2A(z)}{A(z)}\right],$$

into a dyadic iterated exponential Diophantine relation. This transformation is not nearly as much work as reaching the point we've just attained, i.e., the two halves of the proof are not equal halves. Nevertheless, at such a natural resting place as this, the reader might be too tired to go on. For the truly lazy rebel, I offer the following easier, but still respectable, goal.

10.14. Theorem. (7-Variable Result). *Every r.e. relation R has a singlefold iterated exponential Diophantine representation of the form,*

$$R(\boldsymbol{a}):\ \exists x_0 \ldots x_6[E(\boldsymbol{a}, \boldsymbol{x}) = F(\boldsymbol{a}, \boldsymbol{x})].$$

Proof. Writing $A(z) = (E_1 - E_2)/(E_3 - E_4)$ and suppressing a few parameters, observe

$$\begin{aligned}
R(\boldsymbol{a}) \;&\Leftrightarrow\; \exists z\left[2^{M(z)} \mid \binom{2A(z)}{A(z)}\right], \text{ by 10.12}\\
&\Leftrightarrow\; \exists za\left[2^{M(z)} \mid \binom{2A(z)}{A(z)} \ \wedge\ a = A(z)\right]\\
&\Leftrightarrow\; \exists za\left[2^{M(z)} \mid \binom{2A(z)}{A(z)} \ \wedge\ a(E_3 - E_4) = E_1 - E_2\right]\\
&\Leftrightarrow\; \exists zabvw\,\big[2^{M(z)} \mid b \ \wedge\ a(E_3 - E_4) = E_1 - E_2 \ \wedge\\
&\qquad \wedge\ (2^{2a} + 2)^{2a} = v(2^{2a} + 1)^{a+1} + b(2^{2a} + 1)^{a} + w \ \wedge\\
&\qquad \wedge\ w < (2^{2a} + 1)^{a} \ \wedge\ b < 2^{2a} + 1\big], \text{ by 2.9.i}
\end{aligned}$$

$$\Leftrightarrow \ \exists zadvwrs \left[(2^{2a}+2)^{2a} = v(2^{2a}+1)^{a+1} + d2^{M(z)}(2^{2a}+1)^{a} + w \ \wedge \right.$$
$$\wedge \ w+r+1 = (2^{2a}+1)^{a} \ \wedge \ d2^{M(z)} + s = 2^{2a} \ \wedge$$
$$\left. \wedge \ a(E_3 - E_4) = E_1 - E_2 \right]. \qquad \text{QED}$$

10.15. Corollary. (9-Variable Result). *Every r.e. relation has a singlefold exponential Diophantine representation of the form,*

$$R(\boldsymbol{a}):\ \exists x_0 ... x_8 [E(\boldsymbol{a}, \boldsymbol{x}) = F(\boldsymbol{a}, \boldsymbol{x})].$$

Proof. Adding two new variables,

$$x = 2^{2a} + 2 \ \text{ and } \ y = 2^{2a} + 1,$$

results in the system of non-iterated exponential equations,

$$y = 2^{2a}+1 \ \wedge x = y+1 \ \wedge \ x^{2a} = vy^{a+1} + d2^{M(z)}y^{a} + w \ \wedge$$
$$\wedge \ w+r+1 = y^{a} \ \wedge \ d2^{M(z)} + s + 1 = y \ \wedge \ a(E_3 - E_4) = E_1 - E_2. \qquad \text{QED}$$

Theorem 10.14 and Corollary 10.15 do not concern dyadic equations, but they are respectable results and the reader may settle for them.

Corollary 10.15 tells us nothing about the number of variables needed for exponential Diophantine equations when the more restricted Definition 2.7 of "exponential polynomial" is used. Polynomial exponents, like $M(z)$ in $2^{M(z)}$, can be eliminated by the addition of new variables, but the dyadic exponentials E_1, E_2, E_3, E_4 have lots of such polynomial exponents.

The reader who wants the 3-variable dyadic iterated result will have to work a little harder. Matijasevič's original argument to obtain a (non-dyadic) 3-variable iterated exponential Diophantine problem out of

$$\exists z \left[2^{M(z)} \mid \binom{2A(z)}{A(z)}\right] \tag{2}$$

was remarkably simple: Let

$$u > 2^{2A(z)}, \quad 2^{M(z)} \mid u, \quad \text{and} \quad (u+1)^{2A(z)} = tu^{A(z)} + x, \tag{*}$$

where $x < u^{A(z)}$. Then:

$$2^{M(z)} \mid \binom{2A(z)}{A(z)} \ \text{ iff } \ 2^{M(z)} \mid t. \tag{**}$$

Writing $A(z) = (E_1 - E_2)/(E_3 - E_4)$, and choosing

$$U = 2^{(2E_1 + M(z))} \ \text{ and } \ T = (x+y)2^{M(z)},$$

the first two equations of (*) become true and, via (**), (2) is equivalent to

$$\exists zxy \left[(U+1)^{2A(z)} = TU^{A(z)} + x\right]. \tag{16}$$

(Cf. Exercise 5, below.) However, (16) is not dyadic. James Jones joined Matijasevič and, together, they obtained a 3-variable dyadic iterated exponential expression for (2). Their starting point was the following result from the calculus.

10.16. Lemma. *For real numbers x, if* $|x| < 1/4$,

$$\frac{1}{\sqrt{1-4x}} = \sum_{n=0}^{\infty}\binom{2n}{n}x^n. \tag{17}$$

Since this is simple calculus, I leave the proof as an exercise (Exercise 6, below). Replacing x by $1/u$ in (17) and multiplying by u^n, we get (for $|u| > 4$)

$$\frac{u^n}{\sqrt{1-4x}} = u^n + 2u^{n-1} + 6u^{n-2} + \ldots + \binom{2n-2}{n-1}u + \binom{2n}{n} + \\ + \sum_{k=1}^{\infty}\binom{2n+2k}{n+k}u^{-k}. \tag{18}$$

For $u > 0$, an estimate of the fractional part is given by

$$\sum_{k=1}^{\infty}\binom{2n+2k}{n+k}u^{-k} < \sum_{k=1}^{\infty}2^{2n+2k}u^{-k} = 4^n\sum_{k=1}^{\infty}(4/u)^k$$
$$< 4^n(-1/(4/u-1))$$
$$< \frac{4^n \cdot 4}{u-4}, \tag{19}$$

which is < 1 if $u > 4^{n+1} + 4$. Thus, we have the following:

10.17. Lemma. *Let* $n \in \omega$ *be fixed. If* $u > 4^{n+1} + 4$, *then*

$$\left[\frac{u^n}{\sqrt{1-4x}}\right] = u^n + 2u^{n-1} + 6u^{n-2} + \ldots + \binom{2n-2}{n-1}u + \binom{2n}{n}. \tag{20}$$

The proof is simple. The part of (18) missing from (20) is less than 1.

Now, consider what happens if we replace n by $A(z)$ and assume $2^{M(z)} \mid u > 4^{A(z)+1} + 4$. By (20), the condition

$$2^{M(z)} \mid \binom{2A(z)}{A(z)}$$

is equivalent to

$$2^{M(z)} \mid \left[\frac{u^{A(z)}}{\sqrt{1-4x}}\right], \tag{21}$$

which in turn is equivalent to the existence of a (unique) $y \in \omega$ such that

$$2^{M(z)}y = \left[\frac{u^{A(z)}}{\sqrt{1-4x}}\right].$$

For this unique value of y we have

$$\left|\frac{u^{A(z)}}{\sqrt{1-4x}} - 2^{M(z)}y\right| < 1.$$

Looking at (19), we see that we can make this difference less than 1/2 if $u > 2 \cdot 4^{A(z)+1} + 4$.

It might not be a bad idea to summarise what we've got so far:

10.18. Reduction. *Let z be given and let u be any natural number satisfying*

i. $u > 2\cdot 4^{A(z)+1} + 4$

ii. $2^{M(z)} \mid u$.

Then: $2^{M(z)} \mid \binom{2A(z)}{A(z)}$ *iff there is a (unique) y such that*

$$\left| \frac{u^{A(z)}}{\sqrt{1-4x}} \quad 2^{M(z)}y \right| \leq 1/2. \tag{22}$$

We can pretty much coast the rest of the way through the proof. Write (22) as

$$|\alpha - \beta| \leq 1/2$$

and observe the successive equivalences,

$$\begin{aligned} |\alpha - \beta| \leq 1/2 &\Leftrightarrow (\alpha - \beta)^2 \leq 1/4 \\ &\Leftrightarrow \alpha^2 - 2\alpha\beta + \beta^2 \leq 1/4 \\ &\Leftrightarrow \alpha^2 + \beta^2 - 1/4 \leq 2\alpha\beta && (*) \\ &\Leftrightarrow (\alpha^2 + \beta^2 - 1/4)^2 \leq 4\alpha^2\beta^2. && (**) \end{aligned}$$

The point to (*) is that α contains a square root which we want to isolate, square, and thus eliminate. The equivalence of (*) and (**) follows since both sides of the inequality in (*) are positive. Simplifying (**) yields

$$\begin{aligned} |\alpha - \beta| \leq 1/2 &\Leftrightarrow \alpha^4 + 2\alpha^2\beta^2 + \beta^4 - \frac{\alpha^2 + \beta^2}{2} + \frac{1}{16} \leq 4\alpha^2\beta^2 \\ &\Leftrightarrow \alpha^4 + \beta^4 + \frac{1}{16} \leq 2\alpha^2\beta^2 + \frac{\alpha^2 + \beta^2}{2}. && (23) \end{aligned}$$

Suppressing the z, we know

$$\alpha = \frac{u^A}{\sqrt{1 - 4/u}},$$

whence

$$\alpha^2 = \frac{u^{2A}}{1 - 4/u} = \frac{u^{2A+1}}{u - 4}, \quad \alpha^4 = \frac{u^{4A+2}}{(u-4)^2}.$$

Moreover, $\beta = 2^M y$. Substituting these into (23) yields

$$\frac{u^{4A+2}}{(u-4)^2} + 2^{4M}y^4 + \frac{1}{16} \leq 2\cdot \frac{u^{2A+1}}{u-4}\cdot 2^{2M}y^2 + \frac{1}{2}\left(\frac{u^{2A+1}}{u-4} + 2^{2M}y^2\right).$$

Clearing the denominators yields

$$\begin{aligned} &16u^{4A+2} + 16\cdot 2^{4M}y^4(u-4)^2 + (u-4)^2 \leq \\ &\leq 32u^{2A+1}2^{2M}y^2(u-4) + 8u^{2A+1}(u-4) + 8\cdot 2^{2M}y^2(u-4)^2, \end{aligned} \tag{24}$$

as an equivalent to (22), i.e., $|\alpha - \beta| \leq 1/2$.

It remains to determine what value of u to plug into (24), and to simplify the result. Preparatory to doing this, note that, by Lemma 10.12, A is a ratio of differences of dyadic exponential polynomials. Hence, so is $2A + 1$ and we can write

$$2A + 1 = \frac{E_1 - E_2}{E_3 - E_4}, \tag{25}$$

where $E_1 > E_2$ and $E_3 > E_4$ for any $z \in \omega$ (since $2A + 1 \geq 1 > 0$).

For u, we shall choose a large power of 2, say 2^e. By 10.18.i, we want

$$u = 2^e > 2 \cdot 4^{A+1} + 4 = 2^{2A+3} + 4,$$

which is satisfied if we choose

$$u > 2^{2A+4},$$

i.e.,

$$e > 2A + 4 = (2A + 1) + 3.$$

This is satisfied if

$$e > E_1 + 3. \tag{26_1}$$

By 10.18.ii, we want $2^M \mid u$, i.e.,

$$e \geq M. \tag{26_2}$$

Finally, the expression

$$u^{2A+1} = 2^{e(E_1-E_2)/(E_3-E_4)},$$

which occurs in (24), is most simplified if

$$E_3 - E_4 \mid e. \tag{26_3}$$

Combining (26_1) - (26_3), we choose

$$\begin{aligned} u &= 2^{(E_1+4+M)(E_3-E_4)} \\ &= 2^{E_5-E_6}, \end{aligned} \tag{27}$$

say, where E_5, E_6 are dyadic exponential polynomials and $E_5 > E_6$ for all $z \in \omega$. Observe that

$$\begin{aligned} u^{2A+1} &= 2^{(E_1+4+M)(E_1-E_2)} \\ &= 2^{E_7-E_8}, \end{aligned} \tag{28}$$

for some dyadic exponential polynomials E_7 and E_8 for which $E_7 > E_8$ for all $z \in \omega$.

Plugging the values of (27) and (28) of u and u^{2A+1}, respectively, into (24) yields

$$\begin{aligned} &16 \cdot 2^{2E_7-2E_8} + 16 \cdot 2^{4M}y^4(2^{E_5-E_6} - 4)^2 + (2^{E_5-E_6} - 4)^2 \leq \\ &\leq 32 \cdot 2^{E_7-E_8} \cdot 2^{2M}y^2(2^{E_5-E_6} - 4) + 8 \cdot 2^{E_7-E_8}(2^{E_5-E_6} - 4) + 8 \cdot 2^{2M}y^2(2^{E_5-E_6} - 4)^2. \end{aligned} \tag{29}$$

Multiplying both sides of (29) by $2^{2E_6+2E_8}$ will eliminate all subtractions from the exponents. The rest of the subtractions are readily eliminated to yield the following.

10.19. Theorem. *Every r.e. relation can be represented in the form,*

$$R(\boldsymbol{a}): \exists zy\, [E_0(\boldsymbol{a}, z, y) \leq F_0(\boldsymbol{a}, z, y)],$$

where E_0, F_0 are dyadic iterated exponential polynomials. Moreover, z and y, if they exist, are unique.

The proof follows from Reduction 10.18 and the equivalence of (22) with (29). Replacing $E_0 \leq F_0$ in 10.19 by

$$E_0(\boldsymbol{a}, z, y) + x = F_0(\boldsymbol{a}, z, y)$$

yields (finally) Theorem 10.3. Note that neither X nor Y occurs in exponents in the final dyadic iterated exponential polynomials.

Finally, notice that adding two new variables,

$$b = E_7 - E_8, \; c = E_5 - E_6, \tag{30}$$

transforms (29) into

$$16{\cdot}2^{2b} + 16{\cdot}2^{4M}y^4(2^c - 4)^2 + (2^c - 4)^2 \leq$$

$$\leq \; 32{\cdot}2^b{\cdot}2^{2M}y^2(2^c - 4) + 8{\cdot}2^b(2^c - 4) + 8{\cdot}2^{2M}y^2(2^c - 4)^2. \tag{31}$$

Using the system (30) - (31) yields the following.

10.20. Corollary. (5-Variable Result). *Every r.e. relation R has a singlefold dyadic exponential Diophantine representation of the form,*

$$R(\boldsymbol{a}): \; \exists x_0 \ldots x_4 \, [E(\boldsymbol{a}, \boldsymbol{x}) = F(\boldsymbol{a}, \boldsymbol{x})].$$

(Once again, we are using Definition 10.11 rather than 2.8 to define the notion "dyadic exponential".)

Thus, we have the algorithmic unsolvability of the dyadic iterated exponential Diophantine problem in 3 variables and that of the dyadic exponential Diophantine problem in 5 variables. As already remarked several times, the ordinary Diophantine problem is known to be unsolvable in 9 variables. This last result can be found in the paper of Jones cited in the references.

How good is the 3-variable result? That is, what are the complementary positive results? For two variables, nothing is known. For one variable, a great deal is known, but, except in the dyadic case, is not conclusive. Godfrey Harald Hardy proved early in this century that the class of iterated exponential polynomials in one variable is linearly ordered by the eventual dominance relation,

$$E \prec F: \; \exists x \, \forall y > x \, [E(y) < F(y)].$$

In the early 1970s, Andrzej Ehrenfeucht showed that this ordering is, in fact, a well-ordering— every non-empty set of iterated exponential polynomials in one variable has a $\prec$-least element. Despite a great deal of work on the subject, and that fact that, by the linearity of $\prec$, two non-identical such iterated exponential polynomials can share a common value at a common argument only among a finite set of natural numbers, it is not yet known how to bound such a set effectively and it is not even known if the ordering relation is recursive. Among dyadic iterated exponential polynomials, however, the situation is much better understood. Hilbert Levitz has proven the decidability of the ordering in the dyadic case and, therewith, the solvability of the unary dyadic iterated exponential Diophantine problem, i.e., the single variable dyadic iterated exponential Diophantine problem.

Exercises.

1. i. Show: $\sigma(2^r + 2^s - 2) = s$, for $r \geq s \geq 1$

   ii. Show: $\sigma(2^r - 2^s) = r - s$, for $r \geq s \geq 1$.

   [Hint. i. consider separately the cases $s = 1, s > 1$.]

2. i. Prove Lemma 10.4.ii: $\sigma(\sum_{i=0}^{n} t_i 2^{s_i}) = \sum_{i=0}^{n} \sigma(t_i)$, provided that $t_i < 2^{s_{i+1}-s_i}$ for each $i < n$.

   ii. Prove Lemma 10.8:

$$\sigma(B^*(z, w)) = \sum_{x_1=0}^{z-1} \dots \sum_{x_n=0}^{z-1} \sigma(F^*(R(z), Q(w, x))).$$

   iii. Prove:

$$\sigma(A(z)) = \sigma(B^+(z, z)) + \sum_{w=0}^{z-1} \sigma(B^-(z, w)).$$

3. Show that, if one doesn't insist on getting a singlefold representation, one can take

$$A(z) = B^+(z,z)$$
$$M(Z) = Z^n T(Z) + R(Z) - S(Z),$$

   i.e., prove the analogue to Lemma 10.10: z codes a zero of Q iff $\sigma(A(z)) \geq M(z)$.

4. Let $p > 1$ and define, for $i \geq 0$ and $z \in \omega$,

$$S_i(z,p) = \sum_{x=0}^{z-1} x^i p^x.$$

   i. Show: $S_0(z,p) = \dfrac{p^z - 1}{p - 1}$.

   ii. Show: $S_{i+1}(z,p) = \dfrac{1}{p-1}\left(p^z z^{i+1} - p \sum_{k=0}^{i} \binom{i+1}{k} S_k(z,p)\right)$.

   iii. Conclude: $\sum_{x=0}^{z} x^i p^x = \dfrac{G_i(z,p)p^z - H_i(z,p)}{(p-1)^{i+1}}$, for some polynomials G_i, H_i with integral coëfficients.

   iv. Find $S_1(z,p)$.

   [Hint. ii. Write $S_{i+1}(z,p)$ as $\sum_{x=0}^{z-2} (x+1)^{i+1} p^x$. An alternate proof of part iii can be given by appeal to the summation-by-parts formula of Exercise 6 of Chapter I, section 1.]

5. Complete Matijasevič's original proof of Theorem 10.3, i.e., prove the equivalence of (2) with (16).

6. (Calculus). Let $F(x) = (1 - 4x)^{-1/2}$ for real x with $|x| < 1/4$.

   i. Show, by induction on n,

$$F^{(n)}(x) = \frac{(2n)!}{n!}(1 - 4x)^{-(2n+1)/2}.$$

   ii. Prove Lemma 10.16.

7. (Variable Savers). i. Find a polynomial $P(A, B, X)$ such that, if $a \geq 0$,

$$a > 0 \ \wedge \ b > 0 \ \Leftrightarrow \ \exists x\, [P(a, b, x) = 0].$$

ii. Let $b \neq 0$. Show:

$$b \mid c \ \wedge \ d > 0 \ \Leftrightarrow \ \exists x\, [b^2c^2(2d-1) = b^2x + c].$$

iii. Suppose $a_0, \ldots, a_{n-1}$ are pairwise relatively prime. Show:

$$\bigwedge_{i=0}^{n-1} a_i \mid b_i \quad \text{iff} \quad \prod_{i=0}^{n-1} a_i \mid \sum_{i=0}^{n-1} (\prod_{j \neq i} a_j) b_i.$$

iv. Show:

$$\bigwedge_{i=0}^{n-1} \exists x\, (a_i = x^2) \quad \text{iff} \quad \exists x\, [P(a_0, \ldots, a_{n-1}, x) = 0],$$

where $P(A_0, \ldots, A_{n-1}, X)$ is the product over all choices of signs of expressions

$$X \pm \sqrt{A_0} \pm \sqrt{A_1} \cdot Q \pm \ldots \pm \sqrt{A_{n-1}} \cdot Q^{n-1},$$

where $Q = 1 + \Sigma A_i^2$.

11. Reading List

A reference relevant to the entire chapter is:

Martin Davis, Juriĭ Matijasevič, and Julia Robinson, "Hilbert's tenth problem. Diophantine equations: positive aspects of a negative solution", in: Felix Browder, ed., *Mathematical Developments Arising from Hilbert Problems*, AMS, Providence, 1976.

§1.

1. I.G. Bashmakova, *Diophant und diophantische Gleichungen*, UTB-Birkhäuser, Basel, 1974.

2. Eric Temple Bell, *The Development of Mathematics*, McGraw-Hill, New York, 1940.

3. David Hilbert, "Mathematische Probleme", *Arch. f. math. u. Phys.* (3) 1 (1901), 44 - 63; reprinted in: *David Hilbert, Gesammelte Abhandlungen III*, Springer-Verlag, Berlin, 1935; English translation in: Felix Browder, ed., *Mathematical Developments Arising from Hilbert Problems*, AMS, Providence, 1976.

4. François Viète, *The Analytic Art*, Kent State U. Press, Kent (Ohio), 1983.

5. Kurt Vogel, "Diophantus of Alexandria", *Dictionary of Scientific Biography*, Scribners, New York, 1971.

6. André Weil, *Number Theory: An Approach through History; From Hammurapi to Legendre*, Birkhäuser, Boston, 1984.

§2.

1. Martin Davis, Hilary Putnam, and Julia Robinson, "The decision problem for exponential Diophantine equations", *Annals of Math.* 74 (1961), 425 - 436.

The reader might enjoy the following review of this paper.

2. Georg Kreisel, Review # A3061, *Math Reviews* 24 (1962); reprinted in: W.J. LeVeque, ed., *Reviews in Number Theory V*, AMS, Providence, 1974 (review U05-14).

§3.

A great deal of historical material on Pell's equation can be found in Weil's book cited for section 1, above. The next three items offer more on the Pigeon-Hole Principle and its applications. The first reference is fairly concise, but very nice, and gives the historical setting for the Pigeon-Hole Principle. The first chapter of the second reference offers a variety of generalisations of the Pigeon-Hole Principle.

1. Wolfgang M. Schmidt, "Approximation to algebraic numbers", *L'Enseignement mathém.* (2) 17 (1971), 187 - 253.

2. Kenneth B. Stolarsky, *Algebraic Numbers and Diophantine Approximation*, Marcel Dekker, New York, 1974.

3. Juriĭ Matijasevič, "On recursive unsolvability of Hilbert's tenth problem", in: L. Henkin, A. Joja, G. Moisil, and P. Suppes, eds., *Logic, Methodology and Philosophy of Science IV*, North-Holland, Amsterdam, 1973.

Two sources for the more constructive approach to finding solutions to Pell's equation are the following. Old's book is intended for bright high school students and is more elementary; that by Hurwitz and Kritikos also presents a full discussion of the more general theory of binary quadratic forms.

4. C.D. Olds, *Continued Fractions*, MAA, 1963.

5. Adolf Hurwitz and Nikolaos Kritikos, *Lectures on Number Theory*, Springer-Verlag, New York, 1986.

Euler's original proof of Lagrange's Theorem can be found in the next reference; an alternate elementary proof is given in Weil's book cited in the references to section 1.

6. David Eugene Smith, *A Source Book in Mathematics, I*, Dover, New York, 1959.

Finally, the reference on the secant function promised in the text is

7. V. Frederick Rickey and Philip M. Tuchinsky, "An application of geography to mathematics: history of the integral of the secant", *Math. Magazine* 53 (1980), 162 - 166.

§5.

1. Gregory V. Chudnovsky, "Some Diophantine problems", in: Gregory V. Chudnovsky, *Contributions to the Theory of Transcendental Numbers*, AMS, Providence, 1984.

2. Alfred Rupert Hall, *Philosophers at War*, Cambridge, 1980.

3. Raymond Wilder, *Mathematics as a Cultural System*, Pergamon Press, Oxford, 1981.

§6.

1. Dorothy Bollman and Miguel Laplaza, "Some decision problems for polynomial mappings", *Theoretical Comp. Sci.* 6 (1978), 317 - 325.

2. Jan Denef, "Hilbert's tenth problem for quadratic rings", *Proc. AMS* 48 (1975), 214 - 220.

3. Jan Denef and Leonard Lipschitz, "Diophantine sets over some rings of algebraic integers", *J. London Math. Soc.* (2) 18 (1978), 385 - 391.

4. Jan Denef, "Diophantine sets over algebraic number rings II", *Trans. AMS* 257 (1980), 227 - 236.

5. Daniel Richardson, "Some undecidable problems involving functions of a real variable", *J. Symbolic Logic* 33 (1968), 514 - 520.

6. Paul S. Wang, "The undecidability of the existence of zeros of real elementary functions", *J. Assoc. Comp. Mach.* 21 (1984), 586 - 589.

Richardson and Wang offer the most elementary unsolvable problems in calculus. The following two papers offer unsolvable problems from more advanced analysis.

7. Andrew Adler, "Some recursively unsolvable problems in analysis", *Proc. AMS* 22 (1969), 523 - 526.

8. Bruno Scarpellini, "Zwei unentscheidbare Probleme der Analysis", *Zeitschr. f. math. Logik* 9 (1963), 265 - 289.

A reference for Exercise 9 is the following.

9. J.P. Jones, D. Sato, H. Wada, D. Wiens, "Diophantine representation of the set of prime numbers", *Amer. Math. Monthly* 83 (1976), 449 - 464.

§7.

1. Andrew Adler, "A reduction of homogeneous Diophantine problems", *J. London Math. Soc.* (2) 3 (1971), 446 - 448.

2. D.J. Lewis, "Diophantine equations: *p*-adic methods", in: W.J. LeVeque, ed., *Studies in Number Theory*, Prentice-Hall, Englewood Cliffs (N.J.), 1969.

The reference to Siegel's work (for the advanced reader) is the following.

3. Carl Ludwig Siegel, "Zur Theorie der quadratischen Formen", *Göttinger Nachrichten* (1972), 21 - 46.

§8.

1. K. Chandrasekharan, *Introduction to Analytic Number Theory*, Springer-Verlag, New York, 1968.

2. Melvyn Hausner, "Applications of a simple counting technique", *Amer. Math. Monthly* 90 (1983), 127 - 129.

§9.

1. James P. Jones and Juriĭ Matijasevič, "Register machine proof of the theorem on exponential Diophantine representation of enumerable sets", *J. Symbolic Logic* 49 (1984), 818 - 829.

2. C. Smoryński, "A note on the number of zeros of polynomials and exponential polynomials", *J. Symbolic Logic* 42 (1977), 99 - 106.

§10.

The two basic references to this section are:

1. Juriĭ Matijasevič, "Algorithmic undecidability of exponential Diophantine equations in three unknowns", (Russian), in: *Studies in the Theory of Algorithms and Mathematical Logic*, (Russian), Nauka, Moscow, 1979; English translation: *Selecta Math. Sov.* 3 (1983/84), 223 - 232.

2. James P. Jones and Juriĭ Matijasevič, "A new representation for the symmetric binomial coefficient and its applications", *Annales des Sciences Math. du Quebec*

The 13-variable Diophantine result can be found in

3. Juriĭ Matijasevič and Julia Robinson, "Reduction of an arbitrary Diophantine equation to one in 13 unknowns", *Acta Arith.* 27 (1975), 521 - 553.

A proof of the 9-variable Diophantine result and some explicit information on the pairs of numbers of variables and degrees sufficient for unsolvability can be found in

4. James P. Jones, "Universal Diophantine equations", *J. Symbolic Logic* 47 (1982), 549 - 571.

The next two references contain proofs of the linear ordering of the unary exponential polynomials under eventual dominance; the third proves the well-ordering of this relation.

5. Godfrey Harald Hardy, *Orders of Infinity; the 'Infinitärcalcül' of Paul de Bois-Reymond*, 2nd. ed., Cambridge, 1924.

6. Daniel Richardson, "Solution of the identity problem for integral exponential functions", *Zeitschr. f. math. Logik* 15 (1969), 333 - 340.

7. Andrzej Ehrenfeucht, "Polynomial functions with exponentiation are well-ordered", *Alg. Universalis* 3 (1973), 261 - 262.

A proof of the decidability of the single-variable dyadic iterated exponential Diophantine problem is given in reference 8 and some additional relevant information is gathered in reference 9.

8. Hilbert Levitz, "Decidability of some problems pertaining to base 2 exponential Diophantine equations", *Zeitschr. f. math. Logik* 31 (1985), 109 - 115.

9. James P. Jones, Hilbert Levitz, and Alex J. Wilkie, "Classification of quantifier prefixes over exponential Diophantine equations", *Zeitschr. f. math. Logik*

Finally, the following is a short, but simple, survey of positive results available in the ordinary Diophantine case.

10. J.W.S. Cassels, "Explicit results on the arithmetic of curves of higher genus", *Russian Math. Surveys* 40 #4 (1985), 43 - 48.

Chapter III. Weak Formal Theories of Arithmetic

Angling may be said to be so like the mathematics,
that it can never be fully learnt.

—*Izaak Walton*

1. *Ignorabimus*?

There are different themes that run through different branches of mathematics. Arithmetic encoding— the simultaneously astounding breadth and limitations it engenders— has been a theme developed so far in this book. Another theme of logic is formalisation— the creation and study of formal language and logic. Such a theme is taken up in this chapter and ultimately wed to that of encoding and limitation.

In section 2 we will introduce formal language and logic, the so-called *first-order logic*. First-order logic has both syntactic and semantic features, the equivalence of which (a sort of adequacy condition on the syntax) is called the *Completeness Theorem* and is proven in section 3. The reader who has already had a basic course in logic will find nothing new in these sections and can readily skip them.

We have seen in Chapters I and II how, using only addition and multiplication, arbitrary computable properties can be defined arithmetically. Sections 4 and 5 look at addition and multiplication separately. The individual languages of these functions are too weak to allow the definition of the other and each operation has an effectively decidable theory. In the case of addition (section 4), we will see the full details; for multiplication, results will merely be summarised.

Section 6 considers the full language and a weak theory of arithmetic. This theory is shown adequate to derive any true Σ_1-sentence (basically, a sentence asserting formally that a Σ_1-relation holds of some numbers). From this, results of Chapter I readily show that any decent formal extension of this theory is *incomplete* (i.e., it leaves some sentences undecided) and undecidable (derivability is r.e. but not recursive). Section 8 refines our knowledge of undecidability.

In the light of Chapters I and II, such negative results as those of sections 6 and 8 are not too surprising and these sections may appear somewhat anti-climactic. They form, on the contrary, a sort of preview of the deeper and more interesting incompleteness results of the first chapter of the next volume, results that gave birth to the body of knowledge surveyed in these two volumes. The historical background will be discussed in the long-winded introduction to Chapter IV of volume II. For now, I want to discuss the philosophical background to the incompleteness and undecidability results of the present chapter. To do this, we return to Hilbert's problems.

Hilbert's tenth problem was only one of 23 problems compiled by David Hilbert for his address to the Second International Congress of Mathematicians held in Paris in August of 1900. Of possibly greater interest than the 23 problems themselves— which cover a broad spectrum of mathematics and cohere as little as any problem list— are the remarks prefatory to the list. It is here that Hilbert's guiding philosophy is forcefully expressed.

Following a few brief remarks on the coming new century (the year 1900 was the last year of the 19th century, not the first year of the 20th) and how nice it would be to know what the mathematics of the future would be ("If you can look into the seeds of time, And say which grain will grow and which will not..."), Hilbert quickly got to the heart of the matter— problems.

Hilbert began his discussion of problems by focussing on their importance for mathematics— and mathematicians. We can readily exemplify the former by pausing to reflect on the fact that much of the last chapter was the outgrowth of work on one problem— Hilbert's tenth. Hilbert cited several such instances, noting also that, "As long as a branch of science offers an abundance of problems, so long is it alive; a lack of problems foreshadows extinction or the cessation of independent development".

Having explained (in a bit more detail than I have) that mathematics needs problems, Hilbert then addressed the problem of where the problems come from. Initially, they are empirical, arising from geometry, physics, etc. Eventually, the problems arise from mathematics itself. Similarly, logic, which is one level farther removed from the empirical world than mathematics, draws its problems from mathematics (e.g., the last chapter), computer science (I cannot point to Chapter I as this material pre-dated computer science), philosophy (this chapter and Chapter IV of volume II), and ultimately from logic itself (e.g., Chapter V of volume II). This drawing on themselves is a not uncontroversial aspect of mathematics and logic. However, for Hilbert, at least in his lecture on problems, this was not a matter of controversy, but a reäffirmation of the fruitfulness for mathematics of its problems.

The third question Hilbert addresses is a bit more relevant to the present chapter: What constitutes a solution?

> I should say first of all this: that it shall be possible to establish the correctness of the solution by means of a finite number of steps based upon a finite number of hypotheses which are implied by the statement of the problem and which must always be exactly formulated. This requirement of logical deduction by means of a finite number of processes is simply the requirement of rigor in reasoning.

In his second problem, he complements this:

> When we are engaged in investigating the foundations of a science, we must set up a system of axioms which contains an exact and complete description of the relations subsisting between the elementary ideas of that science... no statement within the realm of the science whose foundations we are testing is held to be correct unless it can be derived from those axioms by means of a finite number of logical steps.

This language is partially repeated in the statement of the tenth problem:

> To devise a process according to which it can be determined by a finite number of operations whether the equation is solvable in rational integers.

Other than to note the similarity of language and the common insistence on the finiteness of the desired procedures, I have little to say about the tenth problem here. It stands out among the 23 Hilbert problems in that it is the unique problem on the list calling, not for a proof or disproof of some assertion, but for an algorithm providing such proofs and disproofs (or, more exactly: an algorithm provably providing the answers to be had by such proofs and disproofs).

The desiderata of the preface and of the second problem fit together more closely. They are complementary, like hand and glove. The prefatory requirement is, as Hilbert says, merely that of rigour; the second is the establishment of the relevant notion of rigour. Except for the tenth problem, a problem for Hilbert was (almost) a statement, the truth or falsity of which was to be ferreted out and proven rigorously. The statement concerns a science, e.g., geometry or algebra, and the science must be isolated and *completely* described axiomatically. From this, every problem of the science can be solved: either the statement or its negation will have a proof from the axioms... It is just this that we intend to disprove in general. But I am getting ahead of my story.

Hilbert's fourth point in his discussion of problems was a short enumeration of the difficulties that can arise in trying to find the solution, i.e., in trying to find the proof or disproof desired. It could be the case, for example, that the statement of the given problem is wrong (e.g., the hypotheses are insufficient as when one neglects to require the continuity of the derivative...).

Despite difficulties, Hilbert believed a solution always to be possible. This was the fifth and final point of his general discussion— his expression of faith:

> Take any definite unsolved problem, such as the question as to the irrationality of the Euler-Mascheroni constant C, or the existence of an infinite number of prime numbers of the form $2^n + 1$. However unapproachable these problems may seem to us and however helpless we stand before them, we have, nevertheless, the firm conviction that their solution must follow by a finite number of purely logical processes.
>
> Is this axiom of the solvability of every problem a peculiarity characteristic of mathematical thought alone, or is it possibly a general law inherent in the nature of the mind, that all questions which it asks must be answerable?...
>
> This conviction of the solvability of every mathematical problem is a powerful incentive to the worker. We hear within us the perpetual call: There is the problem. Seek its solution. You can find it by pure reason, for in mathematics there is no *ignorabimus*.

Hilbert's rejection of the ignorabimus was a reaction to the more negative beliefs of Emil du Bois-Reymond and his catchwords "ignoramus" ("We don't know") and "ignorabimus" ("We will never know"). Du Bois-Reymond, elder brother of Paul du Bois-Reymond (the inventor of the diagonal argument later used so fruitfully by Georg Cantor), studied electrophysiology and was also a populariser of science. Du Bois-Reymond believed that there were insoluble problems in science. A lecture on this he gave in 1872 was entitled "Über die Grenzen des Naturerkennens" ("On the limitations of natural philosophy"). In an 1880 lecture, "Die sieben Welträtsel" ("The seven world puzzles"), he cited 7 questions to which we can only answer "ignoramus" or "ignorabimus":

i. the essence of force and matter
ii. the origin of movement
iii. the origin of life
iv. the teleology of nature
v. the origin of sense perception

vi. the origin of thought
vii. free will.

Du Bois-Reymond's views were widely discussed and attained a degree of popularity.

Hilbert's optimism ran counter to du Bois-Reymond's pessimism. His address on problems was given at the turn of the century, a particularly optimistic time, and one may be tempted to attribute Hilbert's faith to the mood of the day. However, the mood changed. There was soon to be the First World War, Heisenberg's Uncertainty Principle, and, in the 1920s, a battle between Hilbert and L.E.J. Brouwer over just such an issue. Yet, on 8 September 1930, an optimistic David Hilbert addressed a radio audience on the occasion of his retirement in his home town of Königsberg. Once again he banished the ignorabimus from mathematics:

> Instead of the foolish Ignorabimus, we call on the contrary our own motto:
> We must know,
> we will know.

The couplet, "We must know, we will know", in its German original, adorns his tombstone.

Both du Bois-Reymond and Hilbert were well aware of the difficulties inherent in the sciences. Du Bois-Reymond declared some to be insurmountable; Hilbert said we just have to be patient. Perhaps the best indication of Hilbert's assessment of the situation is given by an anecdote told by Carl Ludwig Siegel, a 20th century number theorist sometimes referred to in jest as the last of the great 19th century mathematicians. At one lecture in 1920, Hilbert was discussing problems in mathematics. He told the students, who numbered Siegel among them, that much recent work on the Riemann zeta function led him to hope that he'd yet see a proof of the Riemann Hypothesis (locating all the zeroes of the zeta function). The Fermat problem was more difficult, but the youngest in the audience might live to see its solution. However, the transcendence of $2^{\sqrt{2}}$ was so difficult a problem that noöne in the audience would live to see its solution. Within a few years, Siegel had proven this transcendence! (In fact, Siegel was not the first to prove this transcendence result. Today, the Fermat problem is *almost* solved and the Riemann Hypothesis has been overwhelmingly corroborated numerically, but neither problem has been completely solved.)

What this anecdote tells us is that Hilbert's optimism was not a Utopian fantasy. If he did misjudge on specifics, he was nevertheless realistic in recognising the great difficulty of mathematics. But was his appraisal of the non-impossibility correct? Might there be an ignorabimus in mathematics? With our modern knowledge of algorithmic unsolvability, an affirmative answer seems most likely. The evidence available to Hilbert in 1900 seemed to support his rejection of the ignorabimus.

The statement extracted from Hilbert's second problem calls for a resoundingly strong rejection of the ignorabimus, namely the following.

1.1. Programme. Give a complete axiomatic description of mathematics from which any mathematical assertion can be decided by finitely many logical steps.

Evidence that such a programme as 1.1 can be carried out is given, for example, by Richard Dedekind's categorical axiomatic characterisation of the natural numbers: up to isomorphism, $(\omega; S, 0)$ is the unique triple $(X; F, a)$ satisfying

i. $a \in X$ and $F{:}X \to X$ is a function
ii. F is one-one
iii. $a \notin \mathrm{range}(F)$
iv. for any set $Y \subseteq X$, if
  a. $a \in Y$
  b. for all $x \in X$, $x \in Y \Rightarrow F(x) \in Y$,

then $X = Y$.

Another example is the characterisation up to isomorphism of the ordered field of real numbers as a complete Archimedean ordered field. The examples Hilbert cited were his own such axiomatisations of geometry and of the arithmetic of the real numbers.

Such axiomatisations, however completely they characterise the structures being axiomatised, do not completely carry out Programme 1.1. Consider Dedekind's axiomatisation of $(\omega; S, 0)$. From these axioms, addition and multiplication are determined, whence the truth or falsity of any arithmetic assertion, e.g., an assertion of the form,

$$x \in A,$$

where A is an r.e. non-recursive set. If $x \in A$, this fact must follow from Dedekind's axioms by some finite number of logical steps, and indeed it does: the computation putting x into A described in Chapter I depends primarily on our ability to calculate addition and multiplication, which we can do recursively by the recursive definitions of addition and multiplication justified by Dedekind's induction axiom. If $x \notin A$, Dedekind's axioms must yield this information in finitely many logical steps. But how? Indeed, will not the set of all assertions of the form,

$$x \notin A, \tag{*}$$

that can be generated by finitely many logical steps from the given finite set of axioms be recursively enumerable? Will not any decent deductive system have a recursively enumerable set of theorems? After all, mustn't one be able to recognise (recursively) the axioms and the applicability of rules of inference? And, if so, how can the r.e. set of x's for which we can prove (*) coïncide with the non-recursively enumerable complement of A?

What Hilbert didn't realise at the time (1900) was that *second-order* axiomatisations like Dedekind's do not come equipped with any deductive structure and the deductive *first-order* axiomatisations that he later studied in the 1920s do not offer categorical characterisations. Moreover, during the latter work he could not be aware of the future birth and development of recursion theory in the mid- to late 1930s, which development would construct an impassable barrier to the successful completion of Programme 1.1.

The goal of the present chapter is, basically, to study this recursion theoretic obstruction to Programme 1.1. Section 2 introduces first-order logic, the language and deductive structure. Section 3 proves that the deductions capture semantic entailment adequately and verifies informally the effective enumerability of the set of consequences of an effectively enumerable set of non-logical axioms.

Sections 4 and 5 demonstrate that something like Programme 1.1 can be carried out relative to first-order logic for inexpressive fragments of mathematics. Section 6 shows that the Programme must fail once we have both the addition and multiplication of natural numbers in the mathematics under discussion. Following a technical section devoted more-or-less to studying the method of proof in section 6, some applications are discussed in section 8, one such being the incompleteness of second-order logic— second-order entailment is not effectively enumerable.

Hilbert's strong rejection of the ignorabimus— Programme 1.1— is, thus, convincingly refuted. But might there still be an ignorabimus lurking somewhere in mathematics?

1.2. Problem. Is there an absolutely unsolvable problem, i.e., a problem which mathematicians will never be able to solve?

I do not know the answer to Problem 1.2. Perhaps it is an ignorabimus in the philosophy of mathematics. What I can say are two things, conflicting evidence that point in opposite directions for a solution to this Problem. On the one hand is the fact that the bulk of mathematical practice can be coded up in one recursively enumerable theory— axiomatic set theory— and that there are mathematical statements left undecided by this theory; hence there are problems unassailable by the present methods of mathematics. On the other hand, we must bear in mind that mathematics is not a static object like a formal theory; new concepts and new axioms enter mathematics from time to time. Today's unsolved problem may readily fall to tomorrow's axioms.

2. Formal Language and Logic

A *structure*, for us, will consist of a non-empty set, called the *domain* of the structure, together with some relations and functions on the domain, and some distinguished elements of the domain. Thus, for example,

$$(R; <, +, \cdot, 0, 1)$$

(R the set of real numbers) is a structure. The first-order language used to describe a given structure has primitive symbols denoting the specified relations, functions, and elements of the domain, as well as variables ranging over arbitrary elements of the domain. The elements of the domain are often called *individuals*, whence the specified constants are individual constants and the variables are individual variables. There are no collective variables, i.e., variables ranging over subsets of the domain as in Dedekind's *second-order* induction axiom,

$$\forall Y\, [0 \in Y \wedge \forall x\, (x \in Y \rightarrow Sx \in Y) \rightarrow \forall x (x \in Y)].$$

Moreover, the first-order language cannot refer to other structures or elements thereof (as, for example, in Hilbert's axioms for geometry in which there was a maximality axiom asserting that the given structure could not be enlarged and still validate the remaining axioms). The first-order language of a given structure can, thus, be said to be internal.

As in our discussion of primitive recursive functions, we shall give a very explicit and almost excessively formal definition of first-order language and its logic. A few informal observations can, however, be made before we get to the detailed work.

First, a structure is determined by the relations specified, not those possible. Thus,

$$(R; <, +, \cdot, 0, 1)$$

is different from the structure

$$(R; +, \cdot, 0, 1),$$

even though (once we've defined the language fully) the ordering relation is definable in the latter structure:

$$x < y: \ \exists z\,[y = x + z^2 \ \wedge \ z \neq 0].$$

Given this definability, we can eventually relax the stricture and identify the two structures for all practical purposes. However, it can be shown that in neither structure is the set of natural numbers definable, whence the structure,

$$(R; \omega, <, +, \cdot, 0, 1),$$

cannot be identified with either or the two original structures. Similarly, the two structures,

$$(\omega; +, \cdot, 0, 1) \quad \text{and} \quad (\omega; +, 0, 1),$$

are distinct. What is considered part of a structure is largely determined by what can be said about it in its own language.

Second, a structure is not completely determined by its language. An equivalence relation,

$$(X; \equiv),$$

a partial ordering,

$$(P; <),$$

and the graph of a unary function F,

$$(X; G_F),$$

can all be thought of as being given by the same language— that of one binary relation symbol (and, implicitly: equality). The distinction between the various structures is not in this case one of language, but of which expressions in the given language are true in the structures.

As in Programme 1.1, we can set ourselves the following goal. Given a structure, say,

$$(R; <, +, \cdot, 0, 1),$$

find a decent axiomatisation from which all expressions true in the structure can be derived. Setting aside for now the question of how we are to perform the derivations, several axioms immediately present themselves. Ignoring the inequality, we have a field and can list the usual algebraic identities and axioms asserting the existence of additive and multiplicative inverses. The inequality relation gives us axioms of linear order, as well as such order-preservation as holds for the given functions. The reals form, however, a very special ordered field possessing some very special properties: the ordering is complete (any bounded set has a supremum), it is Dedekind complete (every Dedekind cut determines a real), it is real-closed (every positive real has a square root and every polynomial of odd degree has a root). To the extent that these can be expressed in the first-order language, one can use one or more of these special properties as additional axioms. (We shall look at this example again and see how this can be done after we've finally introduced the full language.)

Axioms need not arise from the analysis of a single structure. They can also arise from analysing what is common to a variety of structures. Insofar as this common behaviour is internal to the structures, the axioms are first-order. For example, as already almost touched on in discussing the real umbers above, the axioms for fields are first-order expressible and derive, not from a single structure, but from the isolation of the concept of a field as being that structural aspect of the rational, real, and complex numbers that allows one to perform linear algebra over these domains. The corresponding derivability problem is not to derive everything true in a given structure, but everything true in all fields, that is, everything expressible in the language of fields that is true in all fields.

Let us get down to details.

The description of a first-order language is given via a series of stages. First, we specify the symbols of the language (primitives, variables, propositional connectives, and quantifiers), then we build up the terms (denoting individuals), and finally we build up formulae (asserting things about the individuals). After all this has been accomplished, we have to see how to use the language. This requires an explanation of what it means for a sentence to be true about a structure ("true *in*" the structure), and an explanation of formal provability.

The first step in the descriptive phase is the isolation of primitives.

2.1. Definition. The *primitives* of the full first-order language are the following:
Equality symbol. $\doteq$
Relation symbols (for each $n > 0$). $R_0^n, R_1^n, \ldots$
Function symbols (for each $n > 0$). $F_0^n, F_1^n, \ldots$
Constants. $c_0, c_1, \ldots$

2.2. Definition-Convention. A first-order *language L* is determined by choosing a subset of the set of primitives of the full first-order language given in Definition 2.1. (In this book) it is always assumed that the symbol for equality is in the language L.

A language L is *finite* if it has only finitely many primitives. Note that all first-order languages are countable as they are given by subsets of the countable full first-order language. This restriction, like the insistence that every language include the equality symbol, is a local one enforced in this book; in mathematical logic in general languages need not be countable. Indeed, we shall allow a violation of this restriction shortly when we augment a language by adding constants naming all elements of the domain of a given structure. When the domain is uncountable, the augmented language is uncountable. However, we shall only need such augmented languages as auxiliaries.

2.3. Definition. The other basic symbols of a first-order language are these:
Variables. $v_0, v_1, \ldots$
Propositional connectives. $\neg, \wedge, \vee, \rightarrow$
Quantifiers. $\forall, \exists$.

In informally describing the language of some familiar structure, we will use the usual mathematical symbols for its familiar relations (e.g., = , <) and functions (e.g., +, ·). For the purpose of discussing the correspondence between symbol and object, however, it is convenient to use the unfamiliar symbols R_i^n, F_i^n, and $\doteq$ in the formal language. Once such discussion is over, we can improve readability by treating the language semi-formally by using the familiar <, +, =, etc., as belonging to the formal language. One distinction that is convenient to maintain is the following.

2.4. Convention. We distinguish formal and informal logical symbols as follows:

Formal	*Informal*
$\wedge$	$\wedge$
$\vee$	$\vee$
$\neg$	$\neg$
$\rightarrow$	$\rightarrow$
$\forall$	$\forall$
$\exists$	$\exists$.

Convention 2.4 will be invoked whenever needed throughout the chapter.

Getting back to the present, we have so far only described the generators of the language. The first step in the generation is the construction of terms.

2.5. Definition. Let a language L be given by a suitable choice of primitives. The *terms* of L are inductively defined by

i. variables are terms of L
ii. constants of L are terms of L
iii. if $t_0, \ldots, t_{n-1}$ are terms of L and the n-ary function symbol F_i^n belongs to L, then $F_i^n t_0 \ldots t_{n-1}$ is a term of L.

This last term, $F_i^n t_0 \ldots t_{n-1}$, is just a string of symbols— F_i^n followed by the symbols of t_0, followed by those of t_1, etc. For example, consider the term,

$$F_3^2 F_1^3 c_2 v_4 c_2 c_1. \qquad (*)$$

Parentheses can be informally introduced for ease of readability. Doing this to (*) yields the more intelligible

$$F_3^2(F_1^3(c_2, v_4, c_2), c_1).$$

The parentheses are not necessary: every term can be read in a unique way; there is no ambiguity. This is guaranteed by the fact that all function symbols are written as prefix operators with fixed arities. (Cf. Exercise 1, below.)

Using terms and relation symbols, one quickly defines the atomic formulae of a given language.

2.6. Definition. Let a language L be given. If R_i^n is an n-ary relation symbol of L and $t_0, ..., t_{n-1}$ are terms of L, then

$$R_i^n t_0 ... t_{n-1}$$

is an *atomic formula* of L. Also, if t_0, t_1 are terms of L, then

$$\doteq t_0 t_1$$

is an *atomic formula* of L.

As with function symbols, relation symbols are prefixes and atomic formulae are uniquely readable. For binary relation symbols, infix notation can be used without introducing any ambiguity and one may write

$$t_0 \doteq t_1 \quad \text{or} \quad t_0 R_i^2 t_1$$

*in*formally. As before, parentheses can be introduced for intelligibility, but they are not part of the formal language.

The atomic formulae are, as their name suggests, the building blocks for arbitrary formulae.

2.7. Definition. Let a language L be given. The *formulae* of L are inductively defined by

i. an atomic formula φ is a formula
ii. if φ is a formula, then so is $\neg\varphi$
iii. if φ, ψ are formulae, so are $\wedge\varphi\psi$, $\vee\varphi\psi$, and $\rightarrow\varphi\psi$
iv. if φ is a formula, then $\exists v_i\, \varphi$ and $\forall v_i\, \varphi$ are formulae.

Formulae will generally be denoted by lower-case Greek letters φ, ψ, χ, etc.

Once again, the logical connectives and quantifiers are viewed as prefix operators. As before, the reason is to guarantee the unique readability of formulae. For more humanly readable formulae, it is handy to write $\wedge$, $\vee$, and $\rightarrow$ as infix operators and use parentheses: from φ, ψ, generate $(\varphi \wedge \psi)$, $(\varphi \vee \psi)$, and $(\varphi \rightarrow \psi)$. In a language with parentheses, this will guarantee unique readability. Some parentheses can be eliminated by specifying the usual binding priorities: $\neg$ binds most strongly, then $\wedge$, then $\vee$, and finally $\rightarrow$. Thus,

$$\neg\varphi \wedge \psi \rightarrow \theta \vee \chi$$

reads

$$(((\neg\varphi) \wedge \psi) \rightarrow (\theta \vee \chi))$$

when parentheses are inserted, and it reads

$$\rightarrow\wedge\neg\varphi\psi\vee\theta\chi$$

formally. For fairly obvious reasons, we shall write formulae in the more readable form. In fact, with familiar function symbols (+, ·, etc.) we will use the familiar notation in constructing terms as well.

Consider now the structure $(R; <, +, \cdot, 0, 1)$ mentioned earlier. The language L for this structure includes $\doteq$, R_0^2 (for <), F_0^2 (for +), F_1^2 (for ·), c_0 (for 0), c_1 (for 1). Modulo the conventions for writing formulae in an informal, readable way, we can readily write the field axioms in a first-order way:

$$\forall v_0 \forall v_1 [v_0 + v_1 \;=\; v_1 + v_0]$$
$$\forall v_0 \forall v_1 \forall v_2 [v_0 + (v_1 + v_2) \;=\; (v_0 + v_1) + v_2]$$
$$\forall v_0 [v_0 + 0 \;=\; v_0]$$
$$\forall v_0 \exists v_1 [v_0 + v_1 \;=\; 0]$$

etc.

[Formally, these are

$$\forall v_0 \forall v_1 \doteq F_0^2 v_0 v_1 F_0^2 v_1 v_0$$
$$\forall v_0 \forall v_1 \forall v_2 \doteq F_0^2 v_0 F_0^2 v_1 v_2 F_0^2 F_0^2 v_0 v_1 v_2$$

etc.]

We can express the order axioms for ordered fields:

$$\forall v_0 \neg(v_0 \;<\; v_0)$$
$$\forall v_0 \forall v_1 [v_0 \;<\; v_1 \;\rightarrow\; \neg(v_1 \;<\; v_0)]$$

etc.

We cannot directly express the completeness axiom,

$$\forall X \,\{\exists v_0 \forall v_1 (v_1 \in X \rightarrow v_1 < v_0 \vee v_1 = v_0) \rightarrow$$
$$\rightarrow \exists v_0 [\forall v_1 (v_1 \in X \rightarrow v_1 < v_0 \vee v_1 = v_0) \wedge$$
$$\wedge \forall v_2 [\forall v_1 (v_1 \in X \rightarrow v_1 < v_2 \vee v_1 = v_2) \rightarrow v_0 < v_2 \vee v_0 = v_2]]\},$$

for the language has neither *set variables* X nor a relation symbol for membership. Similarly, we cannot directly express Dedekind's completeness axiom,

$$\forall X \,\forall Y \,\{[\forall v_0 (v_0 \in X \vee v_0 \in Y) \wedge \exists v_0 (v_0 \in X) \wedge \exists v_0 (v_0 \in Y) \wedge$$
$$\wedge \forall v_0 \forall v_1 (v_0 \in X \wedge v_1 \in Y \rightarrow v_0 < v_1)] \rightarrow$$
$$\rightarrow [\exists v_0 [v_0 \in X \wedge \forall v_1 (v_1 \in X \rightarrow v_1 < v_0 \vee v_0 = v_1)] \vee$$
$$\vee \exists v_0 [v_0 \in Y \wedge \forall v_1 (v_1 \in Y \rightarrow v_0 < v_1 \vee v_0 = v_1)]]\}.$$

We can, however, express that positive real numbers are squares,

$$\forall v_0 [0 < v_0 \;\rightarrow\; \exists v_1 (v_0 \;=\; v_1 \cdot v_1)],$$

and we can express schematically that polynomials of odd degree have roots:

$$\forall v_0 \forall v_1 ... \forall v_{2n+1}[v_{2n+1} \neq 0 \rightarrow \exists v_{2n+2}(v_0 + v_1 \cdot v_{2n+2} + ... + v_{2n+1} \cdot v_{2n+2}^{2n+1} = 0)],$$

where v_{2n+2}^{i} abbreviates the i-fold product $v_{2n+2} \cdots v_{2n+2}$.

The completeness axioms for the real line cannot be expressed in a first-order way because they involve quantification over all sets. One can schematically approximate such axioms by asserting each instance expressible in the language. Since these are hideous axioms, let me illustrate this instead with the induction axiom for the structure $(\omega; S, 0)$. In place of the inexpressible,

$$\forall X\ [0 \in X \wedge \forall v_0(v_0 \in X \rightarrow Sv_0 \in X) \rightarrow \forall v_0(v_0 \in X)],$$

we have the schema

$$\varphi(0) \wedge \forall v_0(\varphi(v_0) \rightarrow \varphi(Sv_0)) \rightarrow \forall v_0 \varphi(v_0),$$

for all formulae φv_0 containing the *free variable* v_0, where $\varphi(0)$ and $\varphi(Sv_0)$ denote the results of replacing v_0 by 0 and Sv_0, respectively, in all *free occurrences* of v_0 in φv_0.

The notions of free and *bound* variables are familiar from the calculus, where the latter are termed *dummy* variables. As in calculus, it is generally easy to recognise which occurrences of a variable in a formula are free and which are bound. In, for instance, the formula

$$x^2 + \int 2xy\, dy + y^3 + 1,$$

all occurrences of x are free, as is the final occurrence of y while the occurrence of y in the integrand is bound by the differential dy; in the first-order formula,

$$v_0 = 0 \wedge \forall v_1(v_0 < v_1) \vee v_1 < 0,$$

v_0 similarly has only free occurrences and the last occurrence of v_1 is free, but the occurrence of v_1 in $v_0 < v_1$ is bound by the universal quantification over v_1. What are new in logic are the terminology and, alas, the necessity of being ultra-precise.

The obvious things to give precise definitions of are the notions of *free occurrences* and *bound occurrences* of variables. It turns out, however, that we do not need to define these formally! What we do need to define formally are i. which variables have free occurrences in a formula, ii. which variables have bound occurrences in a formula, iii. how to substitute a term for a free variable in a given formula, and iv. how to relabel bound variables. It turns out these notions can be given simple inductive definitions.

2.8. Definitions. The set $FV(t)$ of *free variables* occurring in a term t is defined inductively on the generation of a term t as follows:

i. $FV(v_i) = \{v_i\}$ and $FV(c_i) = \{\ \}$

ii. $FV(F_i^n t_0 ... t_{n-1}) = FV(t_0) \cup ... \cup FV(t_{n-1})$.

The set $FV(\varphi)$ of free variables occurring in a formula φ is defined inductively on the generation of a formula φ as follows:

i. $FV(\doteq t_0 t_1) = FV(t_0) \cup FV(t_1)$

ii. $FV(R_i^n t_0 ... t_{n-1}) = FV(t_0) \cup ... \cup FV(t_{n-1})$

iii. $FV(\neg\varphi) = FV(\varphi)$

iv. $FV(\wedge\varphi\psi) = FV(\vee\varphi\psi) = FV(\rightarrow\varphi\psi) = FV(\varphi) \cup FV(\psi)$

v. $FV(\exists v_i\varphi) = FV(\forall v_i\varphi) = FV(\varphi) - \{v_i\} = \{v_j\colon v_j \in FV(\varphi) \wedge j \neq i\}$.

2.9. Definition. The set $BV(\varphi)$ of *bound variables* occurring in a formula φ is defined inductively on the generation of formulae as follows:

i. $BV(\varphi) = \{\ \}$, if φ is atomic

ii. $BV(\neg\varphi) = BV(\varphi)$

iii. $BV(\wedge\varphi\psi) = BV(\vee\varphi\psi) = BV(\rightarrow\varphi\psi) = BV(\varphi) \cup BV(\psi)$

iv. $BV(\exists v_i\varphi) = BV(\forall v_i\varphi) = BV(\varphi) \cup \{v_i\}$.

2.10. Definitions. Let t be a term and v_i a variable. The result of substituting t for v_i at all free occurrences of v_i in a term u, written $([t/v_i]u)$, is defined inductively on the generation of terms as follows:

i. $([t/v_i]c_j) = c_j$

ii. $([t/v_i]v_j) = \begin{cases} t, & j = i \\ v_j, & j \neq i \end{cases}$

iii. $([t/v_i]F^n_j t_0 \ldots t_{n-1}) = F^n_j([t/v_i]t_0)\ldots([t/v_i]t_{n-1})$.

The result of substituting t for v_i at all free occurrences of v_i in a formula φ, written $([t/v_i]\varphi)$, is defined inductively on the generation of formulae as follows:

i. $([t/v_i]\dot{=}t_0t_1) = \dot{=}([t/v_i]t_0)([t/v_i]t_1)$

ii. $([t/v_i]R^n_j t_0 \ldots t_{n-1}) = R^n_j([t/v_i]t_0)\ldots([t/v_i]t_{n-1})$

iii. $([t/v_i]\neg\varphi) = \neg([t/v_i]\varphi)$

iv. $([t/v_i]*\varphi\psi) = *([t/v_i]\varphi)([t/v_i]\psi)$, for $* \in \{\wedge, \vee, \rightarrow\}$

v. $([t/v_i]Qv_j\varphi) = \begin{cases} Qv_j\varphi, & j = i \\ Qv_j([t/v_i]\varphi), & j \neq i, \end{cases}$ for $Q \in \{\forall, \exists\}$.

We often write φv_i or $\varphi(v_i)$ to indicate that the variable v_i occurs freely in φ. When we do this, we can also write φt or $\varphi(t)$ for $([t/v_i]\varphi)$.

2.11. Definition. Let v and v_i be variables. The result of relabelling all bound occurrences of v_i in a formula φ by v, written $rlbl(v_i, v, \varphi)$, is defined inductively on the generation of formulae as follows:

i. $rlbl(v_i, v, \varphi) = \varphi$, for atomic formulae φ

ii. $rlbl(v_i, v, \neg\varphi) = \neg rlbl(v_i, v, \varphi)$

iii. $rlbl(v_i, v, *\varphi\psi) = *rlbl(v_i, v, \varphi)rlbl(v_i, v, \psi)$, for $* \in \{\wedge, \vee, \rightarrow\}$

iv. $rlbl(v_i, v, Qv_j\varphi) = \begin{cases} Qv_j\, rlbl(v_i, v, \varphi), & j = i \\ Qv([v/v_i]rlbl(v_i, v, \varphi)), & j \neq i, \end{cases}$

for $Q \in \{\forall, \exists\}$

In earlier writing down the induction schema,

$$\varphi(0) \wedge \forall v_0(\varphi(v_0) \rightarrow \varphi(Sv_0)) \rightarrow \forall v_0\varphi(v_0),$$

we implicitly assumed we knew how to perform the substitutions of 0 and Sv_0 for v_0 in φv_0. With Definition 2.10, we have a formal definition telling us how. Actually, Definition 2.10 is more an explanation to a computer than to a human how to perform such a substitution.

The point of relabelling bound variables is, of course, the fact that we sometimes have to relabel dummy variables before we can make the substitutions we wish for free variables. From the calculus, for example, if we define

$$F(x) = \int_0^x x^2y + 1\,dy + 7, \tag{*}$$

we can write

$$F(2z) = \int_0^{2z} (2z)^2y + 1\,dy + 7;$$

however, we do not have

$$F(2y) = \int_0^{2y} (2y)^2y + 1\,dy + 7,$$

but rather

$$F(2y) = \int_0^{2y} (2y)^2w + 1\,dw + 7. \tag{**}$$

In substituting a term t for free occurrences of a variable v_i in a formula φ, we must be careful not to bind the variables occurring in t. We could relabel all bound occurrences in φ of variables occurring in t to avoid any clash, just as we did for the term $t = 2y$ in going form (*) to (**). Not all such relabelling need be necessary. For example, if

$$G(x) = \int_0^1 x^2z\,dz + \int_0^1 y\,dy,$$

we need not relabel at all before substituting $2y$ for x to get the valid (if not æsthetically pleasing)

$$G(2y) = \int_0^1 (2y^2)z\,dz + \int_0^1 y\,dy.$$

Our next definition inductively defines when a term t may be substituted for all free occurrences of a variable v_i in a formula φ with no relabelling of bound variables necessary.

2.12. Definition. Let t be a term and v_i a variable. We define when t *is free for* v_i *in* a formula φ inductively on the generation of formulae as follows:

i. t is free for v_i in any atomic formula

ii. t is free for v_i in $\neg\varphi$ iff t is free for v_i in φ

iii. t is free for v_i in $*\varphi\psi$ iff t is free for v_i in each of φ, ψ, where $* \in \{\wedge, \vee, \rightarrow\}$

iv. t is free for v_i in $Qv_j\varphi$ iff $i = j$, or $i \neq j$, $v_j \notin FV(t)$, and t is free for v_i in φ, where $Q \in \{\forall, \exists\}$.

The necessity of giving such detailed definitions— of dotting the i's and crossing the t's as it were— stems not from our need to be very explicit in practice. If this were the case, the definitions given would be familiar from the calculus. The fact is that in practice there is no difficulty in relabelling bound variables and performing substitutions, and it is mere pedantry to state every time one writes φt for some given φv that one assumes t to be free for v in φ or that the appropriate relabelling has taken place. In the sequel, I shall be as anti-pedantically lax as anyone in this matter. So why bother with these definitions? There are two reasons. One is that, unlike the situation in the calculus where we are studying *functions* and the difference between

$$F(x) = \int_0^1 x\,y\,dy \quad \text{and} \quad F(x) = \int_0^1 x\,z\,dz$$

is *notational*, whence irrelevant, in logic the notation itself becomes an object of study and rigour requires us to analyse substitution. Further syntactic considerations build on this analysis. A second reason for this excess of precision stems from our desire for effectiveness. We will want, in Chapter IV of the next volume, to encode syntax recursively. The recursive definitions given above serve this purpose well.

There are, however, more immediate, if less compelling, uses for these definitions. We can, for example, illustrate the rôle unique readability plays. If one could read a term t in two ways, say

$$F_i^n t_0 \ldots t_{n-1} \quad \text{and} \quad F_j^m u_0 \ldots u_{m-1},$$

then it would not be immediately obvious that

$$FV(t) = FV(t_0) \cup \ldots \cup FV(t_{n-1}) = FV(u_0) \cup \ldots \cup FV(u_{m-1})$$

would be well-defined. Because terms and formulae are uniquely readable, there is no ambiguity here and we can define functions and relations inductively in this manner without fear of confusion. This will be used again momentarily when we give a precise definition of what it means for a *sentence* of a language to be *true* in a given structure.

We begin by using Definition 2.8 to define the word "sentence".

2.13. Definition. A formula φ is a ***sentence*** if φ has no free variables, i.e., if $FV(\varphi) = \{\ \} = \varnothing$.

Just as algebraic expressions with free variables, e.g.,

$$x^2 + 2x + 13,$$

define functions and do not have fixed values, formulae with free variables define relations and do not have fixed truth values. Sentences, having no free variables, will have definite truth values in structures for the languages in which the sentences lie. It is this notion of truth we now wish to define formally.

Our formal definition of truth will require a few preliminaries. Before getting to these, however, it might be a good idea for me to say a few words about the word "formal". I have been using the word ambiguously in both the mathematical and the more narrow logical senses, and, although this ambiguity has thus far resulted in no loss of clarity, a serious ambiguity can soon arise— particularly with reference to truth. For, on one meaning of "formal", we will give a formal truth definition; and on another meaning either we don't give such a definition or such cannot exist— depending on which formal language is specified.

So what are these two definitions of "formal"? Basically, they are both versions of the same thing. To both the mathematician and the logician, a *formal* language is a precise, unambiguous language— unlike natural languages like Dutch or English. A definition is *formal* if it is given in a formal language. To the mathematician— or anyone but a logician or computer scientist— the usual mathematical language is formal. In logic and computer science, however, formality is of necessity carried to an extreme: precision is not enough; the entire syntax of the language is minutely described. In logic, the word "formal" attaches itself to such artificial languages— in the present book: first-order languages. Thus, mathematically, definitions like 2.5 - 2.13 would be formal definitions; while logically, a formal definition would be one given in a first-order language, e.g.,

$$v_0 < v_1:\ \exists v_2[v_0 + v_2 + 1 = v_1]$$

would formally define < in the language of $(\omega; +, 1)$.

The "formal" definition of truth we shall soon give will be formal in the mathematical sense, not in the logical sense. If we consider the structure $(\omega; +, \cdot, 0, 1)$ and accept the fact (to be established later) that the relations definable in the first-order language of this structure are exactly the arithmetic relations of Chapter I, then Tarski's Theorem I.13.5 tells us that there is no formal truth definition for the language within itself. (Actually, the language does not refer to its syntax and no formal definition can be expected to be given. However, one can numerically encode the syntax and ask for a definition of the set of codes of true sentences. It is this that cannot be given formally. But this is a topic for Chapter IV of the next volume.)

To avoid ambiguity about "formal definitions", "formal proofs", and the like, it will be convenient to make some convention on the use of the word "formal".

2.14. Convention. The word "formal" shall hereafter be limited to its strictly logical use. A formal definition is one given in a first-order language. A formal proof, or derivation, will be one of first-order logic in accordance with the definition thereof to be given. Ordinary mathematical usage shall be deemed "informal", unless we want to emphasise the formality in the mathematical sense, in which case we shall say "precise", "(excessively) detailed", or "semi-formal".

Having said all of this, let me return to the problem at hand— giving a precise, excessively detailed, semi-formal definition of the truth of a sentence in a given language in a structure for the given language.

2.15. Definition. Let a language L be given. A *structure* $\mathfrak{A}$ for L (or, an L-structure $\mathfrak{A}$) is determined by specifying first a non-empty set A, the *domain* of the structure, and then the *interpretations* of the primitives of L (other than the equality symbol, which is always interpreted as identity). These are as follows: Each n-ary relation symbol R_i^n is assigned an n-ary relation $(R^n_i)^{\mathfrak{A}} \subseteq A^n$; each n-ary function symbol F^n_i is assigned an n-ary function $(F^n_i)^{\mathfrak{A}}: A^n \to A$; and each constant symbol c_i is assigned an individual $(c_i)^{\mathfrak{A}} \in A$.

If the interpretations $(R^n_i)^{\mathfrak{A}}$, etc., are fairly obvious, we write $\mathfrak{A}$ as a tuple consisting first of the domain A, then the relations, functions, and constants, as we have been doing.

2.16. Definition. Let $\mathfrak{A}$ be a structure for a language L. The *augmented language* $L(A)$ is defined by adding to L new constants $\overline{a}$ for each element $a \in A$.

2.17. Definitions. i. Let a language L be given. A term t of L is *closed* if t has no free variables (i.e., no variables).
ii. Let L be given and let $\mathfrak{A}$ be a structure for the language L. The *value* $t^{\mathfrak{A}}$ of a closed term of the augmented language $L(A)$ is defined inductively as follows:
a. $(c_i)^{\mathfrak{A}}$ is as defined in 2.15
b. $(\overline{a})$ is a for each new constant of $L(A)$
c. if t is $F^n_i t_0 \ldots t_{n-1}$, then $t^{\mathfrak{A}}$ is $(F^n_i)^{\mathfrak{A}}((t_0)^{\mathfrak{A}}, \ldots, (t_{n-1})^{\mathfrak{A}})$.

2.18. Definitions. Let L be a given language and $\mathfrak{A}$ an L-structure. We define inductively when a sentence φ of the augmented language $L(A)$ is *true* in $\mathfrak{A}$, written

$$\mathfrak{A} \models \varphi,$$

as follows:

i. if φ is atomic:
$\mathfrak{A} \models \doteq t_0 t_1$ iff $(t_0)^{\mathfrak{A}} = (t_1)^{\mathfrak{A}}$
$\mathfrak{A} \models R^n_i t_0 \ldots t_{n-1}$ iff $(R^n_i)^{\mathfrak{A}}((t_0)^{\mathfrak{A}}, \ldots, (t_{n-1})^{\mathfrak{A}})$

ii. $\mathfrak{A} \models \neg\varphi$ iff $\mathfrak{A} \not\models \varphi$ (i.e., $\neg(\mathfrak{A} \models \varphi)$)

iii. $\mathfrak{A} \models \wedge\varphi\psi$ iff $(\mathfrak{A} \models \varphi) \wedge (\mathfrak{A} \models \psi)$
$\mathfrak{A} \models \vee\varphi\psi$ iff $(\mathfrak{A} \models \varphi) \vee (\mathfrak{A} \models \psi)$
$\mathfrak{A} \models \to\varphi\psi$ iff $(\mathfrak{A} \models \varphi) \to (\mathfrak{A} \models \psi)$

iv. $\mathfrak{A} \models \exists v_i \varphi v_i$ iff $\exists a \in A\, (\mathfrak{A} \models \varphi \overline{a})$
$\mathfrak{A} \models \forall v_i \varphi v_i$ iff $\forall a \in A\, (\mathfrak{A} \models \varphi \overline{a})$.

A sentence φ of the language L is deemed true in the structure $\mathfrak{A}$ if it is true in $\mathfrak{A}$ considered as a sentence of $L(A)$. If $\mathfrak{A} \models \varphi$, $\mathfrak{A}$ is said to be a *model* of φ. If $\mathcal{T}$ is a collection of sentences, we write

$$\mathfrak{A} \models \mathcal{T}$$

to mean that every sentence $\varphi \in \mathcal{T}$ is true in $\mathfrak{A}$ and say that $\mathfrak{A}$ is a *model* of $\mathcal{T}$.

By way of example, consider the commutativity of addition in

$$\mathfrak{N} = (\omega; +, \cdot, 0, 1).$$

Observe,

$$\mathfrak{N} \models \forall v_0 \forall v_1 (v_0 + v_1 = v_1 + v_0) \text{ iff } \forall x \in \omega\, \mathfrak{N} \models \forall v_1 (\overline{x} + v_1 = v_1 + \overline{x})$$

$$\text{iff } \forall x \in \omega\ \forall y \in \omega\ \mathfrak{N} \models (\overline{x} + \overline{y} = \overline{y} + \overline{x})$$
$$\text{iff } \forall x \in \omega\ \forall y \in \omega\ \mathfrak{N} \models (\overline{x + y} = \overline{y + x})$$
$$\text{iff } \forall x \in \omega\ \forall y \in \omega\ (x + y = y + x).$$

Thus we see that the formal sentence,

$$\forall v_0 \forall v_1 (v_0 + v_1 = v_1 + v_0),$$

adequately and truly expresses the informal fact,

$$\forall x \in \omega\ \forall y \in \omega\ (x + y = y + x).$$

Following such an example, it may be hard to believe that anything has actually been accomplished by giving a semi-formal truth definition. In mathematics, one often replaces an intuitively clear, informal concept by a precise, semi-formal one. This is what has happened here. Because, however, the concept is TRUTH, some logicians hold that Definition 2.18 is an important philosophical breakthrough, comparable to the isolation of the notion of recursiveness as the proper explication of the notion of effectiveness, and refer to Definition 2.18 as "Tarski's theory of truth". Personally, I don't hold this view.

Having defined truth, we can define *formal* definability.

2.19. Definition. Let L be a language and $\mathfrak{A}$ an L-structure. A relation $R \subseteq A^n$ is *definable in $\mathfrak{A}$ with parameters* (*without parameters*) if there is a formula $\varphi v_0 \ldots v_{n-1}$ of $L(A)$ (respectively, of L) with exactly $v_0, \ldots, v_{n-1}$ free such that, for all $a_0, \ldots, a_{n-1} \in A$,

$$R(a_0, \ldots, a_{n-1}) \text{ iff } \mathfrak{A} \models \varphi\, \overline{a_0} \ldots \overline{a_{n-1}}\,.$$

For the time being, the definition of definability is merely an example of the use of the truth definition. Our immediate goal is axiomatic, to define carefully the notion of logical entailment. A semantic definition is readily given:

2.20. Definition. Let $\mathcal{T}$ be a set of sentences and φ a sentence of some language L. We say that $\mathcal{T}$(semantically) *entails* φ, written

$$\mathcal{T} \models \varphi$$

if, for all L-structures $\mathfrak{A}$ in which $\mathcal{T}$ is true, φ must also be true:

$$\forall \mathfrak{A} (\mathfrak{A} \models \mathcal{T} \rightarrow \mathfrak{A} \models \varphi).$$

An immediate derivative notion to that of entailment is that of validity.

2.21. Definition. Let φ be a sentence of a given language L. We say that φ is *valid*, written

$$\models \varphi,$$

if φ is true in all L-structures.

Note that, by Definition 2.18, the assertion,

$$\mathfrak{A} \models \emptyset,$$

is vacuously true. Thus, for any φ,

$$\models \varphi \text{ iff } \emptyset \models \varphi.$$

The notion of semantic entailment, now precisely defined, is not the notion we want for our discussion of Programme 1.1. Given a set $\mathcal{T}$ of axioms, Hilbert demanded a means of generating any consequences φ of $\mathcal{T}$ in finitely many steps, i.e., he required a notion of formal derivation. Thus, we need a notion of syntactic entailment,

$$\mathcal{T} \vdash \varphi,$$

which agrees with semantic entailment: for any $\mathcal{T}$, φ,

$$\mathcal{T} \vdash \varphi \text{ iff } \mathcal{T} \models \varphi.$$

To this we now turn.

First comes a couple of preliminary definitions.

2.22. Definition. By a *multiset* X, we mean a set X in which multiplicity of occurrence is allowed.

For example,

$$X = \{1, 2, 2, 3\}$$

is a multiset in which 1 and 3 have multiplicity 1, and 2 has multiplicity 2. The usual law of extensionality is not taken to hold: two multisets can have the same elements and fail to be equal; the multiplicities of occurrence must be taken into account in equating two multisets. (The order of the elements is irrelevant. Thus, $\{1, 2, 2, 3\}$ and $\{2, 1, 3, 2\}$ are the same multiset.}

2.23. Definition. Let L be a given language. A *sequent* is an ordered pair of finite multisets of formulae of the language L. We write

$$\Gamma \Rightarrow \Delta$$

for the sequent with multisets Γ and Δ.

Before I start explaining formal derivability, there is a number of things I should say about sequents. The intended meaning of the sequent $\Gamma \Rightarrow \Delta$ is that the conjunction of formulae in Γ implies the disjunction of formulae in Δ, i.e.,

$$\Gamma \Rightarrow \Delta \text{ means } \bigwedge\Gamma \rightarrow \bigvee\Delta.$$

The notations

$$\Gamma \rightarrow \Delta \text{ and } \Gamma \vdash \Delta$$

are also common notations for sequents in the literature.

Other notational matters: One writes $\Gamma, \varphi_0, ..., \varphi_{m-1}$ for $\Gamma \cup \{\varphi_0, ..., \varphi_{m-1}\}$ and $\Delta, \psi_0, ..., \psi_{m-1}$ for $\Delta \cup \{\psi_0, ..., \psi_{m-1}\}$. An empty Γ on the left of the sequent arrow is taken to be vacuously true, and an empty Δ on the right to be vacuously false. The completely empty sequent is false. Thus,

$$\Rightarrow \Delta \text{ means } \bigvee \Delta$$

$$\Gamma \Rightarrow \text{ means } \neg \bigwedge \Gamma$$

$$\Rightarrow \text{ means } \varphi \wedge \neg\varphi.$$

Derivability is defined at the end of a sequence of preliminary definitions— of logical axioms and rules, equality axioms, and non-logical axioms. We assume a fixed language throughout.

2.24. Definition. A sequent of the form,

$$\Gamma, \varphi \Rightarrow \Delta, \varphi$$

is a *logical axiom.*

2.25. Definitions. The *logical rules of inference* by which one infers one sequent from one or two others come in two types— purely logical rules and structural rules. In these rules, the sequents above the lines are called *premises*; those below are *conclusions*. The purely logical rules are:

$$\wedge L \quad \frac{\Gamma, \varphi \Rightarrow \Delta}{\Gamma, \varphi \wedge \psi \Rightarrow \Delta}, \qquad \frac{\Gamma, \psi \Rightarrow \Delta}{\Gamma, \varphi \wedge \psi \Rightarrow \Delta}$$

$$\wedge R \quad \frac{\Gamma \Rightarrow \Delta, \varphi \qquad \Gamma \Rightarrow \Delta, \psi}{\Gamma \Rightarrow \Delta, \varphi \wedge \psi}$$

$$\vee L \quad \frac{\Gamma, \varphi \Rightarrow \Delta \qquad \Gamma, \psi \Rightarrow \Delta}{\Gamma, \varphi \vee \psi \Rightarrow \Delta}$$

$$\vee R \quad \frac{\Gamma \Rightarrow \Delta, \varphi}{\Gamma \Rightarrow \Delta, \varphi \vee \psi} \qquad \frac{\Gamma \Rightarrow \Delta, \psi}{\Gamma \Rightarrow \Delta, \varphi \vee \psi}$$

$$\to L \quad \frac{\Gamma \Rightarrow \Delta, \varphi \qquad \Gamma, \psi \Rightarrow \Delta}{\Gamma, \varphi \to \psi \Rightarrow \Delta} \qquad \to R \quad \frac{\Gamma, \varphi \Rightarrow \Delta, \psi}{\Gamma \Rightarrow \Delta, \varphi \to \psi}$$

$$\neg L \quad \frac{\Gamma \Rightarrow \Delta, \varphi}{\Gamma, \neg\varphi \Rightarrow \Delta} \qquad \neg R \quad \frac{\Gamma, \varphi \Rightarrow \Delta}{\Gamma \Rightarrow \Delta, \neg\varphi}$$

$$\forall L \quad \frac{\Gamma, \varphi t \Rightarrow \Delta}{\Gamma, \forall v_i \varphi v_i \Rightarrow \Delta} \ (**) \qquad \forall R \quad \frac{\Gamma \Rightarrow \Delta, \varphi v}{\Gamma \Rightarrow \Delta, \forall v_i \varphi v_i} \ (*)$$

$$\exists L \quad \frac{\Gamma, \varphi v \Rightarrow \Delta}{\Gamma, \exists v_i \varphi v_i \Rightarrow \Delta} \ (*) \qquad \exists R \quad \frac{\Gamma \Rightarrow \Delta, \varphi t}{\Gamma \Rightarrow \Delta, \exists v_i \varphi v_i} \ (**)$$

where

(*) the variable v is free for v_i in φv_i and does not occur free in any formula in Γ or Δ in the rules $\forall R$ and $\exists L$

and (**) t is free for v_i in φv_i in $\forall L$ and $\exists R$.

The structural rules are:

Weakening
$$\frac{\Gamma \Rightarrow \Delta}{\Gamma, \varphi \Rightarrow \Delta} \qquad \frac{\Gamma \Rightarrow \Delta}{\Gamma \Rightarrow \Delta, \varphi}$$

Contraction
$$\frac{\Gamma, \varphi, \varphi \Rightarrow \Delta}{\Gamma, \varphi \Rightarrow \Delta} \qquad \frac{\Gamma \Rightarrow \Delta, \varphi, \varphi}{\Gamma \Rightarrow \Delta, \varphi}$$

Cut
$$\frac{\Gamma_1, \varphi \Rightarrow \Delta_1 \qquad \Gamma_2 \Rightarrow \Delta_2, \varphi}{\Gamma_1, \Gamma_2 \Rightarrow \Delta_1, \Delta_2}.$$

2.26. Definition. The *equality axioms* are the sequents of the forms

E1. $\Gamma \Rightarrow \Delta, t = t$

E2. $\Gamma, s_0 = t_0, \ldots, s_{n-1} = t_{n-1} \Rightarrow \Delta, F^n_i s_0 \ldots s_{n-1} = F^n_i t_0 \ldots t_{n-1}$

E3. $\Gamma, s_0 = t_0, \ldots, s_{n-1} = t_{n-1}, R^n_i s_0 \ldots s_{n-1} \Rightarrow \Delta, R^n_i t_0 \ldots t_{n-1}$

E4. $\Gamma, s_0 = t_0, s_1 = t_1, s_0 = s_1 \Rightarrow \Delta, t_0 = t_1,$

where F^n_i and R^n_i are in the language under consideration.

Lastly, there is the notion of non-logical axioms. These depend on the theory one wishes to axiomatise. In ordinary mathematics, new axioms are just formulae, e.g., the group axioms:

$$v_0 \cdot (v_1 \cdot v_2) = (v_0 \cdot v_1) \cdot v_2$$
$$v_0 \cdot e = v_0$$
$$\exists v_1 \, (v_0 \cdot v_1 = e).$$

Formally, we can represent such axioms φ by sequents,

$$\Rightarrow \varphi, \tag{1}$$

or

$$\Gamma \Rightarrow \Delta, \varphi, \tag{2}$$

or as rules of inference,

$$\frac{\Gamma, \varphi \Rightarrow \Delta}{\Gamma \Rightarrow \Delta}. \tag{3}$$

The presentation chosen for equality axioms is similar to (2). We could have used axioms in the form (1):

$$\Rightarrow t = t$$
$$\Rightarrow s_0 = t_0 \wedge \ldots \wedge s_{n-1} = t_{n-1} \rightarrow F^n_i s_0 \ldots s_{n-1} = F^n_i t_0 \ldots t_{n-1}$$
$$\Rightarrow s_0 = t_0 \wedge \ldots \wedge s_{n-1} = t_{n-1} \wedge R^n_i s_0 \ldots s_{n-1} \rightarrow R^n_i t_0 \ldots t_{n-1}$$
$$\Rightarrow s_0 = t_0 \wedge s_1 = t_1 \wedge s_0 = s_1 \rightarrow t_0 = t_1.$$

For the sake of definiteness, we assume the convention that non-logical axioms be presented in the form (1).

2.27. Definition. A *theory* is determined by specifying a set $\mathcal{T}$ of formulae to serve as its non-logical axioms. A *non-logical axiom* of $\mathcal{T}$ is a sequent,

$$\Rightarrow \varphi,$$

where $\varphi \in \mathcal{T}$. (Informally, we will refer to φ itself as the axiom.)

Having defined axioms and rules of inference, we can finally put them together and obtain *formal proofs* or *derivations*. We think of such derivations as finite *trees*. The root of the tree contains the sequent to be derived. Immediately above the node of any given sequent is (are) the sequent(s) from which the given sequent is to be derived by some rule of inference. Axioms occur at the top nodes.

2.28. Examples. i. Here is a derivation of $\forall v_i \varphi v_i \rightarrow \forall v_j \varphi v_j$, where v_i is free for v_j in φv_j:

$$\begin{array}{cl}
\varphi v_j \Rightarrow \varphi v_j & \text{Axiom} \\ \hline
\forall v_i \varphi v_i \Rightarrow \varphi v_j & \forall L \\ \hline
\forall v_i \varphi v_i \Rightarrow \forall v_j \varphi v_j & \forall R \\ \hline
\Rightarrow \forall v_i \varphi v_i \rightarrow \forall v_j \varphi v_j \, . & \rightarrow R
\end{array}$$

ii. Here is a derivation of $\varphi \wedge \psi \rightarrow \neg(\neg\varphi \vee \neg\psi)$:

$$\begin{array}{ccl}
\varphi \Rightarrow \varphi & \psi \Rightarrow \psi & \text{Axioms} \\ \hline
\varphi \wedge \psi \Rightarrow \varphi & \varphi \wedge \psi \Rightarrow \psi & \wedge L \\ \hline
\varphi \wedge \psi, \neg\varphi \Rightarrow & \varphi \wedge \psi, \neg\psi \Rightarrow & \neg L \\ \hline
\multicolumn{2}{c}{\varphi \wedge \psi, \neg\varphi \vee \neg\psi \Rightarrow} & \vee L \\ \hline
\multicolumn{2}{c}{\varphi \wedge \psi \Rightarrow \neg(\neg\varphi \vee \neg\psi)} & \neg R \\ \hline
\multicolumn{2}{c}{\Rightarrow \varphi \wedge \psi \rightarrow \neg(\neg\varphi \vee \neg\psi) \, .} & \rightarrow R
\end{array}$$

iii. As a third example, here is a derivation of the "missing" equality axiom $\Gamma, s = t \Rightarrow \Delta, t = s$:

$$\begin{array}{ccl}
\Rightarrow s = s & s = t, s = s, s = s \Rightarrow t = s & E1, E4 \\ \hline
\Gamma \Rightarrow \Delta, s = s & s = t, s = s \Rightarrow t = s & E1, Cut \\ \hline
\multicolumn{2}{c}{\Gamma, s = t \Rightarrow \Delta, t = s \, .} & Cut
\end{array}$$

iv. Finally, let us demonstrate that equality axiom *E3* on the substitutability of equals for equals in atomic formulae holds also for negations of atomic formulae. For the sake of space, we do this for binary atomic formulae. Let

$$(1) \quad \Gamma, s_0 = t_0, t_1 = s_1, Rt_0t_1 \Rightarrow t_0 = s_0, \Delta, Rs_0s_1$$

be provable by iii,

$$(2)\quad \Gamma, t_0 = s_0, t_1 = s_1, Rt_0t_1 \Rightarrow \Delta, Rs_0s_1$$

be an instance of *E3*,

and $$(3)\quad \Gamma, s_0 = t_0, s_1 = t_1, Rt_0t_1 \Rightarrow t_1 = s_1, \Delta, Rs_0s_1$$

be derivable by iii. Then, putting demonstrations of (1) and (3) above these sequents, the following tree yields a derivation of $\Gamma, s_0 = t_0, s_1 = t_1, \neg Rs_0s_1 \Rightarrow \Delta, \neg Rt_0t_1$:

$$\dfrac{\dfrac{\dfrac{\dfrac{\dfrac{\vdots\ (1) \quad \vdots\ (2)}{\Gamma, s_0=t_0, t_1=s_1, Rt_0t_1 \Rightarrow \Delta, Rs_0s_1} \quad \vdots\ (3)}{\Gamma, s_0=t_0, s_1=t_1, Rt_0t_1 \Rightarrow \Delta, Rs_0s_1}\ Cut\ \ Cut}{\Gamma, s_0=t_0, s_1=t_1, Rt_0t_1, \neg Rs_0s_1 \Rightarrow \Delta}\ \neg L}{\Gamma, s_0=t_0, s_1=t_1, \neg Rs_0s_1 \Rightarrow \Delta, \neg Rt_0t_1}\ \neg R}{}$$

If we agree to write

$$\begin{array}{c}\mathcal{D}\\ \Gamma \Rightarrow \Delta\end{array}$$

to indicate that $\mathcal{D}$ is a derivation of $\Gamma \Rightarrow \Delta$, we can readily define the notion of a formal derivation.

2.29. Definition. The class of *formal derivations* (or, *formal proofs*) from a set $\mathcal{T}$ of axioms is inductively defined as follows:

i. every (logical, equality, or non-logical) axiom is a formal derivation of itself,

ii. if $\begin{array}{c}\mathcal{D}\\ \Gamma_1 \Rightarrow \Delta_1\end{array}$ is a formal derivation of $\Gamma_1 \Rightarrow \Delta_1$ and $\Gamma_2 \Rightarrow \Delta_2$ follows from $\Gamma_1 \Rightarrow \Delta_1$ by a one-premise rule of inference ($\wedge L$, $\vee R$, $\rightarrow R$, $\neg L$, $\neg R$, $\forall L$, $\forall R$, $\exists L$, $\exists R$, weakening, or contraction), then

$$\dfrac{\begin{array}{c}\mathcal{D}\\ \Gamma_1 \Rightarrow \Delta_1\end{array}}{\Gamma_2 \Rightarrow \Delta_2}$$

is a formal derivation of $\Gamma_2 \Rightarrow \Delta_2$,

iii. if $\begin{array}{c}\mathcal{D}_1\\ \Gamma_1 \Rightarrow \Delta_1\end{array}$ and $\begin{array}{c}\mathcal{D}_2\\ \Gamma_2 \Rightarrow \Delta_2\end{array}$ are formal derivations of $\Gamma_1 \Rightarrow \Delta_1$ and $\Gamma_2 \Rightarrow \Delta_2$, respectively, and $\Gamma \Rightarrow \Delta$ follows from $\Gamma_1 \Rightarrow \Delta_1$ and $\Gamma_2 \Rightarrow \Delta_2$ by a two-premise rule of inference ($\wedge R$, $\vee L$, $\rightarrow L$, or cut), then

$$\dfrac{\begin{array}{c}\mathcal{D}_1\\ \Gamma_1 \Rightarrow \Delta_1\end{array} \qquad \begin{array}{c}\mathcal{D}_2\\ \Gamma_2 \Rightarrow \Delta_2\end{array}}{\Gamma \Rightarrow \Delta}$$

is a formal derivation of $\Gamma \Rightarrow \Delta$.

2.30. Definitions. Let $\mathcal{T}$ be a set of formulae. We write

$$\mathcal{T} \vdash \Gamma \Rightarrow \Delta$$

if there is a formal derivation of the sequent $\Gamma \Rightarrow \Delta$ from the set $\mathcal{T}$ of axioms. For a formula φ, we write

$$\mathcal{T} \vdash \varphi$$

for $\mathcal{T} \vdash \ \Rightarrow \varphi$.

This, at long last, completes our definition of logic. At this point it is customary in logic texts to give formal derivations of all sorts of things. The reader will find a couple of examples in the exercises. The following theorem is useful in producing such formal derivations.

2.31. Deduction Theorem. *Let $\mathcal{T}$ be a theory, φ a sentence, and $\Gamma \Rightarrow \Delta$ a sequent. Then:*

$$\mathcal{T} \cup \{\varphi\} \vdash \Gamma \Rightarrow \Delta \ \textit{ iff } \ \mathcal{T} \vdash \Gamma, \varphi \Rightarrow \Delta.$$

I leave the proof to the reader as an exercise as I am not much taken with such an excess of syntax. In this, I side with that eminent logician of the University of Wittenberg, Professor Johann Faustus, who said

> *Bene disserere est finis logicis—*
> Is to dispute well logic's chiefest end?
> Affords this art no greater miracle?
> Then read no more, thou hast attained the end.
> A greater subject fitteth Faustus' wit.

If I continue, it is because logic does afford a greater miracle.

Exercises.

1. Define a parity function P on the individual symbols used to construct the terms of a first-order language L as follows:

$$P(v_i) = P(c_i) = -1$$
$$P(F_i^n) = n - 1.$$

Extend P to arbitrary strings $\sigma_0 ... \sigma_{m-1}$ of these symbols by setting

$$P(\sigma_0 ... \sigma_{m-1}) = P(\sigma_0) + ... + P(\sigma_{m-1}).$$

i. Prove: A string $\sigma_0 ... \sigma_{m-1}$ is a term iff

a. for $k < m$, $P(\sigma_0 ... \sigma_{k-1}) > -1$

and b. $P(\sigma_0 ... \sigma_{m-1}) = -1$.

ii. Prove the unique readability of terms: if $F_i^n t_0 ... t_{n-1}$ is $F_j^m s_0 ... s_{m-1}$, then $m = n$, $i = j$, and the terms t_0 and s_0, t_1 and s_1, ... coïncide.

iii. Show that recursions,

$$F(v_i) = G_0(v_i)$$
$$F(c_i) = G_1(c_i)$$
$$F(F_i^n t_0 ... t_{n-1}) = H(F_i^n, (F(t_0), ..., F(t_{n-1}))),$$

yield well-defined functions for given functions G_0, G_1, H.

[Remarks. Part iii involves both the unique readability of terms and a recursion on the lengths (i.e., numbers of symbols) of the terms. Because n varies, we use the sequence $(F(t_0), ..., F(t_{n-1}))$ of values of F as argument.]

2. As in Exercise 1, think of formulae a strings of symbols.
   i. State and prove a unique readability lemma for atomic formulae.
   ii. State and prove such a lemma for formulae in general.
   iii. State and prove a result on recursion analogous to that of Exercise 1.iii.
3. Prove the transitivity of equality:

$$\vdash \Gamma, s = t, t = u \Rightarrow \Delta, s = u.$$

4. Prove: For any formula $\varphi v_0...v_{n-1}$ and any terms $t_0, ..., t_{n-1}, s_0, ..., s_{n-1}$ with each t_i, s_i free for v_i in φ,

$$\vdash \Gamma, s_0 = t_0, ..., s_{n-1} = t_{n-1}, \varphi s_0...s_{n-1} \Rightarrow \varphi t_0...t_{n-1}.$$

[Hint. Induction on the construction of φ.]

5. Prove Theorem 2.31. [Hint. Use the cut and weakening rules.]
6. Let $\mathcal{T}$ be a set of axioms and $\Gamma \Rightarrow \Delta$ any sequent. Show that $\mathcal{T} \vdash \Gamma \Rightarrow \Delta$ iff $\vdash \mathcal{T}_0, \Gamma \Rightarrow \Delta$ for some finite $\mathcal{T}_0 \subseteq \mathcal{T}$.
7. i. Let φv_0 have v_0 free. Consider the two theories obtained by taking

$$\Rightarrow \varphi v_0 \quad \text{and} \quad \forall v_0 \varphi v_0,$$

respectively, as axioms. Show that the same sequents $\Gamma \Rightarrow \Delta$ are derivable in the two systems.

   ii. Repeat part i for φ having n free variables, i.e., consider the axioms,

$$\Rightarrow \varphi v_0...v_{n-1} \quad \text{and} \quad \forall v_0...v_{n-1}\varphi v_0...v_{n-1},$$

(using the obvious abbreviation).

   iii. Show that the non-logical axioms of any theory can all be assumed to be sentences.

[Hint. iii. Apply Exercise 6 and induction on the number of axioms in a finitely axiomatised theory.]

8. Define $\leftrightarrow$ to be the abbreviation

$$\leftrightarrow \varphi\psi: \quad \wedge \rightarrow \varphi\psi \rightarrow \psi\varphi$$

(i.e., $\varphi \leftrightarrow \psi$: $(\varphi \rightarrow \psi) \wedge (\psi \rightarrow \varphi)$).

   i. Derive the rules:

$$\leftrightarrow L \ \frac{\Gamma \Rightarrow \Delta, \varphi \quad \Gamma, \psi \Rightarrow \Delta}{\Gamma, \varphi \leftrightarrow \psi \Rightarrow \Delta} \qquad \frac{\Gamma, \varphi \Rightarrow \Delta \quad \Gamma \Rightarrow \Delta, \psi}{\Gamma, \varphi \leftrightarrow \psi \Rightarrow \Delta}$$

$$\leftrightarrow R \ \frac{\Gamma, \varphi \Rightarrow \Delta, \psi \quad \Gamma, \psi \Rightarrow \Delta, \varphi}{\Gamma \Rightarrow \Delta, \varphi \leftrightarrow \psi}.$$

   ii. Consider an expanded language containing a new 0-ary relation symbol P (in fancy parlance, a propositional constant). Give an inductive definition of $[\varphi/P]\chi$—the result of substituting φ for P throughout χ.

   iii. Prove (by induction on the generation of χ) that, if φ and ψ have no free variables bound in χ,

$$\vdash \Gamma, \varphi \leftrightarrow \psi, [\varphi/P]\chi \Rightarrow \Delta, [\psi/P]\chi.$$

Conclude

$$\vdash \varphi \leftrightarrow \psi \Rightarrow [\varphi/P]\chi \leftrightarrow [\psi/P]\chi.$$

3. The Completeness Theorem

The "greater subject" fitting our Faustian wits is not the formal manipulation of symbols, but the proof of the adequacy of such formal manipulation. That is, we wish to prove for any theory $\mathcal{T}$ and any sentence φ,

$$\mathcal{T} \vdash \varphi \text{ iff } \mathcal{T} \vDash \varphi.$$

Since the notion of entailment concerns sentences, while that of derivability concerns sequents of formulae with free variables, we must first extend our semantic notions to include such formulae and sequents.

3.1. Definitions. Let a language L be given. If $\mathfrak{A}$ is an L-structure and $\varphi v_{i_0}...v_{i_{k-1}}$ is a formula of L with free variables as shown, we say that φ is *true* in $\mathfrak{A}$, written

$$\mathfrak{A} \vDash \varphi v_{i_0}...v_{i_{k-1}},$$

if, for all $a_0, ..., a_{k-1} \in A$,

$$\mathfrak{A} \vDash \varphi\, \overline{a_0} \ldots \overline{a_{k-1}},$$

i.e.,

$$\mathfrak{A} \vDash \varphi v_{i_0}...v_{i_{k-1}} \text{ iff } \mathfrak{A} \vDash \forall v_{i_0}...v_{i_{k-1}}\varphi.$$

A set $\mathcal{T}$ of formulae is *true* in $\mathfrak{A}$ if every formula $\varphi \in \mathcal{T}$ is true in $\mathfrak{A}$, i.e.,

$$\mathfrak{A} \vDash \mathcal{T} \text{ iff } \forall \varphi \in \mathcal{T}(\mathfrak{A} \vDash \varphi).$$

A sequent $\Gamma \Rightarrow \Delta$ is *true* in $\mathfrak{A}$ if the formula $\bigwedge\!\!\bigwedge\Gamma \to \bigvee\!\!\bigvee\Delta$ is true in $\mathfrak{A}$, i.e.,

$$\mathfrak{A} \vDash \Gamma \Rightarrow \Delta \text{ iff } \mathfrak{A} \vDash \bigwedge\!\!\bigwedge\Gamma \to \bigvee\!\!\bigvee\Delta.$$

3.2. Definitions. Let $\mathcal{T}$ be a set of formulae of a given language L. $\mathcal{T}$ (semantically) *entails* a sequent $\Gamma \Rightarrow \Delta$ of L if $\Gamma \Rightarrow \Delta$ is true in all L-structures in which $\mathcal{T}$ is true, i.e.,

$$\mathcal{T} \vDash \Gamma \Rightarrow \Delta \text{ iff } \forall \mathfrak{A}(\mathfrak{A} \vDash \mathcal{T} \to \mathfrak{A} \vDash \Gamma \Rightarrow \Delta).$$

$\mathcal{T}$ *entails* a formula φ iff $\mathcal{T}$ *entails* the sequent $\Rightarrow \varphi$,

$$\mathcal{T} \vDash \varphi \text{ iff } \mathcal{T} \vDash\ \Rightarrow \varphi.$$

It is a routine exercise to show that, if φ is a sentence and $\mathcal{T}$ a set of sentences, the definition of entailment just given agrees with that of Definition 2.20.

The goal of the present section is the proof of the following theorem.

3.3. Completeness Theorem. *For any set $\mathcal{T}$ of sentences and any formula φ, $\mathcal{T} \vDash \varphi$ iff $\mathcal{T} \vdash \varphi$.*

This can be stated more generally for formulae, and φ can even be replaced by an arbitrary sequent. Indeed, half of this will have to be proven in such generality. However, the given statement is the form of greatest interest, so I state it thus.

Half of the Theorem is fairly routine and can be proven directly.

3.4. Soundness Lemma. *Let $\mathcal{T}$ be a set of formulae and $\Gamma \Rightarrow \Delta$ a sequent. Then*:

$$\mathcal{T} \vdash \Gamma \Rightarrow \Delta \;\rightarrow\; \mathcal{T} \models \Gamma \Rightarrow \Delta.$$

Proof. By induction on the number of inferences applied in a derivation of $\Gamma \Rightarrow \Delta$ from $\mathcal{T}$.

The basis consists of the axioms. These are of three kinds. If $\Gamma \Rightarrow \Delta$ is a logical axiom, then some formula ψ occurs in both Γ and Δ, whence

$$\models \bigwedge\Gamma \rightarrow \bigvee\Delta,$$

whence $\mathcal{T} \models \bigwedge\Gamma \rightarrow \bigvee\Delta$, i.e.,

$$\mathcal{T} \models \Gamma \Rightarrow \Delta.$$

If $\Gamma \Rightarrow \Delta$ is a non-logical axiom, it is of the form $\Rightarrow \psi$ for some $\psi \in \mathcal{T}$. But, for any model $\mathfrak{A}$,

$$\mathfrak{A} \models \mathcal{T} \;\rightarrow\; \forall\chi \in \mathcal{T}(\mathfrak{A} \models \chi) \;\rightarrow\; \mathfrak{A} \models \psi,$$

whence $\mathcal{T} \models \psi$, i.e.,

$$\mathcal{T} \models \;\Rightarrow \psi.$$

The most "difficult" case, i.e., the case actually involving some effort, is that in which the axiom is an equality axiom. I will treat one of these axioms and leave the others to the reader. Consider

$$\Gamma, s_0 = t_0, \ldots, s_{n-1} = t_{n-1}, R_i^n s_0 \ldots s_{n-1} \;\Rightarrow\; \Delta, R_i^n t_0 \ldots t_{n-1}. \qquad (*)$$

Let $\mathfrak{A}$ be a structure making the premise true under some interpretation $a_{i_0}, \ldots, a_{i_{k-1}}$ of the free variables $v_{i_0}, \ldots, v_{i_{k-1}}$ occurring in the sequent. If we assume the premise true in $\mathfrak{A}$, then the truth of the equations $s_j = t_j$ means that the values $(s_j)^{\mathfrak{A}}$ and $(t_j)^{\mathfrak{A}}$ of these terms are the same— say, b_j. But:

$$\begin{aligned}
\mathfrak{A} \models R_i^n s_0 \ldots s_{n-1} \;&\rightarrow\; (R_i^n)^{\mathfrak{A}}((s_0)^{\mathfrak{A}}, \ldots, (s_{n-1})^{\mathfrak{A}}) \\
&\rightarrow\; (R_i^n)^{\mathfrak{A}}(b_0, \ldots, b_{n-1}) \\
&\rightarrow\; (R_i^n)^{\mathfrak{A}}((t_0)^{\mathfrak{A}}, \ldots, (t_{n-1})^{\mathfrak{A}}) \\
&\rightarrow\; \mathfrak{A} \models R_i^n t_0 \ldots t_{n-1}.
\end{aligned}$$

Thus,

$$\mathfrak{A} \models \bigwedge\Gamma \wedge \bigwedge s_j = t_j \wedge R_i^n s_0 \ldots s_{n-1} \;\rightarrow\; \bigvee\Delta \vee R_i^n t_0 \ldots t_{n-1},$$

and the sequent (*) is true in all models, whence in all models of $\mathcal{T}$.

As mentioned, I leave the other equality axioms as an exercise to the reader.

The induction step has even more cases than the basis; there is a case for each rule of inference. In each case one must prove that if $\mathcal{T}$ entails the premise(s) of an instance of a given rule of inference, then $\mathcal{T}$ entails the conclusion of the instance. Counting $\wedge L$ and $\vee R$ as two rules apiece, there are 19 rules to handle, i.e., 19 cases in all. I shall treat only two— one single-premise rule and one two-premise rule— and leave the rest to the reader as additional exercises.

The single premise rule I shall treat is $\forall R$ as it brings out most clearly the reason for the choice made in Definitions 3.1 and 3.2 that

$$\mathfrak{A} \models \varphi v_0 \dots v_{n-1}$$

should mean the same as

$$\mathfrak{A} \models \forall v_0 \dots v_{n-1} \varphi v_0 \dots v_{n-1}.$$

Suppose

$$\mathcal{T} \vdash \Gamma \Rightarrow \Delta, \varphi v_j, \tag{1}$$

where v_j does not occur free in any formula of Γ or Δ. Let $\mathfrak{A} \models \mathcal{T}$ and suppose

$$\mathfrak{A} \not\models \Gamma \Rightarrow \Delta, \forall v_i \varphi v_i. \tag{2}$$

This means that

$$\mathfrak{A} \not\models \Gamma(\overline{a_0}, \dots, \overline{a_{k-1}}) \Rightarrow \Delta(\overline{a_0}, \dots, \overline{a_{k-1}}), \forall v_i \varphi(v_i, \overline{a_0}, \dots, \overline{a_{k-1}}), \tag{3}$$

for some replacement $\overline{a_0}, \dots, \overline{a_{k-1}}$ for the free variables of the sequent. Now, (3) consists of sentences of $L(A)$, whence it simply means

$$\mathfrak{A} \models \bigwedge\!\!\!\bigwedge \Gamma(\overline{a_0}, \dots, \overline{a_{k-1}}) \tag{4}$$

and

$$\mathfrak{A} \not\models \bigvee\!\!\!\bigvee \Delta(\overline{a_0}, \dots, \overline{a_{k-1}}) \vee \forall v_i \varphi(v_i, \overline{a_0}, \dots, \overline{a_{k-1}}), \tag{5}$$

whence, in particular,

$$\mathfrak{A} \not\models \bigvee\!\!\!\bigvee \Delta(\overline{a_0}, \dots, \overline{a_{k-1}}). \tag{6}$$

By (1) and (4),

$$\mathfrak{A} \models \bigvee\!\!\!\bigvee \Delta(\overline{a_0}, \dots, \overline{a_{k-1}}) \vee \varphi(v_j, \overline{a_0}, \dots, \overline{a_{k-1}}),$$

i.e.,

$$\mathfrak{A} \models \bigvee\!\!\!\bigvee \Delta(\overline{a_0}, \dots, \overline{a_{k-1}}) \vee \varphi(\overline{a}, \overline{a_0}, \dots, \overline{a_{k-1}}) \tag{7}$$

for any $a \in A$. By (6) and (7), we conclude

$$\forall a \in A \, (\mathfrak{A} \models \varphi(\overline{a}, \overline{a_0}, \dots, \overline{a_{k-1}})),$$

i.e.,

$$\mathfrak{A} \models \forall v_i \varphi(v_i, \overline{a_0}, \dots, \overline{a_{k-1}}),$$

which contradicts (5). Thus, our assumption (2) was incorrect and

$$\mathfrak{A} \models \Gamma \Rightarrow \Delta, \forall v_i \varphi v_i.$$

As a representative two-premise rule, I take $\vee L$. Suppose

$$\mathcal{T} \models \Gamma, \varphi \Rightarrow \Delta \tag{8}$$

and $$\mathcal{T} \models \Gamma, \psi \Rightarrow \Delta. \tag{9}$$

Let $\mathfrak{A} \models \mathcal{T}$ and choose some replacement of free variables by constants so that

$$\mathfrak{A} \models \bigwedge\Gamma \wedge (\varphi \vee \psi). \tag{10}$$

Because $\mathfrak{A} \models \varphi \vee \psi$, we have $\mathfrak{A} \models \varphi$ or $\mathfrak{A} \models \psi$. In the first case, (8) yields

$$\mathfrak{A} \models \bigwedge\Gamma \wedge \varphi \rightarrow \bigvee\Delta,$$

whence

$$\mathfrak{A} \models \bigvee\Delta.$$

In the second case, (9) yields

$$\mathfrak{A} \models \bigwedge\Gamma \wedge \psi \rightarrow \bigvee\Delta,$$

whence again

$$\mathfrak{A} \models \bigvee\Delta.$$

Either way, we have $\mathfrak{A} \models \bigvee\Delta$ on our assumption (10), i.e., we have shown

$$(\mathfrak{A} \models \bigwedge\Gamma \wedge (\varphi \vee \psi)) \rightarrow (\mathfrak{A} \models \bigvee\Delta),$$

i.e., $$\mathfrak{A} \models \bigwedge\Gamma \wedge (\varphi \vee \psi) \rightarrow \bigvee\Delta,$$

i.e., $$\mathfrak{A} \models \Gamma, \varphi \vee \psi \Rightarrow \Delta.$$ QED

With the Soundness Lemma, we quickly derive one implication of the Completeness Theorem:

$$\begin{aligned} \mathcal{T} \vdash \varphi &\rightarrow \mathcal{T} \vdash \Rightarrow \varphi, \text{ by definition} \\ &\rightarrow \mathcal{T} \models \Rightarrow \varphi, \text{ by Soundness} \\ &\rightarrow \mathcal{T} \models \varphi, \end{aligned}$$

by the definition of entailment for sequents. The converse implication,

$$\mathcal{T} \models \varphi \rightarrow \mathcal{T} \vdash \varphi,$$

is proven contrapositively. One assumes $\mathcal{T} \nvdash \varphi$ and constructs a model of $\mathcal{T} + \neg\varphi$ (i.e., $\mathcal{T} \cup \{\neg\varphi\}$), thereby showing $\mathcal{T} \not\models \varphi$. The construction of the model proceeds in three steps. First, one saturates the theory $\mathcal{T} + \neg\varphi$ with constants so that any *completion* of it will yield a model. Second, one completes the newly saturated theory. Finally, one transforms the completion into a model.

As is usually the case with important constructions, the present ones begin with some definitions.

3.5. Definitions. Let $\mathcal{S}$ be a theory (i.e., a set of axioms— cf. Definition 2.27) in a language L. A constant c of L is said to *witness* a sentence $\exists v_i \varphi v_i$ of L if

$$\mathcal{S} \vdash \exists v_i \varphi v_i \rightarrow \varphi c.$$

$\mathcal{S}$ is said to *have witnesses* if every sentence of the form $\exists v_i \varphi v_i$ has a witnessing constant.

The first construction underlying the proof of the Completeness Theorem is the extension of a given theory to one which has witnesses.

3.6. Definitions. Let $L_1 \subseteq L_2$ be two languages and let $\mathcal{S}_1, \mathcal{S}_2$ be theories in these languages. $\mathcal{S}_2$ is an *extension* of $\mathcal{S}_1$ if, for any sentence φ of L_1,

$$\mathcal{S}_1 \vdash \varphi \rightarrow \mathcal{S}_2 \vdash \varphi.$$

An extension $\mathcal{S}_2$ of $\mathcal{S}_1$ is a *conservative extension* of $\mathcal{S}_1$ if the converse holds: for any sentence φ of L_1,

$$\mathcal{S}_2 \vdash \varphi \rightarrow \mathcal{S}_1 \vdash \varphi.$$

3.7. Witnessing Lemma. *Let $\mathcal{S}$ be a theory. $\mathcal{S}$ has a conservative extension which has witnesses.*

Proof. Let $\mathcal{S}$ in a language L be given. Define a new language $L(E)$ by adding to L a denumerable set E of new constants,

$$e_0, e_1, \ldots$$

By the denumerability of the set of primitives of $L(E)$, there is an enumeration,

$$\varphi_0, \varphi_1, \ldots$$

of sentences of the form $\exists v_i \psi v_i$ of $L(E)$.

We will define a sequence of languages,

$$L = L_0 \subseteq L_1 \subseteq \ldots \subseteq L(E) = \cup L_n,$$

each L_i being of the form $L(E_i)$ with

$$\varnothing = E_0 \subseteq E_1 \subseteq \ldots \subseteq E = \cup E_n,$$

a sequence of finite sets of constants. Along with this we will define a sequence of theories,

$$\mathcal{S} = \mathcal{S}_0 \subseteq \mathcal{S}_1 \subseteq \ldots,$$

with each $\mathcal{S}_i$ in the language L_i. We begin with

$$E_0 = \varnothing \text{ and } \mathcal{S}_0 = \mathcal{S}.$$

Given $L_n = L(E_n)$ and $\mathcal{S}_n$, consider $\varphi_n = \exists v_i \psi_n v_i$. Let e_m be the first element of E that is not in E_n and does not occur in $\psi_n v_i$. Define E_{n+1} to be the union of E_n, the constants of E occurring in $\psi_n v_i$, and $\{e_m\}$, and define $\mathcal{S}_{n+1}$ to be

$$\mathcal{S}_n \cup \{\exists v_i \psi_n v_i \rightarrow \psi_n e_m\}.$$

Finally, let

$$\mathcal{S}_\omega = \cup_n \mathcal{S}_n.$$

$\mathcal{S}_\omega$ is the conservative extension of $\mathcal{S}$ that we are looking for. That it is an extension of $\mathcal{S}$ is obvious. To see that the extension is conservative over $\mathcal{S}$, note first that each $\mathcal{S}_{n+1}$ is conservative over $\mathcal{S}_n$: if $\Gamma \Rightarrow \Delta$ is a sequent of L_n, then

$$\mathcal{S}_{n+1} \vdash \Gamma \Rightarrow \Delta \;\rightarrow\; \vdash \mathcal{S}'_{n+1}, \Gamma \Rightarrow \Delta,$$

for some finite $\mathcal{S}'_{n+1} \subseteq \mathcal{S}_{n+1}$ (by Exercise 6 of the preceding section). Letting $\mathcal{S}'_n$ = $\mathcal{S}'_{n+1} - \{\exists v_i \psi_n v_i \rightarrow \psi_n e_m\}$, this means

$$\vdash \mathcal{S}'_n, \Gamma, \exists v_i \psi_n v_i \rightarrow \psi_n e_m \Rightarrow \Delta.$$

Replacing e_m throughout the derivation— say, $\mathcal{D}(e_m)$— of this sequent by a variable v_k occurring nowhere in the derivation results in a new derivation $\mathcal{D}(v_k)$ of the new sequent

$$\vdash \mathcal{S}'_n, \Gamma, \exists v_i \psi_n v_i \rightarrow \psi_n v_k \Rightarrow \Delta.$$

(The reason is that every logical axiom of $\mathcal{D}(e_m)$ applies as well to v_k as to e_m and every rule of inference applicable to constants is applicable to variables.) But one can apply $\exists L$ to this last sequent to derive

$$\vdash \mathcal{S}'_n, \Gamma, \exists v_k (\exists v_i \psi_n v_i \rightarrow \psi_n v_k) \Rightarrow \Delta. \tag{11}$$

On the other hand, it is possible to derive the sequent

$$\Rightarrow \exists v_k (\exists v_i \psi_n v_i \rightarrow \psi_n v_k), \tag{12}$$

(cf. Exercise 4, below). Applying cut to (11) and (12) gives us a derivation of

$$\vdash \mathcal{S}'_n, \Gamma \Rightarrow \Delta,$$

whence (the already cited Exercise yields)

$$\mathcal{S}_n \vdash \Gamma \Rightarrow \Delta.$$

We can now readily verify that $\mathcal{S}_\omega$ is conservative over $\mathcal{S} = \mathcal{S}_0$. Let $\Gamma \Rightarrow \Delta$ be a sequent of $L = L_0$ such that

$$\mathcal{S}_\omega \vdash \Gamma \Rightarrow \Delta.$$

Applying Exercise 6 of the preceding section a third time, let $\mathcal{S}'_\omega$ be a finite subset of $\mathcal{S}_\omega$ such that

$$\vdash \mathcal{S}'_\omega, \Gamma \Rightarrow \Delta.$$

Then $\mathcal{S}'_\omega \subseteq \mathcal{S}'_n$ for some n, i.e.,

$$\mathcal{S}_n \vdash \Gamma \Rightarrow \Delta.$$

Let n be minimum such that this holds. I claim that $n = 0$. For, otherwise $n = m + 1$ for some m and, as we have just seen,

$$\mathcal{S}_{m+1} \vdash \Gamma \Rightarrow \Delta \;\rightarrow\; \mathcal{S}_m \vdash \Gamma \Rightarrow \Delta,$$

since $\Gamma \Rightarrow \Delta$ is in $L_0 \subseteq L_m$.

To complete the proof, it remains only to show that $\mathcal{S}$ has witnesses. But this is fairly easy. Let $\varphi = \exists v_i \psi v_i$ be an existentially quantified sentence of the language $L(D)$ of $\mathcal{S}_\omega$. φ is φ_n for some n, whence there is an m such that

$$\exists v_i \psi v_i \rightarrow \psi e_m \in \mathcal{S}_{n+1} \subseteq \mathcal{S}_\omega. \qquad \text{QED}$$

The second task is to show how a given theory can be completed. This requires a couple of preliminary definitions.

3.8. Definition. A theory $\mathcal{S}$ in a language L is *consistent* if for no sentence $\varphi \in L$ does $\mathcal{S} \vdash \varphi \wedge \neg\varphi$. $\mathcal{S}$ is *inconsistent* if it is not consistent.

The following simple result may provide some orientation.

3.9. Fact. *Let $\mathcal{S}$ be a theory in a language L. The following are equivalent:*
i. *$\mathcal{S}$ is inconsistent*
ii. *for some sentence $\varphi \in L$, $\mathcal{S} \vdash \varphi \wedge \neg\varphi$*
iii. *for some sentence $\varphi \in L$, $\mathcal{S} \vdash \varphi$ and $\mathcal{S} \vdash \neg\varphi$*
iv. *for all sentences $\varphi \in L$, $\mathcal{S} \vdash \varphi$*
v. *$\mathcal{S}$ proves the empty sequent: $\mathcal{S} \vdash \Rightarrow$.*

I leave the proof as an easy exercise for the reader.
A second definition is that of a *complete* theory.

3.10. Definition. Let $\mathcal{S}$ be a theory in a given language L. $\mathcal{S}$ is *complete* if, for every sentence φ of L, either $\mathcal{S} \vdash \varphi$ or $\mathcal{S} \vdash \neg\varphi$.

The obvious examples of complete theories are theories of given models:

$$Th(\mathfrak{A}) = \{\varphi: \mathfrak{A} \models \varphi\}.$$

The Completeness Theorem will tell us that these are the only consistent, complete theories (albeit this is not the only way of *presenting* such theories).

The following lemma is the key formal step in constructing a *completion* (i.e., complete extension) of a given theory.

3.11. Lemma. *Let $\mathcal{S}$ be a consistent theory in a language L and let φ be a sentence of L. Then: At least one of $\mathcal{S} + \varphi$ and $\mathcal{S} + \neg\varphi$ is consistent.*

Proof. Let $\perp$ be any convenient contradiction of the form $\psi \wedge \neg\psi$. By the Deduction Theorem (2.31), if both $\mathcal{S} + \varphi$ and $\mathcal{S} + \neg\varphi$ are inconsistent, there are derivations from the axioms of $\mathcal{S}$ of the forms

$$\begin{array}{c}\mathcal{D}_1\\ \varphi \Rightarrow \perp\end{array} \qquad\qquad \begin{array}{c}\mathcal{D}_2\\ \neg\varphi \Rightarrow \perp\end{array}.$$

We can put these together to obtain a derivation,

$$\dfrac{\dfrac{\begin{array}{c}\mathcal{D}_1\\ \varphi \Rightarrow \perp\end{array} \quad \begin{array}{c}\mathcal{D}_2\\ \neg\varphi \Rightarrow \perp\end{array}}{\varphi \vee \neg\varphi \Rightarrow \perp} \qquad \dfrac{\dfrac{\varphi \Rightarrow \varphi}{\Rightarrow \varphi, \neg\varphi}}{\Rightarrow \varphi \vee \neg\varphi}}{\Rightarrow \perp},$$

of $\perp$ from the axioms of $\mathcal{S}$, concluding the inconsistency of $\mathcal{S}$ (where the double horizontal lines indicate the omission of several steps). This contrapositively establishes the Lemma. QED

With this, we can now perform the second step in the proof of the Completeness Theorem.

3.12. Completion Lemma. *Let $\mathcal{S}$ be a consistent theory. Then: $\mathcal{S}$ has a consistent, complete extension.*

Proof. Let L be the language of $\mathcal{S}$ and let $\varphi_0, \varphi_1, \ldots$ be an enumeration of all sentences of L. Define a sequence of theories in L,

$$\mathcal{S} = \mathcal{S}_0 \subseteq \mathcal{S}_1 \subseteq \ldots \subseteq \cup_n \mathcal{S}_n = \mathcal{S}_\omega,$$

by

$$\mathcal{S}_0 = \mathcal{S}$$

$$\mathcal{S}_{n+1} = \begin{cases} \mathcal{S}_n \cup \{\varphi_n\}, & \text{if } \mathcal{S}_n + \varphi_n \text{ is consistent} \\ \mathcal{S}_n \cup \{\neg\varphi_n\}, & \text{otherwise.} \end{cases}$$

The claim is that $\mathcal{S}_\omega = \cup_n \mathcal{S}_n$ is a consistent, complete extension of $\mathcal{S}$.

By assumption, $\mathcal{S}_0 = \mathcal{S}$ is consistent. By Lemma 3.11, consistency is preserved in the step from $\mathcal{S}_n$ to $\mathcal{S}_{n+1}$, whence an induction yields the consistency of each $\mathcal{S}_n$. $\mathcal{S}_\omega$ must also be consistent since any derivation of $\perp$ would involve only finitely many axioms of $\mathcal{S}_\omega$, whence it would be a derivation from some sub-theory $\mathcal{S}_n$— which derivation cannot exist by the consistency of $\mathcal{S}_n$.

Finally, to see the completeness of $\mathcal{S}_\omega$, let φ be any sentence of L. φ must occur in the list $\varphi_0, \varphi_1, \ldots$ of all sentences of L, whence φ is φ_n for some n. But then either $\varphi \in \mathcal{S}_{n+1} \subseteq \mathcal{S}_\omega$ or $\neg\varphi \in \mathcal{S}_{n+1} \subseteq \mathcal{S}_\omega$, whence

$$\mathcal{S}_\omega \vdash \varphi \quad \text{or} \quad \mathcal{S}_\omega \vdash \neg\varphi.$$

QED

The next lemma is almost exactly what is needed in the final step in constructing a model out of a consistent complete theory having witnesses.

3.13. Lemma. *Let $\mathcal{S}$ be a consistent, complete theory in a language L. For any sentences φ, ψ of L,*

i. $\mathcal{S} \vdash \neg\varphi$ *iff* $\mathcal{S} \nvdash \varphi$

ii. $\mathcal{S} \vdash \varphi \wedge \psi$ *iff* $(\mathcal{S} \vdash \varphi) \wedge (\mathcal{S} \vdash \psi)$

iii. $\mathcal{S} \vdash \varphi \vee \psi$ *iff* $(\mathcal{S} \vdash \varphi) \vee (\mathcal{S} \vdash \psi)$

iv. $\mathcal{S} \vdash \varphi \to \psi$ *iff* $(\mathcal{S} \vdash \varphi) \to (\mathcal{S} \vdash \psi)$.

If $\mathcal{S}$ has witnesses, then, for φv_i with only v_i free,

v. $\mathcal{S} \vdash \exists v_i \varphi v_i$ *iff* $\exists$ *constant* $c \in L\ (\mathcal{S} \vdash \varphi c)$

vi. $\mathcal{S} \vdash \forall v_i \varphi v_i$ *iff* $\forall$ *constants* $c \in L\ (\mathcal{S} \vdash \varphi c)$.

Proof sketch. Since there are so many clauses to prove, I shall only treat a few (i, iii, and v) and leave the rest to the reader.

i. $\Leftarrow$. Trivial by completeness: $\mathcal{S} \nvdash \varphi \;\to\; \mathcal{S} \vdash \neg\varphi$.

→. Suppose $\mathcal{S} \vdash \varphi$ and $\mathcal{S} \vdash \neg\varphi$. Then we have derivations,

$$\begin{array}{c}\mathcal{D}_1\\ \Rightarrow \neg\varphi\end{array} \quad \text{and} \quad \begin{array}{c}\mathcal{D}_2\\ \Rightarrow \varphi\end{array},$$

which by $\wedge R$ yield $\mathcal{S} \vdash \varphi \wedge \neg\varphi$, contrary to the consistency of $\mathcal{S}$.

iii. ←. Again, this is the easy direction. $\mathcal{S} \vdash \varphi$ or $\mathcal{S} \vdash \psi$ yields $\mathcal{S} \vdash \varphi \vee \psi$ by an application of $\vee R$.

→. Suppose $\mathcal{S} \vdash \varphi \vee \psi$, but $\mathcal{S} \nvdash \varphi$ and $\mathcal{S} \nvdash \psi$. By completeness, $\mathcal{S} \vdash \neg\varphi$ and $\mathcal{S} \vdash \neg\psi$ with derivations $\mathcal{D}_1$, $\mathcal{D}_2$, respectively. Let $\mathcal{D}$ be the derivation of $\varphi \vee \psi$, and $\perp$ the unwanted contradiction. We derive the inconsistency of $\mathcal{S}$ as follows:

$$\dfrac{\dfrac{\dfrac{\dfrac{\varphi \Rightarrow \varphi, \perp}{\varphi, \neg\varphi \Rightarrow \perp} \qquad \begin{array}{c}\mathcal{D}_1\\ \Rightarrow \neg\varphi\end{array}}{\varphi \Rightarrow \perp} \qquad \dfrac{\dfrac{\psi \Rightarrow \psi, \perp}{\psi, \neg\psi \Rightarrow \perp} \qquad \begin{array}{c}\mathcal{D}_2\\ \Rightarrow \neg\psi\end{array}}{\psi \Rightarrow \perp}}{\varphi \vee \psi \Rightarrow \perp} \qquad \begin{array}{c}\mathcal{D}\\ \Rightarrow \varphi \vee \psi\end{array}}{\Rightarrow \perp}$$

v. ←. One application of $\exists R$ transforms any proof of $\Rightarrow\ \varphi c$ from $\mathcal{S}$ into one of $\exists v_i \varphi v_i$.

→. Let $\mathcal{D}_1$ be a proof of $\exists v_i \varphi v_i$ and $\mathcal{D}_2$ a proof of a witnessing assertion $\exists v_i \varphi v_i \rightarrow \varphi c$ for some constant c. The following is a proof of φc from $\mathcal{S}$.

$$\dfrac{\begin{array}{c}\mathcal{D}_1\\ \Rightarrow \exists v_i \varphi v_i\end{array} \qquad \dfrac{\dfrac{\exists v_i \varphi v_i, \varphi c \Rightarrow \varphi c \qquad \exists v_i \varphi v_i \Rightarrow \exists v_i \varphi v_i, \varphi c}{\exists v_i \varphi v_i \rightarrow \varphi c, \exists v_i \varphi v_i \Rightarrow \varphi c} \qquad \begin{array}{c}\mathcal{D}_2\\ \Rightarrow \exists v_i \varphi v_i \rightarrow \varphi c\end{array}}{\exists v_i \varphi v_i \Rightarrow \varphi c}}{\Rightarrow\ \varphi\, c}$$

This completes the proof of the lemma. QED

The various clauses of Lemma 3.13 read like those of a truth definition for a structure. The only thing missing is the description of the structure— the designation of domain and the interpretation of the function and relation symbols of the underlying language. Clauses v and vi suggest taking the set of constants of the language as the domain. This almost works; indeed, if one does not insist on interpreting the equality symbol of the language by actual identity, it does work. If equality is to be interpreted by identity, one must identify some constants.

3.14. Definition. Let $\mathcal{S}$ be a consistent, complete theory which has witnesses in a language L. The *term model* $\mathfrak{M}_{\mathcal{S}}$ of $\mathcal{S}$ is defined as follows. The domain M of $\mathfrak{M}_{\mathcal{S}}$ consists of the equivalence classes of constants,

$$[c] = \{d \in L\colon\ \mathcal{S} \vdash c = d\},$$

modulo the relation of provable equality. The interpretation of the equality symbol is identity. Given a function symbol F^n_i, its interpretation $(F^n_i)^{\mathfrak{M}}$, is the function defined by

$$(F_i^n)^{\mathfrak{M}}([d_0],\ldots,[d_{n-1}]) = [d] \text{ iff } \mathcal{S} \vdash F_i^n d_0 \ldots d_{n-1} = d.$$

The interpretation of a relation symbol R_i^n is the relation defined by

$$(R_i^n)^{\mathfrak{M}}([d_0],\ldots,[d_{n-1}]) \text{ iff } \mathcal{S} \vdash R_i^n d_0 \ldots d_{n-1}.$$

3.15. Lemma. *Let $\mathcal{S}$ be a consistent, complete theory with witnesses in a given language L.*

i. *The functions $(F_i^n)^{\mathfrak{M}}$ and relations $(R_i^n)^{\mathfrak{M}}$ of the term model are well-defined.*

ii. *For any variable-free terms s, t of the language L,*

$$\mathfrak{M}_{\mathcal{S}} \models s = t \text{ (i.e., } s^{\mathfrak{M}} = t^{\mathfrak{M}}) \text{ iff } \mathcal{S} \vdash s = t.$$

iii. *For any sentence φ of the language L,*

$$\mathfrak{M}_{\mathcal{S}} \models \varphi \text{ iff } \mathcal{S} \vdash \varphi;$$

in particular, $\mathfrak{M}_{\mathcal{S}}$ is a model of $\mathcal{S}$.

This is the last lemma to be proven before we return to the proof of the Completeness Theorem, which latter proof will merely consist of putting together all that we have done on proving this lemma. The use of function symbols leads to terms more complicated than mere constants and, consequently, complicates the proof of the Lemma. On a first reading, the reader might wish to skip the steps corresponding to function symbols and thereby study the simpler proof for theories with only relation symbols.

Proof of 3.15. i. The well-definedness of the interpretations of relation symbols is merely the requirement that the holding or non-holding of a relation of certain equivalence classes not depend on the particular representatives chosen: if

$$[d_0] = [e_0], \ldots, [d_{n-1}] = [e_{n-1}],$$

then the conditions defining

$$(R_i^n)^{\mathfrak{M}}([d_0],\ldots,[d_{n-1}]): \ \mathcal{S} \vdash R_i^n d_0 \ldots d_{n-1}$$
$$(R_i^n)^{\mathfrak{M}}([e_0],\ldots,[e_{n-1}]): \ \mathcal{S} \vdash R_i^n e_0 \ldots e_{n-1}$$

either both hold or fail to hold. But this follows from the equality axioms,

$$\mathcal{S} \vdash d_0 = e_0, \ldots, d_{n-1} = e_{n-1}, R_i^n d_0 \ldots d_{n-1} \Rightarrow R_i^n e_0 \ldots e_{n-1}$$
$$\mathcal{S} \vdash d_0 = e_0, \ldots, d_{n-1} = e_{n-1}, R_i^n e_0 \ldots e_{n-1} \Rightarrow R_i^n d_0 \ldots d_{n-1}.$$

The corresponding aspect of well-definedness of the function interpretations similarly rests on equality axioms. For functions, there is also the question of their being defined at all, i.e., of there being a value. Equality axiom *E1* and $\exists R$ yield

$$\mathcal{S} \vdash \exists v_j (F_i^n d_0 \ldots d_{n-1} = v_j).$$

If d witnesses this, i.e., if

$$\mathcal{S} \vdash \exists v_j (F_i^n d_0 \ldots d_{n-1} = v_j) \rightarrow F_i^n d_0 \ldots d_{n-1} = d,$$

then we readily see that

$$\mathcal{S} \vdash F_i^n d_0 \ldots d_{n-1} = d,$$

whence

$$(F_i^n)^{\mathfrak{M}}([d_0],...,[d_{n-1}]) = [d]$$

exists.

ii. This is an induction on the total number of function symbols occurring in the terms s and t.

The basis step treats equations involving constants. This case follows by definition:

$$c^{\mathfrak{M}} = d^{\mathfrak{M}} \text{ iff } [c] = [d] \text{ iff } S \vdash c = d.$$

For the induction step, one of the terms s, t of the given equation $s = t$ has an outermost function symbol. Assume for the sake of definiteness that it is s, i.e., s is a term,

$$F_i^n t_0 ... t_{n-1},$$

and we are looking at an equation,

$$F_i^n t_0 ... t_{n-1} = t.$$

As in part i, there are constants d_0, ..., d_{n-1}, d such that

$$S \vdash t_i = d_i, \quad S \vdash t = d.$$

By part i,

$$\begin{aligned} S \vdash F_i^n d_0 ... d_{n-1} = d \quad &\text{iff} \quad (F_i^n)^{\mathfrak{M}}([d_0],...,[d_{n-1}]) = [d] \\ &\text{iff} \quad \mathfrak{M}_S \models F_i^n d_0 ... d_{n-1} = d. \end{aligned}$$

By induction hypothesis,

$$\begin{aligned} S \vdash t_i = d_i \quad &\text{iff} \quad \mathfrak{M}_S \models t_i = d_i \\ S \vdash t = d \quad &\text{iff} \quad \mathfrak{M}_S \models t = d. \end{aligned}$$

Finally, playing with the equality axioms yields

$$\begin{aligned} S \vdash F_i^n d_0 ... d_{n-1} = d \quad &\text{iff} \quad S \vdash F_i^n t_0 ... t_{n-1} = t \\ \mathfrak{M}_S \models F_i^n d_0 ... d_{n-1} = d \quad &\text{iff} \quad \mathfrak{M}_S \models F_i^n t_0 ... t_{n-1} = t. \end{aligned}$$

Thus,

$$\begin{aligned} S \vdash F_i^n t_0 ... t_{n-1} = t \quad &\text{iff} \quad S \vdash F_i^n d_0 ... d_{n-1} = d \\ &\text{iff} \quad \mathfrak{M}_S \models F_i^n d_0 ... d_{n-1} = d \\ &\text{iff} \quad \mathfrak{M}_S \models F_i^n t_0 ... t_{n-1} = t. \end{aligned}$$

iii. This is an induction on the complexity of φ. The basis is handled by i and ii. If φ is an equation, ii directly yields

$$S \vdash \varphi \text{ iff } \mathfrak{M}_S \models \varphi.$$

If φ is of the form $R_i^n t_0 ... t_{n-1}$, let d_0, ..., d_{n-1} be constants such that

$$S \vdash t_i = d_i.$$

Observe,

$$S \vdash R_i^n t_0 ... t_{n-1} \quad \text{iff} \quad S \vdash R_i^n d_0 ... d_{n-1}, \quad \text{by equality axioms}$$

$$\text{iff} \quad \mathfrak{M}_{\mathcal{S}} \models R_i^n d_0 \ldots d_{n-1}, \text{ by part i}$$
$$\text{iff} \quad \mathfrak{M}_{\mathcal{S}} \models R_i^n t_0 \ldots t_{n-1},$$

again by equality axioms.

The induction step follows directly by Lemma 3.13. QED

The Completeness Theorem is now within easy reach.

Proof of the Completeness Theorem. Assume $\mathcal{T} \nvdash \varphi$.

I claim first that $\mathcal{T} + \neg\varphi$ is consistent. Otherwise, by the Deduction Theorem (2.31), $\mathcal{T} \vdash \neg\varphi \Rightarrow$. Letting $\mathcal{D}$ be this derivation, we get the following derivation of φ from $\mathcal{T}$.

$$\dfrac{\dfrac{\dfrac{\varphi \Rightarrow \varphi}{\Rightarrow \varphi, \neg\varphi}}{\neg\neg\varphi \Rightarrow \varphi} \qquad \dfrac{\begin{matrix}\mathcal{D}\\ \neg\varphi \Rightarrow\end{matrix}}{\Rightarrow \neg\neg\varphi}}{\Rightarrow \varphi.}$$

Apply the Witnessing Lemma to $\mathcal{T} + \neg\varphi$ to construct a consistent extension $\mathcal{S}_0 \supseteq \mathcal{T} + \neg\varphi$ which has witnesses. Apply the Completion Lemma to construct a consistent completion $\mathcal{S}$ of $\mathcal{S}_0$. The theory $\mathcal{S}$ still has witnesses, whence we can apply the most recent lemma (Lemma 3.15) to obtain a model $\mathfrak{M}_{\mathcal{S}}$ of $\mathcal{S}$. Since $\mathcal{S}$ contains $\mathcal{T} + \neg\varphi$, we have

$$\mathfrak{M}_{\mathcal{S}} \models \mathcal{T}, \quad \mathfrak{M}_{\mathcal{S}} \not\models \varphi.$$

Thus, $\mathcal{T} \not\models \varphi$. QED

We have at long last accomplished what we set out to do at the beginning of the last section: we have given syntactic notions of language and proof semi-effectively adequate for the corresponding semantic notions of entailment and validity. By "semi-effective" I mean the obvious property of the proof-system that it offers an effective means of generating (enumerating) all logical consequences of any effectively given set of non-logical axioms. [In the first chapter of the next volume, we shall make this more precise: we will assign numerical codes to syntactic objects and show that theories with recursively enumerable sets of (codes of) axioms have recursively enumerable sets of (codes of) theorems.]

The adjectives "semi-effective" and "adequate" hardly appear to be fair descriptions of the proof procedure. The rules are, after all, extremely natural and, as the reader who attempts some formal derivations will readily testify, easy to use. In proof theory, the naturalness of the system is a major fact (indeed, this will be of use in Chapter IV of volume II). For the purposes of the present volume, however, it is the semi-effectiveness and adequacy that are most relevant.

Two immediate corollaries of the Completeness Theorem will shed some light on the situation.

3.16. Compactness Theorem. *Let $\mathcal{T}$ be a set of sentences of a given language L. Then: $\mathcal{T}$ has a model iff every finite subset $\mathcal{T}_0 \subseteq \mathcal{T}$ has a model.*

Proof. Obviously, any model of $\mathcal{T}$ is a model of every finite subset $\mathcal{T}_0$. Hence, the real work of the proof is in proving the converse. Suppose every finite $\mathcal{T}_0 \subseteq \mathcal{T}$ has a model.

By the Soundness Lemma, every such $\mathcal{T}_0$ is thus consistent. But then $\mathcal{T}$ is consistent since any derivation of the empty sequent from $\mathcal{T}$ uses only finitely many axioms. By Completeness, $\mathcal{T}$ has a model. QED

An equivalent formulation of Compactness is the following.

3.17. Theorem. *Let $\mathcal{T}$ be a set of sentences of a given language L and let φ be a sentence of L. Then:*

$$\mathcal{T} \models \varphi \quad \textit{iff} \quad \exists \textit{ finite } \mathcal{T}_0 \subseteq \mathcal{T}\ (\mathcal{T}_0 \models \varphi).$$

Proof. Observe

$$\begin{aligned} \mathcal{T} \models \varphi \quad &\text{iff} \quad \mathcal{T} \vdash \varphi \\ &\text{iff} \quad \exists \text{ finite } \mathcal{T}_0 \subseteq \mathcal{T}\ (\mathcal{T}_0 \vdash \varphi) \\ &\text{iff} \quad \exists \text{ finite } \mathcal{T}_0 \subseteq \mathcal{T}\ (\mathcal{T}_0 \models \varphi). \end{aligned}$$

QED

I leave it to the reader to prove the equivalence of 3.16 and 3.17, and turn instead to the second consequence of the Completeness Theorem.

3.18. Löwenheim-Skolem Theorem. *Let $\mathcal{T}$ be a theory in a language L. If $\mathcal{T}$ has a model, then it has a countable model.*

Proof. If $\mathcal{T}$ has any model at all, then $\mathcal{T}$ is consistent and the construction given in the proof of the Completeness Theorem applies. Let $\mathcal{S}$ be a completion of $\mathcal{T}$ which has witnesses from the countable new set of constants $E = \{e_0, e_1, ...\}$. The model $\mathfrak{M}_{\mathcal{S}}$ of $\mathcal{T}$ is countable because its domain consists of equivalence classes of elements of E. QED

These two theorems— the Compactness Theorem and the Löwenheim-Skolem Theorem— are the beginnings of a branch of logic called model theory. In model theory, the Compactness Theorem is generally the only model-existence theorem one needs, although occasionally model theorists might have to use the semi-effectiveness of the set of logically valid sentences. This situation has a theoretical explanation in two beautiful results of Per Lindström.

3.19. First Lindström Theorem. *No proper extension of first-order logic satisfies both the Compactness Theorem and the Löwenheim -Skolem Theorem.*

3.20. Second Lindström Theorem. *No proper extension of first-order logic both has a semi-effective notion of validity and satisfies the Löwenheim-Skolem Theorem.*

Although these theorems lie beyond the intended scope of the present work, the First Lindström Theorem falls readily to the methods of Chapter V, volume II, and a proof of it will be given there. For now I haven't the space to prove these theorems or even to explain exactly what a "proper extension of first-order logic" is. For this last, suffice it to say that there are natural extensions such as second-order logic, where one allows quantification over subsets of the domain of the model, and extensions by the addition of *cardinality quantifiers*— quantifiers like "there are uncountably many" or "there are infinitely many".

Both the Compactness Theorem and the semi-effectiveness of validity are natural consequences of a Completeness Theorem for any decent notion of derivability which would proceed by the finite number of purely logical steps called for by Hilbert's Programme 1.1. Lindström's assert that, modulo the Löwenheim-Skolem Theorem, each of these consequences is characteristic of first-order logic. I would love to be able to report that the Löwenheim-Skolem Theorem can be dropped and that, thus, first-order logic is *the* logic called for by Programme 1.1. However, this is not the case: the extension of first-order logic by the quantifier "there are uncountably many" satisfies both the Compactness Theorem and semi-effectivity requirement. Fortunately, this won't matter for the refutation of Programme 1.1. It will show that the Programme cannot be carried out for any semi-effective extension of first-order logic. With respect to this Programme, our emphasis on first-order logic and all the work of these past two sections can be viewed as the mere construction of an example of a logic with a semi-effective notion of validity. Thus, the refutation of Programme 1.1 will not be vacuous!

Of course, there is more to first-order logic than the semi-effectiveness of its validity and much of the rest of the book will be devoted to showing this in the case of arithmetic. This first-order study of arithmetic begins mildly in Exercises 7 and 9, below, and in earnest in the next section with the theory of addition. The reader in a hurry to refute Programme 1.1 can safely skip the next two sections and proceed directly to section 6.

Exercises.

1. Complete the proof of the Soundness Lemma (3.4).
2. (Completeness for Formulae). Let $\mathcal{T}$ be a set of formulae in a language L. For each formula $\varphi \in L$, let its *universal closure*,
$$\forall v_{i_0} \ldots v_{i_{n-1}} \varphi, \text{ where } FV(\varphi) = \{v_{i_0}, \ldots, v_{i_{n-1}}\},$$
be denoted by φ'. Prove: For any formula ψ,
$$\mathcal{T} \vdash \psi \text{ iff } \mathcal{T} \models \psi,$$
by reducing the result to that for $\mathcal{T}' = \{\varphi' : \varphi \in \mathcal{T}\}$.
3. Let $\mathcal{D}$ be a derivation involving no non-logical axioms and suppose the constant d occurs in the derivation. Let $[v/d]\mathcal{D}$ denote the result of substituting a variable v for d at all occurrences of d in the derivation. Show: If the variable v does not occur at all in $\mathcal{D}$, then $[v/d]\mathcal{D}$ is a derivation. [Hint. Use induction on the number of inferences occurring in the derivation.]
4. In the proof of the Witnessing Lemma, it was mentioned that $\exists v_k (\exists v_i \psi v_i \to \varphi v_k)$ is derivable.

   i. Derive each of the following using only purely logical rules of inference (i.e., no Cuts or Contractions):

   a. $\vdash (\varphi \to \exists v_k \psi v_k) \Rightarrow \exists v_k (\varphi \to \psi v_k)$, v_k not free in φ

   b. $\vdash \Rightarrow \exists v_i \psi v_i \to \exists v_k \psi v_k$, v_k not free in ψv_i.

   Using Cut, derive
$$\vdash \Rightarrow \exists v_k (\exists v_i \psi v_i \to \psi v_k).$$

   ii. Replacing $\exists R$ by

$$\exists R' \quad \frac{\Gamma \Rightarrow \Delta, \exists v_i \varphi v_i, \varphi t}{\Gamma \Rightarrow \Delta, \exists v_i \varphi v_i},$$

where t is free for v_i in φv_i, derive

$$\vdash \Rightarrow \exists v_k (\exists v_i \psi v_i \rightarrow \psi v_k),$$

without using Cut or Contraction.

iii. Using the original $\exists R$ and Contraction, derive

$$\vdash \Rightarrow \exists v_k (\exists v_i \psi v_i \rightarrow \psi v_k)$$

without using Cut.

[Remark. In pure logic— without equality or non-logical axioms— the Cut rule can be eliminated. The remaining rules of inference all have the *sub-formula property*— the formulae of the premises are (up to considering φt a sub-formula of $Qv_i \varphi v_i$) sub-formulae of those in the conclusion. This is an important property— one that will be applied in Chapter IV of volume II. However, for completeness, one must either retain the contraction rule or i. build contraction into $\exists R$ and $\forall L$ as was done in $\exists R'$ above, and ii. replace $\wedge L$ and $\vee R$ by

$$\wedge L' \quad \frac{\Gamma, \varphi, \psi \Rightarrow \Delta}{\Gamma, \varphi \wedge \psi \Rightarrow \Delta} \qquad \vee R' \quad \frac{\Gamma \Rightarrow \Delta, \varphi, \psi}{\Gamma \Rightarrow \Delta, \varphi \vee \psi}.$$

As an additional exercise, one might derive these rules of inference for the notion of derivability at hand. Alternatively, one could fill in the missing steps in the derivation of $\Rightarrow \varphi \vee \neg \varphi$ occurring in the proof of Lemma 3.11.]

5. Complete the proof of Lemma 3.13.

6. Let $\mathfrak{A}, \mathfrak{B}$ be structures for a given language L.

   a. A function $F{:}A \rightarrow B$ is an *isomorphism* if

      i. F is one-one and onto

      ii. for every function symbol F_i^n of L and all $a_0, \ldots, a_{n-1}$ in A,

      $$F((F_i^n)^{\mathfrak{A}}(a_0, \ldots, a_{n-1})) = (F_i^n)^{\mathfrak{B}}(Fa_0, \ldots, Fa_{n-1})$$

      iii. for every relation symbol R_i^n of L and all $a_0, \ldots, a_{n-1}$ in A,

      $$(R_i^n)^{\mathfrak{A}}(a_0, \ldots, a_{n-1}) \text{ iff } (R_i^n)^{\mathfrak{B}}(Fa_0, \ldots, Fa_{n-1})$$

      iv. if c_i is a constant of L then $(c_i)^{\mathfrak{B}} = F(c_i)^{\mathfrak{A}}$.

   b. $\mathfrak{A}$ and $\mathfrak{B}$ are *elementarily equivalent* if, for any sentence φ of L,

      $$\mathfrak{A} \models \varphi \text{ iff } \mathfrak{B} \models \varphi.$$

   Prove: If there is an isomorphism $F{:}\mathfrak{A} \rightarrow \mathfrak{B}$, then $\mathfrak{A}$ and $\mathfrak{B}$ are elementarily equivalent. [Hint. a.ii is the basis of an induction showing that all terms are preserved. Once this is proven, it and a.iii form a basis for an induction showing

   $$\mathfrak{A} \models \varphi(\overline{a_0}, \ldots, \overline{a_{n-1}}) \text{ iff } \mathfrak{B} \models \varphi(\overline{b_0}, \ldots, \overline{b_{n-1}}),$$

   where $b_i = Fa_i$.]

7. (Theory of Successor). Let $\mathcal{T}$ be the following set of axioms in the language with constant $\overline{0}$ and unary function symbol S:

   a. $\forall v_0(\neg Sv_0 = \overline{0})$

   b. $\forall v_0 v_1(Sv_0 = Sv_1 \rightarrow v_0 = v_1)$

   c. $\forall v_0(\neg v_0 = \overline{0} \rightarrow \exists v_1(v_0 = Sv_1))$

d. $\forall v_0(\neg S^k v_0 = v_0)$, all $k > 0$,

where S^k means the k-fold application of S.

i. Prove that every model $\mathfrak{A}$ of $\mathcal{T}$ can be decomposed into a single ω-chain,

$$0, S0, SS0, \ldots,$$

and a family of Z-chains,

$$\ldots, a_{-1}, a_0, a_1, \ldots,$$

where $a_{n+1} = Sa_n$ for all integers n.

ii. Show: Any two countable models of $\mathcal{T}$ with infinitely many Z-chains are isomorphic.

iii. Show: Any countable model of $\mathcal{T}$ is elementarily equivalent to one with infinitely many Z-chains.

iv. Show: $\mathcal{T}$ is complete.

v. Show: $\mathcal{T} = Th(\omega; S, 0)$.

[Hints. iii. Add new constants $c_0, c_1, \ldots$ and augment $Th(\mathfrak{A})$ by axioms asserting the c_i's to lie on different Z-chains and apply the Compactness Theorem; and iv. don't forget the Löwenheim-Skolem Theorem!]

8. Let $\mathfrak{A}, \mathfrak{B}$ be structures for a given language L. We say $\mathfrak{A}$ is a *substructure* of $\mathfrak{B}$, written $\mathfrak{A} \subseteq \mathfrak{B}$, if

a. $A \subseteq B$

b. $\mathfrak{A} \models \varphi$ iff $\mathfrak{B} \models \varphi$

for every atomic sentence φ of $L(A)$ (i.e., the interpretations of the function symbols of L in $\mathfrak{A}$ are the restrictions to $\mathfrak{A}$ of those of $\mathfrak{B}$, and the interpretations of the relation symbols of L in $\mathfrak{A}$ are the restrictions of those of $\mathfrak{B}$). We say that $\mathfrak{A}$ is an *elementary substructure* of $\mathfrak{B}$, written $\mathfrak{A} \preccurlyeq \mathfrak{B}$ if $\mathfrak{A} \subseteq \mathfrak{B}$ and condition b holds for all sentences of $L(A)$.

i. Let $2\omega = \{0, 2, 4, \ldots\}$. Show that

$$(2\omega; <, 0) \subseteq (\omega; <, 0),$$

but the substructure is not elementary.

ii. Let $\mathfrak{A}_0, \mathfrak{A}_1, \ldots$ satisfy

$$\mathfrak{A}_0 \preccurlyeq \mathfrak{A}_1 \preccurlyeq \ldots,$$

and define $\mathfrak{A} = \cup \mathfrak{A}_n$ in the obvious way. Show: For all n,

$$\mathfrak{A}_n \preccurlyeq \mathfrak{A}.$$

iii. Let $\mathfrak{A}$ be a given L-structure. Define $Th(\mathfrak{A}; a)_{a \in A}$ to be $\{\varphi: \mathfrak{A} \models \varphi \wedge \varphi \in L(A)\}$. Show: Modulo an identification, any model $\mathfrak{B} \models Th(\mathfrak{A}; a)_{a \in A}$ is an elementary extension of $\mathfrak{A}$, i.e., $\mathfrak{A} \preccurlyeq \mathfrak{B}$.

9. (Theory of Successor and Order). Extend the language L of Exercise 7 to include $<$, and extend the theory $\mathcal{T}$ to $\mathcal{T}_<$ by adding the axioms:

a. $\forall v_0 v_1 v_2(v_0 < v_1 \wedge v_1 < v_2 \rightarrow v_0 < v_2)$

b. $\forall v_0 \neg(v_0 < v_0)$

c. $\forall v_0 v_1(v_0 < v_1 \vee v_0 = v_1 \vee v_1 < v_0)$

d. $\forall v_0(v_0 = \overline{0} \vee \overline{0} < v_0)$

e. $\forall v_0 v_1(v_0 < v_1 \rightarrow Sv_0 < Sv_1)$

f. $\forall v_0 v_1 (v_0 < Sv_1 \leftrightarrow v_0 = v_1 \lor v_0 < v_1)$

i. Prove that every model $\mathfrak{A}$ of $\mathcal{T}_<$ can be viewed as an ordered ω-chain followed by an ordered family of ordered Z-chains.

ii. Show: Any two countable models of $\mathcal{T}_<$ in which the Z-chains are grouped into a dense linear order with no first or last element are isomorphic.

iii. Show: Let $\mathfrak{A}$ be a countable model of $\mathcal{T}_<$ with elements a, b such that $\mathfrak{A} \models \overline{a} < \overline{b}$, but a, b are not in the same Z-chain. Then $\mathfrak{A}$ can be elementarily embedded in a model $\mathfrak{B}$ of $\mathcal{T}_<$ with elements c_0, c_1, c_2 such that

$$\mathfrak{B} \models \overline{c_0} < \overline{a} \land \overline{a} < \overline{c_1} \land \overline{c_1} < \overline{b} \land \overline{b} < \overline{c_2},$$

and a, b, c_0, c_1, c_2 all lie on different Z-chains.

iv. Show: Let $\mathfrak{A}$ be a countable model of $\mathcal{T}_<$. $\mathfrak{A}$ can be elementarily embedded in a model $\mathfrak{B}$ of $\mathcal{T}_<$ in which the ordering of Z-chains is dense with no first or last elements.

v. Show: $\mathcal{T}_<$ is complete.

vi. Show: $\mathcal{T}_< = Th(\omega; <, S, 0)$.

[Hints. ii. Any countable dense linear ordering with no first or last element is isomorphic to the ordering of the rationals; iv. iterate step iii and apply Exercise 8.ii.]

[Remark. Exercises 7 and 9 show that Programme 1.1 can be carried out for some structures. But notice how much more difficult Exercise 9 was than Exercise 7. Augmenting the language still further by an addition operation makes the execution of the Programme much more difficult, and involves even deeper model theory. Nonetheless, we shall see that Programme 1.1 can be carried out for $Th(\omega; <, +, S, 0)$; but we shall see this in the next section via a completely different method.]

4. Presburger-Skolem Arithmetic; The Theory of Addition

An extremely (and, perhaps, surprisingly) often cited result of logical number theory is the decidability of the first-order theory of the addition of the natural numbers. This result was proven independently by M. Presburger in 1929 and Thoralf Skolem in 1930 by the method of quantifier-elimination, first introduced in 1919 by Skolem in proving the completeness and decidability of the first-order theory of a special class of boolean algebras. By the time Presburger and Skolem applied the method to the theory of addition, the method was fairly well-known to logicians. Arithmetically, Cooper Harold Langford applied the method to the theories of dense and discrete linear orders (i.e., those of the rationals, integers, and natural numbers) in 1926/27; and, about the time Presburger and Skolem were handling addition, Jacques Herbrand applied the method to the lowly theory of successor. In Warsaw, where Presburger was a student, Alfred Tarski had applied the method and also lectured on it. Thus, when Presburger obtained his result, the method was fairly widely known and Tarski did not feel Presburger's work original enough to merit a doctorate; Presburger was granted a Master's degree. This was one of the ironically bad judgments in the history of logic: of all the quantifier-eliminations in the literature, Presburger's is probably the most widely cited.

Presburger published his result the following year in the proceedings of a conference held in Warsaw. In the published text, he only provided the details for the quantifier-elimination for the structure $(Z; +, 0, 1)$, i.e., the addition of the integers. In an addendum, however, he announced that the proof could be extended to the case in which the order relation was added. Since the non-negative integers are definable in $(Z; <, +, 0, 1)$, his decision procedure covered both structures, $(Z; <, +, 0, 1)$ and $(\omega; <, +, 0, 1)$. The full details were first published by Paul Bernays in *Grundlagen der Mathematik I* in 1934.

Whereas Presburger never got a doctorate, Skolem, who originally had no intention of getting one, did. Thus, it was *Dr.* Skolem who, in 1931, published his quantifier-elimination for the structure $(Z;<,+,0,1)$. Skolem's quantifier-elimination differs substantially from Presburger's, but I cannot explain how until I have carefully explained what "quantifier-elimination" means.

The basic idea of quantifier-elimination is very simple. Given a theory $\mathcal{T}$ in a language L, a quantifier-elimination for $\mathcal{T}$ is a means of associating with each formula φ of L a *quantifier-free* formula ψ of L (i.e., a formula ψ with no quantifiers) such that

$$\mathcal{T} \vdash \varphi \leftrightarrow \psi.$$

(Recall Exercise 7 of section 2 for the definition of $\leftrightarrow$.) In actual practice, one must often first extend L to a larger language L' obtained by making primitive a few defined relations of L. One then shows that every formula φ of L' is equivalent to a quantifier-free formula ψ of L'. For the structure at hand, $(Z;<,+,0,1)$, the language

$$L: \quad =, <, +, \overline{0}, \overline{1},$$

does not admit an elimination of quantifiers; Presburger obtained one for the expanded language,

$$L': \quad =, <, +, \overline{0}, \overline{1}, (\cdot) \equiv (\cdot) \pmod{n} \text{ for } n = 2, 3, \ldots,$$

where $(\cdot) \equiv (\cdot) \pmod{n}$ is defined by

$$v_0 \equiv v_1 \pmod{n}: \ \exists v_2(v_0 = v_1 + (v_2 + \ldots + v_2)),$$

with an n-fold sum of v_2's, i.e., $(\cdot) \equiv (\cdot) \pmod{n}$ expresses congruence modulo the fixed divisor n.

Skolem's quantifier-elimination for $Th(Z;<,+,0,1)$ differs from Presburger's first in Skolem's choice of L'. Skolem violates the rules by adding new multipliers $q\cdot$ for all rational numbers q, i.e., new function symbols $q\cdot$ denoting multiplication by the numbers q. The violation is that Z is not closed under multiplication by rational numbers. To get back into the integers, Skolem also introduces the greatest-integer function $[\cdot]$.

Skolem's treatment also differs from Presburger's in that it is a bit easier. In fact, in as much space as Presburger's proof requires, one can present Skolem's result and derive Presburger's as a corollary. Moreover, Skolem's proof does not require an application of the Chinese Remainder Theorem like Presburger's does and is, thus, a little more direct.

I suppose I have just explained why I intend to follow Skolem's approach instead of Presburger's. The next thing to explain is why I intend to treat the integers instead of the natural numbers. There are two reasons for this. The simple one is that the details are a little easier; the failure of closure under subtraction in the natural numbers creates some awkward circumlocutions (like the necessity in Chapter II of dealing with equations $P = Q$ instead of $P = 0$ in the exponential case). Second, the theory of $(\omega; <, +, 0, 1)$ reduces to that of $(Z; <, +, 0, 1)$ because ω is definable in the latter structure:

$$v_0 \in \omega: \ v_0 = \overline{0} \ \vee \ \overline{0} < v_0.$$

Thus, the quantifier-elimination for $(Z; <, +, 0, 1)$ simultaneously produces decision procedures for $Th(Z; <, +, 0, 1)$ and $Th(\omega; <, +, 0, 1)$.

The latter point above is important and deserves to be explained in detail. I shall do this in some generality. Let $\mathfrak{B} \subseteq \mathfrak{A}$ (cf. Exercise 8 of the preceding section for the definition) be L-structures for a language L. Suppose further that B is definable in $\mathfrak{A}$ by a formula, say $v_0 \in B$. For any formula φ of $L(B)$, we can define its *relativisation* $\varphi^{(B)}$ to B inductively as follows:

i. if φ is atomic, $\varphi^{(B)}$ is φ

ii. $(\neg\varphi)^{(B)}$ is $\neg(\varphi^{(B)})$

iii. $(\varphi * \psi)^{(B)}$ is $\varphi^{(B)} * \psi^{(B)}$ for $* \in \{\wedge, \vee, \rightarrow\}$

iv. $(\exists v_i \varphi v_i)^{(B)}$ is $\exists v_i [v_i \in B \wedge (\varphi v_i)^{(B)}]$

$(\forall v_i \varphi v_i)^{(B)}$ is $\forall v_i [v_i \in B \rightarrow (\varphi v_i)^{(B)}]$,

where we relabel any bound occurrences of v_i in $v_0 \in B$ if necessary (i.e., if v_i is not free for v_0 in $v_i \in B$). An easy induction on the length of a sentence φ of $L(B)$ shows

$$\mathfrak{B} \models \varphi \ \text{ iff } \ \mathfrak{A} \models \varphi^{(B)}.$$

This can be generalised further to the notion of a *relative interpretation* by allowing the primitive functions and relations of $\mathfrak{B}$ to be arbitrary definable relations of $\mathfrak{A}$ (cf. Exercise 1, below).

Since the addition function and order relation on ω are just the restrictions of those on Z to ω, relativisation of quantifiers to the formula $v_0 \in \omega$ reduces $Th(\omega; <, +, 0, 1)$ to $Th(Z; <, +, 0, 1)$. In fact, the application of the quantifier-elimination to a formula $\varphi^{(\omega)}$ will result in a quantifier-free formula ψ with no reference to Z, i.e., a quantifier-elimination for $(\omega; <, +, 0, 1)$ is also obtained by the process,

$$\varphi \ \mapsto \ \varphi^{(\omega)} \ \mapsto \ \text{quantifier-free equivalent.}$$

The same device of relativisation can explain (or, justify) Skolem's unorthodox procedure of stepping outside Z in his quantifier-elimination. What Skolem does amounts to stepping into the structure,

$$(Q; Z, <, +, \{q\cdot : q \in Q\}, [\cdot], 0, 1),$$

with language including a new predicate $Z(v_0)$ defining the set of integers. Skolem then eliminates quantifiers from formulae of the form,

$$Z(v_{i_0}) \wedge \ldots \wedge Z(v_{i_{k-1}}) \wedge \varphi^{(Z)}(v_{i_0}, \ldots, v_{i_{k-1}}), \qquad (*)$$

where φ has exactly $v_{i_0}, ..., v_{i_{k-1}}$ free and Z does not occur as a sub-formula of φ. Since all variables in (*) range over the integers, he can, for the most part, delete reference to the relativising formula $Z(v_0)$. That is, he can write (*) simply as

$$\varphi(v_{i_0}, ..., v_{i_{k-1}}),$$

provided he remembers that all variables free and bound are to be variables over the integers. Perhaps I go too far in this formalism, and probably I introduce it to early. The point is simply this: with terms ranging over integers, the logical rules of inference and the semantic notion of truth both so carefully defined in section 2, above, just won't apply to the case at hand unless I offer some such device to explain how they apply. This will become more apparent in the course of the quantifier-elimination. [In point of fact, all quantifiers can be eliminated in this language, whence there is no conceptual problem at all. The elimination proceeds, however, by cases depending on whether the quantifiers are integral or rational! Cf. Exercise 15, below.]

There is only one thing standing between us and Skolem's quantifier-elimination for the addition and order of the integers. This is the following general lemma on quantifier-eliminations.

4.1. Lemma. *Let $\mathcal{T}$ be a theory in a language L. In order to show that $\mathcal{T}$ admits an elimination of quantifiers in L, it suffices to show that every formula of the form,*

$$\exists v_i \bigwedge \varphi_j v_i, \qquad (*)$$

where each φ_j is an atomic formula or the negation of an atomic formula, is equivalent in $\mathcal{T}$ to a quantifier-free formula.

Proof. By induction on the complexity of φ, we show that there is a quantifier-free formula ψ such that

$$\mathcal{T} \vdash \varphi \leftrightarrow \psi.$$

If φ is atomic, take $\psi = \varphi$.

If φ_1, φ_2 have quantifier-free equivalents ψ_1, ψ_2, respectively, then use the equivalences,

$$\mathcal{T} \vdash \varphi_1 * \varphi_2 \leftrightarrow \psi_1 * \psi_2, \text{ for } * \in \{\wedge, \vee, \rightarrow\} \text{ and } \mathcal{T} \vdash \neg\varphi_1 \leftrightarrow \neg\psi_1,$$

to construct the new quantifier-free equivalents in the propositional cases.

If φ is of the form $\exists v_i \varphi_1$ and φ_1 has ψ_1 as a quantifier-free equivalent, first transform ψ_1 into a *disjunctive normal form,*

$$\bigvee_j \bigwedge_k \psi_{jk},$$

where each ψ_{jk} is atomic or the negation of an atomic formula. (Cf. Exercise 2, below.) Then apply the logically valid equivalence,

$$\exists v_i (\bigvee_j \bigwedge_k \psi_{jk}) \leftrightarrow \bigvee_j \exists v_i (\bigwedge_k \psi_{jk}).$$

(Exercise: Prove this equivalence.) Now, each formula $\exists v_i (\bigwedge_k \psi_{jk})$ is equivalent to a quantifier-free formula χ_j by hypothesis. Thus,

$$\mathcal{T} \vdash \varphi \leftrightarrow \exists v_i \bigvee \bigwedge \psi_{jk}$$

$$\leftrightarrow \bigvee \exists v_i \bigwedge \psi_{jk}$$
$$\leftrightarrow \bigvee \chi_j,$$

and the formula $\psi = \bigvee \chi_j$ is a quantifier-free equivalent to φ.

If φ is of the form $\forall v_i \varphi_1$ and φ_1 has ψ_1 as a quantifier-free equivalent, observe

$$\mathcal{T} \vdash \varphi \leftrightarrow \forall v_i \varphi_1$$
$$\leftrightarrow \neg \exists v_i \neg \varphi_1$$

and the above proofs for negation and the existential quantifier yield the result. QED

Using the equivalences,

$$(\forall v_i \varphi)^{(Z)} \leftrightarrow \forall v_i (Zv_i \rightarrow \varphi^{(Z)})$$
$$\leftrightarrow \neg \exists v_i \neg (Zv_i \rightarrow \varphi^{(Z)})$$
$$\leftrightarrow \neg \exists v_i (Zv_i \wedge \neg \varphi^{(Z)})$$
$$\leftrightarrow \neg (\exists v_i \neg \varphi)^{(Z)},$$

the Lemma yields the result in the relativised case we will now consider.

4.2. Skolem's Quantifier-Elimination. *$Th(Z; <, +, 0, 1)$ admits a quantifier-elimination relative to the augmented language containing rational multipliers $q\cdot$ for all rational numbers q and the greatest integer function* $[\cdot]$.

A few small notational remarks before launching into the proof: We let $\overline{0}$, $\overline{1}$ denote the constants naming 0, 1, respectively. For any term t and any rational number q, we write qt for the term constructed by applying the multiplier $q\cdot$ to t. The special term $q\overline{1}$ will be denoted $\overline{q}$ and $t_1 + (-1)t_2$ will be abbreviated $t_1 - t_2$. Finally, for the sake of clarity, we will use the following:

Formal variables for integers. v, w

Rational numbers. q

Terms. s, t, u

Natural numbers. m, k (and eventually n, i).

These will occasionally be subscripted.

Given all of this, we can start the proof.

Proof of Theorem 4.2. Let a formula φ in the form

$$\exists v \, (\bigwedge_i \varphi_i), \tag{1}$$

with each φ_i atomic or negated atomic, be given.

The elimination proceeds in three stages.

I. The first step is to simplify (1). Using the equivalences,

$$\neg s < t \leftrightarrow (s = t \vee t < s)$$
$$\neg s = t \leftrightarrow (s < t \vee t < s),$$

every negated atomic formula can be replaced by a disjunction of atomic formulae. Repeatedly using the distributive law,

$$\alpha \wedge (\beta \vee \gamma) \leftrightarrow (\alpha \wedge \beta) \vee (\alpha \wedge \gamma),$$

formula (1) can be put into the form,

$$\exists v\, (\bigvee_i \bigwedge_j \chi_{ij}),$$

where each χ_{ij} is atomic. As this is logically equivalent to

$$\bigvee_i \exists v\, (\bigwedge_j \chi_{ij}),$$

we see that we can eliminate the existential quantifier from (1) provided we can do so for all formulae of the form,

$$\exists v\, (\bigwedge_i \chi_i), \quad \text{each } \chi_i \text{ atomic.} \tag{2}$$

II. The second step is to simplify (2) by simplifying the occurrences of the variable v in the atomic formulae χ_i. Consider for example the simple formula:

$$\exists v\, (\tfrac{2}{3}[\frac{\overline{1}}{5} + [\tfrac{1}{2}v + \frac{\overline{1}}{4}]] > \overline{15}). \tag{3}$$

Note that v occurs inside the scopes of multipliers 1/2 and 2/3. The product of their denominators is 6. Now, $v = 6w + k$ for some w and some $0 \le k < 6$. Thus, (3) is successively equivalent to

$$\bigvee_{k=0}^{5} \exists w\, (\tfrac{2}{3}[\frac{\overline{1}}{5} + [\tfrac{1}{2}(6w + \overline{k}) + \frac{\overline{1}}{4}]] > \overline{15})$$

$$\bigvee_{k=0}^{5} \exists w\, (\tfrac{2}{3}[\frac{\overline{1}}{5} + [3w + \tfrac{1}{2}\overline{k} + \frac{\overline{1}}{4}]] > \overline{15})$$

$$\bigvee_{k=0}^{5} \exists w\, (\tfrac{2}{3}[\frac{\overline{1}}{5} + 3w + [\tfrac{1}{2}\overline{k} + \frac{\overline{1}}{4}]] > \overline{15})$$

$$\bigvee_{k=0}^{5} \exists w\, (2w + \tfrac{2}{3}[\frac{\overline{1}}{5} + [\tfrac{1}{2}\overline{k} + \frac{\overline{1}}{4}]] > \overline{15}),$$

and w no longer occurs inside any $[\cdot]$. This can be done in general.

In each χ_i, v occurs inside the scopes of various multipliers. Let these be $q_0, \ldots, q_{k-1}$ (including repetitions) with denominators $m_0, \ldots, m_{k-1}$, respectively. Let $m = m_0 \cdots m_{k-1}$ (1, if $k = 0$), and observe that

$$\exists v\, (\bigwedge \chi_i(v)) \leftrightarrow \bigvee_{k=0}^{m-1} \bigwedge \chi_i(mw + \overline{k}).$$

Now, m has been chosen to contain enough factors to cancel the denominators of any multipliers it meets in attempting to step outside the scope of any $[\cdot]$ it may happen to lie in. Thus, the variable w can be pulled outside all $[\cdot]$'s and w occurs linearly in each resulting term. Ordinary algebra allows us to simplify each atomic formula by collecting the w's and, if the resulting coëfficients are non-zero, to turn these coëfficients into 1. Thus, (2) is equivalent to

$$\bigvee_{k=0}^{m-1} \exists w\, (\bigwedge_i w < s_i \;\wedge\; \bigwedge_j t_j < w \;\wedge\; \bigwedge_n w = u_n \;\wedge\; \psi),$$

where w does not occur in ψ or in any s_i, t_j, or u_n. If we pull ψ outside the scope of the quantifier, we see that we have reduced our problem to that of eliminating the existential quantifier from (several) formulae of the form,

$$\exists w \left(\bigwedge_i w < s_i \ \wedge\ \bigwedge_j t_j < w \ \wedge\ \bigwedge_k w = u_k\right), \tag{4}$$

where w does not occur in any s_i, t_j, or u_k.

III. We can now eliminate the quantifier from (4).

Case 1. The conjunction of equations is non-vacuous. Then an equation $w = u_1$ occurs. Since the variable w runs over integers and the term u_1 can denote a non-integral rational number, we cannot simply substitute u_1 for w in (4), but must also conjoin the assertion that u_1 is integral. This can be done in a simple way, yielding the following quantifier-free equivalent to (4):

$$\bigwedge_i u_1 < s_i \ \wedge\ \bigwedge_j t_j < u_1 \ \wedge\ \bigwedge_{k \neq 1} u_1 = u_k \ \wedge\ u_1 = [u_1].$$

Case 2. No equations are present and only one kind of inequality occurs, i.e., (4) is of one of the two forms,

$$\exists w \left(\bigwedge_i w < s_i\right) \quad \text{or} \quad \exists w \left(\bigwedge_j t_j < w\right). \tag{5}$$

The unboundedness of Q and cofinality of Z in Q readily yields the truth of (5), i.e., the equivalence of (5) with $\overline{0} < \overline{1}$.

Case 3. Formula (4) is of the form,

$$\exists w \left(\bigwedge_i w < s_i \ \wedge\ \bigwedge_j t_j < w\right), \tag{6}$$

where neither conjunction is vacuous. Obviously, we must have each $t_j < s_i$ in order to fit anything between them. Since we want to fit an integer between the t_j's and the s_i's, we cannot simply take

$$\bigwedge_{i,j} t_j < s_i\,,$$

as a quantifier-free equivalent to (6). However, we can take

$$\bigwedge_{i,j} [t_j] + \overline{1} < s_i\,,$$

since $w \geq [t_j] + 1 > t_j$ whenever $w > t_j$. QED

Æsthetically, Skolem's quantifier-elimination is not quite so nice as Presburger's because the outcome is expressed in terms of rational numbers instead of integers. We will shortly remedy this defect by deriving Presburger's result from Skolem's. First, however, let us note that Skolem's result admits some nice applications.

4.3. Application. *Th*(Z; <, +, 0, 1) *is decidable, i.e., there is an effective procedure which, given any sentence* φ *of the language of this theory, will decide the truth or falsity of* φ *in* (Z; <, +, 0, 1). *The same holds for Th*(ω; <, +, 0, 1).

The proof is simple. Given φ, find its quantifier-free equivalent ψ by the above procedure. The truth or falsity of ψ is merely a matter of computation. [As for

$Th(\omega; <, +, 0, 1)$, recall that a sentence φ holds in $(\omega; <, +, 0, 1)$ iff its relativisation $\varphi^{(Z)}$ holds in $(Z; <, +, 0, 1)$. Thus, the decidability of the latter theory yields that of the former.]

The procedure of a quantifier-elimination, although theoretically effective, is generally practically unfeasible. Consider, for example, a formula,

$$\varphi: \ \forall v \, \exists w \, \psi,$$

with ψ quantifier-free. First, one must put ψ into disjunctive normal form $\bigvee\!\!\bigwedge \psi_{ij}$ and consider

$$\forall v \bigvee \exists w \bigwedge \psi_{ij} \, .$$

Applying the elimination to each of the existentially quantified disjuncts, we get

$$\forall v \bigvee \chi_i, \tag{*}$$

each χ_i quantifier-free. Each χ_i is a disjunction of conjunctions, whence (*) itself looks like

$$\forall v \, (\bigvee \bigwedge \chi_{ij}),$$

which Lemma 4.1 tells us to rewrite as

$$\neg \exists v \, (\neg \bigvee \bigwedge \chi_{ij}),$$

i.e.,

$$\neg \exists v \, (\bigwedge \bigvee \neg \chi_{ij}).$$

The quantifier-free part here must be changed to a disjunction of conjunctions before the quantifier can be eliminated. In short, in addition to all the steps outlined in the proof of Theorem 4.2, one must put quantifier-free formulae into disjunctive normal form over and over again. It can, in fact, be shown that any decision procedure for $Th(Z; <, +, 0, 1)$ is more than exponentially bad: there is a positive real constant c such that, for any decision procedure for $Th(Z; <, +, 0, 1)$ and all large natural numbers n, there are sentences φ of size $\le n$ (under a natural measure of size) which require more than $2^{2^{cn}}$ steps to be decided by the given procedure.

The pessimism of the last paragraph concerns the worst possible cases. In the good cases, the procedure can occasionally be used. The next application to an old result of J.J. Sylvester offers a good example.

4.4. Application. *Let m, n be relatively prime positive integers and let $c_0 = mn - m - n$. Let c be a natural number.*

i. *if $c > c_0$, c is representable over the set of natural numbers by the linear form $mX + nY$;*

ii. *if $c \le c_0$, exactly one of c and $c_0 - c$ is so representable.*

Proof. Let c be a given natural number. The assertion that c can be represented by the form $mX + nY$ over ω is expressed in the language of $(Z; <, +, 0, 1)$ by the formula,

$$\exists v \, \exists w \, (\overline{-1} < v \wedge \overline{-1} < w \wedge mv + nw = \overline{c}). \tag{7}$$

If we apply the elimination procedure to (7), we obtain successively (omitting the overlines):

$$\exists v\,(-1 < v \wedge \exists w\,(-1 < w \wedge w = \frac{c - mv}{n}))$$

$$\exists v\,(-1 < v \wedge -1 < \frac{c - mv}{n} \wedge \frac{c - mv}{n} = [\frac{c - mv}{n}])$$

$$\bigvee_{i=0}^{n-1} \exists w(-1 < nw + i \wedge -1 < \frac{c - m(nw + i)}{n} \wedge \frac{c - m(nw + i)}{n} = \left[\frac{c - m(nw + i)}{n}\right])$$

$$\bigvee_{i=0}^{n-1} \exists w(-\frac{i + 1}{n} < w \wedge w < \frac{c + n - mi}{mn} \wedge \frac{c - mi}{n} = \left[\frac{c - mi}{n}\right])$$

$$\bigvee_{i=0}^{n-1} (\frac{c - mi}{n} = \left[\frac{c - mi}{n}\right] \wedge \exists w(-\frac{i + 1}{n} < w \wedge w < \frac{c + n - mi}{mn}))$$

$$\bigvee_{i=0}^{n-1} (\frac{c - mi}{n} = \left[\frac{c - mi}{n}\right] \wedge \left[-\frac{i + 1}{n}\right] + 1 < \frac{c + n - mi}{mn})$$

$$\bigvee_{i=0}^{n-1} (\frac{c - mi}{n} = \left[\frac{c - mi}{n}\right] \wedge 0 < \frac{c + n - mi}{mn}), \tag{8}$$

since, for $0 \leq i \leq n - 1$, $[-(i + 1)/n] = -1$.

Now, the equation in (8) simply asserts that $(c - mi)/n$ is integral, i.e., that $n \mid c - mi$. Further, the denominator mn being positive, it can be omitted from the inequality. Thus (8), and hence (7), is equivalent to

$$\bigvee_{i=0}^{n-1} (n \mid c - mi \;\wedge\; mi - n < c). \tag{9}$$

The quantity $mi - n$ is maximised by taking $i = n - 1$, yielding $m(n - 1) - n = mn - m - n = c_0$. Hence, for $c > c_0$, the inequality holds automatically and (9) is equivalent to

$$\bigvee_{i=0}^{n-1} n \mid c - mi\;,$$

which is just the quantifier-free equivalent to

$$\exists v\, \exists w(mv + nw = c),$$

as can be seen by deleting the inequalities in the reduction of (7) to (9), and which is true in the integers since m, n are relatively prime. This proves assertion i.

To prove ii, note that the assertion that $c_0 - c$ is representable is equivalent (as seen by substituting $c_0 - c$ for c in (9)) to

$$\bigvee_{i=0}^{n-1} (n \mid m(n - 1) - n - c - mi \;\wedge\; mi - n < m(n - 1) - n - c),$$

which is equivalent to

$$\bigvee_{i=0}^{n-1} (n \mid m(n - 1 - i) - c \;\wedge\; c < m(n - 1 - i)),$$

which in turn can be rewritten as

$$\bigvee_{j=0}^{n-1} (n \mid mj - c \ \wedge \ c < mj),$$

by taking $j = n-1-i$. Relabelling j by i, this finally yields

$$\bigvee_{i=0}^{n-1} (n \mid mi - c \ \wedge \ c < mi). \tag{10}$$

Disjoining (9) and (10) to obtain the assertion that one of c and $c_0 - c$ is representable, we get successively

$$\bigvee_{i=0}^{n-1}(n \mid c - mi \ \wedge \ mi - n < c) \ \vee \ \bigvee_{i=0}^{n-1} (n \mid mi - c \ \wedge \ c < mi)$$

$$\bigvee_{i=0}^{n-1}(n \mid c - mi \ \wedge \ (c < mi \ \vee \ mi - n < c)).$$

The inner disjunction trivially being true, this yields

$$\bigvee_{i=0}^{n-1}(n \mid c - mi\ ),$$

which, as remarked above, is true. Thus, at least one of c and $c_0 - c$ is representable by the form in question over the natural numbers.

Finally, to see that one cannot have both c and $c_0 - c$ representable by $mX + nY$ for $c \le c_0$, note that it suffices to show c_0 not to be so representable. To establish this last, plug c_0 into (9) to obtain, after simplifying,

$$\bigvee_{i=0}^{n-1}(n \mid -m(i+1) \ \wedge \ i < n-1),$$

i.e.,

$$\bigvee_{i=0}^{n-2} n \mid -m(i+1). \tag{11}$$

Since m, n are relatively prime, (11) can only be satisfied if $n \mid i+1$ for some $0 \le i \le n-2$, which is impossible. Thus, (11) and hence the assertion of the representability of c_0 are false. QED

The reader might wish to verify Application 4.4 with a numerical example. The equation,

$$4X + 5Y = c,$$

is simple enough to do by hand. For more along these lines, cf. Exercise 5, below.

Application 4.4 is atypical. Most applications of a quantifier-elimination are, like Application 4.3, *meta*mathematical in character; they tell us something about the theory itself. In addition to decidability, the two most common metamathematical applications are an axiomatisation and an analysis of definability. First, we consider the problem of axiomatisation.

4.5. Application. *A sentence* φ *is true in* $(\mathbb{Z};<,+,0,1)$ *iff it is a consequence of the following set* $\mathcal{T}_+$ *of axioms*:

A1. $\forall v_0 v_1 v_2 (v_0 + (v_1 + v_2) = (v_0 + v_1) + v_2)$

A2. $\forall v_0 (v_0 + \overline{0} = v_0)$

A3. $\forall v_0 \exists v_1 (v_0 + v_1 = \overline{0})$

A4. $\forall v_0 v_1 (v_0 + v_1 = v_1 + v_0)$

O1. $\forall v_0 v_1 (v_0 < v_1 \rightarrow \neg(v_1 < v_0))$

O2. $\forall v_0 v_1 v_2 (v_0 < v_1 \wedge v_1 < v_2 \rightarrow v_0 < v_2)$

O3. $\forall v_0 v_1 (v_0 < v_1 \vee v_0 = v_1 \vee v_1 < v_0)$

O4. $\forall v_0 v_1 (v_0 < v_1 + \overline{1} \leftrightarrow v_0 = v_1 \vee v_0 < v_1)$

OA. $\forall v_0 v_1 v_2 (v_0 < v_1 \rightarrow v_0 + v_2 < v_1 + v_2)$

D_n. $\forall v_0 \exists v_1 (\bigvee_{k=0}^{n-1} (v_0 = n v_1 + \overline{k})),$

where n is an integer > 1, nt denotes the n-fold sum $t + \ldots + t$ for any term t, and $\overline{k}$ the term $k\overline{1}$ (unless $k = 0$, when $\overline{k}$ denotes the constant $\overline{0}$).

Proof sketch. The reader familiar with algebra will recognise *A1 - A4* as the axioms for an *abelian group*. The additive inverse provided by *A3* is unique and, for any element a of an abelian group, we can denote this inverse by $-a$. Subtraction is definable by

$$a - b = a + (-b).$$

All the usual laws of addition and subtraction hold.

Axioms *O1 - O4* are the axioms for *discrete linear order* ($a + 1$ being the successor of an element a); and axiom *OA* asserts the *preservation of order* under addition. These axioms can be exploited to show, by induction on $n > 1$, that the only things between 0 and n are 1, 2, ..., $n - 1$ (i.e., 1, 1 + 1, ..., 1 + 1 + ... + 1):

$$\mathcal{T}_+ \vdash \forall v_0 (\overline{0} < v_0 \wedge v_0 < \overline{n} \rightarrow v_0 = \overline{1} \vee \ldots \vee v_0 = \overline{n-1}). \qquad (12)$$

The final axiom D_n is a schematisation of the division algorithm: For each integer $n > 1$, one can divide an element a by n to get a quotient b and remainder k:

$$a = nb + k, \ 0 \leq k < n.$$

Using (12) we can prove the unicity of quotient and remainder:

$$\mathcal{T}_+ \vdash \forall v_0 v_1 \bigwedge_{i,k=0}^{n-1} (n v_0 + \overline{i} = n v_1 + \overline{k} \rightarrow v_0 = v_1 \wedge \overline{i} = \overline{k}).$$

To see this, suppose for $a, b \in A$ in a model $\mathfrak{A}$ of $\mathcal{T}_+$ and for some $0 \leq i, k < n$,

$$na + i = nb + k. \qquad (13)$$

Without loss of generality, we can assume $k \geq i$. Manipulating (13) yields

$$n(a - b) = k - i \geq 0. \qquad (14)$$

By *OA*,

$$\begin{aligned} a < b &\rightarrow na < nb \\ &\rightarrow n(a - b) < 0. \end{aligned}$$

Thus, $a \geq b$. But *OA* again yields

$$b \leq a \ \rightarrow\ 0 \leq a-b \leq n(a-b) = k-i < n.$$

Therefore, (12) yields

$$a-b = j$$

for some $0 \leq j < n$. But $n(a-b) = nj < n$ (true by (14)) implies $j = 0$. Thus, $a = b$ and, by (14), it follows that $k-i = n0 = 0$, i.e., $k = i$.

So much for preliminaries! We can now embedd any model $\mathfrak{A} \models \mathcal{T}_+$ into a structure $\mathfrak{B}$ in which we can multiply by arbitrary rationals. The construction of $\mathfrak{B}$ is familiar from algebra: it is basically the same as the construction of the rationals from the integers. First, one defines

$$B' = \{\langle a, n\rangle: a \in A \ \wedge n \in \omega^+ = \{1, 2, \ldots\}\}.$$

On B', define an equivalence relation,

$$\langle a, n\rangle \sim \langle b, m\rangle \ \text{ iff } \ ma = nb.$$

[Think of $\langle a, n\rangle$ and $\langle b, m\rangle$ as a/n and b/m.] The domain of $\mathfrak{B}$ is the set of equivalence classes modulo this equivalence relation:

$$B = \{[\langle a, n\rangle]_\sim: \langle a, n\rangle \in B'\},$$

where

$$[\langle a, n\rangle]_\sim = \{\langle b, m\rangle \in B': \langle a, n\rangle \sim \langle b, m\rangle\}.$$

For convenience, we will denote the equivalence class $[\langle a, n\rangle]_\sim$ by a/n. This has the two-fold advantage of reminding us what the classes are to represent and of not looking like the greatest integer function to be introduced. Addition is defined on B by

$$a/n + b/m = (ma + nb)/(mn),$$

and the 0, 1 of B are the classes 0/1, 1/1, respectively. Finally, the order relation is defined by

$$a/n < b/m \ \text{ iff } \ ma < nb.$$

The addition operation and order relation can be shown to be well-defined, i.e., not to depend on the choice of representatives of the equivalence classes. I illustrate this by proving the result for the order relation. Let $\langle a, n\rangle \sim \langle a', n'\rangle$ and $\langle b, m\rangle \sim \langle b', m'\rangle$. We must show:

$$ma < nb \ \leftrightarrow\ m'a' < n'b'.$$

To see this, observe

$$\begin{aligned} ma < nb \ &\leftrightarrow\ mn'a < nn'b && (*) \\ &\leftrightarrow\ mna' < nn'b \\ &\leftrightarrow\ m'mna' < m'nn'b && (*) \\ &\leftrightarrow\ mnm'a' < nn'mb' \\ &\leftrightarrow\ m'a' < n'b', && (*) \end{aligned}$$

where the steps labelled (*) follow by *OA* and cancellation (i.e., uniqueness of the quotient), and the other steps follow by the definition of $\sim$.

I leave the rest of the proof of the well-definedness of $\mathfrak{B}$ to the reader, who may also verify that $\mathfrak{B}$ is an ordered abelian group (i.e., $\mathfrak{B}$ satisfies axioms *A1* - *A4*, *OA*, and *O1* - *O3*, but not *O4*). The reader may also verify that the map $\mathfrak{A} \rightarrow \mathfrak{B}$ given by mapping a to $a/1$ is an isomorphic embedding identifying $\mathfrak{A}$ with an ordered abelian subgroup of $\mathfrak{B}$.

It is not $\mathfrak{B}$ *per se* we are interested in, but the structure

$$\mathfrak{B}^* = (B; A, <, +, \{q\cdot : q \in Q\}, [\cdot], 0, 1).$$

For this, we must define the rational multipliers $q\cdot$ for $q \in Q$ and the greatest integer function. The first of these is direct: if $q = m/k$ (in the usual arithmetic sense), define

$$q\cdot(a/n) = (ma)/(kn).$$

As for the greatest integer function, given a/n, find c, k such that

$$a = nc + k,$$

and define

$$[a/n] = c.$$

These operations can be shown to be well-defined.

To complete our picture of the algebraic laws holding in $\mathfrak{B}$, note first that, because of the rational multipliers, $\mathfrak{B}^*$ is more-or-less a vector space over the rationals— a very special one: $\mathfrak{B}^*$ is an ordered vector space in which the scalar multiplication interacts with the order as follows:

$$q \text{ positive} \rightarrow (a < b \leftrightarrow qa < qb)$$

$$q \text{ negative} \rightarrow (a < b \leftrightarrow qb < qa).$$

Moreover, $\mathfrak{B}^*$ has a designated discrete subgroup A and a function $[\cdot]$ mapping an element b of the space to the greatest element a of A less than or equal to b.

If one now looks at Skolem's quantifier-elimination, one will find that, for eliminating quantifiers with variables ranging over A, all the steps can be carried out in $\mathfrak{B}^*$. Thus, for any sentence φ of the original language in $\mathfrak{A}$, if ψ is the quantifier-free equivalent, one has

$$\mathfrak{A} \models \varphi \quad \text{iff} \quad \mathfrak{B}^* \models \psi.$$

Now, $(Z; <, +, 0, 1)$ is directly embeddable in $\mathfrak{A}$, whence $Q^* = (Q; Z, <, +, \{q\cdot : q \in Q\}, [\cdot], 0, 1)$ is embeddable in $\mathfrak{B}^*$ (map q to $q1$) and one has, for any quantifier-free sentence ψ,

$$Q^* \models \psi \quad \text{iff} \quad \mathfrak{B}^* \models \psi.$$

Thus,

$$\begin{aligned} Z \models \varphi \quad & \text{iff} \quad Q^* \models \psi \\ & \text{iff} \quad \forall \mathfrak{B}^* (\mathfrak{B}^* \models \psi) \\ & \text{iff} \quad \forall \mathfrak{A} (\mathfrak{A} \models \varphi) \\ & \text{iff} \quad \mathcal{T}_+ \vdash \varphi. \end{aligned}$$

QED

The proof of Application 4.5 is an argument in favour of Presburger's proof of a quantifier-elimination for $(Z; <, +, 0, 1)$. Presburger has to expand the language, but he does not have to move into a larger structure and verify (seemingly) infinitely many defined relations to be well-defined: he stays within an arbitrarily given model of $\mathcal{T}_+$. On the other hand, the reader familiar with the construction of the rationals from the integers will find the present proof routine; and the reader who is not already familiar with such ought to become so and there is no chance like the present... (If this justification of my choice of treatment strikes the reader as just a little specious, I can only say that the reader has entirely the wrong attitude and would probably mistrust his own mother!)

Speaking of Presburger, it is high time we considered his quantifier-elimination for $Th(Z; <, +, 0, 1)$. As announced, we will derive Presburger's result from Skolem's.

4.6. Presburger's Quantifier-Elimination. *$Th(Z; <, +, 0, 1)$ admits a quantifier-elimination relative to the augmented language containing, for each $n > 1$, the mod-n congruence relation $(\cdot) \equiv (\cdot) \pmod{n}$.*

Proof. Let us begin by relabelling the two languages:

$$L_S: \ <, +, q\cdot, [\cdot], \overline{0}, \overline{1} \quad (q \text{ ranging over all rationals})$$

$$L_P: \ <, +, \overline{0}, \overline{1}, (\cdot) \equiv (\cdot) \pmod{n} \quad (n = 2, 3, \ldots).$$

Let us also introduce an intermediate language,

$$L_I: \ <, +, \overline{0}, \overline{1}, q\cdot, [\cdot], n \mid (\cdot), \quad (q \in Q \wedge n = 2, 3, \ldots)$$

where "$n \mid t$" reads "n divides t". L_I is introduced because the transformation of a quantifier-free formula of L_S into one of L_P must go through several steps.

I. We begin with a preliminary simplification of terms. Every term t of L_S can be written in the form $\Sigma q_i t_i$, where each t_i is a variable, a constant $\overline{0}$ or $\overline{1}$, or a term of the form $[s]$. Thus, each t_i is integral and, if we let m be the greatest common denominator of the q_i's, then we can write $t = (1/m) \Sigma n_i t_i$, each summand now being integral.

II. Suppose we have an atomic formula,

$$q[t] + u \sim s,$$

where $\sim$ is one of $=$, $<$, or $>$. We eliminate the given application of the greatest integer function as follows: Let t be written as $(1/m)\Sigma n_i t_i = (1/m)t'$ as in I. Then:

$$\begin{aligned}
q[t] + u \sim s \ &\leftrightarrow \ \bigvee_{k=0}^{m-1} \exists v \left(t' = mv + \overline{k} \wedge q\left[\frac{mv + \overline{k}}{m}\right] + u \sim s\right) \\
&\leftrightarrow \ \bigvee_{k=0}^{m-1} \exists v \left(t' - \overline{k} = mv \wedge qv + u \sim s\right) \\
&\leftrightarrow \ \bigvee_{k=0}^{m-1} \exists v \left(t' - \overline{k} = mv \wedge qmv + mu \sim ms\right) \\
&\leftrightarrow \ \bigvee_{k=0}^{m-1} \left(m \mid t' - \overline{k} \ \wedge \ q(t' - \overline{k}) + mu \sim ms\right).
\end{aligned}$$

Note that each summand of t' is integral, whence the divisibility assertion $m \mid t' - \overline{k}$ is meaningful.

In the final expression, we have reduced the total nestings of $[\cdot]$'s by one. However, if t' has a summand of the form $n_i[s_i]$, then we have introduced an occurrence of $[\cdot]$ into a new context— inside a divisibility statement. To eliminate such occurrences, we must introduce another reduction.

III. We must eliminate occurrences of $[\cdot]$ from contexts of the form

$$n \mid k[t] + u.$$

Let $t = (1/m)\Sigma n_i t_i = (1/m)t'$ as in I. Then:

$$\begin{aligned}
n \mid k[t] + u \quad &\leftrightarrow \quad \bigvee_{k=0}^{m-1} \exists v \left(n \mid k\left[\frac{mv + \overline{i}}{m}\right] + u \wedge t' = mv + \overline{i}\right) \\
&\leftrightarrow \quad \bigvee_{k=0}^{m-1} \exists v\,(n \mid kv + u \wedge mv = t' - \overline{i}) \\
&\leftrightarrow \quad \bigvee_{k=0}^{m-1} \exists v\,(mn \mid (km)v + mu \wedge mv = t' - \overline{i}) \\
&\leftrightarrow \quad \bigvee_{k=0}^{m-1} (mn \mid k(t' - \overline{i}) + mu \wedge m \mid t' - \overline{i}).
\end{aligned}$$

Again, we have reduced the possible depth of nestings of applications of the greatest integer function by one.

Iterating steps II and III, we can successively eliminate all applications of the greatest integer function. The result, a quantifier-free formula of L_I, is almost a quantifier-free formula of L_P. There is still the easy task of eliminating the rational multipliers. First, one replaces rational multipliers by integral ones by simply multiplying both sides of any equation or inequality by the positive greatest common denominator of all multipliers occurring in the equation or inequality. Negative integral multipliers can then be eliminated from equations and inequalities by simply adding their opposites to both sides of the given equations or inequalities. Similarly, a term in a divisibility statement can be split into positive and negative parts so that $n \mid t$ becomes $n \mid t_1 + (-1)t_2$, where t_1, t_2 have only positive multipliers. Replace this latter by $t_1 \equiv t_2 \pmod{n}$. One last step completes the transformation: rewrite every term mt with a positive integral multiplier m as the m-fold sum $t + \ldots + t$.

If one faithfully applies the above procedure to a quantifier-free formula in L_S, one will end up with a quantifier-free equivalent in L_P. Since every formula of the language of $(Z;<,+,0,1)$ has a quantifier-free equivalent in L_S, it has such in L_P. QED

4.7. Corollary. *$Th(\omega; <, +, 0, 1)$ admits a quantifier-elimination relative to the augmented language containing, for each $n > 1$, the mod-n congruence relation $(\cdot) \equiv (\cdot) \pmod{n}$.*

Proof. Let φ be a formula of the language of $(\omega; <, +, 0, 1)$. Passing successively from φ to $\varphi^{(\omega)}$ to its Skolem quantifier-eliminant φ_S to its Presburger quantifier-eliminant φ_P, we have, for any $x_0, \ldots, x_{n-1} \in \omega$ (letting $\equiv_n$ denote the mod-n congruence):

$$(\omega; <, +, 0, 1) \models \varphi\,\overline{x_0} \ldots \overline{x_{n-1}} \quad \leftrightarrow \quad (Z; <, +, 0, 1) \models \varphi^{(\omega)}\,\overline{x_0} \ldots \overline{x_{n-1}}$$

$$\leftrightarrow (Q; Z, <, +, \ldots) \models \varphi_S\, \overline{x_0} \ldots \overline{x_{n-1}}$$
$$\leftrightarrow (Z; <, +, \{\equiv_n: n > 1\}, 0, 1) \models \varphi_P\, \overline{x_0} \ldots \overline{x_{n-1}}$$
$$\leftrightarrow (\omega; <, +, \{\equiv_n: n > 1\}, 0, 1) \models \varphi_P\, \overline{x_0} \ldots \overline{x_{n-1}},$$

this last equivalence holding because the quantifier-free formula $\varphi_P\, \overline{x_0} \ldots \overline{x_{n-1}}$ lies entirely in the language of $(\omega; <, +, \{\equiv_n: n > 1\}, 0, 1)$ and the evaluations of truth in ω and Z are thus identical. Since $x_0, \ldots, x_{n-1}$ were arbitrary,

$$(\omega; <, +, \{\equiv_n: n > 1\}, 0, 1) \models \forall v_0 \ldots v_{n-1}(\varphi \leftrightarrow \varphi_P). \qquad \text{QED}$$

I mentioned in the opening paragraph of this section that Presburger's quantifier-elimination for $Th(\omega; <, +, 0, 1)$ is the most widely cited quantifier-elimination in the literature. One of the reasons for this is the expressive power of the language. For many structures admitting a quantifier-elimination (e.g., $(\omega; S, 0)$, $(\omega; <, S, 0)$, $(Q; +, 0)$, $(C; +, \cdot, 0, 1)$, where C is the set of complex numbers), the definable sets are either finite or co-finite (i.e., complements of finite sets); the definable sets in the language of $(\omega; <, +, 0, 1)$, although still simple, are more extensive and coïncide with a class of sets that arise in automata theory.

4.8. Definition. Let $X \subseteq \omega$ be given. X is *ultimately periodic* iff there are a positive integer p (a *period* of X) and a natural number x_0 such that, for $x \geq x_0$,

$$x \in X \;\text{ iff }\; x + p \in X. \qquad (15)$$

4.9. Definitions. An *arithmetic progression* is a function $F: \omega \to \omega$ of the form,

$$F(x) = m + nx,$$

for some $m, n \in \omega$. (If $n = 0$, one has a degenerate progression $F(x) = m$.) A set $X \subseteq \omega$ is *semi-linear* if X is a union of the ranges of finitely many arithmetic progressions.

4.10. Theorem. (Ginsburg-Spanier). *Let* $X \subseteq \omega$. *The following are equivalent:*
i. *X is definable in the language of* $(\omega; <, +, 0, 1)$, *i.e., there is a formula* φv_0 *with only* v_0 *free such that, for any* $x \in \omega$,

$$x \in X \;\textit{ iff }\; (\omega; <, +, 0, 1) \models \varphi \overline{x}$$

ii. *X is ultimately periodic*
iii. *X is semi-linear.*

Proof. The easy implications are ii $\to$ iii $\to$ i.

ii $\to$ iii. Let X be ultimately periodic and choose $p > 0$, x_0 so that (15) holds. For those numbers $x_i = x_0 + i$ for $0 \leq i < p$ such that $x_i \in X$, define an arithmetic progression,

$$F_i(x) = x_i + px.$$

For those $j < x_0$ such that $j \in X$, define

$$G_j(x) = j + 0x.$$

A moment's thought should convince the reader that X is the union of the ranges of the F_i's and G_j's that have been defined.

iii $\rightarrow$ i. Given an arithmetic progression,

$$F(x) = m + nx,$$

observe that the formula,

$$\varphi v_0:\ \exists v_1(v_0 = \overline{m} + nv_1),$$

defines the range of F. If $\varphi_0 v_0, ..., \varphi_{n-1} v_0$ define the ranges of progressions $F_0, ..., F_{n-1}$, respectively, then

$$\varphi v_0:\ \varphi_0 v_0 \vee ... \vee \varphi_{n-1} v_0$$

defines the union of these ranges.

To prove i $\rightarrow$ ii, we consider first an atomic formula φv_0 of L_P. Performing some algebraic simplifications, φv_0 assumes one of the forms,

$$mv_0 = \overline{k} \tag{16}$$

$$mv_0 < \overline{k} \tag{17}$$

$$\overline{k} < mv_0 \tag{18}$$

$$mv_0 \equiv \overline{k} \pmod{n}, \tag{19}$$

where m, k are non-negative integers and $n > 1$. If m is 0, φv_0 reduces to a sentence that is either true or false, whence φv_0 defines ω or $\emptyset$— in either case an ultimately periodic set. Thus, we can assume $m > 0$.

If φv_0 assumes either form (16) or (17), it defines a finite— hence ultimately periodic— set; if φv_0 assumes form (18), it defines the co-finite— hence ultimately periodic— set of natural numbers greater than k/m.

Thus, assume φv_0 takes the form (19). Let d be the greatest common divisor of m, n. If $d \nmid k$, (19) cannot be true of any natural number number, whence φv_0 defines $\emptyset$. If $d \mid k$, we can divide m, k and n in (19) to obtain a similar form,

$$m_0 v_0 \equiv \overline{k_0} \pmod{n_0}, \tag{20}$$

with m_0, n_0 relatively prime. Then m_0 has a multiplicative inverse m' modulo n_0, whence (20) is equivalent to

$$m' m_0 v_0 \equiv m' \overline{k_0} \pmod{n_0},$$

i.e., to

$$v_0 \equiv \overline{k_1} \pmod{n_0}, \tag{21}$$

for $k_1 = m'k_0$. The form (21) of φv_0 clearly defines an ultimately periodic set (in fact, a periodic set).

Thus, we have seen that every atomic formula of L_P with only one free variable defines an ultimately periodic set. Since every formula of the language of $(\omega; <, +, 0, 1)$ is equivalent to a quantifier-free formula of L_P with no additional free variables, and since every such quantifier-free formula of L_P is equivalent to a formula of L_P generated from the atomic formulae by disjunction, conjunction, and negation, it suffices to prove the class of ultimately periodic sets to be closed under finite union, finite intersection, and complementation. That is, the Theorem is proven by proving the following lemma.

4.11. Lemma. *The family of ultimately periodic sets is closed under finite union, finite intersection, and complementation.*

Proof. First observe that, if X is ultimately periodic with x_0, p satisfying (15),

$$(15): \quad \text{for } x \geq x_0,\ x \in X \leftrightarrow x+p \in X,$$

then,

$$\text{for } x \geq x_0,\ x \notin X \leftrightarrow x+p \notin X.$$

In other words, the family of ultimately periodic sets is closed under complementation. By de Morgan's Law, closure under intersection thus reduces to closure under union.

Let X, Y be ultimately periodic with x_0, p, y_0, q such that

$$\forall x \geq x_0 (x \in X \leftrightarrow x+p \in X)$$

$$\forall y \geq y_0 (y \in Y \leftrightarrow y+q \in Y).$$

Let $z_0 = \max\{x_0, y_0\}$ and r = least common multiple of p, q. I claim,

$$\forall x \geq z_0 (x \in X \cup Y \leftrightarrow x+r \in X \cup Y).$$

To see this, first observe that, for any $x \geq z_0$,

$$x \in X \cup Y \leftrightarrow x \in X \vee x \in Y. \tag{22}$$

Then observe:

$$\begin{aligned} x \in X &\leftrightarrow x+p \in X \\ &\leftrightarrow x+2p \in X \\ &\quad \vdots \\ &\leftrightarrow x+r \in X. \end{aligned} \tag{23}$$

Similarly,

$$x \in Y \leftrightarrow x+r \in Y. \tag{24}$$

Combining (22) - (24), we have

$$\begin{aligned} x \in X \cup Y &\leftrightarrow x \in X \vee x \in Y \\ &\leftrightarrow x+r \in X \vee x+r \in Y \\ &\leftrightarrow x+r \in X \cup Y. \end{aligned}$$

QED

A small remark: The direct proof of implication iii $\rightarrow$ ii of Theorem 4.10 reduces to showing the class of ultimately periodic sets to be closed under finite union. For, the range of an arithmetic progression is clearly ultimately periodic. Hence, every semi-linear set is the finite union of ultimately periodic sets.

Theorem 4.10 finishes our discussion of the quantifier-elimination and its consequences. The only thing left is the exercise set— and just before it, one quick and obvious application of Theorem 4.10. In proving the quantifier-eliminations, we have freely used multiplication by *fixed* scalars and divisibility by *fixed* divisors. Theorem 4.10 has as immediate application the non-definability of multiplication and the divisibility predicate.

4.12. Application. *The following numerical sets and relations are not definable in the language of* $(\omega; <, +, 0, 1)$:

i. *x is a square*
ii. *x is prime*
iii. $z = x \cdot y$
iv. $x \mid y$.

For the proof, cf. Exercise 11, below.

Exercises.

1. (Relative interpretability). Let $\mathfrak{A}$ be an L_1-structure for a language L_1 and let $\mathfrak{B}$ be an L_2-structure for a language L_2, and suppose

a. $B \subseteq A$ is definable in $\mathfrak{A}$ by an L_1-formula $\varphi_B v$, i.e.,

$$B = \{a \in A\colon \mathfrak{A} \models \varphi_B \overline{a}\};$$

b. for every n-ary function symbol F of L_2 there is a term t_F of L_1 with exactly $v_0, \ldots, v_{n-1}$ occurring in t_F and such that, for all $b_0, \ldots, b_{n-1} \in B$,

$$(F)^{\mathfrak{B}}(b_0, \ldots, b_{n-1}) = (t_F \overline{b_0} \ldots \overline{b_{n-1}})^{\mathfrak{A}};$$

c. for every n-ary relation symbol R of L_2, there is a formula φ_R of L_1 with exactly $v_0, \ldots, v_{n-1}$ free and such that, for all $b_0, \ldots, b_{n-1} \in B$,

$$\mathfrak{B} \models R\,\overline{b_0} \ldots \overline{b_{n-1}} \quad \text{iff} \quad \mathfrak{A} \models \varphi\, \overline{b_0} \ldots \overline{b_{n-1}};$$

d. for every constant c of L_2 there is a constant d_c of L_1 such that, if $b = c^{\mathfrak{B}}$, then $\mathfrak{A} \models \overline{b} = d_c$, i.e., $b = (d_c)^{\mathfrak{A}}$.

i. Define an interpretation I of terms of $L_2(B)$ into $L_1(A)$ as follows:

$$(v_i)^I \text{ is } v_i, \quad \overline{b}^I \text{ is } \overline{b}, \quad c^I \text{ is } d_c \ (\text{for } c \in L_2),$$
$$(Ft_0 \ldots t_{n-1})^I \text{ is } t_F(t_0)^I \ldots (t_{n-1})^I.$$

Show: For any closed terms t_1, t_2 of $L_2(B)$,

$$\mathfrak{B} \models t_1 = t_2 \quad \text{iff} \quad \mathfrak{A} \models (t_1)^I = (t_2)^I.$$

ii. Extend the interpretation I from terms to formulae as follows:

$$(t_1 = t_2)^I \text{ is } (t_1)^I = (t_2)^I$$
$$(Rt_0 \ldots t_{n-1})^I \text{ is } \varphi_R((t_0)^I, \ldots, (t_{n-1})^I)$$
$$(\varphi * \psi)^I \text{ is } \varphi^I * \psi^I, \text{ for } * \in \{\wedge, \vee, \rightarrow\}$$
$$(\neg\varphi)^I \text{ is } \neg(\varphi^I)$$
$$(\exists v_i\, \varphi v_i)^I \text{ is } \exists v^*[\varphi_B(v^*) \wedge (\varphi v^*)^I]$$
$$(\forall v_i\, \varphi v_i)^I \text{ is } \forall v^*[\varphi_B(v^*) \rightarrow (\varphi v^*)^I],$$

where v^* is any variable not occurring already in φ_B or φ^I. Show: For any sentence φ of $L_2(B)$,

$$\mathfrak{B} \models \varphi \quad \text{iff} \quad \mathfrak{A} \models \varphi^I.$$

iii. Prove the corresponding assertion for relativisations $\varphi^{(B)}$.

iv. Suppose the choices $\varphi_B, t_F, \varphi_R, d_c$ do not depend on $\mathfrak{A}$, i.e., suppose we have two theories $\mathcal{T}_1, \mathcal{T}_2$ in languages L_1, L_2, respectively, and that $\varphi_B, t_F, \varphi_R, d_c$ define, for each model $\mathfrak{A} \models \mathcal{T}_1$ a model $\mathfrak{B} \models \mathcal{T}_2$. Show: For any sentence φ of L_2,

$$\mathcal{T}_2 \vdash \varphi \;\rightarrow\; \mathcal{T}_1 \vdash \varphi^I.$$

[Hint. iv. Apply the Completeness Theorem, not induction on the length of a formal proof.]

[Remark. The treatment of function symbols given in this Exercise is not as general as it could be. Functions can also be represented by their graphs. Thus, in place of t_F interpreting F, one could have a formula φ_F such that:

a. $\mathfrak{A} \models \forall v_0 \ldots v_{n-1}(\bigwedge\!\!\!\bigwedge \varphi_B v_i \rightarrow \exists! v_n\, \varphi_F v_0 \ldots v_{n-1} v_n)$

b. for all $b_0, \ldots, b_{n-1}, b_n \in B$,

$$\mathfrak{B} \models F\overline{b_0} \ldots \overline{b_{n-1}} = \overline{b_n} \;\rightarrow\; \mathfrak{A} \models \varphi_F \overline{b_0} \ldots \overline{b_{n-1}},$$

where $\exists! v_n \psi$ abbreviates $\exists v_n \psi v_n \wedge \forall v_n v[\psi v_n \wedge \psi v \rightarrow v_n = v]$ in a. I leave to the reader the task of adapting Exercise 1 to this case.]

2. (Disjunctive Normal Form). Using the valid equivalences,

$$\neg(\alpha \wedge \beta) \leftrightarrow \neg\alpha \vee \neg\beta$$
$$\neg(\alpha \vee \beta) \leftrightarrow \neg\alpha \wedge \neg\beta$$
$$\alpha \rightarrow \beta \leftrightarrow \neg\alpha \vee \beta$$
$$\alpha \wedge (\beta \vee \gamma) \leftrightarrow (\alpha \wedge \beta) \vee (\alpha \wedge \gamma),$$

as well as the more obvious,

$$\alpha \vee (\beta \vee \gamma) \leftrightarrow (\alpha \vee \beta) \vee \gamma, \quad \alpha \wedge (\beta \wedge \gamma) \leftrightarrow (\alpha \wedge \beta) \wedge \gamma$$
$$\alpha \vee \beta \leftrightarrow \beta \vee \alpha, \quad \alpha \wedge \beta \leftrightarrow \beta \wedge \alpha,$$

show by induction on the number of logical connectives occurring in a quantifier-free formula φ that there is a formula ψ of the form,

$$\bigvee\!\!\!\bigvee \bigwedge\!\!\!\bigwedge \psi_{ij}, \quad \psi_{ij} \text{ atomic or negated atomic,}$$

such that $\vdash \varphi \leftrightarrow \psi$.

3. (Prenex Normal Form).

i. Prove the equivalences:

a. $\exists v\varphi \vee \exists v\psi \;\leftrightarrow\; \exists v(\varphi \vee \psi)$

b. $\forall v\varphi \wedge \forall v\psi \;\leftrightarrow\; \forall v(\varphi \wedge \psi)$

c. $\varphi \wedge \exists v\psi \;\leftrightarrow\; \exists v_0(\varphi \wedge \psi v_0)$

d. $\varphi \vee \forall v\psi \;\leftrightarrow\; \exists v_0(\varphi \vee \psi v_0)$

e. $\neg\exists v\varphi \;\leftrightarrow\; \forall v \neg\varphi$

f. $\neg\forall v\varphi \;\leftrightarrow\; \exists v \neg\varphi,$

where v_0 is free for v in ψv and does not occur free in φ in equivalences c, d.

ii. Prove the equivalences:

a. $(\exists v\varphi \rightarrow \psi) \;\leftrightarrow\; \forall v_0\, (\varphi v_0 \rightarrow \psi)$

b. $(\forall v\varphi \rightarrow \psi) \;\leftrightarrow\; \exists v_0\, (\varphi v_0 \rightarrow \psi)$

c. $(\psi \to \exists v\varphi) \leftrightarrow \exists v_0(\psi \to \varphi v_0)$

d. $(\psi \to \forall v\varphi) \leftrightarrow \forall v_0(\psi \to \varphi v_0)$,

where v_0 is free for v in φv and does not occur free in ψ.

iii. Prove: Every formula φ is equivalent to one of the form,

$$Q_0 v_{i_0} \ldots Q_{k-1} v_{i_{k-1}} \psi,$$

where each Q_i is a quantifier and ψ is quantifier-free. ψ can be taken to be in disjunctive normal form.

4. In each of the following parts, give a quantifier-elimination for one of the given structures. (You will not need to extend the given languages.) Give axiomatisations of the theories of the structures.

i. $(\omega; S, 0)$, $(Z; S, 0)$

ii. $(\omega; <, S, 0)$, $(Z; <, S, 0)$

iii. $(Q; <, +, 0, 1)$, $(R; <, +, 0, 1)$ (R the set of real numbers).

5. (Sylvester, Frobenius).

i. Let m, n be relatively prime positive integers. Show that every positive integer $c > mn$ is representable over the set of positive integers by the form $mX + nY$, i.e., the equation

$$mX + nY = c$$

is solvable in positive integers for $c > mn$. Show that this bound is optimal.

ii. Let m, n, p be relatively prime positive integers. Show: There is a number c_0 such that every natural number $c > c_0$ is representable over the natural numbers by the form $mX + nY + pZ$.

iii. Generalise ii to the form,

$$m_0 X_0 + \ldots + m_{k-1} X_{k-1},$$

with $m_0, \ldots, m_{k-1}$ relatively prime.

[Remark. The bound c_0 of part ii given by the quantifier-elimination, $c_0 = \min\{m(n-1) + p(n-1) - n,\ n(m-1) + p(m-1) - m,\ m(p-1) + n(p-1) - p\}$, is not best possible.]

6. Let $\mathcal{T}$ be a theory in a given language L. $\mathcal{T}$ is said to be *model complete* if, for any models $\mathfrak{A}, \mathfrak{B}$ of $\mathcal{T}$,

$$\mathfrak{A} \subseteq \mathfrak{B} \to \mathfrak{A} \preccurlyeq \mathfrak{B}.$$

(Recall Exercise 8 of section 3, above.)

i. Show: If $\mathcal{T}$ admits a quantifier-elimination in L, then $\mathcal{T}$ is model complete.

ii. Show: $(Q; +, 0, 1) \preccurlyeq (R; +, 0, 1)$.

iii. Suppose $\mathcal{T}$ is model complete and has a model $\mathfrak{A}_0$ that is isomorphically embeddable into every model $\mathfrak{B}$ of $\mathcal{T}$ (i.e., $\mathfrak{A}_0$ is isomorphic to a submodel of $\mathfrak{B}$). Show: $\mathcal{T}$ is complete.

[Remark. The notion of model completeness is due to Abraham Robinson, who showed how it could be used to simplify many proofs of completeness. Together with Elias Zakon, he applied his method to $Th(Z; <, +, 0, 1)$— cf. the Reading List for the reference.]

7. Fill in the missing details of the proof of Application 4.5; in particular, verify the crucial properties of $\mathfrak{B}$ and $\mathfrak{B}^*$.

8. Fix $n > 0$ and define A_n to be the set of polynomials,

$$a_0 + a_1X + ... + a_nX^n,$$

where $a_0 \in Z$ and $a_1, ..., a_n \in Q$. Define 0, 1, + on A_n in the obvious way and define < by

$$P(X) < Q(X) \text{ iff } Q \text{ eventually dominates } P,$$

i.e., iff $\exists x_0 \forall x > x_0(P(x) < Q(x))$.

i. Show: $\mathfrak{A}_n = (A_n; <, +, 0, 1) \models \mathcal{T}_+$, with $\mathcal{T}_+$ as in Application 4.5.

ii. Describe the structures $\mathfrak{B}_n$ and $\mathfrak{B}_n{}^*$ arising from $\mathfrak{A}_n$ by applying the construction of Application 4.5. Given an element $b \in B_n$, what is $[b]$?

9. i. Show that $Th(\omega:<,+,0,1)$ can be axiomatised by

a. replacing axiom $A3$ of $\mathcal{T}_+$ by

$$\forall v_0 v_1 [v_0 < v_1 \rightarrow \exists v_2(v_0 + v_2 = v_1)]$$

and b. adding the axiom,

$$\forall v_0(v_0 = \overline{0} \vee \overline{0} < v_0).$$

ii. Show that < is definable in the structure $(\omega; +, 0, 1)$.

iii. Describe polynomial models of $Th(\omega; <, +, 0, 1)$ analogous to those of Exercise 8.

[Hint. i. Show how to embedd any model of the proposed axioms for $Th(\omega; <, +, 0, 1)$ into a model of $\mathcal{T}_+$.]

10. i. State and prove the analogue to Theorem 4.10 for the structure $(Z; <, +, 0, 1)$.

ii. Assuming a quantifier-elimination for $Th(Z; +, 0, 1)$, i.e., the addition of the integers without order, in the language augmented by the congruence relations, state and prove the analogue to Theorem 4.10 for $(Z; +, 0, 1)$.

iii. Applying ii, show that < is not definable in the language of $(Z; +, 0, 1)$.

iv. Recalling Exercise 6 of section 3, give a simpler proof that < is not definable in the language of $(Z; +, 0, 1)$.

[Hints. i. Given any $X \subseteq Z$, split it into two pieces,

$$X^+ = \{x \in \omega: x \in X\}, \; X^- = \{x \in Z - \omega: x \in X\},$$

and apply 4.10; ii. mimick the proof of 4.10.]

11. Prove Application 4.12: The sets of squares and primes, the graph of the multiplication function, and the divisibility relation are not definable in the structure $(\omega;<,+,0,1)$.

12. Consider the model $\mathfrak{A}_1 \models \mathcal{T}_+$ described in Exercise 8, above:

$$A_1 = \{a + bX: a \in Z \wedge b \in Q\}.$$

i. Show: There is no way of defining an order-preserving multiplication on $\mathfrak{A}_1$, i.e., there is no operation $\cdot$ on A_1 such that $\mathfrak{A}_1$ satisfies

$$m \cdot (a + bX) = (ma) + (mbX), \quad \text{for all } m \in \omega$$

$$\forall v_0 v_1 v_2(\overline{0} < v_0 \wedge v_1 < v_2 \rightarrow v_0 \cdot v_1 < v_0 \cdot v_2).$$

ii. The graph of the multiplication function is not definable in $(Z; <, +, 0, 1)$.

[Hints. i. Consider X^2; ii. if one could define multiplication in $(Z; <, +, 0, 1)$, there would be a multiplication function on $\mathfrak{A}_1$!]

13. Let $n > 1$ be fixed and consider the set of polynomials,

$$A^n = \{a + (m/n^k)X:\ a, m \in Z \wedge k \in \omega\}.$$

Define <, +, 0, 1 as in Exercise 8 and let $\mathfrak{A}^n$ denote the resulting structure.

i. Show that $\mathfrak{A}^n$ is a model of axioms *A1* - *A4*, *O1* - *O3* and *OA*.

ii. Show that, for any divisor m of n, $\mathfrak{A}^n$ satisfies axiom D_m.

iii. Let p be a prime number that doesn't divide n. Show that $\mathfrak{A}^n$ does not satisfy D_p.

iv. Show that no finite subset of the set of axioms of $\mathcal{T}_+$ entails all the axioms of $\mathcal{T}_+$.

[Hints. iii. Try dividing X by p; iv. any finite subset of $\mathcal{T}_+$ is valid in some $\mathfrak{A}^n$.]

14. Let $\mathcal{T}$ be a theory in a language L. Suppose $\mathcal{T}$ has two axiomatisations given by sets X, Y of sentences. Show: If Y is finite, then $\mathcal{T}$ can be axiomatised by a finite subset of X. Using this and the result of Exercise 13, conclude that $\mathcal{T}_+$ has no finite axiomatisation.

15. Give a quantifier-elimination for the structure

$$(Q;\ Z, <, +, 0, 1, \{q\cdot :\ q \in Q\}, [\cdot]).$$

[Hint-sketch. The predicate Zv can be replaced by the equation $v = [v]$; thus, one needs only consider formulae of the form (2). If the quantified variable is integral, apply Skolem's procedure. Otherwise, replace $\exists x(\bigwedge\!\!\!\bigwedge \alpha_i)$ by $\exists z_0 ... z_{n-1} \exists x\ (\bigwedge\!\!\!\bigwedge \alpha_i' \wedge z_j = [t_j])$, where α_j' and t_j have no occurrences of $[\cdot]$'s. Change this to

$$\exists z_0 ... z_{n-1} (\bigwedge\!\!\!\bigwedge Z(z_j) \wedge \exists x\ (\bigwedge\!\!\!\bigwedge \alpha_i' \wedge \bigwedge\!\!\!\bigwedge z_j \le t_j \wedge \bigwedge\!\!\!\bigwedge t_j < z_j + \overline{1}))$$

which has no $[\cdot]$'s. Eliminate the quantifier $\exists x$ as in Exercise 4.iii, then replace each $Z(z_j)$ by $z_j = [z_j]$ and apply Skolem's procedure to eliminate the remaining quantifiers.]

*5. Skolem Arithmetic; The Theory of Multiplication

In the same paper in which he provided his quantifier-elimination for the theory of addition, Skolem proved the decidability of the theory $Th(\omega^+; \cdot, 1)$ of multiplication of the positive integers. He did not prove this latter result by a quantifier-elimination; in fact, he didn't really *prove* the result: Skolem chose five examples illustrating his decision method and declared the method completely general.

The first complete proof of the decidability of $Th(\omega^+; \cdot, 1)$ published was given by Andrzej Mostowski in 1952. Mostowski considered the general problem of relating the decidability of the theory of a product of structures to the decidabilities of the factor structures. As an illustration of one of his results, he reduced the decidability of $Th(\omega^+; \cdot, 1)$ to $Th(\omega; +, 0, 1)$. This reduction is based on the Fundamental Theorem of Arithmetic: Let $p_0 = 2, p_1, ...$ be the sequence of prime numbers. Every positive integer x can be written uniquely in the form,

$$x = (p_0)^{n_0}(p_1)^{n_1}...,$$

where each $n_i \in \omega$ and all but finitely many n_i's are 0. If, moreover,

$$y = (p_0)^{m_0}(p_1)^{m_1}\cdots,$$

then

$$xy = (p_0)^{n_0+m_0}(p_1)^{n_1+m_1}\cdots;$$

and

$$1 = (p_0)^0(p_1)^0\cdots.$$

In short, $(\omega^+; \cdot, 1)$ is (isomorphic to) the *weak direct power* of $(\omega; +, 0)$.

5.1. Definition. Let $\mathfrak{A}$ be a structure for a language L with a unique constant c_0, and let $a = (c_0)^{\mathfrak{A}}$. The *weak direct power* $\mathfrak{B}$ of $\mathfrak{A}$ is defined as follows: The domain of $\mathfrak{B}$ is

$$B = \{F:\omega \to A \mid \exists x_0 \in \omega\, \forall x > x_0\, (Fx = a)\}.$$

The interpretation of the constant c_0 is the constant sequence $F_a(n) = a$. For any n-ary function symbol F_i^n of L, $(F_i^n)^{\mathfrak{B}}$ is defined by,

$$(F_i^n)^{\mathfrak{B}}(F_0,\ldots,F_{n-1}) = F,$$

where

$$F(m) = (F_i^n)^{\mathfrak{A}}(F_0(m),\ldots,F_{n-1}(m)).$$

Finally, for any n-ary relation symbol R_i^n of L, one defines

$$(R_i^n)^{\mathfrak{B}}(F_0, \ldots, F_{n-1}) \text{ iff } \forall m\, (R_i^n)^{\mathfrak{A}}(F_0(m), \ldots, F_{n-1}(m)).$$

In words, functions and relations are defined component-wise on the elements of B.

What Mostowski observed was that $(\omega^+; \cdot, 1)$ is the weak direct power of $(\omega; +, 0)$, with 0 becoming 1 and + becoming · (and < becoming the divisibility relation when < is added to the latter structure). The relevant theorem of Mostowski was the following.

5.2. Theorem. *Let $\mathcal{T}$ be a decidable theory with a unique distinguished constant. The theory of weak direct powers of models of $\mathcal{T}$ is decidable.*

I won't prove this here. Indeed, as I announced in section 1, I intent not to prove anything in this section, but merely to outline the results.

In the late 1950s, Solomon Feferman and Robert Vaught generalised Mostowski's results, but did not specifically apply their work— an elaborate quantifier-elimination— to $Th(\omega^+; \cdot, 1)$. This was done in 1980 by Patrick Cegielski, who transformed the quantifier-elimination for the theory of addition into one for the theory of multiplication. He was also able to supply an axiomatisation for $Th(\omega^+; \cdot, 1)$. Both the quantifier-elimination and the axiomatisation are best understood by thinking of $(\omega^+; \cdot, 1)$ as the weak direct power of $(\omega; +, 0)$.

Let me begin by presenting the axioms for $Th(\omega^+; \cdot, 1)$.

First, there are axioms asserting that $(\omega^+; \cdot, 1)$ is a torsion-free, cancellative, commutative semi-group with identity and no other unit:

S1. $\forall v_0 v_1 v_2 [v_0 \cdot (v_1 \cdot v_2) = (v_0 \cdot v_1) \cdot v_2]$

S2. $\forall v_0 v_1 (v_0 \cdot v_1 = v_1 \cdot v_0)$

$S3 \quad \forall v_0(\overline{1} \cdot v_0 = v_0)$

$S4. \quad \forall v_0 v_1 v_2 (v_0 \cdot v_2 = v_1 \cdot v_2 \rightarrow v_0 = v_1)$

$S5. \quad \forall v_0 v_1 (v_0 \cdot v_1 = \overline{1} \rightarrow v_0 = \overline{1} \vee v_1 = \overline{1})$

$S6_n. \quad \forall v_0 v_1 [(v_0)^n = (v_1)^n \rightarrow v_0 = v_1],$

where t^n abbreviates 1 for $n = 0$, and the n-fold product $t \cdots t$ for $n > 0$.

Note the partial correspondence of with the axioms $\mathcal{T}_+$ of $Th(Z; <, +, 0, 1)$:

$$S1 - A1, \quad S2 - A4, \quad S3 - A2.$$

S4 corresponds to *A3*, but not quite adequately because $\mathcal{T}_+$ allows negative integers— $(Z; +, 0)$ is a group and not merely a semi-group. The correspondence with the subtraction axiom (Exercise 9 of the preceding section) is closer, but still not exact because $(\omega^+; \cdot, 1)$ is *partially* rather than totally ordered. *S5* corresponds similarly to the axiom asserting that 0 is the least element of ω— if one replaces $<$ by its definition in terms of $+$, the correspondence is clearer. Axiom schema $S6_n$ is best associated with *OA*, which guarantees against torsion in $(\omega; <, +, 0, 1)$.

Divisibility is readily definable in the multiplicative language:

$$v_0 \mid v_1: \exists v_2 (v_1 = v_0 \cdot v_2).$$

Using this, we can write the multiplicative version of the division algorithm axiom D_n:

$MD_n. \quad \forall v_0 \exists v_1 v_2 (v_0 = (v_1)^n v_2 \wedge \forall v_3 v_4 [v_0 = (v_3)^n v_4 \rightarrow v_2 \mid v_4]).$

Here, the point is that, if we take "logs", we have

$$l_0 = n l_1 + l_2,$$

with a *minimum* l_2— inequality in $(\omega; +, 0)$ corresponding to divisibility in $(\omega^+; \cdot, 1)$.

Except for $S6_n$, we haven't mentioned any analogues to the order axioms. This is a much more complicated matter. The order on $(\omega^+; \cdot, 1)$ is given by the divisibility relation. The minimum non-trivial elements of this order (corresponding to 1 in $(\omega; <, +, 0, 1)$) are the primes. Observe that one can define the notion of primeness:

$$v_0 \text{ is prime: } \forall v_1 (v_1 \mid v_0 \rightarrow v_1 = \overline{1} \vee v_1 = v_0) \wedge \neg v_0 = \overline{1}.$$

For any prime p, the p-primary numbers are those of the form p^n, or, put more readily expressibly, those divisible by no primes other than p:

$$PR(v_0, v_1): v_0 \text{ is prime} \wedge \forall v_2 (v_2 \text{ is prime} \wedge v_0 \neq v_2 \rightarrow \neg(v_2 \mid v_1)).$$

Two simple axioms assert the existence of primes and the totality of the ordering given by $\mid$ on the p-primary numbers:

$P1. \quad \forall v_0 \exists v_1 (v_1 \text{ is prime} \wedge \neg(v_1 \mid v_0))$

$P2. \quad \forall v_0 v_1 v_2 (PR(v_0, v_1) \wedge PR(v_0, v_2) \rightarrow (v_1 \mid v_2 \vee v_2 \mid v_1)).$

If we imagine the individual factors of the weak direct product as being indexed by the primes, we see that the p-primary numbers under multiplication are a copy of the p-th factor. The next definition and group of axioms are designed to allow us to decompose any element of the product into its components in the various factors. From a different perspective, it embodies the Fundamental Theorem of Arithmetic.

First, the definition of the *p-adic valuation* $V(p, x)$ of a number x:

$$v_2 = V(v_0, v_1): \; PR(v_0, v_1) \wedge v_2 \mid v_1 \wedge \forall v_3(PR(v_0, v_3) \wedge v_3 \mid v_1 \rightarrow v_3 \mid v_2).$$

In words, $V(p, x)$ is the largest power of p dividing x, for a prime p. The valuation axioms begin with existence:

V1. $\forall v_0 v_1(v_0 \text{ is prime} \rightarrow \exists v_2(v_2 = V(v_0, v_1)))$

V2. $\forall v_1 v_2(v_1 = v_2 \leftrightarrow \forall v_0[v_0 \text{ is prime} \rightarrow V(v_0, v_1) = V(v_0, v_2)])$

V3. $\forall v_0 v_1 v_2(v_0 \text{ is prime} \rightarrow V(v_0, v_1 \cdot v_2) = V(v_0, v_1) \cdot V(v_0, v_2))$

V4. $\forall v_1 v_2(\forall v_0[v_0 \text{ is prime} \rightarrow V(v_0, v_1) \mid V(v_0, v_2)] \rightarrow v_1 \mid v_2)$.

There are only three axioms left to discuss. The first allows a truncation operation. Given two numbers x, y, the truncation of y relative to x is that number z obtained by deleting from the prime factorisation of y all factors corresponding to primes not dividing x.

T. $\forall v_1 v_2 \exists v_3 \forall v_0[v_0 \text{ is prime} \rightarrow (v_0 \mid v_1 \rightarrow V(v_0, v_3) = V(v_0, v_2)) \wedge$
$\wedge (\neg(v_0 \mid v_1) \rightarrow V(v_0, v_3) = \overline{1})]$.

The number asserted to exist is unique (by *V2*) and we denote it by $T(v_1, v_2)$.

Another function-introducing axiom is the incrementation axiom, which allows an increase of each non-zero exponent in the prime factorisation of a number by 1:

$$I((p_{i_0})^{n_0} \cdots (p_{i_{k-1}})^{n_{k-1}}) = (p_{i_0})^{n_0+1} \cdots (p_{i_{k-1}})^{n_{k-1}+1}.$$

The axiom,

I. $\forall v_1 \exists v_2 \forall v_0[v_0 \text{ is prime} \rightarrow (v_0 \mid v_1 \rightarrow V(v_0, v_2) = v_0 \cdot V(v_0, v_1)) \wedge$
$\wedge (\neg(v_0 \mid v_1) \rightarrow V(v_0, v_2) = \overline{1})]$.

asserts the existence of $I(v_1)$. As with T, the element asserted to exist is unique and we can introduce a new function symbol I.

Preparatory to introducing the last axiom, define, for each $n \in \omega$, the relation "y/x is an n-th power":

$$v_0 \mid_n v_1: \; \exists v_2(v_1 = (v_2)^n \cdot v_0).$$

Our final axiom is a schema:

SP_n. $\forall v_1 v_2 \exists v_3 \forall v_0[v_0 \text{ is prime} \rightarrow$
$\rightarrow (v_0 \mid v_1 \cdot v_2 \wedge V(v_0, v_1) \mid_n V(v_0, v_2) \rightarrow V(v_0, v_3) = v_0) \wedge$
$\wedge (\neg(v_0 \mid v_1 \cdot v_2) \vee \neg(V(v_0, v_1) \mid_n V(v_0, v_2)) \rightarrow V(v_0, v_3) = \overline{1})]$.

To explain schema SP_n, first note that we can add a new function symbol SP_n to the language yielding the postulated v_3. For each prime p, and any x, y such that $p \mid xy$,

$$V(p, x) \mid_n V(p, y) \text{ iff } V(p, SP_n(x, y)) = p,$$

whence we get a characteristic function of $\mid_n$ restricted to values $V(p, x)$. The axiom itself is a limited *comprehension*, or *separation*, axiom asserting in a coded way the existence of the "set" $SP_n(x, y)$ of all "elements" x of y which n-divide y.

This completes the list of Cegielski's axioms for $Th(\omega^+; \cdot, 1)$.

5.3. Theorem. *$Th(\omega^+; \cdot, 1)$ is axiomatised by the axioms S1 - S5, $S6_n$, MD_n, P1 - P2, V1 - V4, T, I, and SP_n.*

This is, of course, only half of Cegielski's result. We must also state that quantifiers can be eliminated in the appropriate language.

For any model $\mathfrak{A}$ of the given axioms, and any prime p of the model, one can define

$$A_p = \{a \in A\colon\ a \text{ is } p\text{-primary}\},$$

and consider the structure,

$$\mathfrak{A}_p = (A_p; \cdot, 1).$$

The axioms given for $Th(\omega^+; \cdot, 1)$ are strong enough to guarantee each $\mathfrak{A}_p$ to be a model of $Th(\omega; +, 0)$. In fact, for the model $(\omega^+; \cdot, 1)$, each $\mathfrak{A}_p$ is isomorphic to $(\omega; +, 0)$.

Equally important is the fact that A_p is definable in $\mathfrak{A}$ in terms of the parameter p:

$$v_0 \in A_p\colon\ PR(\overline{p}, v_0).$$

For any formula $\varphi v_1...v_n$ of the language of $\mathfrak{A}$, we can find a formula φ^p such that, for all $a_0, ..., a_{n-1} \in A$,

$$\mathfrak{A} \models \varphi^p(\overline{a_0}, ..., \overline{a_{n-1}}) \text{ iff } \mathfrak{A}_p \models \varphi(\overline{b_0}, ..., \overline{b_{n-1}}),$$

where $b_i = V(p, a_i)$. To define φ^p, first relativise φ to A_p and then replace each free variable v_i in φ by $V(\overline{p}, v_i)$.

The construction of φ^p is uniform in the constants $\overline{p}$, i.e., for each φ, there is a single formula $\varphi^{v_0}v_1...v_n$ from which each φ^p is obtained by substituting the constant $\overline{p}$ for the variable v_0.

The Feferman-Vaught Theorem tells us that every formula of the language in question (with primitives $\cdot$, 1, =) is provably equivalent to a propositional combination of formulae of the form,

$$\exists v_0...v_{k-1}[\bigwedge_{i<j} v_i \neq v_j \wedge \bigwedge_i (v_i \text{ is prime } \wedge \varphi^{v_i})]. \qquad (*)$$

In $\mathfrak{A}_p$, the formula is equivalent to a quantifier-free formula if we add the definable constant 1, and relations $<$, $(\cdot) \equiv (\cdot) \pmod n$ for $n > 1$. This allows some simplification of the formulae (*). With the addition of the function symbols I, T, and SP_n ($n > 0$), it turns out the only formulae of the form (*) that need to be simulated by new primitives to obtain a quantifier-elimination are the unary predicates,

$$E_n(v_0)\colon\ \exists v_0...v_{k-1}[\bigwedge_{1\le i<j\le n} v_i \neq v_j \wedge \bigwedge_{i\le n}(v_i \text{ is prime } \wedge v_i \mid v_0)],$$

for $n \geq 1$. Basically, the reason for this is that the other new primitives of the augmented language (interpreted in the multiplicative structures $\mathfrak{A}_p$) are replaced by the functions I, T, and SP_n. Consider, for example, the additive formula $\overline{b_0} < \overline{b_1}$ for $b_i = V(p, a_i)$:

$$\begin{aligned}\mathfrak{A}_p \models \overline{b_0} < \overline{b_1} \quad &\leftrightarrow \quad \mathfrak{A} \models V(\overline{p}, \overline{a_0}) \mid V(\overline{p}, \overline{a_1}) \\ &\leftrightarrow \quad \mathfrak{A} \models V(\overline{p}, \overline{a_0}) \mid_1 V(\overline{p}, \overline{a_1}) \\ &\leftrightarrow \quad \mathfrak{A} \models V(\overline{p}, SP_1(\overline{a_0}, \overline{a_1})) = \overline{p}\end{aligned}$$

$$\leftrightarrow \quad \mathfrak{A} \models \overline{p} \mid SP_1(\overline{a_0}, \overline{a_1}).$$

Thus, a formula like

$$\exists v_0 v_1 [v_0 \neq v_1 \wedge (v_0 \text{ is prime} \wedge (\overline{b_0} < \overline{b_1})^{v_0}) \wedge (v_1 \text{ is prime} \wedge (\overline{b_0} < \overline{b_1})^{v_1}),$$

is equivalent to

$$\exists v_0 v_1 [v_0 \neq v_1 \wedge (v_0 \text{ is prime} \wedge v_0 \mid SP_1(\overline{a_0}, \overline{a_1})) \wedge$$
$$\wedge (v_1 \text{ is prime} \wedge v_0 \mid SP_1(\overline{a_0}, \overline{a_1})),$$

i.e., to $E_2(SP_1(\overline{a_0}, \overline{a_1}))$.

Obviously, this sketch is far from a proof, but it should at least make plausible the full result of Cegielski's:

5.4. Theorem. *$Th(\omega^+; \cdot, 1)$ admits a quantifier-elimination when the language is augmented by the function symbols I, T, SP_n ($n \geq 0$), and the unary relation symbols E_n ($n \geq 1$).*

A full proof of Theorem 5.4, as well as of Theorem 5.3, can be found in Cegielski's paper cited in the Reading List for this Chapter.

I have only a small closing comment to make on the theory of multiplication: as with the theory of addition, the quantifier-elimination yields a decision procedure for the theory. The decision procedure is, however, much more complicated than that for addition. First, the new primitives are more complicated. The computationally most difficult primitives for the theory of addition were the congruence relations, which depended on division. In the theory of multiplication, the new primitives I, T, SP_n, and E_n all depend on the inherently more difficult task of factoring. Second, the quantifier-elimination procedure outlined for the theory of multiplication requires multiple application of that for the theory of addition. Computer scientists studying the complexities of decision procedures have shown that $Th(\omega; +, 0)$ has complexity of order $2^{2^{cn}}$, i.e., there are constants c, c' such that
i. every sentence φ of the given language of size $\leq n$ can be decided in at most $2^{2^{cn}}$ steps,
and ii. for any decision procedure there are, for infinitely many n, formulae of size $\leq n$ which require more than $2^{2^{c'n}}$ steps of the procedure to be decided.
The lower complexity for $Th(\omega^+; \cdot, 1)$ has been shown to be of order $2^{2^{2^{cn}}}$.

When one mixes addition and multiplication, the complexity becomes infinite, i.e., there is no decision procedure. This we shall see in the next section.

6. Theories with + and · ; Incompleteness and Undecidability

The decidability of the theory of addition, as well as that of the theory of multiplication, is due in no small part to the limited expressive power of the underlying language. In a language with both addition and multiplication we will be able to express a great deal more, whence undecidability rather than decidability will confront us. The present section offers a first round of undecidability results. It also, of necessity, takes a first look at syntactic notions of definability. Indeed, it is easy to lose track of our goals of proving incompleteness and undecidability results, treating them as afterthoughts— remarks of the form "Oh, by the way, ...". What readily comes to the fore is the study of these various weak notions of definability. The emphasis on definability, rather than on incompleteness and undecidability, is the result of an evolution in research and exposition, and somewhat distorts the picture. In the next chapter (Chapter IV of volume II), a more historically correct approach will be followed.

The first requisite for a logical notion of definability is a language, in the present case *a* language of arithmetic. I say "*a*" rather than "the" language of arithmetic because it is convenient to vary the language from application to application. Moreover, for greater generality in some results (especially in Chapter V of volume II), one often only wants the language to "contain" a language of arithmetic in some sense of the word "contain". In the case of axiomatic set theory, for instance, "containment" means the existence of a relative interpretation; in the present work, "containment" will usually mean that "the" arithmetic primitives are among those of the given language.

And what are these arithmetic primitives? For this and the next section, they shall be the following.

6.1. Definition. The language L_R has as primitives, in addition to the equality symbol, the following:
Numerical constants. $\overline{0}$, $\overline{1}$, $\overline{2}$, ...
Function symbols. $+, \cdot$
Relation symbol. $\leq$.

Some other familiar variations of the arithmetic language drop most of the numerical constants (leaving only $\overline{0}$ or $\overline{0}$ and $\overline{1}$), add a symbol for the successor function, and either drop the weak inequality symbol or replace it by the symbol for strong inequality. Sometimes a great deal more is added, e.g., function symbols for all primitive recursive functions. Throughout the remainder of this chapter we shall stick with L_R and call it *the* language of arithmetic; in the next volume, *the* language of arithmetic will be slightly different.

Oddly enough, the possession of a language is not enough to define "definability". For, what is one defining? Formulae? No, one usually wants to define *relations* and these must be defined in domains of various kinds. The most straightforward definition of definability in L_R is the following.

6.2. Definition. Let $\mathfrak{N} = (\omega; \leq, +, \cdot, 0, 1, 2, ...)$. A formula $\varphi v_0...v_{n-1}$ of L_R with exactly $v_0, ..., v_{n-1}$ free *defines* a relation $R \subseteq \omega^n$ if, for all $x_0, ..., x_{n-1} \in \omega$,

$$R(x_0, ..., x_{n-1}) \text{ iff } \mathfrak{N} \models \varphi\, \overline{x_0} \ldots \overline{x_{n-1}}\,.$$

A relation R is *definable in* L_R, or *arithmetically definable*, if R is defined by some formula of L_R.

In the next volume (specifically, in Chapters V and VI), we will consider nonstandard models of arithmetic— structures for the language L_R which satisfy many properties of $\mathfrak{N}$— and will consider some additional semantic notions of definability over these models.

Although 6.2 is the most straightforward such definition, it is not the only one. However, as our first theorem shows, it is a very special one.

6.3. Theorem. *A relation $R \subseteq \omega^n$ is arithmetically definable iff it is an arithmetic relation in the sense of Definition* I.10.4.

Before saying anything about the proof, let me note a quick corollary.

6.4. Corollary. *$Th(\omega; \leq, +, \cdot, 0, 1, 2, ...)$ is undecidable, i.e., there is no algorithm to decide which sentences φ of L_R are true in the natural numbers.*

Proof. Let ψv_0 define a non-recursive set X. Any algorithm deciding the truth or falsity of the sentences $\varphi \overline{x}$ would decide the membership or non-membership of numbers x in X, thereby rendering X recursive. QED

The proof of Theorem 6.3 itself is fairly unremarkable. It consists of two routine inductions. The one, like several of Chapter I, is an induction on the inductive generation of arithmetic relations. One shows the initial polynomial relations to be definable by atomic formulae of L_R and the closure conditions on the class of arithmetic relations (closure under conjunction, disjunction, complementation, and quantification) to be handled by the propositional connectives and quantifiers of L_R. This shows every arithmetic relation to be arithmetically definable. In the converse direction, the induction is on the inductive construction of formulae (i.e., on their lengths). One shows equations to define polynomial relations and inequalities to define primitive recursive relations, and then observes that the application of propositional connectives other than implication, as well as the application of quantifiers correspond to closure conditions of the class of arithmetic relations. For implications, $\varphi \rightarrow \psi$, one simply observes

$$(\varphi \rightarrow \psi) \leftrightarrow (\neg\varphi \vee \psi).$$

I leave the details to the reader.

The weak inequality need not have been included in L_R. Indeed, the reader who supplies all the details of the proof of Theorem 6.3 will not use the inequality in showing the definability in the formal language of the arithmetic relations. The reason for including $\leq$ in L_R will emerge slowly in the sequel. One reason that can be given quickly is the use of this relation in bounded quantifiers. If $\leq$ is a primitive, then a bounded quantifier, say

$$\exists v_0 \leq v_1\, \varphi\text{:}\ \exists v_0(v_0 \leq v_1 \wedge \varphi), \qquad (*)$$

will not conceal an extra real quantifier like an abbreviation,

$$v_0 \leq v_1\text{:}\ \exists v_2(v_0 + v_2 = v_1), \qquad (**)$$

would: using (**), (*) reads,

$$\exists v_0 \leq v_1\, \varphi\text{:}\ \exists v_0(\exists v_2(v_0 + v_2 = v_1) \wedge \varphi).$$

Such æsthetic ground is not the true reason for choosing inequality as a primitive, but it will have to do for now.

Bounded quantifiers play a rôle in the arithmetic language comparable to that they play in recursion theory. To begin with we can use them to define syntactic counterparts to the notions of Σ_1-relations, Δ_0-relations, and strict-Σ_1-relations.

6.5. Definition. Let φ be a formula, v_0 a variable, and t a term not containing v_0. We introduce into L_R the following abbreviations:

$$\forall v_0 \leq t\ \varphi\text{: } \forall v_0(v_0 \leq t \rightarrow \varphi)$$
$$\exists v_0 \leq t\ \varphi\text{: } \exists v_0(v_0 \leq t \wedge \varphi).$$

We say that t *bounds the quantifier* in these formulae.

6.6. Definition. A formula φ of L_R is a Δ_0-*formula* if every quantifier in φ is bounded by a variable. Equivalently, the class of Δ_0-formulae is generated inductively as follows:
i. every atomic formula is a Δ_0-formula
ii. if φ, ψ are Δ_0-formulae, then so are $\neg\varphi$, $\varphi \wedge \psi$, $\varphi \vee \psi$, and $\varphi \rightarrow \psi$
iii. if φ is a Δ_0-formula, then so are $\forall v_i \leq v_j\ \varphi$ and $\exists v_i \leq v_j\ \varphi$ (v_j distinct from v_i).

As in Chapter I, section 10, where we defined the notion of a Δ_0-relation, we have only allowed variable bounds to bounded quantifiers in the definition of a Δ_0-formula. One can be more liberal and allow more general terms. For the purposes of the present book, it doesn't matter which choice is made.

6.7. Definition. A formula φ of L_R is a *strict*-Σ_1-*formula* if φ is of the form $\exists v\psi$, where ψ is a Δ_0-formula. If φ is of the form $\forall v\psi$, where ψ is a Δ_0-formula, then φ is a *strict*-Π_1-*formula.*

Note that strict-Π_1-formulae are logically equivalent to negations of strict-Σ_1-formulae, and vice versa.

6.8. Definitions. The class of Σ_1-*formulae* is generated inductively as follows:
i. atomic formulae and their negations are Σ_1-formulae
ii. if φ, ψ are Σ_1-formulae, then so are $\varphi \wedge \psi$ and $\varphi \vee \psi$
iii. if φ is a Σ_1-formula, then so are $\forall v_i \leq v_j\ \varphi$, $\exists v_i \leq v_j\ \varphi$, and $\exists v_i\ \varphi$ (v_j distinct from v_i).

The class of Π_1-*formulae* is generated inductively similarly, but with $\exists v_i\ \varphi$ replaced by $\forall v_i\ \varphi$ in clause iii.

Once again, the Π_1-formulae are logically equivalent to negations of Σ_1-formulae, and vice versa. The proof is not quite as trivial as in the "strict case" and is an induction on the length of a formula.

Except for including negations of atomic formulae in the initial clause, the definition of Σ_1-formulae is analogous to the definition of the class of Σ_1-relations (I.10.5). The expected results hold.

6.9. Theorem. *Let* Γ *be one of:* Δ_0, Σ_1, Π_1, *strict*-Σ_1, *strict*-Π_1. *A relation* $R \subseteq \omega^n$ *is definable by a* Γ-*formula iff* R *is a* Γ-*relation.*

I leave the proof to the reader.

The result can be generalised to the various levels of the Arithmetic Hierarchy: one can define Σ_n- and Π_n- formulae for $n > 1$ and prove that they define exactly the Σ_n- and Π_n-relations. This is, however, hardly exciting and we change course here.

Our new goal is the study of definability in *theories* (as opposed to definability in structures). We begin this study by introducing a few variants of the theory $\mathcal{R}$ of Raphael Robinson. $\mathcal{R}$ is a more-or-less minimal theory with which to work.

6.10. Definitions. Consider the following axiom schemata:

R1. $\overline{x} + \overline{y} = \overline{z}$, if $x + y = z$

R2 $\overline{x} \cdot \overline{y} = \overline{z}$, if $x \cdot y = z$

R3. $\neg\, \overline{x} = \overline{y}$, if $x \neq y$

R4. $\forall v_0(v_0 \leq \overline{x} \rightarrow v_0 = \overline{0} \vee v_0 = \overline{1} \vee \ldots \vee v_0 = \overline{x})$

R4'. $\forall v_0(v_0 \leq \overline{x} \leftrightarrow v_0 = \overline{0} \vee v_0 = \overline{1} \vee \ldots \vee v_0 = \overline{x})$

R5. $\forall v_0(v_0 \leq \overline{x} \vee \overline{x} \leq v_0)$

R5'. $\forall v_0(v_0 \leq v_1 \vee v_1 \leq v_0)$.

The theories $\mathcal{R}^-$, $\mathcal{R}$, and $\mathcal{R}^+$ are obtained by choosing for them the axioms:

$\mathcal{R}^-$: *R1 - R3, R4'*

$\mathcal{R}$: *R1 - R4, R5*

$\mathcal{R}^+$: *R1 - R4, R5'*.

These theories are best explained by their applications. The theory $\mathcal{R}^-$ naturally proves every true Σ_1-sentence; the theory $\mathcal{R}$ almost allows one to prove the disjointness of pairs of "obviously disjoint" Σ_1-definable sets; and $\mathcal{R}^+$ can prove the functionality of certain representations of partial recursive functions.

If I cannot convincingly offer the best explanation yet, I can offer some partial explanation of these axioms. *R1 - R2* will allow the correct formal computation of polynomials (6.11, below). This guarantees the provability of any true equational sentence, and *R3* will add to this the refutability of any false such sentence. This yields the basis for an inductive proof that $\mathcal{R}^-$ proves all true Σ_1-sentences. Axiom schema *R4'* handles the bounded quantifiers (6.12, below) in the induction step of this proof. Schema *R5*, or axiom *R5'*, is necessary in handling the comparison of witnesses in the more advanced recursion theoretic constructions.

Let us begin with the application of *R1 - R3*.

6.11. Lemma. *Let s, t be terms of L_R with variables from the list $v_0, \ldots, v_{n-1}$, and let them define polynomials P, Q, respectively. Then, for any $x_0, \ldots, x_{n-1} \in \omega$,*

i. $P(x_0, \ldots, x_{n-1}) = Q(x_0, \ldots, x_{n-1}) \rightarrow \mathcal{R}^- \vdash s\,\overline{x_0} \ldots \overline{x_{n-1}} = t\,\overline{x_0} \ldots \overline{x_{n-1}}$

ii. $P(x_0, \ldots, x_{n-1}) \neq Q(x_0, \ldots, x_{n-1}) \rightarrow \mathcal{R}^- \vdash \neg s\,\overline{x_0} \ldots \overline{x_{n-1}} = t\,\overline{x_0} \ldots \overline{x_{n-1}}$.

The "polynomial defined by a term t" is not one of the subtle notions of definability to be studied in this section. I mean merely the obvious. E.g., the term,

$$\overline{2} \cdot v_0 + \overline{3} \cdot v_0 \cdot v_1 + \overline{4},$$

defines the polynomial,

$$2X_0 + 3X_0X_1 + 4.$$

Proof of Lemma 6.11. By induction on the sum of the numbers of function symbols occurring in s and t.

The basis of the induction is given by choosing s and t to be variables or constants. After substituting numerals for the variables, the equation $s = t$ becomes an equation $\overline{x} = \overline{y}$. If $x = y$, the numerals $\overline{x}$ and $\overline{y}$ are identical and logic alone yields $\mathcal{R}^- \vdash \overline{x} = \overline{y}$; if $x \neq y$, axiom *R3* yields $\mathcal{R}^- \vdash \neg\, \overline{x} = \overline{y}$.

The induction step occurs when one or more of the terms is composite. By symmetry, we can assume this term to be s.

Case 1. s is $s_1 + s_2$ with s, s_1, s_2 defining polynomials P, P_1, P_2, respectively. Substitute $\overline{x_i}$ for v_i in the terms and let

$$y = P(x_0, ..., x_{n-1}),\ y_1 = P_1(x_0, ..., x_{n-1}),\ y_2 = P_2(x_0, ..., x_{n-1}).$$

By induction hypothesis,

$$\mathcal{R}^- \vdash \overline{y_1} = s_1\, \overline{x_0} \ldots \overline{x_{n-1}}$$

$$\mathcal{R}^- \vdash \overline{y_2} = s_2\, \overline{x_0} \ldots \overline{x_{n-1}},$$

and, by *R1*,

$$\mathcal{R}^- \vdash \overline{y} = \overline{y_1} + \overline{y_2}.$$

Thus,

$$\mathcal{R}^- \vdash \overline{y} = s_1\, \overline{x_0} \ldots \overline{x_{n-1}} + s_2\, \overline{x_0} \ldots \overline{x_{n-1}},$$

i.e.,

$$\mathcal{R}^- \vdash \overline{y} = s\, \overline{x_0} \ldots \overline{x_{n-1}}. \qquad (*)$$

Again, the equation $\overline{y} = t$ has fewer function symbols than $s = t$, whence the induction hypothesis yields, assuming t defines a polynomial Q,

$$y = Q(x_0, \ldots, x_{n-1}) \rightarrow \mathcal{R}^- \vdash \overline{y} = t$$

$$\rightarrow \mathcal{R}^- \vdash s = t, \text{ by } (*)$$

where we suppress mention of the x_i's in s and t. Similarly,

$$y \neq Q(x_0, \ldots, x_{n-1}) \rightarrow \mathcal{R}^- \vdash \neg\, s = t, \text{ by } (*).$$

Case 2. s is $s_1 \cdot s_2$. The proof is similar. QED

As mentioned, the following explains the choice of *R4'*.

6.12. Lemma. *Let* $x \in \omega$.

i. $\mathcal{R} \vdash \forall v_0 (v_0 \leq \overline{x} \leftrightarrow v_0 = \overline{0} \vee v_0 = \overline{1} \vee \ldots \vee v_0 = \overline{x})$

ii. $\mathcal{R}^- \vdash \forall v_0 \leq \overline{x}\ \varphi v_0 \leftrightarrow \varphi \overline{0} \wedge \ldots \wedge \varphi \overline{x}$

iii. $\mathcal{R}^- \vdash \exists v_0 \leq \overline{x}\ \varphi v_0 \leftrightarrow \varphi \overline{0} \vee \ldots \vee \varphi \overline{x}$.

Proof. Parts ii and iii, which explain the choice of *R4'* and are the interesting parts of the Lemma are trivial. Part i shows that $\mathcal{R}^-$ is a sub-theory of $\mathcal{R}$ (which is clearly a sub-theory of $\mathcal{R}^+$) and is not as trivial.

By *R5*,

$$\mathcal{R} \vdash v \le \overline{x} \vee \overline{x} \le v. \tag{1}$$

Substituting $\overline{x}$ for v in (1) yields

$$\mathcal{R} \vdash \overline{x} \le \overline{x} \vee \overline{x} \le \overline{x},$$

whence

$$\mathcal{R} \vdash \overline{x} \le \overline{x}. \tag{2}$$

For $y < x$, *R4* yields

$$\mathcal{R} \vdash \overline{x} \le \overline{y} \rightarrow \overline{x} = \overline{0} \vee \ldots \vee \overline{x} = \overline{y}. \tag{3}$$

But *R3* yields $\neg\, \overline{x} = \overline{z}$ for $z \le y < x$, whence (3) yields

$$\mathcal{R} \vdash \neg\, \overline{x} \le \overline{y}. \tag{4}$$

Substituting $\overline{y}$ for v in (1), this last yields

$$\mathcal{R} \vdash \overline{y} \le \overline{x},$$

i.e.,

$$\mathcal{R} \vdash v = \overline{y} \rightarrow v \le \overline{x}, \text{ for } 0 \le y < x. \tag{5}$$

Combining (2) and (5),

$$\mathcal{R} \vdash v = \overline{0} \vee \ldots \vee v = \overline{x} \rightarrow v \le \overline{x}.$$

With *R4*, this yields

$$\mathcal{R} \vdash v \le \overline{x} \leftrightarrow v = \overline{0} \vee \ldots \vee v = \overline{x}.$$

QED

We are now ready to prove an important theorem.

6.13. Theorem. (Σ_1-Completeness Theorem). *Let* φ *be a* Σ_1*-formula with free variables from* $v_0, \ldots, v_{n-1}$*, and let* $x_0, \ldots, x_{n-1} \in \omega$*. Then*

$$\mathfrak{N} \models \varphi\, \overline{x_0} \ldots \overline{x_{n-1}} \rightarrow \mathcal{R}^- \vdash \varphi\, \overline{x_0} \ldots \overline{x_{n-1}},$$

where $\mathfrak{N}$ *denotes the structure* $(\omega; \le, +, \cdot, 0, 1, \ldots)$.

Proof. By induction on the complexity of φ.
Basis. φ is either an atomic formula or the negation of an atomic formula. If the atomic formula is an equation, Lemma 6.11 yields the result immediately. Thus, suppose the atomic formula is an inequality, say, $s \le t$, with s, t defining polynomials P, Q, respectively. Let

$$y = P(x_0, \ldots, x_{n-1}), \quad z = Q(x_0, \ldots, x_{n-1}).$$

By Lemma 6.11 we have (suppressing mention of the x's)

$$\mathcal{R}^- \vdash \overline{y} = s \text{ and } \mathcal{R}^- \vdash \overline{z} = t,$$

whence,

$$\mathcal{R}^- \vdash s \le t \leftrightarrow \overline{y} \le \overline{z}. \tag{*}$$

But,

$$\begin{aligned}\mathfrak{N} \models s \leq t \;&\rightarrow\; y \leq z\\ &\rightarrow\; \mathcal{R}^- \vdash \overline{y} = \overline{0} \vee \ldots \vee \overline{y} = \overline{z}\\ &\rightarrow\; \mathcal{R}^- \vdash \overline{y} \leq \overline{z}, \text{ by } R4'\\ &\rightarrow\; \mathcal{R}^- \vdash s \leq t, \text{ by } (*);\end{aligned}$$

while

$$\begin{aligned}\mathfrak{N} \models \neg s \leq t \;&\rightarrow\; \neg y \leq z\\ &\rightarrow\; \mathcal{R}^- \vdash \neg(\overline{y} = \overline{0} \vee \ldots \vee \overline{y} = \overline{z}), \text{ by } R3\\ &\rightarrow\; \mathcal{R}^- \vdash \neg\, \overline{y} \leq \overline{z}, \text{ by } R4'\\ &\rightarrow\; \mathcal{R}^- \vdash \neg s \leq t, \text{ by } (*).\end{aligned}$$

Induction step. φ is a compound formula. Once again, I suppress mention of the x's.

If φ is $\psi \wedge \chi$, then

$$\begin{aligned}\mathfrak{N} \models \varphi \;&\rightarrow\; \mathfrak{N} \models \psi \wedge \chi\\ &\rightarrow\; (\mathfrak{N} \models \psi) \wedge (\mathfrak{N} \models \chi)\\ &\rightarrow\; (\mathcal{R}^- \vdash \psi) \wedge (\mathcal{R}^- \vdash \chi)\\ &\rightarrow\; \mathcal{R}^- \vdash \psi \wedge \chi\\ &\rightarrow\; \mathcal{R}^- \vdash \varphi.\end{aligned}$$

The case when φ is $\psi \vee \chi$ is treated similarly.

Suppose φ is $\forall v \leq v_j \psi v$. Then,

$$\begin{aligned}\mathcal{R}^- \vdash \varphi \;&\leftrightarrow\; \forall v \leq \overline{x_j}\, \psi v\\ &\leftrightarrow\; \psi \overline{0} \wedge \ldots \wedge \psi \overline{x_j}, \text{ by 6.12.}\end{aligned}$$

We can now appeal to an iteration of the conjunction case.

The treatment of $\exists v \leq v_j \psi v$ is similar.

Finally, consider the case where φ is $\exists v\, \psi v$:

$$\begin{aligned}\mathfrak{N} \models \varphi \;&\rightarrow\; \mathfrak{N} \models \exists v\, \psi v\\ &\rightarrow\; \exists x\, (\mathfrak{N} \models \psi \overline{x})\\ &\rightarrow\; \exists x\, (\mathcal{R}^- \vdash \psi \overline{x})\\ &\rightarrow\; \mathcal{R}^- \vdash \exists v \psi v\\ &\rightarrow\; \mathcal{R}^- \vdash \varphi.\end{aligned}$$

QED

6.14. Corollary. *Let φ be a Δ_0-formula with free variables from $v_0, \ldots, v_{n-1}$, and let $x_0, \ldots, x_{n-1} \in \omega$. Then:*

i. $\mathfrak{N} \models \varphi \overline{x_0} \ldots \overline{x_{n-1}} \;\rightarrow\; \mathcal{R}^- \vdash \varphi \overline{x_0} \ldots \overline{x_{n-1}}$

ii. $\mathfrak{N} \models \neg\varphi \overline{x_0} \ldots \overline{x_{n-1}} \;\rightarrow\; \mathcal{R}^- \vdash \neg\varphi \overline{x_0} \ldots \overline{x_{n-1}}$,

where $\mathfrak{N}$ denotes the structure $(\omega; \leq, +, \cdot, 0, 1, \ldots)$.

This reduces fairly easily to Theorem 6.13: by an induction on the complexity of $\varphi \in \Delta_0$, one can easily prove each of φ and $\neg\varphi$ to be logically equivalent to Σ_1-formulae. (This makes a good exercise for those not tired of such inductions.)

6.15. Corollary. *Let φ be a strict-Σ_1-formula with free variables from $v_0, ..., v_{n-1}$, and let $x_0, ..., x_{n-1} \in \omega$. Then:*

$$\mathfrak{N} \models \varphi\,\overline{x_0}\ldots\overline{x_{n-1}} \quad\rightarrow\quad \mathcal{R}^- \vdash \varphi\,\overline{x_0}\ldots\overline{x_{n-1}},$$

where $\mathfrak{N}$ denotes the structure $(\omega; \leq, +, \cdot, 0, 1, ...)$.

To prove this, either prove φ logically equivalent to a Σ_1-formula or apply the argument for the case of the unbounded existential quantifier in the proof of Theorem 6.13 to Corollary 6.14.

Theorem 6.13, together with its converse, has a couple of more interesting corollaries. Each of them requires an introductory definition.

6.16. Definition. A theory $\mathcal{T}$ in L_R is Σ_1-*sound* if, for any Σ_1-sentence φ,

$$\mathcal{T} \vdash \varphi \quad\rightarrow\quad \mathfrak{N} \models \varphi.$$

The theories $\mathcal{R}^-$, $\mathcal{R}$, and $\mathcal{R}^+$ are all Σ_1-sound.

6.17. Corollary. (Gödel's Incompleteness Theorem). *Let $\mathcal{T}$ be a Σ_1-sound extension of $\mathcal{R}^-$ with an effectively enumerable set of axioms. Then: $\mathcal{T}$ is incomplete; in fact, there is a true* (*strict-*) *Π_1-sentence ψ such that $\mathcal{T} \nvdash \psi$ and $\mathcal{T} \nvdash \neg\psi$.*

Proof. Let φv be a (strict-) Σ_1-formula defining an r.e. non-recursive set X (recall I.10.10 and 6.9) and consider,

$$Y = \{x \in \omega\colon\ \mathcal{T} \vdash \neg\varphi\overline{x}\}.$$

By Theorem 6.13,

$$x \in X \quad\rightarrow\quad \mathcal{T} \vdash \varphi\overline{x},$$

i.e.,

$$\mathcal{T} \vdash \neg\varphi\overline{x} \quad\rightarrow\quad x \notin X,$$

and $Y \subseteq \omega - X$.

By the effective enumerability of the axioms of $\mathcal{T}$ and the effectively enumerable nature of first-order derivability, Y is an r.e. set. (This will be more formally demonstrated in Chapter IV of volume II.) Because X is r.e. and not recursive, $\omega - X$ is not r.e. In particular, $Y \neq \omega - X$ and there is some $x_0 \in \omega - (X \cup Y)$. But,

$$\mathcal{T} \nvdash \neg\varphi\overline{x_0}, \text{ because } x_0 \notin Y$$

$$\mathcal{T} \nvdash \varphi\overline{x_0}, \text{ because } x_0 \notin X \text{ and } \mathcal{T} \text{ is } \Sigma_1\text{-sound.}$$

Thus, the sentence $\psi = \neg\varphi\overline{x_0}$ is undecided by $\mathcal{T}$. It is true because $x_0 \notin X$. QED

A variant of Gödel's Incompleteness Theorem is the following:

6.18. Corollary. *Let $\mathcal{T}$ be a Σ_1-sound extension of $\mathcal{R}^-$. Then: $\mathcal{T}$ is undecidable, i.e., there is no algorithm for deciding which sentences are derivable in $\mathcal{T}$ and which are not.*

Proof. As before, let φv be a Σ_1-formula defining an r.e. non-recursive set X and appeal to 6.13 and Σ_1-soundness to conclude

$$X = \{x \in \omega\colon \mathcal{T} \vdash \varphi \overline{x}\}.$$

A decision procedure for $\mathcal{T}$ would decide X. QED

Before commenting on these, let me introduce our first syntactic notion of definability and state the corollary concerning it.

6.19. Definition. Let $R \subseteq \omega^n$ and let $\mathcal{T}$ be a theory in L_R. A formula $\varphi v_0 ... v_{n-1}$ with exactly $v_0, ..., v_{n-1}$ free is said to *semi-represent* R in $\mathcal{T}$ if, for all $x_0, ..., x_{n-1} \in \omega$,

$$R(x_0, ..., x_{n-1}) \quad \text{iff} \quad \mathcal{T} \vdash \varphi \overline{x_0} \ldots \overline{x_{n-1}}.$$

6.20. Corollary. *Let $\mathcal{T}$ be a Σ_1-sound extension of $\mathcal{R}^-$ with an effectively enumerable set of axioms. Then: A relation $R \subseteq \omega^n$ is semi-representable in $\mathcal{T}$ iff R is recursively enumerable.*

The proof is trivial and I omit it.

It may be instructive to compare Corollaries 6.17 and 6.18 with what is obtainable from Theorem 6.3 and Corollary 6.4. By Theorem 6.3, $Th(\mathfrak{N})$ semi-represents all arithmetic sets. Since any effectively enumerable theory can semi-represent only r.e. sets, no such theory can coïncide with $Th(\mathfrak{N})$. If, in particular, an effectively enumerable theory $\mathcal{T}$ is a sub-theory of $Th(\mathfrak{N})$ (i.e., $\mathcal{T}$ has only true axioms), $\mathcal{T}$ must be incomplete. This refutes Programme 1.1. What Corollary 6.17 does is to generalise this slightly: $\mathcal{T}$ need not lie completely inside $Th(\mathfrak{N})$; only its Σ_1-consequences are required to be true. To the extent that we are only interested in true theories of arithmetic, this generalisation may appear insignificant. If, however, we are interested in theories of structures similar to, but different from $\mathfrak{N}$, Corollary 6.17 does not go far enough. Such a theory need not be Σ_1-sound.

The comparison of Corollary 6.18 with Corollary 6.4 parallels that of the incompleteness results. Corollary 6.4 tells us that $Th(\mathfrak{N})$ is undecidable. A more careful look at Theorem 6.3 will tell us that any consistent theory containing all true Σ_1- and Π_1-sentences is undecidable; but, by Gödel's Incompleteness Theorem 6.17, this does not cover theories with effectively enumerable sets of axioms. Corollary 6.18 is thus much more general: all Σ_1-sound extensions of $\mathcal{R}^-$ are undecidable. The advantage here is the incredible weakness of $\mathcal{R}^-$; the disadvantage, again, is the restriction to Σ_1-sound theories.

That the restriction to Σ_1-sound theories is not a vacuous restriction is readily established.

6.21. Example. *There are consistent, non-Σ_1-sound theories.*

Proof. Let φ be a Σ_1-sentence such that $\mathcal{R}^- \nvdash \varphi$ and $\mathcal{R}^- \nvdash \neg\varphi$. Then we can choose $\mathcal{T}$ to be $\mathcal{R}^- + \varphi$.

$\mathcal{T}$ is consistent because (for $\perp$ any convenient refutable sentence),

$$\begin{aligned} \mathcal{T} \vdash \perp \quad &\rightarrow \quad \mathcal{R}^- + \varphi \vdash \perp \\ &\rightarrow \quad \mathcal{R}^- \vdash \neg\varphi, \end{aligned}$$

contrary to the choice of φ.

$\mathcal{T}$ is not Σ_1-sound because $\mathcal{T} \vdash \varphi$ and, if φ were true we would have $\mathcal{R}^- \vdash \varphi$ (by 6.13) and we chose φ so that $\mathcal{R}^- \nvdash \varphi$. QED

We might as well also note the following.

6.22. Example. *There is a Σ_1-sound theory $\mathcal{T}$ which is not sound.*

Proof sketch. Let $\mathcal{T}_0$ be the sound theory obtained by adding to $\mathcal{R}^-$ all true Π_1-sentences, i.e.,

$$\mathcal{T}_0 = \mathcal{R}^- + \{\varphi \in \Pi_1 \colon \mathfrak{N} \models \varphi\}.$$

The proof of Theorem 6.13 relativises: $\mathcal{T}_0$ proves all true Σ_2-sentences (i.e., sentences of the form $\exists v\varphi$ with $\varphi \in \Pi_1$). Starting with a Σ_2-set which is not Π_2, the proof of Corollary 6.17 yields a Σ_2- sentence φ such that

$$\mathcal{T}_0 \nvdash \varphi \quad \text{and} \quad \mathcal{T}_0 \nvdash \neg\varphi.$$

Let $\mathcal{T} = \mathcal{R}^- + \varphi$ (or even $\mathcal{T}_0 + \varphi$). QED

The constructions underlying Examples 6.21 and 6.22 yield more than stated. In 6.21, for example, we can replace $\mathcal{R}^-$ by any effectively enumerable Σ_1-sound extension $\mathcal{T}_0$ of $\mathcal{R}^-$ and obtain a consistent, non-Σ_1-sound extension $\mathcal{T}$ of $\mathcal{T}_0$.

These constructions are also a bit abstract. Because $\mathcal{R}^-$ is so weak to begin with, we can actually give explicit examples of consistent, non-Σ_1-sound extensions.

6.23. More Explicit Example. Let $Pol(\omega)$ be the set of all unary polynomials with non-negative integral coëfficients. Define

$$\mathfrak{Pol} = (Pol(\omega); \leq, +, \cdot, 0, 1, \ldots),$$

where

i. $+, \cdot$ are the usual operations on polynomials
ii. $0, 1, \ldots$ are the constant polynomials, and
iii. $\leq$ is eventual domination:

$$P \leq Q \ \text{ iff } \ \exists x\, \forall y > x\, (P(y) \leq Q(y)).$$

$\mathfrak{Pol}$ is readily seen to be a model of $\mathcal{R}^-$ (even of $\mathcal{R}^+$), but it is not Σ_1-sound:

$$\mathfrak{Pol} \models \exists v_0(\neg\, v_0 = \overline{0} \ \wedge\ \forall v_1 \leq v_0 \neg\, (v_0 = v_1 + \overline{1})).$$

(Let v_0 be the polynomial $P(X) = X$.)

Some other explicit examples of non-Σ_1-sound extensions of $\mathcal{R}^-$ can be found in the Exercises (specifically in Exercise 1), below. In each of these examples we can conclude, say, the undecidability of the theory in question, a theory of a given structure, by defining ω in the structure and thereby reducing $Th(\mathfrak{N})$ to the given theory. However, not every consistent, non-Σ_1-sound extension of $\mathcal{R}^-$ is as amenable as this and it would be nice to conclude such undecidability results as that of $Th(\mathfrak{Pol})$ directly from the fact that these theories extend $\mathcal{R}^-$.

It is thus to the extension of Corollaries 6.17 and 6.18, and eventually 6.20, to the non-Σ_1-sound case to which we next turn. How do we avoid the problem of blowing an r.e. non-recursive set up into a recursive set? I.e., how can we guarantee that, if X is an r.e. non-recursive set, φv_0 defines X in $\mathfrak{N}$, and

$$A = \{x: \mathcal{T} \vdash \varphi\overline{x}\},$$

for some consistent extension $\mathcal{T}$ of $\mathcal{R}^-$, then A, although possibly larger than X, is not recursive? So stated, the problem has an easy solution: don't use any old X, but use an X that belongs to a pair of effectively inseparable sets X, Y, and show

$$X \subseteq A = \{x: \mathcal{T} \vdash \varphi\overline{x}\}$$
$$Y \subseteq \{x: \mathcal{T} \vdash \neg\varphi\overline{x}\}. \qquad (*)$$

The set A will certainly not be recursive and the analogues to Corollaries 6.17 and 6.18 will follow. The analogue to Corollary 6.20, asserting that the sets A are exactly the r.e. sets, would also yield analogues to Corollary 6.17 and 6.18, but is a bit deeper. However, the solution to this deeper problem also begins with the pair of effectively inseparable sets and the inclusions cited, as the reader will see in the next section.

Conceptual clarity and simplicity of execution do not always go hand-in-hand, particularly if the means of execution are limited. $\mathcal{R}^-$ is too weak to allow a decent proof of inclusion (*); $\mathcal{R}$ is just strong enough to allow derivation of (*), but not of the expected $\forall v\,(\psi v \rightarrow \neg\varphi v)$ (where ψv defines Y), which would quickly yield (*); and $\mathcal{R}^+$ gives us this last implication— the provable disjointness of X and Y. I shall present these proofs for $\mathcal{R}$.

Evidently, it is for the purpose of proving (*) that we postulated *R5*. In point of fact, it is really for this purpose that we need the weak inequality. For Corollaries 6.17, 6.18, and 6.20 we could have used *R1* - *R2* to derive all true Diophantine sentences, semi-represent Diophantine (= r.e.) relations, and dealt with Diophantine-correct theories (of which, incidentally, $Th(\mathfrak{Pol})$ of Example 6.23 is one).

As for the construction, we needn't worry about the construction of the pair of effectively inseparable r.e. sets; we have just to assume given such a pair from Chapter I. We can also assume given Σ_1-formulae φ, ψ defining them— we will need strict-Σ_1-formulae, but Theorem 6.9 yields these. What we need to do is to find special definitions which are almost provably disjoint in $\mathcal{R}$. And this is not too difficult— we use the trick used in Chapter I to calculate a function F defined by

$$F(x) \simeq \begin{cases} G_1(x), & x \in X_1 \\ G_2(x), & x \in X_2, \end{cases}$$

where G_1, G_2 are partial recursive and X_1, X_2 disjoint r.e. sets. This trick was to look for the first *witness* to the membership of x in X_1 or X_2 before attempting to calculate $G_1(x)$ or

$G_2(x)$. [The reader who has worked out Exercise 5 of Chapter I, section 12, will find in that Exercise an even closer parallel with what we are about to do.]

In order to apply this trick, we need a preliminary definition and a couple of preliminary remarks.

6.24. Definitions. Define

$$v_0 < v_1:\ v_0 \leq v_1 \wedge \neg v_1 = v_0$$
$$\forall v_0 < v_1\, \varphi:\ \forall v_0 (v_0 < v_1 \rightarrow \varphi)$$
$$\exists v_0 < v_1\, \varphi:\ \exists v_0 (v_0 < v_1 \wedge \varphi).$$

Observe that $v_0 < v_1$ is Δ_0. Also, note the logical equivalences:

$$\forall v_0 < v_1 \varphi \leftrightarrow \forall v_0 \leq v_1 (\neg v_1 = v_0 \rightarrow \varphi)$$
$$\exists v_0 < v_1 \varphi \leftrightarrow \exists v_0 \leq v_1 (\neg v_1 = v_0 \wedge \varphi).$$

By these, the classes of Δ_0- and Σ_1-formulae are provably closed under application of these new bounded quantifiers. [Forgive me if I seem to emphasise trivialities: $\leq$ and $<$, although written as orderings, need not satisfy all the obvious properties of orderings. For example,

$$\mathcal{R} \nvdash v_0 \leq v_1 \leftrightarrow v_0 < v_1 \vee v_0 = v_1,$$

because, although

$$\mathcal{R} \vdash v_0 \leq v_1 \rightarrow v_0 < v_1 \vee v_0 = v_1,$$

it happens that

$$\mathcal{R} \nvdash v_0 = v_1 \rightarrow v_0 \leq v_1.]$$

Following the little Definition 6.24 is a major definition.

6.25. Definition. Given existentially quantified formulae $\varphi = \exists v \chi v$ and $\psi = \exists v \theta v$, define their *witness comparison formulae* as follows:

$$\varphi \preccurlyeq \psi:\ \exists v\, [\chi v \wedge \forall v_0 < v \neg \theta v_0]$$
$$\varphi \prec \psi:\ \exists v\, [\chi v \wedge \forall v_0 \leq v \neg \theta v_0].$$

In words: Call any number x such that $\chi \overline{x}$ a *witness* to φ. Then $\varphi \preccurlyeq \psi$ ($\varphi \prec \psi$) asserts that, not only is φ witnessed, but that some witness to φ occurs no later than (definitely earlier than) any witness to ψ. Moreover, $\varphi \preccurlyeq \varphi$ asserts there to be a *least* witness to φ. The *Least Number Principle* can thus be written,

$$\varphi \rightarrow (\varphi \preccurlyeq \varphi),$$

for existentially quantified φ. The reader should bear this in mind as it will make the *truths* of various assertions more obvious. He should also bear in mind that $\mathcal{R}$ has no induction and can prove very little about $<$ and $\leq$, whence the *proofs* in $\mathcal{R}$ of obvious truths may not be obvious, or may not even be.

The informal version of what we want to prove is the following.

6.26. Informal Reduction Lemma. *Let φ, ψ be strict-Σ_1-sentences and define*

$$\varphi_1: \ \varphi \preccurlyeq \psi$$
$$\psi_1: \ \psi \prec \varphi.$$

Then: i. $\mathfrak{N} \models \varphi_1 \to \varphi, \ \mathfrak{N} \models \psi_1 \to \psi$

ii. $\mathfrak{N} \models \varphi \vee \psi \ \to \ \varphi_1 \vee \psi_1$

iii. $\mathfrak{N} \models \neg(\varphi_1 \wedge \psi_1)$.

Formally, we would like to replace truth in $\mathfrak{N}$ by provability in $\mathcal{R}$ in this Lemma. For 6.26.i, this is trivial as the implications $\varphi_1 \to \varphi$ and$\psi_1 \to \psi$ are logically valid. The second implication cannot be proven in $\mathcal{R}$as it makes use of the Least Number Principle (as the industrious reader who supplies a proof of this Lemma will verify), which $\mathcal{R}$ is too weak to prove. And, alas, even 6.26.iii is only easily proven in $\mathcal{R}^+$. The formal counterpart is the following.

6.27. Formal Reduction Lemma. *Let φ, ψ be strict-Σ_1-sentences and define*

$$\varphi_1: \ \varphi \preccurlyeq \psi$$
$$\psi_1: \ \psi \prec \varphi.$$

Then: i. $\mathcal{R} \vdash \varphi_1 \to \varphi, \ \mathcal{R} \vdash \psi_1 \to \psi$

ii. a. $\mathfrak{N} \models \varphi_1 \to \mathcal{R} \vdash \varphi_1$

b. $\mathfrak{N} \models \psi_1 \to \mathcal{R} \vdash \psi_1$

iii. a. $\mathfrak{N} \models \varphi_1 \to \mathcal{R} \vdash \neg\psi_1$

b. $\mathfrak{N} \models \psi_1 \to \mathcal{R} \vdash \neg\varphi_1$

iv. $\mathcal{R}^+ \vdash \neg(\varphi_1 \wedge \psi_1)$.

Proof. i. Simple logic.

ii. By Corollary 6.15. Observe that, if φ, ψ are strict-Σ_1-formulae, then so are φ_1, ψ_1.

iii. Write

$$\varphi_1: \ \exists v_0[\chi v_0 \wedge \forall v < v_0 \neg\theta v]$$
$$\psi_1: \ \exists v_1[\theta v_1 \wedge \forall v \le v_1 \neg\chi v].$$

a. Let $\mathfrak{N} \models \varphi_1$. Choose x such that

$$\mathfrak{N} \models \chi\overline{x} \wedge \forall v < \overline{x}\, \neg\theta v.$$

Then

$$\mathcal{R} \vdash \chi\overline{x} \wedge \forall v < \overline{x}\, \neg\theta v. \tag{6}$$

Consider

$$\theta\overline{x} \wedge \forall v \le \overline{x}\, \neg\chi v. \tag{7}$$

Now,

$$\mathcal{R} \vdash \forall v \le v_1 \neg\chi v \ \to \ \forall v \le v_1 \neg\chi v \wedge \chi\overline{x}, \text{ by (6)}$$
$$\to \neg\, \overline{x} \le v_1 \tag{8}$$
$$\to v_1 \le \overline{x}, \text{ by } R5$$
$$\to v_1 < \overline{x} \vee v_1 = \overline{x}$$

$$\to\ v_1 < \overline{x}\,, \tag{9}$$

as follows from (8) and

$$\mathcal{R} \vdash v_1 = \overline{x}\ \to\ \overline{x} \le v_1$$

(which follows by *R5*: $\mathcal{R} \vdash \overline{x} \le \overline{x}\ \vee\ \overline{x} \le \overline{x}$, whence $\mathcal{R} \vdash \overline{x} \le \overline{x}$, whence $\mathcal{R} \vdash v_1 = \overline{x}\ \to\ \overline{x} \le v_1$). But

$$\mathcal{R} \vdash v_1 < \overline{x}\ \to\ v_1 < \overline{x}\ \wedge\ \forall v < \overline{x}\ \neg\theta v, \text{ by (6)}$$
$$\to\ \neg\theta v_1.$$

Combining this with (9),

$$\mathcal{R} \vdash \forall v \le v_1 \neg\chi v\ \to\ \neg\theta v_1,$$

whence

$$\mathcal{R} \vdash \neg(7)$$
$$\vdash \neg\exists v \neg[\theta v_1\ \wedge\ \forall v \le v_1 \neg\chi v\,]$$
$$\vdash \neg\psi_1.$$

b. Let $\mathfrak{N} \models \psi_1$. Choose x such that

$$\mathfrak{N} \models \theta\overline{x}\ \wedge\ \forall v \le \overline{x}\ \neg\chi v.$$

Again,

$$\mathcal{R} \vdash \theta\overline{x}\ \wedge\ \forall v \le \overline{x}\ \neg\chi v. \tag{10}$$

Consider,

$$\chi v_0\ \wedge\ \forall v < v_0 \neg\theta v. \tag{11}$$

Now,

$$\mathcal{R} \vdash \forall v < v_0 \neg\theta v\ \to\ \forall v < v_0 \neg\theta v\ \wedge\ \theta\overline{x}, \text{ by (10)}$$
$$\to\ \neg\,\overline{x} < v_0$$
$$\to\ \neg(\overline{x} \le v_0\ \wedge\ \neg\,\overline{x} = v_0)$$
$$\to\ \neg\,\overline{x} \le v_0\ \vee\ \overline{x} = v_0$$
$$\to\ v_0 \le \overline{x}\ \vee\ \overline{x} = v_0, \text{ by } R5$$
$$\to\ v_0 \le \overline{x}\,, \tag{12}$$

since $\mathcal{R} \vdash \overline{x} = v_0\ \to\ v_0 \le \overline{x}$. But,

$$\mathcal{R} \vdash v_0 \le \overline{x}\ \to\ v_0 \le \overline{x}\ \wedge\ \forall v \le \overline{x}\ \neg\chi v, \text{ by (10)}$$
$$\to\ \neg\chi v_0.$$

Combining this with (12),

$$\mathcal{R} \vdash \forall v < v_0 \neg\theta v\ \to\ \neg\chi v_0,$$

which yields

$$\mathcal{R} \vdash \neg(11)$$
$$\vdash \neg\exists v_0[\chi v_0\ \wedge\ \forall v < v_0 \neg\theta v]$$

$$\vdash \neg\varphi_1.$$

iv. Consider

$$\chi v_0 \wedge \forall v < v_0 \neg\theta v \wedge \theta v_1 \wedge \forall v \leq v_1 \neg\chi v. \tag{13}$$

Using *R5′*,

$$\mathcal{R}^+ \vdash v_0 \leq v_1 \vee v_1 \leq v_0,$$

one quickly concludes

$$\mathcal{R}^+ \vdash v_0 \leq v_1 \vee v_1 < v_0 \vee v_1 = v_0$$
$$\vdash v_0 \leq v_1 \vee v_1 < v_0.$$

Formula (13) is contradicted in each case:

$$\mathcal{R}^+ \vdash v_0 \leq v_1 \wedge \forall v \leq v_1 \neg\chi v \rightarrow \neg\chi v_0$$
$$\mathcal{R}^+ \vdash v_1 < v_0 \wedge \forall v < v_0 \neg\theta v \rightarrow \neg\theta v_1.$$

QED

We are now in position to obtain the non-Σ_1-sound analogue to Corollaries 6.17 and 6.18.

6.28. Corollary. (Rosser's Theorem). *Let $\mathcal{T}$ be a consistent theory extending $\mathcal{R}$. Then:*
i. *$\mathcal{T}$ is undecidable, and*
ii. *if $\mathcal{T}$ has an effectively enumerable set of axioms, $\mathcal{T}$ is incomplete.*

Proof. Let X, Y be a pair of effectively inseparable r.e. sets, with strict Σ_1-formulae φv, ψv defining them, and let

$$\varphi_1 v:\ \varphi v \preccurlyeq \psi v$$
$$\psi_1 v:\ \psi v \prec \varphi v,$$

be their witness-comparison formulae.

Observe,

$$x \in X \rightarrow \mathfrak{N} \models \varphi\overline{x} \wedge \neg\psi\overline{x}$$
$$\rightarrow \mathfrak{N} \models \varphi_1\overline{x}$$
$$\rightarrow \mathcal{R} \vdash \varphi_1\overline{x}, \text{ by 6.27.ii}$$
$$x \in Y \rightarrow \mathfrak{N} \models \psi\overline{x} \wedge \neg\varphi\overline{x}$$
$$\rightarrow \mathfrak{N} \models \psi_1\overline{x}$$
$$\rightarrow \mathcal{R} \vdash \neg\varphi_1\overline{x}, \text{ by 6.27.iii.}$$

Thus, if

$$A = \{x:\ \mathcal{T} \vdash \varphi_1\overline{x}\},\ B = \{x:\ \mathcal{T} \vdash \neg\varphi_1\overline{x}\},$$

then

$$X \subseteq A \text{ and } Y \subseteq B.$$

The disjointness of A and B is an immediate consequence of the consistency of $\mathcal{T}$. By the effective inseparability of X and Y, A is not recursive, whence $\mathcal{T}$ is undecidable.

If, additionally, $\mathcal{T}$ is effectively enumerable, then the sets A and B are r.e., whence the effective inseparability of X, Y yields the existence of some $x_0 \notin A \cup B$: $\mathcal{T} \nvdash \varphi_1 \overline{x_0}$, $\mathcal{T} \nvdash \neg \varphi_1 \overline{x_0}$. QED

The undecidability of any consistent extension of $\mathcal{R}$ offers a powerful refinement of the refutation of Programme 1.1. Yet, it is only the beginning of the story. Decidability of theories is the exception, not the rule, and many logic textbooks tend at this point to illustrate the pervasiveness of undecidability through the enumeration of examples and techniques. This is pleasant material and we should not shun the pleasure. I shall, however, postpone it and we shall only take of it in moderation. There are, as I've already hinted, greater subjects suiting our wits— what John Stuart Mill called "higher susceptibilities". I don't imagine, however, that the reader will quite agree with my judgment of what is higher.

Before coming to further undecidability results in section 8, below, I wish to discuss further the question of which relations are semi-representable in consistent, but not Σ_1-sound, effectively enumerable extensions of $\mathcal{R}$. Although a Σ_1-definition φv of an r.e. set X could semi-represent a proper superset X' of X, the obvious conjecture is that the semi-representable sets are exactly the r.e. ones. Why is this an obvious conjecture? Well, for an effectively enumerable theory, the semi-representable sets are recursively enumerable, and we have already seen in the proof of Corollary 6.28 that some r.e. non-recursive sets are semi-representable. Given the (admittedly irrational, but generally justified) belief that some reasonable characterisation must exist, we must ask ourselves: what reasonable characterisations can we think of? The best of these is the obvious conjecture.

The obvious conjecture is true. It is not, however, a trivial fact and we will have to work to prove it. This, along with some refinements, is the task of the next section. For now, let us content ourselves with deriving what we can from the Reduction Lemma.

6.29. Definition. Let $R \subseteq \omega^n$ and let $\mathcal{T}$ be a theory in L_R. A formula $\varphi v_0 \ldots v_{n-1}$ with exactly $v_0, \ldots, v_{n-1}$ free is said to *represent* R in $\mathcal{T}$ if, for all $x_0, \ldots, x_{n-1} \in \omega$,

i. $R(x_0, \ldots, x_{n-1}) \rightarrow \mathcal{T} \vdash \varphi \overline{x_0} \ldots \overline{x_{n-1}}$

ii. $\neg R(x_0, \ldots, x_{n-1}) \rightarrow \mathcal{T} \vdash \neg \varphi \overline{x_0} \ldots \overline{x_{n-1}}$.

6.30. Corollary. *Let $R \subseteq \omega^n$ be recursive and let $\mathcal{T}$ be a consistent extension of $\mathcal{R}$. Then: R is representable in $\mathcal{T}$.*

Proof. Recall that R is recursive iff both R and $\omega^n - R$ are recursively enumerable. Thus, if R is recursive, there are strict-Σ_1-formulae φv, ψv which define R, $\omega^n - R$, respectively. Let $\varphi_1 v$, $\psi_1 v$ be their witness comparisons and observe:

$$\begin{aligned} R(x_0, \ldots, x_{n-1}) &\rightarrow \mathfrak{N} \models \varphi \overline{x_0} \ldots \overline{x_{n-1}} \wedge \neg \psi \overline{x_0} \ldots \overline{x_{n-1}} \\ &\rightarrow \mathfrak{N} \models \varphi_1 \overline{x_0} \ldots \overline{x_{n-1}} \\ &\rightarrow \mathcal{R} \vdash \varphi_1 \overline{x_0} \ldots \overline{x_{n-1}} \\ &\rightarrow \mathcal{T} \vdash \varphi_1 \overline{x_0} \ldots \overline{x_{n-1}}; \end{aligned}$$

while

$$\begin{aligned}\neg R(x_0, ..., x_{n-1}) &\rightarrow \mathfrak{N} \models \psi \overline{x_0} \dots \overline{x_{n-1}} \wedge \neg\varphi \overline{x_0} \dots \overline{x_{n-1}} \\ &\rightarrow \mathfrak{N} \models \varphi_1 \overline{x_0} \dots \overline{x_{n-1}} \\ &\rightarrow \mathcal{R} \vdash \neg\varphi_1 \overline{x_0} \dots \overline{x_{n-1}}, \quad \text{by 6.27.iii} \\ &\rightarrow \mathcal{T} \vdash \neg\varphi_1 \overline{x_0} \dots \overline{x_{n-1}}.\end{aligned}$$

QED

In the effectively enumerable case, there is the following converse.

6.31. Fact. *Let $\mathcal{T}$ be a consistent recursively enumerable extension of $\mathcal{R}$, and let $R \subseteq \omega^n$.*

i. *if R is semi-representable in $\mathcal{T}$, then R is recursively enumerable*

ii. *if R is representable in $\mathcal{T}$, then R is recursive.*

I leave the proof to the reader.

Following the Exercises, we will resume the study of semi-representability in section 7 and that of undecidability in section 8.

Exercises.

1. (Models of $\mathcal{R}$). We begin with a catalogue of structures.

   a. $\mathfrak{Pol}$ is defined in Example 6.23.

   b. $\mathfrak{Pol}_Z$ extends $\mathfrak{Pol}$ by allowing polynomials with negative non-leading coëfficients. That is, the domain of $\mathfrak{Pol}_Z$ consists of polynomials,

$$a_n X^n + a_{n-1} X^{n-1} + ... + a_0,$$

   where the coëfficients $a_0, ..., a_n$ are integers and either $n = 0$ and $a_n \geq 0$, or $n > 0$ and $a_n > 0$.

   c. $\mathfrak{Mat}_n$ has the set of $n \times n$ matrices over ω as domain. The addition and multiplication operations on $\mathfrak{Mat}_n$ are just the usual matrix operations; the denotation of a numeral $\overline{x}$ is the scalar matrix,

$$xI = \begin{bmatrix} x & 0 & \dots & 0 \\ 0 & x & \dots & 0 \\ & & & \\ 0 & 0 & \dots & x \end{bmatrix}$$

   I the identity matrix; and $\leq$ is defined by

   $A \leq B$ iff $\exists x, y \in \omega[A = xI \wedge B = yI \wedge x \leq y]$ or B is not a scalar matrix.

   d. $\mathfrak{Mat}_n'$ differs from $\mathfrak{Mat}_n$ only in the interpretation of $\leq$:

   $A \leq B$ iff $\exists x, y \in \omega[A = xI \wedge B = yI \wedge x \leq y]$ or A is and B is not a scalar matrix.

   e. $\mathfrak{N}_\infty$ is obtained from $\mathfrak{N}$ by adjoining a new element ∞ to ω and defining, for $a \in \omega \cup \{\infty\}$,

$$a + \infty = \infty + a = \infty; \; a \cdot \infty = \infty \cdot a = \infty; \; a \leq \infty.$$

f. $\mathfrak{N}_{2\infty}$ extends $\mathfrak{N}$ by the adjunction of two new elements ∞_1, ∞_2 and defining, for $x \in \omega$ and $i, j \in \{1, 2\}$,

$$x + \infty_i = \infty_i + x = \infty_i;\ x \cdot \infty_i = \infty_i \cdot x = \infty_i$$

$$\infty_i + \infty_j = \infty_i;\ \infty_i \cdot \infty_j = \infty_i$$

$$x \le \infty_i;\ \infty_1 \le \infty_2.$$

g. For any ordinal number α closed under $+,\cdot$, $\mathfrak{Ord}_\alpha$ is given by choosing $\alpha = \{\beta: \beta < \alpha\}$ as domain, ordinal addition and multiplication as $+,\cdot$, and the usual weak ordering on ordinals as $\le$.

h. (For those who know some set theory). For any ordinal number α closed under $+,\cdot$, $\mathfrak{Set}_\alpha$ is given by choosing as domain the collection of all sets of rank less than α, and extending the operations on $\mathfrak{Ord}_\alpha$ trivially to sets that are not ordinals, i.e., for x, y sets other than ordinals, one defines

$$x + y = x \cdot y = 0;$$

and $\le$ is never defined for non-ordinals. Now for the work:

i. Show that $\mathfrak{Set}_\alpha$ is a model of $\mathcal{R}^-$, $\mathfrak{Mat}_n'$ is a model of $\mathcal{R}$, and all the other structures are models of $\mathcal{R}^+$.

ii. For which structures $\mathfrak{A}$ from the list a - h is $Th(\mathfrak{A})$ Σ_1-sound?

iii. Show that $\le$ is definable in $L_R - \{\le\}$ in the structures $\mathfrak{Pol}_Z$, $\mathfrak{Mat}_n$, $\mathfrak{N}_\infty$, and $\mathfrak{Ord}_\alpha$. Show that $\le$ is not definable in the restricted language in $\mathfrak{N}_{2\infty}$. What about $\mathfrak{Pol}$?

iv. In which structures is $\le$ defined by

$$v_0 \le v_1:\ \exists v_2(v_0 + v_2 = v_1)$$

$$v_0 \le v_1:\ \exists v_2(v_2 + v_0 = v_1)?$$

v. Show that the following are not provable in $\mathcal{R}^+$:

$$\forall v_0 v_1(v_0 + v_1 = v_1 + v_0)$$

$$\forall v_0 v_1(v_0 \cdot v_1 = v_1 \cdot v_0)$$

$$\forall v_0 v_1 v_2[v_0 + (v_1 + v_2) = (v_0 + v_1) + v_2]$$

$$\forall v_0 v_1 v_2[v_0 \cdot (v_1 \cdot v_2) = (v_0 \cdot v_1) \cdot v_2]$$

$$\forall v_0 v_1 v_2[v_0 \cdot (v_1 + v_2) = (v_0 \cdot v_1) + (v_0 \cdot v_2)]$$

$$\forall v_0(v_0 \le v_0)$$

$$\forall v_0 v_1 v_2[v_0 \le v_1 \wedge v_1 \le v_2 \rightarrow v_0 \le v_2].$$

[Hints: iii. Show that ω is definable in terms of $+$ and $\cdot$ (for $\mathfrak{Mat}_n$, look for a multiplicative characterisation of the scalar matrices); v. you will have to construct some new models.]

2. (Interpretability I). Recall Exercise 1 of section 4, above.

i. Define

$$N_0(v):\ \forall v_0(v_0 \le v \vee v \le v_0)$$

$$v_0 \le' v_1:\ (N_0(v_1) \wedge v_0 \le v_1) \vee \neg N_0(v_1).$$

a. Show: $\mathcal{R} \vdash N_0(\overline{x})$ for each $x \in \omega$

b. Show: $\mathcal{R} \vdash N_0(v_1) \rightarrow (v_0 \le' v_1 \leftrightarrow v_0 \le v_1)$

c. Show $\mathcal{R}^+$ is interpretable in $\mathcal{R}$ by interpreting $+, \cdot, \overline{0}, \overline{1}, \ldots$ by $+, \cdot, \overline{0}, \overline{1}, \ldots$ and $\le$ by $\le'$.

ii. Define

$$N_1(v)\colon\ \overline{0} \le v \wedge \forall v_0 \le v\,(v_0 + \overline{1} \le v \vee v \le v_0 + \overline{1})$$

$$v_0 \le' v_1\colon\ (N_1(v_1) \wedge v_0 \le v_1) \vee \neg N_1(v_1)$$

a. Show: $\mathcal{R}^- \vdash N_0(\overline{x})$ for each $x \in \omega$

b. Show: $\mathcal{R}^- \vdash N_0(v_1) \to (v_0 \le' v_1 \leftrightarrow v_0 \le v_1)$

c. Show $\mathcal{R}$ is interpretable in $\mathcal{R}^-$ by interpreting $+, \cdot, \overline{0}, \overline{1}, \ldots$ by $+, \cdot, \overline{0}, \overline{1}, \ldots$ and $\le$ by $\le'$.

3. (Interpretability II). Define $\mathcal{R}^{LO}$ to be the extension of $\mathcal{R}$ by the axioms,

$R5'$. $\forall v_0 v_1 (v_0 \le v_1 \vee v_1 \le v_0)$

$R6$. $\forall v_0 v_1 (v_0 \le v_1 \wedge v_1 \le v_0 \to v_0 = v_1)$

$R7$. $\forall v_0 v_1 v_2 (v_0 \le v_1 \wedge v_1 \le v_2 \to v_0 \le v_2)$.

i. Show: There is no formula $\varphi v_0 v_1$ in L_R such that

a. $\mathcal{R} \vdash \forall v_0 v_1 (\varphi v_0 v_1 \vee \varphi v_1 v_0)$

b. $\mathcal{R} \vdash \forall v_0 v_1 (\varphi v_0 v_1 \wedge \varphi v_1 v_0 \to v_0 = v_1)$

c. $\mathcal{R} \vdash \forall v_0 v_1 v_2 (\varphi v_0 v_1 \wedge \varphi v_1 v_2 \to \varphi v_0 v_2)$.

ii. Define

$N_0(v)$: $\forall v_0 (v_0 \le v \vee v \le v_0)$

$N_1(v)$: $\forall v_0 \le v\,(v \le v_0 \to v = v_0)$

$N_2(v)$: $\forall v_1 \le v\, \forall v_0 \le v_1 (v_0 \le v)$

$N(v)$: $N_0(v) \wedge N_1(v) \wedge N_2(v)$.

Show:

a. $\mathcal{R} \vdash \forall v_0 v_1 (N(v_0) \wedge N(v_1) \to v_0 \le v_1 \vee v_1 \le v_0)$

b. $\mathcal{R} \vdash \forall v_0 v_1 [N(v_0) \wedge N(v_1) \to (v_0 \le v_1 \wedge v_1 \le v_0 \to v_0 = v_1)]$

c. $\mathcal{R} \vdash \forall v_0 v_1 v_2 [N(v_0) \wedge N(v_1) \to (v_0 \le v_1 \wedge v_1 \le v_2 \to v_0 \le v_2)]$

iii. Define

$$v_0 \oplus v_1 = v_2 \leftrightarrow (v_2 = v_0 + v_1 \wedge N(v_0 + v_1)) \vee \\ \vee (v_2 = \overline{0} \wedge \neg N(v_0 + v_1))$$

$$v_0 \otimes v_1 = v_2 \leftrightarrow (v_2 = v_0 \cdot v_1 \wedge N(v_0 \cdot v_1)) \vee \\ \vee (v_2 = \overline{0} \wedge \neg N(v_0 \cdot v_1)).$$

Recalling the definition of relative interpretability given in Exercise 1 of section 4 and the comment immediately following it, show that $\mathcal{R}^{LO}$ is interpretable in $\mathcal{R}$. [Hint: i. Construct a model of $\mathcal{R}$ with two elements a and b in the domain and an automorphism F defined by

$$F(c) = \begin{cases} b, & c = a \\ a, & c = b, \\ c & \text{otherwise.}] \end{cases}$$

4. (Interpretability III). Define a consistent theory $\mathcal{T}_0$ to be *essentially undecidable* if every consistent extension $\mathcal{T}$ of $\mathcal{T}_0$ is undecidable.

i. Suppose $\mathcal{T}_1$ and $\mathcal{T}_2$ are consistent theories and I is an effective interpretation of $\mathcal{T}_1$ in $\mathcal{T}_2$. Show: If $\mathcal{T}_1$ is essentially undecidable, then so is $\mathcal{T}_2$.

ii. Show: $\mathcal{R}^-$ is essentially undecidable.

5. Let $\mathcal{R}^{WO}$ be the extension of $\mathcal{R}^{LO}$ of Exercise 3 by the schema,

$$\mathrm{LNP}\Sigma_1: \quad \varphi \rightarrow (\varphi \preccurlyeq \varphi)$$

for strict-Σ_1-formulae φ.

i. Show: For arbitrary strict-Σ_1 φ, ψ,

$$\mathcal{R}^{LO} \vdash (\varphi \prec \psi) \rightarrow (\varphi \preccurlyeq \psi).$$

ii. Show: $\mathcal{R}^{LO} \nvdash \mathrm{LNP}\Sigma_1$, i.e., for some strict-$\Sigma_1$ φ, $\mathcal{R}^{LO} \nvdash \varphi \rightarrow (\varphi \preccurlyeq \varphi)$.

iii. Show: For some strict-Σ_1 φ, ψ,

$$\mathcal{R}^{LO} \nvdash \varphi \wedge \psi \rightarrow (\varphi \preccurlyeq \psi) \vee (\psi \preccurlyeq \varphi).$$

iv. Show: For any strict-Σ_1 φ,

$$\mathcal{R}^{WO} \vdash \varphi \vee \psi \rightarrow (\varphi \preccurlyeq \psi) \vee (\psi \prec \varphi).$$

6. Let $\mathcal{R}^0$ be the sub-theory of $\mathcal{R}^-$ with $\leq$ dropped from the language and only schemata *R1* and *R2* as axioms.

i. Define Diophantine formulae and show that in $(\omega; +, \cdot, 0, 1, ...)$ they define exactly the Diophantine relations.

ii. Show the Diophantine completeness of $\mathcal{R}^0$, i.e. prove the Diophantine analogue of Theorem 6.13.

iii. Show that the relations semi-representable in $\mathcal{R}^0$ are exactly the recursively enumerable relations.

iv. Show that any Diophantine correct extension of $\mathcal{R}^0$ is undecidable.

v. Show that $\mathcal{R}^0$ is not essentially undecidable.

[Hint: v. Find a finite model of $\mathcal{R}^0$. Note that this trick will not work if one augments $\mathcal{R}^0$ by the schema *R3*. The resulting theory is, nonetheless, not essentially undecidable. Its extensions include $Th(R; +, \cdot, 0, 1, ...)$ and $Th(C; +, \cdot, 0, 1, ...)$, both of which theories admit eliminations of quantifiers in appropriate languages.]

7. Semi-Representability of Functions

In the last section we considered several notions of the definability of numerical relations: definability in $\mathfrak{N}$, and semi-representability and representability in theories. There are also several notions of the definability of functions, beginning with the mere definability of the graphs of the given functions. Semantically, this is all one really needs: if a numerical function F has an arithmetically definable graph, then all the usual functional properties of F are true in $\mathfrak{N}$ for any formula φ defining the graph of F. Syntactically, this is not the case— as we should expect from incompleteness.

7.1. Example. The formula,

$$\varphi v_0 v_1: \quad v_1 = v_0 + \overline{1} \ \vee\ \forall v_2 < v_1 (v_2 + \overline{1} < v_1),$$

represents the graph of the successor function in $\mathcal{R}$, but it does not define a function in every model of $\mathcal{R}$. For example,

$$\mathfrak{N}_\infty \models \varphi(\overline{0}, \overline{1}) \wedge \varphi(\overline{0}, \overline{\infty}),$$

where $\mathfrak{N}_\infty$ is as described in Exercise 1 of the last section.

Clearly we cannot consider the formula φ of this Example as a good *functorial* representation of the successor function even though it is an acceptable representation of the graph. Before we will state that a formula φ represents a numerical function in a theory $\mathcal{T}$, we will insist that $\mathcal{T}$ proves φ to define a function— in some sense. The question is: in what sense? There are several possible requirements to impose. The goal of the present section is to impose the weakest usable requirement— unicity. Stronger requirements are matters of greater subtlety and will be considered in later chapters (i.e., they are deferred to volume II).

There are two notions of unicity, depending on whether or not one cares about the behaviour of the function outside its intended domain.

7.2. Definition. A formula $\varphi v_0 \ldots v_n$ *represents* a function $F{:}\omega^n \to \omega$ in a theory $\mathcal{T}$ if

i. φ represents the graph of F in $\mathcal{T}$

ii. φ satisfies the unicity condition: for all $x_0, \ldots, x_{n-1} \in \omega$,

$$\mathcal{T} \vdash \varphi(\overline{x_0}, \ldots, \overline{x_{n-1}}, v_0) \wedge \varphi(\overline{x_0}, \ldots, \overline{x_{n-1}}, v_1) \rightarrow v_0 = v_1.$$

7.3. Definition. A formula $\varphi v_0 \ldots v_n v$ *strongly represents* a function $F{:}\omega^n \to \omega$ in a theory $\mathcal{T}$ if

i. φ represents the graph of F in $\mathcal{T}$

ii. φ satisfies the strong unicity condition:

$$\mathcal{T} \vdash \varphi v_0 \ldots v_{n-1} v \wedge \varphi v_0 \ldots v_{n-1} v_n \rightarrow v = v_n.$$

7.4. Example. The formula,

$$\varphi v_0 v_1{:}\ v_1 = v_0 + \overline{1} \vee \forall v_2 < v_0 (v_2 + \overline{1} < v_0) \wedge v_1 = \overline{0},$$

represents the graph of the successor function in $\mathcal{R}$, but it does not strongly represent this function as

$$\mathfrak{N}_\infty \models \varphi(\overline{\infty}, \overline{0}) \wedge \varphi(\overline{\infty}, \overline{\infty}).$$

Despite the obvious preferability of the notion of strong representability over that of representability, it is the latter, weaker notion that has traditionally made its appearance in recursion theoretic expositions of incompleteness and undecidability. There are two reasons for this. First, the applications only require representability, not strong representability. Second, these traditional expositions stem from the 1950s, when the problem of finding undecidable *weak* theories was particularly stressed. Thus, the theory $\mathcal{R}$ was to be preferred to $\mathcal{R}^+$ and it happens that the most direct proof establishes the representability of recursive functions in $\mathcal{R}$ and the strong representability of these functions in $\mathcal{R}^+$. Exactly as in the proof of the Reduction Theorem in the last section, the difference between representability and strong representability more-or-less hinges on the difference between the pointwise comparability of $\leq$ of axiom schema *R5* and the uniform comparability given by *R5'*. It would seem that the most reasonable thing to do would be to replace $\mathcal{R}$ by a stronger theory— $\mathcal{R}^+$, for example. However, regardless of the choice of theory, the overall approach is the same and there may be some mild axiomatic interest in seeing exactly what is used.

The construction of a functional representation of a recursive function from a representation of a graph of the function will be the same sort of application of minimum witnesses that we used in proving the Reduction Theorem. A slight complication arises because we want to compare *witnesses* and get unique *values*. Basically, this means that two comparisons— one of witnesses and one of values— must occur. The following lemma precludes any conflict between the two comparisons.

7.5. Lemma. *Let* $\varphi v_0...v_n$ *be* $\exists v\, \psi v v_0...v_n$, *with* $\psi \in \Delta_0$. *Define,*

$$\psi' v v_0...v_n:\ \exists v_{n+1} \le v\, [\psi v_{n+1} v_0...v_n \wedge v_n \le v]$$
$$\varphi' v_0...v_n:\ \exists v\, \psi' v v_0...v_n.$$

Then:

i. $\mathcal{R} \vdash \psi' v v_0...v_n \rightarrow v_n \le v$

ii. a. $\mathcal{R} \vdash \varphi' v_0...v_n \rightarrow \varphi v_0...v_n$

b. *for any* $x_0,...,x_n \in \omega$,

$$\mathcal{R} \vdash \varphi(\overline{x_0},..., \overline{x_n}) \leftrightarrow \mathcal{R} \vdash \varphi'(\overline{x_0},..., \overline{x_n})$$

iii. $\mathcal{R}^+ \vdash \forall v_0...v_n[\varphi v_0...v_n \leftrightarrow \varphi' v_0...v_n]$.

Proof. i. Clear.

ii. a. By pure logic,

$$\begin{aligned} \varphi' v_0...v_n &\rightarrow \exists v\, \exists v_{n+1} \le v\, [\psi v_{n+1} v_0...v_n \wedge v_n \le v\,] \\ &\rightarrow \exists v_{n+1}\, \psi v_{n+1} v_0...v_n \\ &\rightarrow \varphi v_0...v_n\,. \end{aligned}$$

b. φ and φ' are strict-Σ_1-formulae. Any instance of φ is true iff the corresponding instance of φ' is true (apply iii, below, to prove this), whence the Σ_1-completeness and Σ_1-soundness of $\mathcal{R}$ yield the result:

$$\begin{aligned} \mathcal{R} \vdash \varphi(\overline{x_0},..., \overline{x_n}) &\leftrightarrow \mathfrak{N} \models \varphi(\overline{x_0},..., \overline{x_n}) \\ &\leftrightarrow \mathfrak{N} \models \varphi'(\overline{x_0},..., \overline{x_n}) \\ &\leftrightarrow \mathcal{R} \vdash \varphi'(\overline{x_0},..., \overline{x_n}). \end{aligned}$$

iii. By ii.a, it remains only to prove

$$\mathcal{R}^+ \vdash \varphi v_0...v_n \rightarrow \varphi' v_0...v_n.$$

Suppose $\varphi v_0...v_n$, i.e., suppose there is v_{n+1} such that

$$\psi v_{n+1} v_0...v_n\,.$$

By *R5'*, either $v_n \le v_{n+1}$ or $v_{n+1} \le v_n$. In the first case, let v be v_{n+1}; in the second choose v to be v_n. Either way we have

$$\psi v_{n+1} v_0 \ldots v_n \ \wedge \ v_{n+1} \leq v \ \wedge \ v_n \leq v,$$

whence

$$\exists v_{n+1} \leq v \ [\psi v_{n+1} v_0 \ldots v_n \ \wedge \ v_n \leq v],$$

i.e. $\psi' v v_0 \ldots v_n$. Thus,

$$\mathcal{R}^+ \vdash \varphi v_0 \ldots v_n \ \rightarrow \exists v \ \psi' v v_0 \ldots v_n$$
$$\rightarrow \varphi' v v_0 \ldots v_n .$$

QED

7.6. Corollary. *Let φ and φ' be as in Lemma 7.5. If φ represents a relation R in $\mathcal{R}$, then φ' also represents R in $\mathcal{R}$.*

Proof. If we were working in $\mathcal{R}^+$, we could appeal to 7.5.iii to conclude φ and φ' to represent the same relation. In $\mathcal{R}$, we have only 7.5.ii. By 7.5.ii.b, if φ represents R in $\mathcal{R}$, φ' *semi*-represents R in $\mathcal{R}$. To conclude representability, apply 7.5.ii.a:

$$\neg R(x_0, \ldots, x_n) \ \rightarrow \ \mathcal{R} \vdash \neg\varphi(\overline{x_0}, \ldots, \overline{x_n}), \text{ by representability}$$
$$\rightarrow \ \mathcal{R} \vdash \neg\varphi'(\overline{x_0}, \ldots, \overline{x_n}), \text{ by 7.5.ii.a.}$$

QED

Corollary 7.6 is far less relevant for our immediate purpose than it seems. For, the full representability of the graph of a function in the definition of the representability of a function can, in the face of the unicity condition, be replaced by the weaker condition of the semi-representability of the graph. (Cf. Exercise 2, below.) Nevertheless, Corollary 7.6 is worth noting as a comment on Lemma 7.5.

As for our immediate purpose, its fulfillment is at hand:

7.7. Theorem. *Let $F{:}\omega^n \rightarrow \omega$ be a recursive function. There are strict-Σ_1- and strict-Π_1-formulae which represent F in $\mathcal{R}$ and strongly represent F in $\mathcal{R}^+$.*

Proof. Let $\varphi = \exists v \ \psi v v_0 \ldots v_n$ be a strict-Σ_1-formula (semi-)representing the graph of F in $\mathcal{R}$ and suppose, by the Lemma, that ψ satisfies

$$\mathcal{R} \vdash \psi v v_0 \ldots v_n \ \rightarrow \ v_n \leq v. \tag{1}$$

Define

$$\varphi' v_0 \ldots v_n{:}\ \exists v \ [\psi v v_0 \ldots v_n \ \wedge \forall v' \leq v \ \forall v'' \leq v' \ (\psi v' v_0 \ldots v_{n-1} v'' \ \rightarrow \ v'' = v_n)].$$

The formula φ' is strict-Σ_1 and it semi-represents the graph of F in $\mathcal{R}$. (For, it *defines* the graph of F and it is Σ_1.) As already noted, the unicity condition will ensure that φ' represents the graph of F in $\mathcal{R}$. Thus, to show that φ' (strongly) represents F functionally, it suffices to show that φ' satisfies the (strong) unicity condition in $\mathcal{R}$ (respectively, in $\mathcal{R}^+$).

Let $F(x_0, \ldots, x_{n-1}) = x_n$. Let $y \in \omega$ be the minimum witness to $\varphi'(\overline{x_0}, \ldots, \overline{x_{n-1}}, \overline{x_n})$. Then, $\mathfrak{R}$ satisfies

$$\psi(\overline{y}, \overline{x_0}, \ldots, \overline{x_n}) \ \wedge \ \forall v' \leq \overline{y} \ \forall v'' \leq v' \ [\psi(v', \overline{x_0}, \ldots, \overline{x_{n-1}}, v'') \rightarrow v'' = \overline{x_n}] \tag{2}$$

and, the sentence being Δ_0, $\mathcal{R}$ proves it. Now suppose $\varphi'(\overline{x_0},\ldots,\overline{x_{n-1}},v_1)$, i.e., for some v_0,

$$\psi(v_0,\overline{x_0},\ldots,\overline{x_{n-1}},v_1) \wedge \forall v' \le v_0 \forall v'' \le v' [\psi(v',\overline{x_0},\ldots,\overline{x_{n-1}},v'') \to v'' = v_1]. \quad (3)$$

By *R5*,

$$v_0 \le \overline{y} \vee \overline{y} \le v_0.$$

The first disjunct, (1), and (2) yield

$$\psi(v_0,\overline{x_0},\ldots,\overline{x_{n-1}},v_1) \to v_1 = \overline{x_n}.$$

The second disjunct, (1), and (3) yield

$$\psi(\overline{y},\overline{x_0},\ldots,\overline{x_n}) \to \overline{x_n} = v_1.$$

Thus,

$$\mathcal{R} \vdash \varphi'(\overline{x_0},\ldots,\overline{x_{n-1}},v_1) \to \overline{x_n} = v_1,$$

and φ represents F in $\mathcal{R}$.

The proof of the strong unicity condition in $\mathcal{R}^+$ is similar and I omit it.

As for the strict-Π_1-representation of F, let $\varphi = \exists v\, \psi$ be a strict-Σ_1-representation of F (not just of the graph of F) such that (1) holds for ψ. [Note that this property is preserved in the above construction of φ'.] Then choose

$$\varphi' v_0 \ldots v_n\colon\ \forall v\, \forall v' \le v\, [\psi v v_0 \ldots v_{n-1} v' \to v' = v_n].$$

I leave the verification that this works as an exercise to the reader (Exercise 3, below). QED

7.8. Corollary. *Let $\mathcal{T}$ be an effectively enumerable consistent extension of $\mathcal{R}$. Then:*
i. *the functions representable in $\mathcal{T}$ are exactly the recursive functions; and*
ii. *if $\mathcal{T}$ also extends $\mathcal{R}^+$, the recursive functions are strongly representable in $\mathcal{T}$.*

The proof is an easy application of the known and the obvious: the known is the recursiveness of any relation (here: the graph of a representable function) representable in $\mathcal{T}$ (by 6.31) and the recursiveness of any function with a recursive graph. The obvious is the fact that a (strong) representation of a function in a theory $\mathcal{T}_0$ is also a (strong) representation of the given function in any consistent extension $\mathcal{T}_1 \supseteq \mathcal{T}_0$.

At the moment, we have only an intuitive notion of "effective enumerability" of a theory. In the next chapter we will develop an encoding of syntax and replace this intuitive notion by the precise one of "recursive enumerability" of a theory. With such precision, Corollary 7.8 seems almost tautological and of only mild interest: the functions representable in r.e. theories are exactly the recursive ones. We can, however, replace "effective enumerability" by another precise notion— that of finiteness. In the next section, we will see that $\mathcal{R}$ can be naturally interpreted in a finitely axiomatised theory $\mathcal{Q}$ in a finite language. Corollary 7.8 transforms into: the functions representable in finitely axiomatised theories containing $\mathcal{Q}$ are precisely the recursive functions. This affords us with another characterisation of recursiveness: the recursive functions are precisely those functions numeralwise computable in sufficiently strong finitely axiomatised theories. The invariance of the class of functions under the choice of theories may be taken as further evidence of the correctness of the explication of the intuitive notion of effective computability by the precise notion of recursiveness. Indeed, in the late 1960s, the *uniform* representability of recursive relations and functions in consistent *complete* extensions of $\mathcal{R}$ was a matter of some heuristic importance in generalised recursion theory— the study of "effectiveness" in some non-numerical contexts.

However, I do not wish to discuss weighty issues in the present section, but to present a fairly technical development. I wish now to discuss the semi-representability of *partial* recursive functions. There are, again, two definitions.

7.9. Definition. A formula $\varphi v_0 \ldots v_n$ *semi-represents* a partial function $F{:}\subseteq\omega^n \to \omega$ in a theory $\mathcal{T}$ if

i. φ semi-represents the graph of F in $\mathcal{T}$

ii. φ satisfies the unicity condition: for all $x_0,\ldots,x_{n-1}$ in the domain of F,

$$\mathcal{T} \vdash \varphi(\overline{x_0},\ldots,\overline{x_{n-1}},v_0) \wedge \varphi(\overline{x_0},\ldots,\overline{x_{n-1}},v_1) \to v_0 = v_1.$$

7.10. Definition. A formula $\varphi v_0 \ldots v_n$ *strongly semi-represents* a partial function $F{:}\subseteq\omega^n \to \omega$ in a theory $\mathcal{T}$ if

i. φ semi-represents the graph of F in $\mathcal{T}$

ii. φ satisfies the strong unicity condition:

$$\mathcal{T} \vdash \varphi v_0 \ldots v_{n-1} v \wedge \varphi v v_0 \ldots v_{n-1} v_n \to v = v_n.$$

These Definitions are entirely analogous to Definitions 7.2 and 7.3 of representability of total functions and require no particular explanation. Indeed, they may help explain why we did not include provable totality as a condition in these earlier Definitions: totality is a different issue from that of functionality, which is what we are considering here.

For Σ_1-sound theories we can obtain semi-representations of partial recursive functions in the same way we obtained representations of recursive functions. We have but to look at the proof of Theorem 7.7 and see what it really proves. This will be the very next thing we do and the result is probably the most important of this section. Following that we will turn to the non-Σ_1-sound case, where some additional recursion theoretic tools must be introduced.

To avoid messy formulations and overmuch subtlety, we shall restrict our attention throughout the rest of the section to the theory $\mathcal{R}^+$ and its extensions.

7.11. Selection Theorem. *For any strict-Σ_1-formula $\varphi v_0 \ldots v_n$ with exactly $v_0, \ldots, v_n$ free, there is another strict-Σ_1-formula* $\mathsf{Sel}[\varphi]$ *with the same free variables and is such that*

i. a. $\mathfrak{N} \models \forall v_0 \ldots v_n (\mathsf{Sel}[\varphi] \to \varphi)$

b. $\mathfrak{N} \models \forall v_0 \ldots v_{n-1}(\exists v_n \varphi \to \exists v_n \mathsf{Sel}[\varphi])$

c. $\mathfrak{N} \models \forall v_0 \ldots v_{n-1} v v_n (\mathsf{Sel}[\varphi] v_0 \ldots v_{n-1} v \wedge \mathsf{Sel}[\varphi] v_0 \ldots v_{n-1} v_n \to v = v_n)$

ii. a. $\mathcal{R}^+ \vdash \mathsf{Sel}[\varphi] \to \varphi$

b. *for any* $x_0, \ldots, x_{n-1} \in \omega$,

$$\mathfrak{N} \models \exists v_n \varphi(\overline{x_0},\ldots,\overline{x_{n-1}},v_n) \to \exists z\,(\mathcal{R}^+ \vdash \mathsf{Sel}[\varphi](\overline{x_0},\ldots,\overline{x_{n-1}},\overline{z}))$$

c. $\mathcal{R}^+ \vdash \mathsf{Sel}[\varphi] v_0 \ldots v_{n-1} v \wedge \mathsf{Sel}[\varphi] v_0 \ldots v_{n-1} v_n \to v = v_n$).

The Selection Theorem is a sort of effective version of the Axiom of Choice, one form of which asserts every relation to include a function. The recursion theoretic half, 7.11.i, asserts that every recursively enumerable relation (a strict-Σ_1-relation) contains a partial recursive function (a function with a strict-Σ_1-graph) with the same domain as the original relation (7.11.i.b). The syntactic half, 7.11.ii, asserts that $\mathcal{R}^+$ can prove the functoriality of the subfunction, and that $\mathcal{R}^+$ *schematically* proves the domain of the function equal to the domain of the relation: each *instance* is provable. To derive the uniform assertion, one would have to add the Least Number Principle (i.e., induction) for Δ_0-formulae to $\mathcal{R}^+$. This is done in the next chapter.

Proof of Theorem 7.11. Let $\varphi = \exists v\, \psi v v_0 \ldots v_n$ be strict-Σ_1. By Lemma 7.5, we can assume

$$\mathcal{R} \vdash \psi v v_0 \ldots v_n \ \rightarrow\ v_n \leq v\,.$$

Let, as in the proof of Theorem 7.7, $\mathsf{Sel}[\varphi]$ be

$$\exists v\, [\psi v v_0 \ldots v_n \ \wedge\ \forall v' \leq v\ \forall v'' \leq v'\, (\psi v' v_0 \ldots v_{n-1} v'' \ \rightarrow\ v'' = v_n)]. \tag{4}$$

Assertions i.a and ii.a are trivial.

To see the truth of i.b and ii.b, let

$$\mathfrak{N} \models \exists v_n\, \varphi(\overline{x_0}, \ldots, \overline{x_{n-1}}, v_n),$$

i.e.,

$$\mathfrak{N} \models \exists v\ \exists v_n\, \psi(v, \overline{x_0}, \ldots, \overline{x_{n-1}}, v_n).$$

Choose y minimal so that

$$\mathfrak{N} \models \exists v_n\, \psi(\overline{y}, \overline{x_0}, \ldots, \overline{x_{n-1}}, v_n),$$

and z minimal so that

$$\mathfrak{N} \models \psi(\overline{y}, \overline{x_0}, \ldots, \overline{x_{n-1}}, \overline{z}). \tag{5}$$

By these combined minimalities,

$$\mathfrak{N} \models \forall v' \leq \overline{y}\ \forall v'' \leq v'\, (\psi(v', \overline{x_0}, \ldots, \overline{x_{n-1}}, v'') \ \rightarrow\ v'' = \overline{z}). \tag{6}$$

By (4), (5), and (6), we see

$$\mathfrak{N} \models \mathsf{Sel}[\varphi](\overline{x_0}, \ldots, \overline{x_{n-1}}, \overline{z}),$$

whence

$$\mathcal{R}^+ \vdash \mathsf{Sel}[\varphi](\overline{x_0}, \ldots, \overline{x_{n-1}}, \overline{z}).$$

Thus, i.b and ii.b are satisfied.

Finally, i.c and ii.c are proven exactly as in the proof of Theorem 7.7, where the details for the strong unicity result were left to the reader as an exercise. The reader has but to xerox his solution to that exercise to solve the present one. QED

7.12. Corollary. *Let $\mathcal{T}$ be any Σ_1-sound extension of $\mathcal{R}^+$. Then: Every partial recursive function is strongly semi-representable in $\mathcal{T}$ by a strict-Σ_1-formula.*

Proof. By Corollary 6.20, the graph of every partial recursive function is semi-representable in $\mathcal{T}$ by a strict-Σ_1-formula. Applying the construction of the Selection Theorem to such a semi-representation yields a strong semi-representation of the function itself.

QED

Note that Corollary 7.12 doesn't use the full power of the Selection Theorem. We will make such a use in the next Chapter.

In the non-Σ_1-sound case, the full analogue to Corollary 7.12 holds for consistent effectively enumerable extensions of $\mathcal{R}^+$. The effective enumerability condition is essential. Without it, there is no guarantee that a semi-representable function is partial recursive— or even that a representable function is recursive: in $\mathcal{T} = Th(\mathfrak{N})$, every arithmetic function is representable. More seriously, in the non-effectively enumerable case, partial recursive functions which are not recursive can fail to be semi-represented. The fact is that the desired analogue to Corollary 7.12 is a recursion theoretic fact, a characterisation of the class of partial recursive functions among its subclasses:

7.13. Visser's Theorem. *Let $\mathcal{F}$ be a family of partial recursive unary functions such that*

i. *if $F \in \mathcal{F}$ and G is recursive, then $FG \in \mathcal{F}$*

ii. *if G is partial recursive, then, for some $F \in \mathcal{F}$, $G \subseteq F$.*

Then: Every partial recursive unary function is in $\mathcal{F}$.

Before proving this theorem, let us draw the desired conclusions from it.

7.14. Corollary. *Let $\mathcal{T}$ be a consistent effectively enumerable extension of $\mathcal{R}^+$. The functions strongly semi-representable in $\mathcal{T}$ are exactly the partial recursive functions. Moreover, the strong semi-representations can be taken to be given by strict-Σ_1-formulae.*

Proof. If F is semi-representable in $\mathcal{T}$, then the graph of F is semi-representable and, by Fact 6.31, is recursively enumerable. Thus, F is partial recursive.

For the converse, let $\mathcal{F}$ be the class of unary functions strongly representable in $\mathcal{T}$ by strict-Σ_1-formulae. We will first apply Visser's Theorem to show that $\mathcal{F}$ coincides with the class of partial recursive functions and then appeal to the strong representability of recursive n-tupling functions within $\mathcal{T}$ to obtain the full n-ary result.

We have already proven that $\mathcal{F}$ consists of partial recursive functions.

i. Let $\exists v\, \psi_1(v, v_0, v_1)$ be a strict-Σ_1-formula strongly semi-representing a function $F \in \mathcal{F}$ in $\mathcal{T}$, and let G be a recursive function strongly represented in $\mathcal{T}$ by a strict-Σ_1-formula $\exists v\, \psi_2(v, v_0, v_1)$. It is easy to see that the formula,

$$\exists v v' v_2[\psi_2(v, v_0, v_2) \wedge \psi_1(v', v_2, v_1)], \qquad (*)$$

strongly semi-represents the composition $F \circ G$ in $\mathcal{T}$. (Exercise.) The formula is, however, not strict-Σ_1 as it has three unbounded quantifiers. We can, however, appeal to the trick used in the proof of Lemma 7.5: the formula (*) is equivalent in $\mathcal{R}^+$ to

$$\exists v_3\, \exists v v_4 \leq v_3\, \exists v' v_2 \leq v_4\, [\psi_2(v, v_0, v_2) \wedge \psi_1(v', v_2, v_1)].$$

ii. This is easier: Let F be partial recursive and let $\exists v\, \psi(v, v_0, v_1)$ be a strict-Σ_1 strong semi-representation of F in $\mathcal{R}^+$. Passage to the extension $\mathcal{T} \supseteq \mathcal{R}^+$ does not destroy strong unicity, whence $\exists v\, \psi$ strongly semi-represents a function G in $\mathcal{T}$. Because $\mathcal{R}^+ \subseteq \mathcal{T}$, it follows that $F \subseteq G$. [N.B. Ordinary unicity could be destroyed in passage to $\mathcal{T}$.]

Visser's Theorem applies and we can conclude $\mathcal{F}$ to coincide with the class of all partial recursive unary functions, i.e., all partial recursive unary functions are strongly semi-represented in $\mathcal{T}$. If F is an n-ary partial recursive function, then

$$F(x_0, ..., x_{n-1}) \simeq G(\langle x_0, ..., x_{n-1}\rangle_n)$$

for some unary partial recursive function G and a recursive n-tupling function $\langle\cdot,...,\cdot\rangle_n$. Repeating the argument of i shows F to have a strong strict-Σ_1-semi-representation in $\mathcal{T}$. QED

7.15. Corollary. *Let $\mathcal{T}$ be a consistent effectively enumerable extension of $\mathcal{R}^+$. The relations semi-representable in $\mathcal{T}$ are exactly the recursively enumerable relations. Moreover, the semi-representations can be taken to be given by strict-Σ_1-formulae.*

Proof. By Fact 6.31, the semi-representable relations are r.e. To prove the converse, let $R \subseteq \omega^n$ be r.e. and define, for $x_0, ..., x_{n-1} \in \omega$,

$$F(x_0, ..., x_{n-1}) \simeq \begin{cases} 0, & R(x_0, \ldots, x_{n-1}) \\ \text{undefined}, & \text{otherwise.} \end{cases}$$

Let $\varphi v_0...v_n$ semi-represent F in $\mathcal{T}$ and observe

$$R(x_0, ..., x_{n-1}) \text{ iff } \mathcal{T} \vdash \varphi(\overline{x_0}, \ldots, \overline{x_{n-1}}, \overline{0}).$$

Thus, $\varphi(v_0, ..., v_{n-1}, \overline{0})$ semi-represents R in $\mathcal{T}$. QED

Proof of Theorem 7.13. Let $D(x) \simeq \varphi_x(x)$ be the diagonal function and choose $G \in \mathcal{F}$ extending D. Applying the Parameter Theorem, find F recursive such that, for all $x,y \in \omega$,

$$\varphi_{F(x)}(y) \simeq \begin{cases} D(x), & D(x)\!\downarrow\, \preccurlyeq G(y)\!\downarrow \\ G(y) + 1, & G(y)\!\downarrow\, \prec D(x)\!\downarrow, \end{cases}$$

where, in analogy to the formal witness comparisons of Definition 6.24, we write

$$D(x)\!\downarrow \preccurlyeq G(y)\!\downarrow: \quad \exists z\, [T_1(d, x, z) \wedge \forall w < z\, \neg T_1(g, y, w)]$$
$$G(y)\!\downarrow \prec D(x)\!\downarrow: \quad \exists z\, [T_1(g, y, z) \wedge \forall w \le z\, \neg T_1(d, x, w)],$$

where d, g are fixed indices for D, G, respectively.

Claim. For any x,

$$D(x) \simeq \varphi_{F(x)}(Fx) \simeq G(Fx).$$

To prove the Claim, calculate $\varphi_{F(x)}(Fx)$:

$$\varphi_{F(x)}(Fx) \simeq \begin{cases} D(x), & D(x)\downarrow \preccurlyeq G(Fx)\downarrow \\ G(Fx) + 1, & G(Fx)\downarrow \prec D(x)\downarrow. \end{cases} \tag{7}$$

Assuming for the moment that $D(x)$, $\varphi_{F(x)}(Fx)$, and $G(Fx)$ all exist, we can easily obtain the desired equalities. Since G extends D,

$$G(Fx) = D(Fx) = \varphi_{F(x)}(Fx),$$

which is one of the desired equations. For the other, note that the second clause in the calculation cannot apply as it would yield

$$G(Fx) = \varphi_{F(x)}(Fx) = G(Fx) + 1.$$

Thus, the first clause applies and

$$\varphi_{F(x)}(Fx) = D(x).$$

We thus see that the equations of the Claim hold provided all three values exist. We have yet to show that the existence of any one of the values implies that of the other two. If $D(x)$ is defined, then either $G(Fx)$ is defined or it is not. If the former, (7) immediately yields a value for $\varphi_{F(x)}(Fx)$; if the latter, the first clause of (7) holds and $\varphi_{F(x)}(Fx)$ is defined to be $D(x)$. But then $G(Fx)$ must be defined and equal to $\varphi_{F(x)}(Fx)$ because $\varphi_{F(x)}(Fx)$ is just $D(Fx)$ and G extends D. Thus, if $D(x)$ is defined, the other two values are also.

If $G(Fx)$ is defined, then either $D(x)$ is defined, whence so is $\varphi_{F(x)}(Fx)$, or $D(x)$ is not defined. If this latter should happen, (7) would yield

$$\varphi_{F(x)}(Fx) = G(Fx) + 1,$$

and, as G extends D,

$$G(Fx) = D(Fx) = \varphi_{F(x)}(Fx) = G(Fx) + 1,$$

a contradiction. Thus, if $G(Fx)$ is defined, then so is $D(x)$, whence $\varphi_{F(x)}(Fx)$.

Finally, if $\varphi_{F(x)}(Fx)$ is defined, (7) tells us that at least one of $D(x)$ and $G(Fx)$ is defined— but we know that the existence of either implies the existence of the other. This completes the proof of the Claim.

By the Claim, the diagonal function $D = GF$ is in $\mathcal{F}$ (7.13.i). To complete the proof, it suffices to show every partial recursive function H to be of the form DK for some recursive K. To this end, let H be a given partial recursive function and define K recursively such that

$$\varphi_{K(x)}(y) \simeq \begin{cases} H(x), & H(x) \text{ is defined} \\ \text{undefined}, & \text{otherwise.} \end{cases}$$

It is easy to show that

$$H(x) \simeq \varphi_{K(x)}(Kx) \simeq D(Kx). \qquad \text{QED}$$

The proof given of the semi-representability of partial recursive functions and r.e. relations in effectively enumerable extensions of $\mathcal{R}^+$ is only one of three such proofs, and derives, in fact, from the other two. As it may help to shed some light on the proof given here, I shall digress to say a few words about the original proof of Corollary 7.15.

To begin with, there is the recursion theoretic notion of a *complete* r.e. set introduced by Emil Post in 1944 in a paper entitled "Recursively enumerable sets of positive integers and their decision problems". In this paper, Post tried to get a picture of non-recursive r.e. sets and introduced *creative* sets (Definition I.12.9), *simple* and *hypersimple* sets (Dekker's construction of the latter is given in Exercise 7 of Chapter I, section 12), and the complete sets defined below.

7.16. Definition. An r.e. set A is *complete* if for every r.e. set X there is a recursive function F such that, for all $x \in \omega$,

$$x \in X \text{ iff } F(x) \in A .$$

7.17. Example. *The set* $K = \{x\colon x \in W_x\} = \{x\colon \varphi_x(x)$ *is defined*$\}$ *is complete.*

Proof. Define F via the Parameter Theorem so that

$$\varphi_{F(x)}(y) \simeq \begin{cases} 0 & x \in X \\ \text{undefined,} & \text{otherwise,} \end{cases}$$

and observe that

$$\varphi_{F(x)}(Fx) \text{ is defined iff } x \in X . \qquad \text{QED}$$

The completeness of K shows that it is as non-recursive as any r.e. set can be. Not every non-recursive r.e. set is complete. In 1955, however, John Myhill showed that every creative set is complete.

7.18. Theorem. *Every creative set is complete.*

Proof. Let C be creative via H, i.e., assume, for any r.e. set W_e,

$$W_e \subseteq \omega - C \;\Rightarrow\; H(e) \notin C \cup W_e. \tag{8}$$

Let X be an r.e. set and choose G recursive by the Parameter Theorem so that for all x, $y, z \in \omega$,

$$\varphi_{G(x,y)}(z) \simeq \begin{cases} 0, & z = H(x) \text{ and } y \in X \\ \text{undefined,} & \text{otherwise.} \end{cases}$$

Thus,

$$W_{G(x,y)} = \begin{cases} \{H(x)\}, & y \in X \\ \varnothing, & y \notin X. \end{cases}$$

Next apply the parametric form of the Recursion Theorem (cf. Exercise 12 of Chapter I, section 12) to find a recursive function F so that, for all $y \in \omega$,

$$W_{F(y)} = W_{G(F(y),y)} = \begin{cases} \{H(F(y))\}, & y \in X \\ \varnothing, & y \notin X . \end{cases}$$

Now observe:

$$y \in X \;\rightarrow\; H(F(y)) \in W_{F(y)}$$

$$\rightarrow W_{F(y)} \not\subseteq \omega - C \text{ , by (8)}$$
$$\rightarrow H(F(y)) \in C .$$
$$y \notin X \rightarrow W_{F(y)} = \varnothing$$
$$\rightarrow H(F(y)) \notin C \text{, by (8) again.}$$

Thus, the set X reduces to C via the composition HF and C is seen to be complete. QED

In 1960, Andrzej Ehrenfeucht and Solomon Feferman applied Theorem 7.18 to derive Corollary 7.15 by observing that, if $\mathcal{T}$ is any consistent effectively enumerable extension of $\mathcal{R}^+$, then some creative set is semi-representable in $\mathcal{T}$. This observation follows from the proof in the last section of Rosser's Theorem (6.28): Let A, B be a pair of effectively inseparable r.e. sets and let

$$C = \{x \in \omega: \mathcal{T} \vdash \varphi(\overline{x})\}, \quad D = \{x \in \omega: \mathcal{T} \vdash \neg\varphi(\overline{x})\},$$

where φ is constructed so that

$$A \subseteq C, B \subseteq D.$$

The sets C, D are readily seen to be effectively inseparable, whence C is creative. If X is any r.e. set, we can obtain a semi-representation of X in $\mathcal{T}$ as follows. Choose F recursive so that, for $x \in \omega$,

$$x \in X \text{ iff } Fx \in C,$$

and let $\psi v_0 v_1$ strongly semi-represent the function F in $\mathcal{T}$ by Theorem 7.7. Clearly, the formula,

$$\chi v_0: \exists v_1[\psi v_0 v_1 \wedge \varphi v_1],$$

semi-represents X in $\mathcal{T}$.

If we compare the Ehrenfeucht-Feferman proof of Corollary 7.15 with the proof of Theorem 7.13, we see some parallels. The second construction in the proof of Theorem 7.13, for example, displays an evident *completeness* property of the diagonal function $D(x) \simeq \varphi_x(x)$ among the class of partial recursive functions. The first construction (that of the Claim), however, goes a bit beyond providing analogues to Myhill's Theorem on the completeness of the members of a reasonable family of partial recursive functions (say, those extending D) and the existence of such a complete partial function in the given class $\mathcal{F}$. It shows that the partial function D *uniformly* reduces to D and an arbitrary extension $G \supseteq D$: there is a recursive F such that, for all $x \in \omega$,

$$\begin{cases} D(x) \simeq D(F(x)) \\ D(x) \simeq G(F(x)). \end{cases} \tag{9}$$

To better see what is gained by the uniform reduction, consider what such uniformity would mean in the Ehrenfeucht-Feferman approach. Let $A \subseteq C$ be creative sets and φv_0 a formula defining A and semi-representing C in $\mathcal{T}$. A uniform reduction,

$$x \in X \text{ iff } Fx \in A$$
$$\text{iff } Fx \in C,$$

of an r.e. set X to A and C would yield a formula

$$\chi v_0\colon\ \exists v_1[\psi v_0 v_1 \wedge \varphi v_1],\quad (\psi \text{ semi-representing } F)$$

simultaneously defining X in $\mathfrak{N}$ and semi-representing X in $\mathcal{T}$.

7.19. Definition. A semi-representation φ of a relation R (or a function F) in a theory $\mathcal{T}$ is *correct* if φ is also a definition of R (respectively, F) in $\mathfrak{N}$.

7.20. Theorem. *Let $\mathcal{T}$ be a consistent effectively enumerable extension of $\mathcal{R}^+$.*
i. *Every partial recursive function has a correct strict-Σ_1-semi-representation in $\mathcal{T}$;*
ii. *Every r.e. relation has a correct strict-Σ_1-semi-representation in $\mathcal{T}$.*

Because of the uniform reduction (9), the semi-representations constructed in the proofs of Corollaries 7.14 and 7.15 are correct. I leave the details of the proof to the reader. An alternate proof of Theorem 7.20 will be given in full in Chapter IV of volume II.

Visser's Theorem and the present proof of Theorem 7.20 are the end-products of a surprisingly slow development of over two decades. Those most directly involved in this development, in addition to Myhill, Ehrenfeucht, and Feferman, include alphabetically Hilary Putnam, Robert W. Ritchie, W. Ritter, John Shepherdson, C. Smoryński, Raymond Smullyan, Albert Visser, and Paul Young. To go into detail on the individual contributions would require more space than the subject merits. Indeed, I may have already exceeded the subject's just allotment of space. The situation is, however, not untypical of mathematical development and the reader may take it as an indication of why most mathematics texts do not present the history of the subject the way in which texts in other fields do. Even minor developments can be rather complicated and most mathematicians feel the energy put into the scholarly effort of sorting things out could be better used on mathematics itself.

Exercises.

1. Prove the assertions of Examples 7.1 and 7.4.
2. Let φ be a formula semi-representing the graph of a total function F in a given theory $\mathcal{T}$. Suppose φ satisfies the unicity or strong unicity condition of Definitions 7.2 and 7.3. Show: φ represents F, i.e., φ represents the graph of F in $\mathcal{T}$.
3. Complete the proof of Theorem 7.7 by showing that the strict-Π_1-formula,
$$\forall v\, \forall v' \leq v\, [\psi v v_0 \ldots v_{n-1}\, v' \rightarrow v' = v_n],$$
represents the recursive function F in $\mathcal{R}$.
4. (Verena Huber-Dyson). Show that every recursive function $F\colon\omega^n \rightarrow \omega$ has a strong representation $\varphi v_0 \ldots v_n$ in $\mathcal{R}^+$ that satisfies
$$\mathcal{R}^+ \vdash \forall v_0 \ldots v_{n-1}\, \exists v_n\, \varphi v_0 \ldots v_n.$$
[Warning: φ will generally *not* be Σ_1.]
5. Show, by interpreting $\mathcal{R}^+$ in $\mathcal{R}$, that all partial recursive functions are strongly semi-representable in $\mathcal{R}$. [Remark: It is not clear that the proof via Visser's Theorem can be applied to arbitrary consistent effectively enumerable extensions $\mathcal{T}$ of $\mathcal{R}$; for the proof of Corollary 7.14 used the preservation of *strong* unicity under the

extension of $\mathcal{R}^+$ to $\mathcal{T}$. A version of Theorem 7.20 does hold for such extensions of $\mathcal{R}$: Cf. Exercises 9-11, below.]

6. Show that, in general,

$$\mathcal{R} \nvdash \exists v_0 \exists v_1 \varphi v_0 v_1 \leftrightarrow \exists v \exists v_0 \le v \exists v_1 \le v \varphi v_0 v_1,$$

but,

$$\mathcal{R}^+ \vdash \exists v_0 \exists v_1 \varphi v_0 v_1 \leftrightarrow \exists v \exists v_0 \le v \exists v_1 \le v \varphi v_0 v_1.$$

[Hint: Use a variant of the model $\mathfrak{N}_{2\infty}$ of Exercise 1 of section 6.]

7. (Carl Jockusch - Robert Soare). Let $\mathcal{T}_0$ be any consistent effectively enumerable extension of $\mathcal{R}^+$. Let $X \subseteq \omega$ be a non-recursive r.e. relation. Show: There is a consistent completion $\mathcal{T} \supseteq \mathcal{T}_0$ such that X is not semi-representable in $\mathcal{T}$. [Hint: Construct $\mathcal{T}$ as the union of a chain $\mathcal{T}_0 \subseteq \mathcal{T}_1 \subseteq$ ... At odd-numbered stages in the construction, add a sentence or its negation as a new axiom as in the proof of Lemma 3.12; at even numbered stages, destroy a possible semi-representation φv of X by adding an axiom $\varphi(\overline{x})$ for some $x \notin X$. This construction generalises to any countable collection of non-recursive r.e. sets— $\mathcal{T}$ can be chosen to represent no non-recursive r.e. set.]

8. Prove the converse to Theorem 7.18: If A is a complete r.e. set, then A is creative. [Hint: Assume as given some creative set, e.g. K.]

9. Let $\mathcal{T}$ be any consistent extension of $\mathcal{R}$.

   i. Suppose φ is a correct semi-representation in $\mathcal{T}$ of an r.e. relation $R \subseteq \omega^n$ and suppose φ has the form $\exists v_0 v_1 \psi$, with $\psi \in \Delta_0$. Show that $\exists v \exists v_0 \le v \exists v_1 \le v \psi$ is also a correct semi-representation of R in $\mathcal{T}$ if the variable v does not occur in ψ.

   ii. Suppose φ is a correct strict-Σ_1-semi-representation in $\mathcal{T}$ of the graph of a partial recursive function F in $\mathcal{T}$. Show that the Selection Theorem applies to yield a correct strict-Σ_1-semi-representation of F in $\mathcal{T}$.

10. (Smoryński). Let $A = \{x\colon \varphi_x(x) = 0\}$, $B = \{x\colon \varphi_x(x) = 1\}$, and let $A \subseteq C$, $B \subseteq D$ with C, D disjoint r.e. sets. Let X, Y be disjoint r.e. sets and define F recursively such that

$$\varphi_{F(x)}(y) \simeq \begin{cases} 0, & (x \in X \vee y \in D) \preccurlyeq (x \in Y \vee y \in C) \\ 1, & (x \in Y \vee y \in C) \prec (x \in X \vee y \in D) \end{cases}$$

where we write, e.g.,

$$x \in X \vee y \in D: \; \exists z\, [T_1(e_0, x, z) \vee T_1(d, y, z)],$$

for e_0 an index of X and d one of D.

   i. Show: For any $x \in \omega$,
$$x \in X \to F(x) \in C \to \varphi_{F(x)}(Fx) = 0 \to F(x) \in A.$$

   ii. Show: For any $x \in \omega$,
$$F(x) \in C \to x \in X.$$

   iii. Show: For any $x \in \omega$,
$$x \in X \leftrightarrow F(x) \in B \leftrightarrow F(x) \in D.$$

11. Use Exercises 9 - 10 to prove an analogue to Theorem 7.20 for extensions of $\mathcal{R}$. [N.B. In the functional case, the analogue should provide semi-representations, not strong semi-representations.]
12. Prove Theorem 7.20 by the method desired in the text.

8. Further Undecidability Results

We close this Chapter with some additional undecidability results of primarily logical rather than arithmetic interest, but which have arithmetic proofs. One exception will be the arithmetically significant undecidability of the first-order theory of the rational number field, which undecidability result will, however, be discussed and not proved— the proof requires more number theory than can be covered here.

Specifically, the results to be discussed in this section are: the undecidability of a finitely axiomatised theory of arithmetic, the undecidability of first-order logic itself, the undecidability— indeed, the non-effective enumerability of— second-order logic, the undecidability of the first-order theory of finite models, and, as mentioned, the undecidability of the theory of the rational number field (8.28 and following).

The main results of this section will not figure in the sequel (i.e., volume II). Some of the tools will occur again, and some have already occurred, thus making this section partly a review and partly an anticipation.

Our first task is to construct a finitely axiomatised theory of arithmetic. Since one cannot do much to distinguish infinitely many constants with finitely many axioms, we begin with a change of language.

8.1. Definition. The language L_Q has as primitives, in addition to the equality symbol, the following:
Numerical constant. $\overline{0}$
Function symbols. $S, +, \cdot$.

The symbol "S " stands for the successor function and can be used to generate *numerals* $\overline{0}, \overline{1}, \ldots$ replacing the numerical constants $\overline{0}, \overline{1}, \ldots$ of L_R (Definition 6.1). The numerals $\overline{x}$, for $x \in \omega$, are the special terms defined inductively by
i. the numeral $\overline{0}$ is the constant $\overline{0}$
ii. if $y = x + 1$, the numeral $\overline{y}$ is $S\,\overline{x}$, where $\overline{x}$ is the numeral for x.
The numeral $\overline{x}$ will also be written $S^x\,\overline{0}$ when convenient.

The "inequality" relation of L_R will be taken to be an *abbreviation* in L_Q:

$$v_0 \le v_1:\ \exists v_2 (v_2 + v_0 = v_1).$$

Notice that v_2 is on the *left* side of the addition. Unless one wants to take the easy way out and assume commutativity as an axiom, one cannot use the formula,

$$\exists v_2 (v_0 + v_2 = v_1),$$

to interpret $\le$.

The axioms chosen for the finitely axiomatised theory of arithmetic are very natural.

8.2. Definition. The theory Q is the theory in the language L_Q possessing as non-logical axioms the following:

$Q1$. $\forall v_0 (Sv_0 \neq \overline{0})$

$Q2$. $\forall v_0 [v_0 \neq \overline{0} \rightarrow \exists v_1 (v_0 = Sv_1)]$

$Q3$. $\forall v_0 v_1 (Sv_0 = Sv_1 \rightarrow v_0 = v_1)$

$Q4$. $\forall v_0 (v_0 + \overline{0} = v_0)$

$Q5$. $\forall v_0 v_1 [v_0 + Sv_1 = S(v_0 + v_1)]$

$Q6$. $\forall v_0 (v_0 \cdot \overline{0} = \overline{0})$

$Q7$. $\forall v_0 v_1 [v_0 \cdot Sv_1 = (v_0 \cdot v_1) + v_0]$.

As I said, the axioms are quite natural and memorable: $Q1$ - $Q3$ assert that S is a one-one function the range of which includes everything but 0; $Q4$ - $Q5$ are the recursion equations for addition; and $Q6$ - $Q7$ are the recursion equations for multiplication. The obvious missing axiom is a schema of induction. This will be added in the next chapter.

The system Q, like $\mathcal{R}$, is due to Raphael Robinson, who proved the following.

8.3. Theorem. *Q interprets $\mathcal{R}$, i.e.,*

i. $Q \vdash \overline{x} + \overline{y} = \overline{z}$, *whenever* $x + y = z$

ii. $Q \vdash \overline{x} \cdot \overline{y} = \overline{z}$, *whenever* $x \cdot y = z$

iii. $Q \vdash \neg\, \overline{x} = \overline{y}$, *whenever* $x \neq y$

iv. $Q \vdash v \leq \overline{x} \rightarrow v = \overline{0} \vee \ldots \vee v = \overline{x}$, *for all* $x \in \omega$

v. $Q \vdash v \leq \overline{x} \vee \overline{x} \leq v$, *for all* $x \in \omega$.

Proof. i. Induction on y.

Basis. By $Q4$, $Q \vdash \overline{x} + \overline{0} = \overline{x}$.

Induction step. Apply $Q5$ and the induction hypothesis:

$$Q \vdash \overline{x} + S\overline{y} = S(\overline{x} + \overline{y})$$

$$= S\overline{z}, \text{ if } x + y = z.$$

Since $S\overline{z}$ is the numeral for $z + 1 = x + (y + 1)$, we are finished.

ii. Again by induction on y.

Basis. $Q6$ yields, $Q \vdash \overline{x} \cdot \overline{0} = \overline{0}$.

Induction step. Observe,

$$Q \vdash \overline{x} \cdot S\overline{y} = (\overline{x} \cdot \overline{y}) + \overline{x}, \text{ by } Q7$$

$$= \overline{z} + \overline{x}, \text{ if } x \cdot y = z,$$

by the induction hypothesis. But, if $x \cdot (y + 1) = w$, then $w = x \cdot y + x = z + x$ and, by i,

$$Q \vdash \overline{z} + \overline{x} = \overline{w},$$

whence substitution yields

$$Q \vdash \overline{x} \cdot S\overline{y} = \overline{w}.$$

iii. Suppose $x \neq y$. By symmetry we can assume $x = y + z$ with $z \neq 0$. Observe,

$$Q \vdash S^x \overline{0} = S^y \overline{0} \to S^{x-1} \overline{0} = S^{y-1} \overline{0} \quad \text{by Q3}$$

$$\vdots$$

$$\to S^z \overline{0} = \overline{0}.$$

But, by *Q1*, $Q \vdash \neg S^z \overline{0} = \overline{0}$, whence

$$Q \vdash \neg S^x \overline{0} = S^y \overline{0},$$

i.e., $Q \vdash \neg \overline{x} = \overline{y}$.

iv. By induction on x.

Basis. Observe,

$$Q \vdash v \neq \overline{0} \wedge v \leq \overline{0} \to \exists v_1 v_2 (v = Sv_1 \wedge v_2 + v = \overline{0}), \text{ by } Q2$$
$$\to \exists v_1 v_2 (v_2 + Sv_1 = \overline{0})$$
$$\to \exists v v_1 v_2 (S(v_2 + v_1) = \overline{0}).$$

But this last conclusion contradicts *Q1*, whence,

$$Q \vdash \neg(v \neq \overline{0} \wedge v \leq \overline{0}),$$

i.e.,

$$Q \vdash v \leq \overline{0} \to v = \overline{0}.$$

Induction step. Argue by cases.

$$Q \vdash v \leq S^{x+1}\overline{0} \wedge v = \overline{0} \to v = \overline{0} \qquad (1)$$

$$Q \vdash v \leq S^{x+1}\overline{0} \wedge v \neq \overline{0} \to \exists v_1 v_2 (v = Sv_1 \wedge v_2 + v = S^{x+1}\overline{0})$$
$$\to \exists v_1 v_2 (v = Sv_1 \wedge v_2 + Sv_1 = S^{x+1}\overline{0})$$
$$\to \exists v_1 v_2 (v = Sv_1 \wedge S(v_2 + v_1) = S^{x+1}\overline{0})$$
$$\to \exists v_1 v_2 (v = Sv_1 \wedge v_2 + v_1 = S^{x}\overline{0})$$
$$\to \exists v_1 (v = Sv_1 \wedge v_1 \leq S^{x}\overline{0})$$
$$\to \exists v_1 (v = Sv_1 \wedge (v_1 = \overline{0} \vee \ldots \vee v_1 = \overline{x}))$$
$$\to v = S\overline{0} \vee \ldots \vee v = S\overline{x}))$$
$$\to v = S\overline{0} \vee \ldots \vee v = S^{x+1}\overline{0}. \qquad (2)$$

Combining (1) and (2), we have

$$Q \vdash v \leq S^{x+1}\overline{0} \to v = \overline{0} \vee \ldots \vee v = S^{x+1}\overline{0}.$$

v. The proof will be an induction on x. The basis is simple, but the induction step requires a lemma.

Basis. Begin with *Q4*:

$$Q \vdash v + \overline{0} = v$$

$$\vdash \exists v_1 (v_1 + \overline{0} = v)$$
$$\vdash \overline{0} \le v$$
$$\vdash v \le \overline{0} \lor \overline{0} \le v .$$

The lemma needed for the induction step is the following.

8.4. Lemma. i. $Q \vdash v \le S^x \overline{0} \rightarrow v \le S^{x+1}\overline{0}$
ii. $Q \vdash Sv + S^x \overline{0} = v + S^{x+1}\overline{0}$.

Proof. i. By 8.3.iv,

$$Q \vdash v \le \overline{x} \rightarrow \bigvee_{y \le x} v = \overline{y} . \tag{3}$$

However, if $y \le x$ then there is some z such that $z + y = x + 1$ and 8.3.i yields

$$Q \vdash \overline{z} + \overline{y} = S^{x+1}\overline{0}$$
$$\vdash \exists v (v + \overline{y} = S^{x+1}\overline{0})$$
$$\vdash \overline{y} \le S^{x+1}\overline{0} . \tag{4}$$

Combining (3) and (4) we have

$$Q \vdash v \le S^x\overline{0} \rightarrow v \le S^{x+1}\overline{0} .$$

ii. By induction on x.
Basis. $Q \vdash Sv + \overline{0} = Sv = S(v + \overline{0}) = v + S\,\overline{0}$.
Induction step. Observe,

$$\begin{aligned} Q \vdash Sv + S^{x+1}\overline{0} &= S(Sv + S^x \overline{0}), \text{ by } Q5 \\ &= S(v + S^{x+1}\overline{0}), \text{ by induction hypothesis} \\ &= v + S^{x+2}\overline{0} . \end{aligned}$$

QED

Continuation of the proof of Theorem 8.3: Assume as induction hypothesis,

$$Q \vdash \forall v (v \le \overline{x} \lor \overline{x} \le v). \tag{5}$$

Argue by cases.

$$\begin{aligned} Q \vdash v \le S^x\overline{0} &\rightarrow v \le S^{x+1}\overline{0}, \text{ by 8.4.i} \\ &\rightarrow v \le S^{x+1}\overline{0} \lor S^{x+1}\overline{0} \le v . \end{aligned} \tag{6}$$

$$Q \vdash S^x\overline{0} \le v \rightarrow \exists v_1 (v_1 + S^x\overline{0} = v).$$

Argue by subcases. Fix v_1 such that $v_1 + S^x\overline{0} = v$.

$$v_1 = \overline{0} \rightarrow \overline{0} + S^x\overline{0} = v \tag{*}$$

$$\rightarrow S^x\overline{0} = v , \text{ by 8.3.i: } 0 + x = x$$

$$\rightarrow v \le S^x\overline{0} \text{ (switching } v \text{ and } S^x\overline{0} \text{ in (*))}$$

$$\rightarrow v \le S^{x+1}\overline{0} \lor S^{x+1}\overline{0} \le v, \text{ by (6).}$$

$$v_1 \ne \overline{0} \rightarrow \exists v_2 (v_1 = Sv_2 \lor v_1 + S^x\overline{0} = v)$$

$$\rightarrow \exists v_2 (Sv_2 + S^x\overline{0} = v)$$
$$\rightarrow \exists v_2 (v_2 + S^{x+1}\overline{0} = v), \text{ by 8.4.ii}$$
$$\rightarrow S^{x+1}\overline{0} \le v$$
$$\rightarrow v \le S^{x+1}\overline{0} \lor S^{x+1}\overline{0} \le v.$$

QED

An immediate corollary to Theorem 8.3 is the undecidability of $\mathcal{Q}$. If we recall the following definition from Exercise 4 of section 6, we get a nice statement of this undecidability result.

8.5. Definition. A theory $\mathcal{T}_0$ is *essentially undecidable* if every consistent extension $\mathcal{T} \supseteq \mathcal{T}_0$ in the language of $\mathcal{T}_0$ is undecidable.

8.6. Corollary. *$\mathcal{Q}$ is essentially undecidable.*

This is immediate by Theorem 8.3 and the essential undecidability of $\mathcal{R}$ (Rosser's Theorem 6.28).

The importance of the finite axiomatisability of $\mathcal{Q}$ lies in the following corollary proven originally by Alonzo Church.

8.7. Corollary. (Church's Theorem). *First-order logic is undecidable.*

Proof. Let χ be the conjunction of the seven non-logical axioms of $\mathcal{Q}$ and observe: for any sentence φ,

$$\mathcal{Q} \vdash \varphi \text{ iff } \vdash \chi \rightarrow \varphi.$$

Thus, a decision procedure for first-order logic would yield one for $\mathcal{Q}$, which is undecidable. QED

Before proving that validity in second-order logic is not effectively enumerable, we must define precisely what second-order logic is. The idea behind second-order logic is very simple. It is the logic characterised by augmenting structures with the class of all sets of individuals in the domains. Thus, for example, the first-order structure,

$$\mathfrak{N} = (\omega; S, +, \cdot, 0),$$

would be augmented to

$$\mathfrak{N}^2 = (\omega; P(\omega); \in, S, +, \cdot, 0).$$

More precisely, the second-order language extending a given first-order language will have new primitive symbols:

Set variables. $V_0, V_1, \ldots$

Membership. $\in$.

The language is thus enlarged by introducing new atomic formulae $t \in V_i$ for any term t of the original language.

The notion of truth in a second-order model is clear enough and I will not bother giving the detailed definition. Second-order validity and logical consequence are semantically defined concepts: if $\Gamma \cup \{\varphi\}$ is a set of second-order sentences of a language L, we define

$\models^2 \varphi$ iff φ is true in all second-order L-structures

$\Gamma \models^2 \varphi$ iff φ is true in all second-order L-structures in which all sentences of Γ are true.

The power of second-order logic is illustrated by the following.

8.8. Theorem. *The theory $Q^2 + Ind$ in the second-order extension of L_Q with axioms $Q1$ - $Q7$ and,*

$$Ind: \forall V_0[\overline{0} \in V_0 \wedge \forall v_0(v_0 \in V_0 \rightarrow Sv_0 \in V_0) \rightarrow \forall v_0(v_0 \in V_0)],$$

is complete.

This is just a variant of Dedekind's famous characterisation of the structure $(\omega; P(\omega); \in, S, 0)$ up to isomorphism: any second-order model of $Q^2 + Ind$ is isomorphic to $\mathfrak{N}^2$, whence, for any sentence φ,

$$Q^2 + Ind \models^2 \varphi \quad \text{iff} \quad \mathfrak{N}^2 \models^2 \varphi.$$

(Cf. I.8.1 and its following remarks.)

The desired non-enumerability result is an immediate corollary:

8.9. Corollary. *The set of valid sentences of second-order logic is not effectively enumerable.*

Proof. Let φv be a formula of L_Q defining a non-recursively enumerable set X of natural numbers, and let χ be the conjunction of $Q1$ - $Q7$ and Ind. Observe: for any $x \in \omega$,

$$x \in X \quad \text{iff} \quad \models^2 \chi \rightarrow \varphi\overline{x}.$$

An effective enumeration of valid second-order sentences would recursively enumerate X, which cannot be done. QED

That was awfully easy and the reader might suspect I tried to slip something past him. I really haven't. There are subtleties, but they are not relevant to the above proof: I have been brief in describing the second-order language, but the reader who consults a more detailed reference will readily verify that I have not hidden any difficulties there. The omitted proof of Theorem 8.8 is also unproblematic: having had an algebra course, the reader will know what has to be proven and, with all the inductions he has been subjected to so far in this book, he should have no difficulty providing a proof of this Theorem. (Hint: Use Theorem I.8.1 to define the isomorphism.) So what are the subtleties? Well, there is an ambiguity in the literature: some accounts will label as "second-order logic" the axiomatic logical system describing structures with incomplete power sets. Thus, *a* "second-order model of arithmetic" would be given by,

$$\mathfrak{A} = (\omega; A; \in, S, +, \cdot, 0),$$

where $A \subseteq P(\omega)$ and $\mathfrak{A}$ satisfies some closure properties on A, e.g., the comprehension schema,

$$\exists V_0 \forall v_0 [v_0 \in V_0 \leftrightarrow \varphi v_0], \; V_0 \text{ not free in } \varphi.$$

The axiomatic "second-order logic" has the same language as the semantic second-order logic, but is really a *two-sorted* first-order logic. Theorem 8.8 and Corollary 8.9 say nothing about such theories.

A more interesting matter of such subtlety concerns the choice of primitives of the arithmetic language. Dedekind's original characterisation was of the structure,

$$\mathfrak{N}_1^2 = (\omega; P(\omega); \in, S, 0).$$

Induction and the recursion equations extend his result quickly to cover

$$\mathfrak{N}^2 = (\omega; P(\omega); \in, S, +, \cdot, 0).$$

In Corollary 8.9, it was more convenient to refer to the structure $\mathfrak{N}^2$ because we already know the definability in the first-order fragment of its language of non-recursively enumerable sets. We also know the decidability of the first-order theory of the structure $\mathfrak{N}_1^2$, i.e., of $Th(\omega; S, 0)$, as well as that of $Th(\omega; S, +, 0)$. Attempting to use the simpler structure $\mathfrak{N}_1^2$ or,

$$\mathfrak{N}_2^2 = (\omega; P(\omega); \in, S, +, 0),$$

to prove Corollary 8.9 would indeed be non-trivial tasks. In the case of $\mathfrak{N}_1^2$, the task is impossible: J. Richard Büchi has proven the decidability of the second-order theory of successor, whence its second-order language defines no non-recursive sets. (In fact, the unary sets it defines are exactly those definable in the first-order theory of addition!) The structure $\mathfrak{N}_2^2$ has a very undecidable second-order theory: Multiplication is definable in the second-order theory of addition. (Cf. Exercise 4, below.)

If our last task was too easy, our next will require more of an effort. We want to prove the undecidability of the theory of finite structures. This result is due to B. Trahtenbrot, who went on in 1953 to prove an even stronger result, the exact statement of which requires some preliminary definitions.

8.10. Definitions. Let $\mathcal{T}$ be a theory in a first-order language L. A sentence φ is said to be *finitely refutable* over $\mathcal{T}$ if φ is false in some finite model of $\mathcal{T}$. $\mathcal{T}$ has the *finite inseparability* property if the sets of finitely refutable sentences over $\mathcal{T}$ and of theorems of $\mathcal{T}$ are effectively inseparable.

As we haven't yet coded syntax and given a precise definition of what constitutes a recursively enumerable set of sentences, the definition of the finite inseparability property is imprecise. The reader in psychological need of such precision can peek ahead to the next volume; the others can simply accept the fact that proofs of finite inseparability show that, modulo an intuitively effective reduction, the finite refutables and the theorems of $\mathcal{T}$ separate pairs of effectively inseparable sets, whence neither the set of finitely refutable sentences nor the set of theorems of $\mathcal{T}$ is decidable. From the undecidability of the problem of which sentences are false in some finite models follows the undecidability of the complementary problem of which sentences are true in all such models. Thus, the undecidability of the theory of finite models (Corollary 8.12, below) is an immediate consequence of Trahtenbrot's Theorem:

8.11. Trahtenbrot's Theorem. *First-order logic has the finite inseparability property.*

8.12. Corollary. *The first-order theory of finite models, i.e., the set of sentences true in all finite models, is undecidable.*

The proof of Trahtenbrot's Theorem, like the proof of Rosser's Theorem, relies on the provable disjointness of special strict-Σ_1-definitions of two effectively inseparable r.e. sets. This is roughly where the similarity ends, as model theoretic properties of Σ_1-sentences replace the provability of true Σ_1-formulae in $\mathcal{R}$.

The first step in proving Theorem 8.11 is to choose the proper arithmetic theory. Desiderata include the just cited ability to prove disjointness— hence a linear ordering, finite axiomatisability, and a quantity of finite models of a recognisably arithmetic nature. The finite models we will use will simply be the finite initial segments of the natural numbers with their inherited order and inherited *partial* functions of successor, addition, and multiplication. With partial functions, function symbols must be replaced by relation symbols for the graphs of the functions. Thus, we need a different language.

8.13. Definition. The language L_P has as primitives, in addition to the equality symbol, the following:
Numerical constant. $\overline{0}$
Binary relation symbols. $\leq, S$
Ternary relation symbols. A, M.

"S", of course, stands for successor, "A" for addition, and "M" for multiplication. The last position in the predicate is reserved for the value; thus, $S(x, y)$ means $y = Sx$, etc.

Unlike "$\mathcal{Q}$" and "$\mathcal{R}$", which are the standard names of standard theories, the following theory $\mathcal{P}$, as well as its name, is merely of local interest.

8.14. Definition. The theory $\mathcal{P}$ is the theory in the language L_P possessing as non-logical axioms the following:

P1. $\forall v_0(\overline{0} \leq v_0)$

P2. $\forall v_0 v_1 (v_0 \leq v_1 \vee v_1 \leq v_0)$

P3. $\forall v_0 v_1 v_2 (v_0 \leq v_1 \wedge v_1 \leq v_2 \rightarrow v_0 \leq v_2)$

P4. $\forall v_0 v_1 (v_0 \leq v_1 \wedge v_1 \leq v_0 \rightarrow v_0 = v_1)$

P5. $\forall v_0 \neg S(v_0, \overline{0})$

P6. $\forall v_0 (v_0 \neq \overline{0} \rightarrow \exists v_1 \leq v_0\, S(v_1, v_0))$

P7. $\forall v_0 v_1 v_2 (S v_1 v_0 \wedge S v_2 v_0 \rightarrow v_1 = v_2)$

P8. $\forall v_0 v_1 v_2 (S v_0 v_1 \wedge S v_0 v_2 \rightarrow v_1 = v_2)$

P9. $\forall v_0 v_1 v_2 v_3 (v_0 \leq v_1 \wedge S v_0 v_2 \wedge S v_1 v_3 \rightarrow v_2 \leq v_3)$

P10. $\forall v_0 v_1 v_2 [S v_0 v_1 \rightarrow (v_2 \leq v_1 \leftrightarrow v_2 \leq v_0 \vee v_2 = v_1)]$

P11. $\forall v_0 v_1 (v_0 \leq v_1 \wedge v_0 \neq v_1 \rightarrow \exists v_2 \leq v_1\, S v_0 v_2)$

P12. $\forall v_0 A(v_0, \overline{0}, v_0)$

P13. $\forall v_0v_1v_2v_1'v_2'[Av_0v_1v_2 \wedge Sv_1v_1' \wedge Sv_2v_2' \rightarrow Av_0v_1'v_2']$

P14. $\forall v_0v_1v_1'v_2'[Av_0v_1'v_2' \wedge Sv_1v_1' \rightarrow \exists v_2 \leq v_2'(Sv_2v_2' \wedge Av_0v_1v_2)]$

P15 $\forall v_0v_1v_2v_3[Av_0v_1v_2 \wedge Av_0v_1v_3 \rightarrow v_2 = v_3]$

P16. $\forall v_0\, M(v_0, \overline{0}, \overline{0})$

P17. $\forall v_0v_1v_2v_3v_1'[Mv_0v_1v_2 \wedge Sv_1v_1' \wedge Av_2v_0v_3 \rightarrow Mv_0v_1'v_3]$

P18. $\forall v_0v_1v_2v_3v_1'[Mv_0v_1'v_3 \wedge Sv_1v_1' \rightarrow \exists v_2 \leq v_3(Av_2v_0v_3 \wedge Mv_0v_1v_2)]$

P19. $\forall v_0v_1v_2v_3[Mv_0v_1v_2 \wedge Mv_0v_1v_3 \rightarrow v_2 = v_3]$.

The explanation of the axioms is as follows: *P1* - *P4* assert that $\leq$ is a linear order with first element 0; *P5* - *P7* are analogues to *Q1* - *Q3*; *P8* is a strong unicity condition on *S*— thus guaranteeing that *S* defines the graph of a partial function; *P9* - *P11* assert that *S* defines the successor function in the order-theoretic sense; *P12* - *P14* are recursion equations for addition— initial value (*P12*), upward recursion (*P13*), and downward recursion (*P14*); *P15* asserts the strong unicity of *A* ; and *P16* - *P19* are the multiplicative analogues to *P12* - *P15*.

One curious feature of the axioms is the bounding of the existential quantifiers. This is redundant in *P6* and *P11*, but seems not to be in *P14* and *P18*. The main purpose of these bounds is to render all axioms Π_1— once we define the classes of Σ_1- and Π_1-formulae of the language L_P.

Before getting involved in the details of the proof, let me give an overall outline of the proof. In broad terms, the proof consists of proving five assertions:

I. the finite models of $\mathcal{P}$ are exactly the initial segments of $\mathfrak{N}$;

II. every other model of $\mathcal{P}$ has $\mathfrak{N}$ as an initial segment;

III. every Π_1-sentence true in $\mathfrak{N}$ is true in all finite models;

IV. every Σ_1-sentence true in $\mathfrak{N}$ is true in some finite model;

V. if φv, ψv are $\mathcal{P}$-provably disjoint strict-Σ_1-definitions of sets X, Y, respectively, then, for $x \in \omega$,

$$x \in Y \rightarrow \mathfrak{N} \models \psi\overline{x}$$
$$\rightarrow \mathcal{P} \vdash \neg\varphi\overline{x}$$
$$x \in X \rightarrow \varphi\overline{x} \text{ has a finite model}$$
$$\rightarrow \neg\varphi\overline{x} \text{ is false in some finite model.}$$

These assertions are connected by a crucial model theoretic property of Π_1-sentences (or, dually, Σ_1-sentences). Before stating this property, we need some definitions— which we shall give in some generality.

8.15. Definitions. Let L be a language containing a binary relation $\leq$. Bounded quantifiers in L are defined as abbreviations:

$$\forall v_i \leq v_j\, \varphi: \ \forall v_i(v_i \leq v_j \rightarrow \varphi)$$
$$\exists v_i \leq v_j\, \varphi: \ \exists v_i(v_i \leq v_j \wedge \varphi)$$

The class of Σ_1- (Π_1-) formulae of L is defined inductively by:

i. atomic formulae and their negations are Σ_1- (Π_1-) formulae

ii. if φ, ψ are Σ_1- (Π_1-) formulae, then so are $\varphi \vee \psi, \varphi \wedge \psi$
iii. if φ is a Σ_1- (Π_1-) formula and v_i, v_j are distinct variables, then $\exists v_i \le v_j \varphi$, $\forall v_i \le v_j \varphi$ are Σ_1- (Π_1-) formulae
iv. if φ is a Σ_1- (Π_1-) formula, then $\exists v_i \varphi$ ($\forall v_i \varphi$) is a Σ_1-formula (respectively, Π_1-formula).

I leave to the reader the precise definitions of Δ_0- and strict-Σ_1- formulae.

8.16. Definition. Let L be a language containing $\le$ and let $\mathfrak{A} = (A; \le, \ldots)$, $\mathfrak{B} = (B; \le, \ldots)$ be L-structures such that $\mathfrak{A}$ is a substructure of $\mathfrak{B}$. We call $\mathfrak{B}$ an *end extension* of $\mathfrak{A}$ if

$$\forall a \in A\ \forall b \in B\ (b \le a \to b \in A).$$

We also say, particularly if $\le$ is a linear ordering on B, that $\mathfrak{A}$ is an *initial segment* of $\mathfrak{B}$.

Needless to say, we are interested in the linearly ordered case. Stating the definition more generally allows us, should we want to (and which we will not do in this book), to discuss initial segments of models of $\mathcal{Q}$ and $\mathcal{R}$ as well as of $\mathcal{P}$. The definition also makes sense for the $\in$-relation of set theory, which relation is very far from linear.

The crucial model theoretic properties of Π_1- and Σ_1-sentences is the following.

8.17. Theorem. *Let L be a language containing $\le$ and let $\mathfrak{A}$ be an initial segment of $\mathfrak{B}$ for some L-structures $\mathfrak{A}, \mathfrak{B}$. Then:*
i. *for any Π_1-sentence φ of $L(\mathfrak{A})$, $\mathfrak{B} \models \varphi \to \mathfrak{A} \models \varphi$*
ii. *for any Σ_1-sentence φ of $L(\mathfrak{A})$, $\mathfrak{A} \models \varphi \to \mathfrak{B} \models \varphi$.*

Proof. i. By induction on the construction of φ. The interesting case is that in which φ is obtained by application of the bounded existential quantifier. Thus, let φ be $\exists v \le v_i\, \psi v v_0 \ldots v_{n-1}$ and let $a_0, \ldots, a_{n-1} \in A$. Observe,

$$\begin{aligned}
\mathfrak{B} \models \exists v \le \overline{a_i}\ \psi(v, \overline{a_0}, \ldots, \overline{a_{n-1}}) &\to \exists b \in B\ (b \le a_i \wedge \mathfrak{B} \models \psi(\overline{b}, \overline{a_0}, \ldots, \overline{a_{n-1}})) \\
&\to \exists b \in A\ (b \le a_i \wedge \mathfrak{B} \models \psi(\overline{b}, \overline{a_0}, \ldots, \overline{a_{n-1}})) \qquad (*) \\
&\to \exists b \in A\ (b \le a_i \wedge \mathfrak{A} \models \psi(\overline{b}, \overline{a_0}, \ldots, \overline{a_{n-1}})) \qquad (**) \\
&\to \mathfrak{A} \models \exists v \le \overline{a_i}\ \psi(v, \overline{a_0}, \ldots, \overline{a_{n-1}}),
\end{aligned}$$

where (*) follows from the fact that $\mathfrak{A}$ is an initial segment of $\mathfrak{B}$ and (**) by the induction hypothesis.

I leave the other cases to the reader.

ii. Similar— or, observe that Σ_1-sentences are logically equivalent to negations of Π_1-sentences. If $\varphi \leftrightarrow \neg\psi$ and $\psi \in \Pi_1$,

$$\begin{aligned}
\mathfrak{B} \models \neg\varphi &\to \mathfrak{B} \models \psi \\
&\to \mathfrak{A} \models \psi \\
&\to \mathfrak{A} \models \neg\varphi,
\end{aligned}$$

whence $\mathfrak{A} \models \varphi \to \mathfrak{B} \not\models \neg\varphi$, i.e., $\mathfrak{B} \models \varphi$. QED

8.18. Definition. Let $\mathfrak{B} = (B; \leq, S, A, M, 0)$ be a model of $\mathcal{P}$ and let $a \in B$. Define

$$\mathfrak{B}_a = (B_a; \leq_a, S_a, A_a, M_a, 0_a)$$

as follows:

i. $B_a = \{b \in B : b \leq a\}$

ii. $\leq_a, S_a, A_a, M_a$ are the restrictions to B_a of $\leq, S, A, M$, respectively

iii. 0_a is 0.

8.19. Lemma. *Let* $\mathfrak{B} \models \mathcal{P}$ *and* $a \in B$. *Then:* $\mathfrak{B}_a \models \mathcal{P}$.

Proof. The structure $\mathfrak{B}_a$ is an initial segment of $\mathfrak{B}$, and the axioms of $\mathcal{P}$ are (logically equivalent to) Π_1-sentences. Thus, Theorem 8.17 applies. QED

8.20. Corollary. *For each* $n \in \omega$, $\mathfrak{N}_n$ *is a model of* $\mathcal{P}$.

For, $\mathfrak{N}$ is a model of $\mathcal{P}$. What we really need, however, is a sort of converse to Corollary 8.20.

8.21. Lemma. *Let* $\mathfrak{B}$ *be a finite model of* $\mathcal{P}$. *For some* $n \in \omega$, $\mathfrak{B}$ *is isomorphic to* $\mathfrak{N}_n$.

Proof. Let the domain B of $\mathfrak{B}$ have cardinality $n + 1$. By the axioms of order and successor, B consists of 0 and n successive successors. We may write

$$B = \{0, 1, ..., n\}$$

with

$$0 < 1 < ... < n \text{ and } S(0,1), S(1,2), ..., S(n-1,n).$$

We have to show, for $x, y, z \in B$, that

$$\mathfrak{B} \models A(\overline{x}, \overline{y}, \overline{z}) \text{ iff } x+y = z$$
$$\mathfrak{B} \models M(\overline{x}, \overline{y}, \overline{z}) \text{ iff } x \cdot y = z.$$

To do this, we fix x and prove by induction on y that

$$\forall z\, [\mathfrak{B} \models A(\overline{x}, \overline{y}, \overline{z}) \text{ iff } x+y = z]$$
$$\forall z\, [\mathfrak{B} \models M(\overline{x}, \overline{y}, \overline{z}) \text{ iff } x \cdot y = z].$$

First, we consider addition.

Basis. $y = 0$. By *P12*,

$$\mathfrak{B} \models A(\overline{x}, \overline{0}, \overline{x}). \qquad (*)$$

That

$$\mathfrak{B} \models A(\overline{x}, \overline{0}, \overline{z}) \rightarrow z = x$$

follows from (*) by *P15*.

Induction step. Let $x + (y + 1) = (z + 1)$. By induction hypothesis,

$$\mathfrak{B} \models A(\overline{x}, \overline{y}, \overline{z}).$$

But also

$$\mathfrak{B} \models S(\overline{y}, \overline{y+1}) \text{ and } \mathfrak{B} \models S(\overline{z}, \overline{z+1}).$$

Thus, *P13* yields

$$\mathfrak{B} \models A(\overline{x}, \overline{y+1}, \overline{z+1}).$$

Conversely, if

$$\mathfrak{B} \models A(\overline{x}, \overline{y+1}, \overline{z}),$$

for some z, then P14 yields

$$\mathfrak{B} \models A(\overline{x}, \overline{y}, \overline{z'}) \wedge S(\overline{z'}, \overline{z})$$

for some z'. By induction hypothesis,

$$x + y = z'$$

whence

$$x + y + 1 = z' + 1 = z.$$

The multiplication case is handled similarly. I leave the details of the proof to the reader.

QED

With Lemma 8.21 we have accomplished our first task— that of showing the finite models of $\mathcal{P}$ to be exactly the initial segments of $\mathfrak{N}$. Our second task is a corollary to this.

8.22. Corollary. *Let $\mathfrak{B}$ be an infinite model of $\mathcal{P}$. Then: $\mathfrak{N}$ is isomorphic to an initial segment of $\mathfrak{B}$.*

Proof. Map $\mathfrak{N}$ to $\mathfrak{B}$ as follows: First, map $0 \in \omega$ to the 0 of $\mathfrak{B}$ — denote it by b_0. Suppose $0, \ldots, x$ have been mapped to $b_0, \ldots, b_x$, respectively, in such a way that

$$\mathfrak{B} \models S(\overline{b_0}, \overline{b_1}), \ \ldots, \ \mathfrak{B} \models S(\overline{b_{x-1}}, \overline{b_x}).$$

Thus, $b_0 < b_1 < \ldots < b_x$ constitute an initial segment of the ordering of $\mathfrak{B}$. Because $\mathfrak{B}$ is infinite, this means b_x is not the last element of $\mathfrak{B}$ and *P11* says there is some $b \in B$ such that

$$\mathfrak{B} \models S(\overline{b_x}, \overline{b}).$$

Call this (unique) element b_{x+1} and map $x + 1$ to it.

So far we have $(\omega; \leq, S, 0)$ isomorphically embedded as an initial segment of $(B; \leq, S, 0)$. To complete the proof of the Corollary, we have to show addition and multiplication to be preserved. This can be done by a pair of inductions as in the proof of Lemma 8.21, or by a clever appeal to the Lemma as follows: suppose $x, y, z \in \omega$ are such that $x + y = z$ (or, $x \cdot y = z$). Choose $n = \max\{x, y, z\}$ and observe that the isomorphism of $\mathfrak{N}_n$ with $\mathfrak{B}_{b_n}$ is just the restriction of the embedding of $\mathfrak{N}$ into $\mathfrak{B}$. But the finite isomorphism preserves addition and multiplication.

QED

Our third major task is also a corollary to Lemma 8.21:

8.23. Corollary. *Every Π_1-sentence true in $\mathfrak{N}$ is true in all finite models.*

Proof. Let φ be a Π_1-sentence. Observe,

$$\mathfrak{N} \models \varphi \ \rightarrow \forall n \in \omega\, (\mathfrak{N}_n \models \varphi), \text{ by 8.17}$$

$$\rightarrow \forall \text{ finite } \mathfrak{B} \models \mathcal{P}(\mathfrak{B} \models \varphi), \text{ by 8.21.}$$

QED

Our fourth major task is to prove a dual to this.

8.24. Lemma. *Every Σ_1-sentence true in $\mathfrak{N}$ is true in some finite model.*

Proof. We will prove something a bit stronger: for any Σ_1-formula $\varphi v_0 \ldots v_{m-1}$, there is a function $n_\varphi(x_0, \ldots, x_{m-1})$ such that, for all $x_0, \ldots, x_{m-1} \in \omega$, if the sentence $\varphi \overline{x_0} \ldots \overline{x_{m-1}}$ of the augmented language $L(\mathfrak{N})$ is true in $\mathfrak{N}$, then for all $n \geq n_\varphi(x_0, \ldots, x_{m-1})$, $\mathfrak{N}_n \models \varphi \overline{x_0} \ldots \overline{x_{m-1}}$.

The proof of this is by induction on the construction of φ.

If φ is a binary atomic formula,

$$v_i = v_j, \ v_i \leq v_j, \ \text{or} \ S(v_i, v_j),$$

let $n_\varphi(x, y) = \max\{x, y\}$; if φ is a ternary atomic formula,

$$A(v_i, v_j, v_k) \ \text{or} \ M(v_i, v_j, v_k),$$

let $n_\varphi(x, y, z) = \max\{x, y, z\}$. In either case, n_φ is the maximum w such that $\overline{w}$ occurs in φ.

If φ is $\psi \vee \chi$, choose a true disjunct (under some given interpretation of the unexhibited free variables; here and below I shall suppress mention of inactive variables), say ψ, and let $n_\varphi = \max\{n_\psi, \max\{x : \overline{x} \text{ occurs in } \varphi\}\}$.

If φ is $\psi \wedge \chi$, choose $n_\varphi = \max\{n_\psi, n_\chi\}$.

If φ is $\forall v_i \leq v_j \, \psi v_i$ or if φ is $\exists v_i \leq v_j \, \psi v_i$, choose $n_\varphi(x) = \max\{x, \max\{n_{\psi\overline{y}} = n_\psi(y) : y \leq x\}\}$, where x is the number whose numeral is to replace the variable v_j.

Finally, if φ is $\exists v \, \psi v$, choose (assuming some assignment of numerals for the free variables of φ making φ true in $\mathfrak{N}$) x such that $\psi \overline{x}$ is true in $\mathfrak{N}$ and let $n_\varphi = n_{\psi\overline{x}} = n_\psi(x)$.

I leave to the reader the inductive verifications that these values work. QED

We are now only a few minor, but tedious, steps away from completing our fifth and final major task in proving Trahtenbrot's Theorem.

First, there is the notational nuisance. We are going to want to choose formulae φv, ψv defining effectively inseparable r.e. sets X, Y and ask when $\neg\varphi\overline{x}$ is provable and when it has a finite countermodel. For $x \neq 0$, however, when we are not using the augmented language $L(\mathfrak{N})$ (or one of the smaller $L(\mathfrak{N}_n)$'s), there is no constant naming x. What do we mean by $\neg\varphi\overline{x}$? The answer is very simple: for any formula χv, and any $0 \neq x \in \omega$, we could take $\chi\overline{x}$ to be an abbreviation for one of the formulae,

$$\exists v_0 \ldots v_{x-1}[S(\overline{0}, v_0) \wedge \bigwedge S(v_i, v_{i+1}) \wedge \chi v_{x-1}],$$

$$\forall v_0 \ldots v_{x-1}[S(\overline{0}, v_0) \wedge \bigwedge S(v_i, v_{i+1}) \rightarrow \chi v_{x-1}].$$

The truth or falsity of $\chi\overline{x}$ would therefore be a meaningful question even in models $\mathfrak{N}_n$ with $n < x$, i.e., models in which x does not exist. Since we wish $\neg\varphi\overline{x}$ to be Π_1, we shall assume for it the second of these abbreviations. (Note that $\neg\varphi\overline{x}$ is automatically true in models $\mathfrak{N}_n$ with $n < x$; choosing the alternative abbreviation would make the sentence automatically false.)

The second point is the definability of every r.e. set by a strict-Σ_1-formula of L_P. Note that strict-Σ_1-formulae of L_R translate naturally into Σ_1-formulae, not *strict*-Σ_1-formulae, of L_P. For example,

$$\exists v_0 (v_0 + v_1 = v_2 + v_2)$$

translates to

$$\exists v_0 v_3 [A(v_0, v_1, v_3) \wedge A(v_2, v_2, v_3)].$$

This is equivalent in $\mathfrak{N}$ to

$$\exists v \, \exists v_0 \leq v \, \exists v_3 \leq v \, [A(v_0, v_1, v_3) \wedge A(v_2, v_2, v_3)],$$

which is strict-Σ_1. Every r.e. set has a strict-Σ_1-definition in L_P, but this is a fact requiring proof anew. I leave the details to the reader. (Exercise 8, below.) Note that, for φv strict-Σ_1 it does not follow that the instances $\varphi \overline{x}$ will be *strict*-Σ_1 under the abbreviations used above. For the first abbreviation, $\varphi \overline{x}$ will be Σ_1 but not strict-Σ_1; for the second, the sentence will not even be Σ_1. This does not matter, for the strictness is only used at one point to construct special strict-Σ_1-formulae, the instances and negations of which need only be Σ_1 and Π_1, respectively.

The third point— the one where we use the strictness just established— is that if φv, ψv are strict-Σ_1-formulae defining sets X, Y in $\mathfrak{N}$, respectively, then

$$\varphi_1 v: \quad \varphi \preccurlyeq \psi$$

$$\psi_1 v: \quad \psi \prec \varphi$$

are strict-Σ_1-formulae defining X, Y in $\mathfrak{N}$, and

$$\mathcal{P} \vdash \forall v \neg (\varphi_1 v \wedge \psi_1 v). \tag{7}$$

The proof of this— that is, the analogous portion of the Reduction Theorem (6.27.iv)— carries over readily as all that was used in the earlier proof was the comparability of order. I leave the details to the reader (Exercise 9, below).

Given all of this, it is now possible to complete the proof of Trahtenbrot's Theorem.

Proof of Theorem 8.11. Let X, Y be a pair of effectively inseparable r.e. sets and let φv, ψv be strict-Σ_1-formulae defining X, Y in $\mathfrak{N}$ and such that

$$\mathcal{P} \vdash \forall v (\psi v \rightarrow \neg \varphi v) \tag{8}$$

as in (7), above. Let $x \in \omega$, and observe:

$$\begin{aligned} x \in X \ &\rightarrow \mathfrak{N} \models \varphi \overline{x} \\ &\rightarrow \exists n (\mathfrak{N}_n \models \varphi \overline{x}) \qquad (9) \\ &\rightarrow \neg \varphi \overline{x} \text{ is false in some finite model of } \mathcal{P}. \end{aligned}$$

Further, observe

$$\begin{aligned} x \in Y \ &\rightarrow \mathfrak{N} \models \psi \overline{x} \qquad (10) \\ &\rightarrow \mathfrak{B} \models \psi \overline{x} \qquad (11) \end{aligned}$$

for all infinite models $\mathfrak{B} \models \mathcal{P}$ by Theorem 8.17 and Corollary 8.22. But (8) and (11) yield,

$$x \in Y \rightarrow \mathfrak{B} \models \neg \varphi \overline{x}$$

for all infinite $\mathfrak{B} \models \mathcal{P}$. Applying (8) and (10) yields

$$\begin{aligned} x \in Y &\rightarrow \mathfrak{N} \models \neg \varphi \overline{x}, \text{ since } \neg \varphi \overline{x} \text{ is } \Pi_1 \\ &\rightarrow \forall n\, (\mathfrak{N}_n \models \neg \varphi \overline{x}) \\ &\rightarrow \mathfrak{B} \models \neg \varphi \overline{x} \end{aligned}$$

for all finite models $\mathfrak{B} \models \mathcal{P}$. Thus we see that

$$\begin{aligned} x \in Y &\rightarrow \mathcal{P} \models \neg \varphi \overline{x} \\ &\rightarrow \mathcal{P} \vdash \neg \varphi \overline{x}. \end{aligned}$$

Finally, let χ be the conjunction of the axioms of $\mathcal{P}$. We have just seen, for $x \in \omega$,

$$x \in X \rightarrow \chi \rightarrow \neg \varphi \overline{x} \text{ is false in some finite } L_P\text{-structure}$$

$$x \in Y \rightarrow \vdash \chi \rightarrow \neg \varphi \overline{x}.$$

Thus, the set of sentences of L_P which can be falsified in some finite structures and the set of valid sentences of L_P separate a pair of effectively inseparable r.e. sets, as was to be proven. (With the encoding of the next volume, one can show that the sets of finitely refutable sentences and theorems form, in fact, a pair of effectively inseparable r.e. sets.) QED

Corollary 8.12, on the undecidability of the set of sentences valid in all finite structures also follows immediately: the set,

$$F = \{x : \chi \rightarrow \neg \varphi \overline{x} \text{ is valid in all finite structures}\}$$

contains Y and is disjoint from X, where X, Y are the given effectively inseparable r.e. sets. Hence, F is not recursive.

The undecidability of first-order logic, i.e., Corollary 8.7, follows anew by a simple shift of perspective: the set of valid sentences, together with the set of finitely refutables, also separates X and Y, whence it cannot be decidable.

The theory $\mathcal{P}$ is undecidable, but not essentially so as it has many finite models and each finite model has a decidable theory (as is readily established by a quantifier-elimination). If one adds to $\mathcal{P}$ the single axiom,

$$\forall v_0 \exists v_1\, S(v_0, v_1),$$

asserting the totality of the successor function, the resulting theory $\mathcal{P}^+$ is essentially undecidable. In fact, by formula (9) of the proof of Trahtenbrot's Theorem,

$$\begin{aligned} x \in X &\rightarrow \forall \text{ infinite } \mathfrak{B} \models \mathcal{P} (\mathfrak{B} \models \varphi \overline{x}) \\ &\rightarrow \forall \mathfrak{B} \models \mathcal{P}^+ (\mathfrak{B} \models \varphi \overline{x}) \\ &\rightarrow \mathcal{P}^+ \vdash \varphi \overline{x}, \end{aligned}$$

while

$$\begin{aligned} x \in Y &\rightarrow \mathcal{P} \vdash \neg \varphi \overline{x} \\ &\rightarrow \mathcal{P}^+ \vdash \neg \varphi \overline{x}. \end{aligned}$$

Thus, as with $\mathcal{Q}$, the theorems and outright refutables of any consistent extension $\mathcal{T}$ of $\mathcal{P}^+$ separate a pair of effectively inseparable r.e. sets.

These are rather minor remarks. There are more important things to be said.

8.25. Remark. Assuming the encoding of syntax by numbers so that we may properly speak of the recursiveness, recursive enumerability, or even arithmetic definability of sets of sentences, we may state Trahtenbrot's Theorem and its Corollary in the following stronger forms:

8.11. *Let V be any class of structures containing all finite structures. Then:*

$$Th(V) = \{\varphi: \forall \mathfrak{K} \in V (\mathfrak{K} \models \varphi)\}$$

is not recursive.

8.12. *Let V_{dec} be the class of structures with decidable theories and V_{fin} the class of finite structures. If V is such that $V_{fin} \subseteq V \subseteq V_{dec}$, then Th(V) is not r.e.*

The restatement of 8.11 is fairly immediate; the statement of 8.12 in this generality is due to Robert Vaught (1960) and follows from the fact that the models of $\mathcal{P}$ with decidable theories are all finite (Exercise 11, below).

8.25. Remark (continued). There are two further results along these lines. The first is due to Georg Kreisel and Andrzej Mostowski, the second to Vaught. I offer Vaught's formulations:

i. *Let $V_{r.e.}$ be the class of structures with r.e. domains and relations, and let $V_{p.r.}$ be the class of structures with primitive recursive domains and relations. If*

$$V_{p.r.} \subseteq V \subseteq V_{r.e.},$$

then Th(V) is not arithmetic.

ii. *If $V_{fin} \subseteq V \subseteq V_{r.e.}$, then Th(V) is not r.e.*

Remark 8.25.i will be proven in Chapter V, of volume II; Remark 8.25.ii will not be proven in this work.

8.26. Remark . All of the logical undecidability results mentioned (first-order logic (8.7), second-order logic (8.9), $Th(V_{fin})$ (8.12), and Remark 8.25) were established for first-order logic *with equality*. This is unnecessary as the languages used in the proofs have only finitely many primitives— whence require only finitely many equality axioms. If χ represents the conjunction of these axioms in, say, the language L_Q, then, for any sentence φ of L_Q, we see that

$$\vdash^{=} \varphi \text{ iff } \vdash \chi \rightarrow \varphi,$$

where we temporarily denote derivability in logic with equality by "$\vdash^{=}$" and derivability in the corresponding logic in which equality is just another non-logical primitive by "$\vdash$". Thus, we see the undecidability of first-order logic without equality.

8.27. Remark. Even when equality has been deprived of its special status, the complexity of the languages involved in the logical undecidability results is too great— in L_P there are three binary predicates (=, ≤, S), two ternary predicates (A, M), and a constant ($\overline{0}$). This can be greatly reduced: none of the undecidability results cited requires more than a single binary relation symbol. A mild familiarity with set theory suggests that ∈ is the right binary relation to use to establish these results. This is, indeed, the case, although it seems not to have been completely recognised until the 1970s and I am not sure if the set theoretic approach to Trahtenbrot's Theorem (first due, I think, to George Boolos) has yet been published. These proofs proceed by formulating set theoretic analogues to $\mathcal{Q}$ and $\mathcal{P}$ for the *hereditarily finite sets*, i.e. the sets obtained from the empty set by finite iterations of the power set. Precisely, define

$$V_0 = \varnothing$$

$$V_{n+1} = \mathrm{P}(V_n).$$

The class of hereditarily finite sets is just

$$V_\omega = \cup_n V_n.$$

If it were not for the arithmetic outlook of this book, and the fact that some of the material of this section is of definite (if not immediately obvious) arithmetic interest, I would have based my discussion of undecidability in logic on the hereditarily finite sets. The curious reader can find the beginnings of such an approach in Exercise 12, below.

With this, we have finished our discussion of purely logical undecidability results. Somewhat anti-climactically, I do not wish quite yet to call it quits and move on to Chapter IV. I wish first to say a few words about Julia Robinson's proof of the undecidability of the theory,

$$Th(\mathfrak{Q}) = Th(Q; +, \cdot, 0, 1),$$

of the field of rational numbers. The starting point is the following result.

8.28. Theorem. *$Th(\omega; +, \cdot, 0, 1)$ is undecidable.*

This is essentially Corollary 6.4. The difference is the change of language.

An easy corollary is the following.

8.29. Corollary. *$Th(Z; +, \cdot, 0, 1)$ is undecidable.*

This is established by showing that the natural numbers are definable by a formula Nv. Then, for any sentence φ of the appropriate language,

$$\mathfrak{N} \models \varphi \text{ iff } \mathfrak{Z} \models \varphi^{(N)}, \text{ for } \mathfrak{Z} = (Z; +, \cdot, 0, 1),$$

whence the undecidable $Th(\mathfrak{N})$ reduces to $Th(\mathfrak{Z})$.

There are several possible choices for the formula Nv. Lagrange's Theorem (II.3.8) yields

$$Nv: \quad \exists v_0 v_1 v_2 v_3 [v = (v_0)^2 + (v_1)^2 + (v_2)^2 + (v_3)^2].$$

Raphael Robinson observed that the number of quantifiers can be reduced by appeal to the Pell equation (II.3.7):

$$Nv:\ \exists v_0 v_1 [v = (v_0)^2 \vee ((v_0)^2 = 1 + v\,(v_1)^2 \wedge (v_0)^3 \neq v_0)].$$

Finally, there is Julia Robinson's Theorem:

8.30. Theorem. *$Th(Q; +, \cdot, 0, 1)$ is undecidable.*

This is proven by finding a formula Zv defining the integers in $\mathfrak{Q}$ and relativising as before:

$$\mathfrak{Z} \models \varphi \text{ iff } \mathfrak{Q} \models \varphi^{(Z)}.$$

Like the construction of the formula Nv above, that of Zv depends on some genuine number theory, in this case the characterisation of the solutions of two quadratic equations in the rationals.

The two quadratic equations are defined in terms of parameters p, q, a and differ only in the choice of q.

8.31. Lemma. *Let p, q be odd prime numbers with p of the form $4k + 1$. Suppose further that q is a quadratic non-residue of p, i.e., the equation,*

$$x^2 \equiv q \pmod p,$$

is unsolvable. Then, for any rational number a, the Diophantine equation,

$$2 + pqa^2 + pZ^2 = X^2 + qY^2,$$

is solvable in rationals iff the denominator of a, when a is reduced to lowest terms, is divisible by neither p nor q.

(The denominator of 0, written in lowest terms, is taken to be ± 1.)

8.32. Lemma. *Let p be an odd prime number with p of the form $4k + 3$. Then, for any rational number a, the Diophantine equation,*

$$2 + pa^2 + pZ^2 = X^2 + Y^2,$$

is solvable in rationals iff the denominator of a, when a is reduced to lowest terms, is odd and not divisible by p.

The proofs of these lemmas ought not to lie beyond the scope of this book, but they do. They rest on results on the p-adic numbers and a proper exposition would entail a new chapter and the inclusion of yet further results of logical number theory dealing with the p-adics. Right or wrong, I have deemed this entire subject to be a topic in advanced rather than introductory logical number theory.

If we do not prove these two lemmas here we can, however, readily derive Theorem 8.30 from them. To this end, first define

$$\varphi v_0 v_1 v:\ \exists v_2 v_3 v_4 [\overline{1} + \overline{1} + v_0 v_1 v^2 + v_0 (v_4)^2 = (v_2)^2 + v_1 (v_3)^2],$$

and then define

$$Zv:\ \forall v_0 v_1[\varphi v_0 v_1 \overline{0} \wedge \forall v'(\varphi v_0 v_1 v' \rightarrow \varphi(v_0, v_1, v' + \overline{1})) \rightarrow \varphi v_0 v_1 v].$$

Proof that Zv defines the integers in $\mathfrak{Q}$. We work with the augmented language $L(Q)$.

We must first show that

$$\mathfrak{Q} \models Z\overline{x}$$

for any $x \in \mathrm{Z}$. First, $Z\overline{0}$ follows trivially from the tautology $\varphi v_0 v_1 \overline{0} \rightarrow \varphi v_0 v_1 \overline{0}$. The fact that Zv looks like induction guarantees that $Z\overline{x} \rightarrow Z\overline{x+1}$: Suppose $Z\overline{x}$ holds. Choose any $a, b \in Q$. If $\varphi(\overline{a}, \overline{b}, \overline{x+1})$ holds, then so does

$$\varphi(\overline{a}, \overline{b}, \overline{0}) \wedge \forall v'(\varphi(\overline{a}, \overline{b}, v') \rightarrow \varphi(\overline{a}, \overline{b}, v' + \overline{1})) \rightarrow \varphi(\overline{a}, \overline{b}, \overline{x+1}). \quad (*)$$

If $\varphi(\overline{a}, \overline{b}, \overline{x+1})$ fails, then either $\varphi(\overline{a}, \overline{b}, \overline{x})$ holds or it fails. In the former case,

$$\forall v'(\varphi(\overline{a}, \overline{b}, v') \rightarrow \varphi(\overline{a}, \overline{b}, v' + \overline{1})) \quad (**)$$

fails, whence (*) holds; in the latter, since $Z\overline{x}$ holds, either $\varphi(\overline{a}, \overline{b}, \overline{0})$ fails or (**) fails— in either case (*) holds. Since a, b were arbitrary, we have shown that $Z\overline{x+1}$ holds.

By induction, we conclude that $Z\overline{x}$ holds for all $x \in \omega$. If we look at φ, we see that the variable v is squared in it, whence $Z\overline{x} \rightarrow Z\overline{-x}$, i.e., $Z\overline{x} \rightarrow Z\overline{y}$ for $y = -x$. Thus, $Z\overline{x}$ holds for all $x \in \mathrm{Z}$.

The interesting half of the proof is the converse, that

$$\mathfrak{Q} \models Z\overline{a} \rightarrow a \in \mathrm{Z}.$$

To this end, suppose $\mathfrak{Q} \models Z\overline{a}$.

Begin by choosing any prime $p \equiv 3 \pmod 4$. By Lemma 8.32,

$$\mathfrak{Q} \models \varphi(\overline{p}, \overline{1}, \overline{0}). \quad (*)$$

Moreover, since b and $b + 1$ have the same denominator when written in lowest terms,

$$\mathfrak{Q} \models \forall v'[\varphi(\overline{p}, \overline{1}, v') \rightarrow \varphi(\overline{p}, \overline{1}, v' + \overline{1})]. \quad (**)$$

(Note: Using this fact would have simplified the first part of this proof.) Now $Z\overline{a}$ yields

$$\mathfrak{Q} \models (*) \wedge (**) \rightarrow \varphi(\overline{p}, \overline{1}, \overline{a}),$$

whence

$$\mathfrak{Q} \models \varphi(\overline{p}, \overline{1}, \overline{a}).$$

Yet another application of Lemma 8.32 now yields: p does not divide the denominator of a and 2 does not divide the denominator of a.

Similarly, Lemma 8.31 yields: If $p \equiv 1 \pmod 4$, then p does not divide the denominator of a. (Cf. Exercise 13, below.) It follows that, the denominator having no prime divisor, a is integral. QED

Exercises.

1. Consider the following seven structures of the form $\mathfrak{A} = (A; \oplus, \otimes, \hat{S}, 0)$.

$\mathfrak{A}_1$: $A = \{0\}$, $\hat{S}0 = 0$, $0 \oplus 0 = 0$, $0 \otimes 0 = 0$.

$\mathfrak{A}_2$: A = non-negative real numbers

$$\hat{S}x = x+1,\ x \oplus y = x+y,\ x \otimes y = x \cdot y.$$

$\mathfrak{A}_3$: $A = \{0,1\}$, $\hat{S}x = 1$,

$$x \oplus y = \begin{cases} 0, & x = y = 0 \\ 1, & \text{otherwise,} \end{cases}$$

$$x \otimes y = \begin{cases} 1, & x = y = 1 \\ 0, & \text{otherwise.} \end{cases}$$

$\mathfrak{A}_4$: $A = \omega$, $\hat{S}x = x+1$, $x \oplus y = y$,

$$x \otimes y = \begin{cases} 0, & y = 0 \\ x, & y \neq 0. \end{cases}$$

$\mathfrak{A}_5$: $A = \omega$, $\hat{S}x = x+1$, $x \oplus y = x$, $x \otimes y = 0$

$\mathfrak{A}_6$: $A = \omega \cup \{\infty\}$, $x \otimes y = \infty$,

$$\hat{S}x = \begin{cases} x+1, & x \in \omega \\ \infty, & \text{otherwise,} \end{cases}$$

$$x \oplus y = \begin{cases} \infty, & x = \infty \ \vee \ y = \infty \\ x+y & \text{otherwise.} \end{cases}$$

$\mathfrak{A}_7$: $A = \omega$, $\hat{S}x = x+1$, $x \oplus y = x+y$, $x \otimes y = 0$.

i. Choose a number n from 1 to 7 and show $\mathfrak{A}_n$ to model all axioms of Q other than Q_n.

ii. Choose a number n other than 2 from 1 to 7 and show that $Th(\mathfrak{A}_n)$ is decidable.

[Hints-Remark: The structures with finite domains are easily shown to have decidable theories; the theories of the structures with domain ω or $\omega \cup \{\infty\}$ can be interpreted in $Th(\omega; +, \leq, 0, 1)$. The structure $\mathfrak{A}_2$ also has a decidable theory, as can be shown by a quantifier-elimination. It follows that Q is *minimally* essentially undecidable in the weak sense that the theories obtained by deleting the individual axioms of Q are not essentially undecidable. This, together with the naturalness of the axioms of Q, explains why we did not trivialise the proof that $Q \vdash R5$ by simply adding $\forall v_0 v_1 (v_0 \leq v_1 \vee v_1 \leq v_0)$ to Q as an axiom. The next Exercise shows how a few additional arithmetic axioms effect what is provable about order.]

2. i. Show that adding the commutativity of addition to Q yields a simplified proof of Lemma 8.4: part i no longer requires a reference to 8.3.iv, and part ii no longer requires an induction.

ii. Show that, if one adds the associativity of addition to Q, the transitivity of $\leq$ becomes provable.

iii. Show that if one adds to Q the commutativity and associativity of addition, as well as the uniqueness of 0 (as an additive identity), the anti-symmetry of $\leq$,

$$\forall v_0 v_1 (v_0 \leq v_1 \wedge v_1 \leq v_0 \rightarrow v_0 = v_1),$$

becomes provable.

iv. Show that left-subtraction,

$$\forall v_0 v_1 \exists v_2 (v_2 + v_0 = v_1 \vee v_2 + v_1 = v_0),$$

yields dichotomy, $\forall v_0 v_1(v_0 \le v_1 \lor v_1 \le v_0)$, over $\mathcal{Q}$.

3. Let $A = \omega \cup \{\infty\}$. On ω, define

$$\hat{S}x = Sx \ (= x + 1),\ x \oplus y = x + y,\ x \otimes y = x \cdot y.$$

We want to extend $\hat{S}$, $\oplus$, $\otimes$ to all of A so that $\mathfrak{A} = (A; \oplus, \otimes, \hat{S}, 0) \models \mathcal{Q}$.

i. How must one define $\hat{S}\infty$?

ii. For $x \in \omega$, what must $\infty \oplus x$ and $\infty \otimes x$ be?

iii. Define $x \oplus \infty$ and $x \otimes \infty$ in such a way that

a. $\mathfrak{A} \models \mathcal{Q}$

b. $\mathfrak{A} \models \neg\forall v_0 v_1(v_0 + v_1 = v_1 + v_0)$

c. $\mathfrak{A} \models \neg\forall v_0 v_1 (v_0 \cdot v_1 = v_1 \cdot v_0)$

d. $\mathfrak{A} \models \exists v\, [v \ne \overline{0} \land \exists v_1(v + v_1 = \overline{0})]$.

[Remark: By part iii.d, Theorem 8.3.iv could not be proven if we were to use the definition,

$$v_0 \le v_1: \quad \exists v_2(v_0 + v_2 = v_1).$$

Note that this definition, but not,

$$v_0 \le v_1: \quad \exists v_2(v_2 + v_0 = v_1),$$

defines the ordering of the ordinal numbers. The interested reader might like to see which of the models of $\mathcal{R}$ given in Exercise 1 of section 6, above, also models $\mathcal{Q}$.]

4. i. Letting Sq denote the set of squares, show that multiplication is definable in the structure $(\omega; +, \le, 0, 1, Sq)$, i.e.,there is a formula $\varphi v_0 v_1 v_2$ of the language of the structure such that, for all $x, y \in \omega$,

$$z = x \cdot y \ \text{ iff } \ (\omega; +, \le, 0, 1, Sq) \models \varphi(\overline{x}, \overline{y}, \overline{z}).$$

Conclude the undecidability of $Th(\omega; +, \le, 0, 1, Sq)$.

ii. Prove the undecidability of the theory obtained from $Th(\omega; +, \le, 0, 1)$ by the adjunction of a new unary predicate symbol R.

iii. Prove the undecidability of $Th(\omega; P(\omega); \in, +, \le, 0, 1)$, i.e., the second-order theory of the addition of the natural numbers.

[Hints: i. First show the definability of the relation $y = x^2$ by seeing how x relates to the next square after x^2; ii. replace Sq by R in part i and write down the recursion equations for multiplication as new axioms. If you are ambitious, find a sentence φ in the augmented language such that, for any $X \subseteq \omega$,

$$(\omega; +, \le, 0, 1, X) \models \varphi \to X = Sq.]$$

5. Show that second-order logic satisfies neither the Löwenheim-Skolem Theorem nor the Compactness Theorem. [Remark: By the First Lindström Theorem (3.19), one of these must fail; according to this Exercise, neither holds. In formulating the Löwenheim-Skolem Theorem for second-order logic, the cardinality of the structure $\mathfrak{A} = (A; P(A); \in, \ldots)$ should be taken to be that of the domain A of individuals. Hint for this part: Don't use the arithmetic of the natural numbers.]

6. Let $\mathcal{P}^+ = \mathcal{P} + \forall v_0 \exists v_1\, S v_0 v_1$. Conclude from results of the present section: If φ is a Σ_1-sentence,

$$\mathfrak{N} \models \varphi \to \mathcal{P}^+ \vdash \varphi.$$

[Hint: All models of $\mathcal{P}^+$ are infinite.]

7. $\mathcal{P}^+$ and $\mathcal{Q}$ are incomparable theories.

i. Show: The structure $\mathfrak{A}$ of Exercise 3 is not a model of $\mathcal{P}^+$ (modulo the obvious linguistic change).

ii. Define a structure $\mathfrak{A}$ as follows:

$$A = \omega \times \{0\} \cup Z \times \{1\}, \ \hat{0} \text{ is } \langle 0, 0\rangle, \ \hat{S}\langle x, i\rangle = \langle x+1, i\rangle,$$
$$\langle x, i\rangle \le \langle z, j\rangle \text{iff } i < j \lor i = j \land x \le z$$
$$\langle x, 0\rangle \oplus \langle y, 0\rangle = \langle x+y, 0\rangle$$
$$\langle z, 1\rangle \oplus \langle x, 0\rangle = \langle z+x, 1\rangle$$
$$\langle x, 0\rangle \otimes \langle y, 0\rangle = \langle x \cdot y, 0\rangle$$
$$\langle z, 1\rangle \otimes \langle 0, 0\rangle = \langle 0, 0\rangle,$$

and all other "additions" and "multiplications" are left undefined. Show: $\mathfrak{A} = (A; \oplus, \otimes, \hat{S}, \hat{0}) \models \mathcal{P}^+$, but

$$\mathfrak{A} \not\models \forall v_0 v_1 \exists v_2 A v_0 v_1 v_2,$$
$$\mathfrak{A} \not\models \forall v_0 v_1 \exists v_2 M v_0 v_1 v_2.$$

[Intuitively, $\mathfrak{A}$ consists of a copy of the natural numbers followed by one of the integers, but with addition and multiplications limited to those instances that are absolutely necessary.]

8. Show that every r.e. relation is definable in $(\omega; \le, S, A, M, 0)$ by a strict-Σ_1-formula of L_P. [Prove this by induction on the complexity of a Σ_1-formula of L_Q or L_R.]

9. Prove the following adjunct to the Reduction Theorem: Let φ, ψ be strict-Σ_1 and

$$\varphi_1 v: \quad \varphi \preccurlyeq \psi$$
$$\psi_1 v: \quad \psi \prec \varphi.$$

Then: $\mathcal{P} \vdash \neg(\varphi_1 \land \psi_1)$.

10. Define a new quantifier Q— "there are infinitely many" by letting,

$$\mathfrak{A} \models Qv\varphi v \ \text{ iff } \ \{a \in A: \mathfrak{A} \models \varphi \overline{a}\} \text{ is infinite.}$$

i. Show that validity for the logic with this new quantifier is not effectively enumerable.

ii. Show that the Compactness Theorem fails for this logic.

[Hint: i. Trahtenbrot's Theorem. The Löwenheim-Skolem Theorem can easily be proven for this logic. Thus, it offers a different sort of illustration of the two Lindström Theorems from that offered by second-order logic.]

11. Prove the second assertion of Remark 8.25: If V is a class of structures satisfying

$$V_{fin} \subseteq V \subseteq V_{dec},$$

then $Th(V)$ is not effectively enumerable. [Hint: What are the models of $\mathcal{P}$ with decidable theories?]

12. (For those knowing some naïve set theory). For $n \in \omega$, define V_n recursively by

$$V_0 = \varnothing$$
$$V_{n+1} = P(V_n) = V_n \cup P(V_n).$$

Then $V_\omega = \cup_n V_n$ is called the class of *hereditarily finite sets*. Also, define $E \subseteq \omega^2$ by,

$$x E y : x \in D_y.$$

i. Show: $(\omega; E) \cong (V_\omega; \in)$.

ii. Let L_E be the language with only $\in$ as non-logical primitive. Show: Every relation on ω that is Σ_1 in L_E is Σ_1 in L_Q, i.e.,Σ_1-relations of $(\omega;E)$ are Σ_1-relations of $(\omega; S, +, \cdot, 0, 1)$.

iii. On the structure $(V_\omega; \in)$, show that the following relations are Δ_0 (without mentioning equality):

$$a = b,\ a \in \omega,\ c = \{a, b\},\ c = \langle a, b\rangle,$$
$$c \text{ is a pair},\ c \text{ is an ordered pair},\ f \text{ is a function}.$$

iv. Show that ordinal addition and multiplication are Σ_1 in $(V_\omega; \in)$.

v. Let F be the isomorphism $F:(\omega; E) \to (V_\omega; \in)$. For $x \in \omega$, let o_x be the x-th finite ordinal in V_ω. Show that the function $\hat{F}$ defined by

$$\hat{F} o_x = Fx$$

is Σ_1 in $(V_\omega; \in)$.

vi. Show: If R is a Σ_1-relation of $(\omega; S, +, \cdot, 0, 1)$, then R is a Σ_1-relation of $(\omega; E)$.

vii. Construct a set theoretic analogue to $\mathcal{P}$, the finite models of which are exactly the structures $(V_n; \in)$ and the infinite models of which are end extensions of $(V_\omega; \in)$.

viii. Prove Trahtenbrot's Theorem for first-order logic with a single binary relation symbol.

[Remarks: This Exercise, especially part vii, is only for the strong-hearted and I won't bother providing any hints.]

13. Complete the derivation of Theorem 8.30 from Lemmas 8.31 and 8.32. Specifically,

i. Let $p \equiv 1 \pmod 4$ be prime. Show that there is an odd prime q such that q is not congruent to a square modulo p.

ii. Assuming $\mathfrak{Q} \models Z\overline{a}$, let $p \equiv 1 \pmod 4$ be prime and choose q by part i of this Exercise. Show: $\mathfrak{Q} \models \varphi(\overline{p}, \overline{q}, \overline{a})$ and conclude p does not divide the denominator of a.

[Hint: i. if x is not a square modulo p, one of x and $x + p$ must be odd.]

9. Reading List

§1.

Hilbert's 1900 address on problems can be found in:

1. David Hilbert, "Mathematische Probleme", *Arch. f. Math. u. Phys.* (3) 1 (1901), 44 - 63; reprinted in: David Hilbert, *Gesammelte Abhandlungen III*, Springer-Verlag, Berlin, 1935; English translation in: F. Browder, ed., *Mathematical Developments Arising from Hilbert Problems*, AMS, Providence, 1976.

For more about Hilbert and du Bois-Reymond, I offer the following two references.

2. Constance Reid, *Hilbert*, Springer-Verlag, New York, 1970.

3. K.E. Rothschuh, "Emil Heinrich du Bois-Reymond", *Dictionary of Scientific Biography*, Scribners, New York, 1971.

Now for a few philosophical references. In the first of these, Abrusci suggests that Problem 1.1 is not the correct interpretation of Hilbert's demand for completeness in the quote from the second Problem. One might take the word "complete" to mean adequacy for actual practice, which, in the case of logical incompleteness, can always be extended by the addition of new axioms. Abrusci may be historically correct that, in 1900, Hilbert was not thinking of the sort of completeness demanded by Problem 1.1; in the 1920s, however, Hilbert was quite explicit about expecting this sort of completeness. We shall go into historical matters in a little more detail in section 1 of the Chapter (in the next volume). For now, I recommend Abrusci's paper for the philosophically minded reader.
For the reader with some background in set theory, the reference by Moore is an excellent account of the adoption of new axioms. For balance, I also recommend the paper by Putnam on non-proof methods in mathematics.

4. Vito Abrusci, "'Proof', 'theory', and 'foundations' in Hilbert's mathematical work from 1885 to 1900", in: M.L. Dalla Chiara, ed., *Italian Studies in the Philosophy of Science*, Reidel, Dordrecht, 1981.

5. Gregory Moore, *Zermelo's Axiom of Choice*, Springer-Verlag, New York, 1982.

6. Hilary Putnam, "What is mathematical truth?", in: H. Putnam, *Philosophical Papers I; Mathematics, Matter and Method*, Cambridge, 1975; reprinted in: T. Tymoczko, ed., *New Directions in the Philosophy of Mathematics*, Birkhäuser, Boston, 1986.

§§2 - 3.

My discussion of first-order logic is rather skeletal. Most logic textbooks can flesh it out a bit. I particularly recommend the following for syntactic matters.

1. Jean Gallier, *Logic for Computer Science: Foundations of Automatic Theorem Proving*, Harper & Row, New York, 1986.

The Completeness Theorem was first proved by Kurt Gödel; the version I have presented is due essentially to Leon Henkin. Several variants can be found in Gallier's book. The background to the proof of the Completeness Theorem— the events leading up to it and the emergence of first-order logic as the dominant one is nicely described in the following paper.

2. Warren Goldfarb, "Logic in the twenties: the nature of the quantifier", *J. Symbolic Logic* 44 (1979), 351 - 368.

Finally, for the reader willing to study a bit more model theory, a good account of both of Lindström's Theorems can be found in the following book already cited in the Reading List of Chapter I.

3. Heinz-Dieter Ebbinghaus, Jörg Flum, and Wolfgang Thomas, *Mathematical Logic*, Springer-Verlag, New York, 1984.

§4.

The original papers of Presburger and Skolem are the following:

1. M. Presburger, "Über die Vollständigkeit eines gewissen Systems der Arithmetik ganzer Zahlen, in welchem die Addition als einzige Operation hervortritt", in: *Comptes Rendus I Congrès des Mathématiciens des Pays Slaves*, Warsaw, 1930.

2. Thoralf Skolem, "Über einige Satzfunktionen in der Arithmetik", Skrifter Vitenskapsakademiet i Oslo, I, No 7 (1931), 1 - 28; reprinted in: Thoralf Skolem, *Selected Works in Logic*, Universitetsforlaget, Oslo, 1970.

Presburger's published proof was given for the integers in the language of addition without <, and does not cover the theory of the addition of the natural numbers. Separate from his paper in the same volume is an abstract announcing the result for the extended language. The first full proof published was given by Bernays in the following reference. English treatments along Presburger's lines can be found in each of the ensuing two references.

3. David Hilbert and Paul Bernays, *Grundlagen der Mathematik I*, Springer-Verlag, Berlin, 1934. (*Vid.* pages 359 - 368, or 368 - 377 of the second edition (1970).)

4. Herbert B. Enderton, *A Mathematical Introduction to Logic*, Academic Press, New York, 1972.

5. Ann Yasuhara, *Recursive Function Theory and Logic*, Academic Press, New York, 1971.

The characterisation of the sets definable in the structure $(\omega; +, \leq, 0, 1)$ is due to Ginsburg and Spanier. They actually characterised all definable relations. For this and an application in formal language theory in computer science, see:

6. S. Ginsburg and E. Spanier, "Semigroups, Presburger formulas, and languages", *Pacific J. Math.* 16 (1966), 285 - 296.

The reader curious about the Sylvester-Frobenius problem (Application 4.4 and Exercise 5) is referred to the following for more material and references.

7. C. Smoryński, "Skolem's solution to a problem of Frobenius", *Math. Intelligencer* 3 (1981), 123 - 132.

Finally, the reader with a good algebraic background can find a model theoretic proof of the completeness of the axioms of Application 4.5 using the notion of *model completeness* (from Exercise 6) in the following paper.

8. Abraham Robinson and Elias Zakon, "Elementary properties of ordered Abelian groups", *Transactions AMS* 96 (1960), 222 - 236; reprinted in: H.J. Keisler, S. Körner, W.A.J. Luxemburg, and A.D. Young, eds., *Selected Papers of Abraham Robinson I; Model Theory and Algebra*, North-Holland, Amsterdam, 1979.

§5.

Skolem's paper is cited as reference 2 for section 4, above. The other major references are:

1. Andrzej Mostowski, "On direct products of theories", *J. Symbolic Logic* 17 (1952), 1 - 31; reprinted in: A. Mostowski, *Foundational Studies; Selected Works II*, North-Holland, Amsterdam, and PWN— Polish Scientific Publishers, Warsaw, 1979.

2. Patrick Cegielski, "Théorie élémentaire de la multiplication des entiers naturels", in: C. Berline, K. McAloon, and J.-P. Ressayre, eds., *Model Theory and Arithmetic*, Springer-Verlag, New York, 1981.

Finally, let me make a tangential reference to the question of complexity of decision problems for theories:

3. Jeanne Ferrante and Charles W. Rackoff, *The Computational Complexity of Logical Theories*, Springer-Verlag, New York, 1979.

§8.

The following reference is pertinent to sections 6 and 7 as well.

1. Alfred Tarski, Andrzej Mostowski, and Raphael Robinson, *Undecidable Theories*, North-Holland, Amsterdam, 1953.

The best overall reference on undecidability available at the time of writing is the following survey of over two decades ago.

2. Yu. L. Ershov, I.A. Lavrov, A.D. Taimanov, and M.A. Taitslin, "Elementary theories", (Russian), *Uspekhi Matem. Nauk* 20 4 (1965), 37 - 108; English translation: *Russian Math. Surveys* 20 4 (1965), 35 - 105.

The theory Q, despite its weakness, has been studied again recently. The mathematical reasons for such a renascence of interest in Q can be gleaned from the following book; the philosophical reasons offered should, however, be dismissed.

3. Edward Nelson, *Predicative Arithmetic*, Princeton, 1986.

The most thorough reference on the decidability of the second-order theory of $(\omega; S, 0)$ is the following.

4. Dirk Siefkes, *Büchi's Monadic Second-Order Successor Arithmetic*, Springer-Verlag, Heidelberg, 1970.

Finally, for the reader desiring a complete proof of Robinson's theorem on the undecidability of the first-order theory of the rational number field, I offer the following two references. The paper by Lewis includes a proof of the Hasse-Minkowski Theorem and the first appendix of Fraïssé's book derives Robinson's lemmas therefrom.

5. D.J. Lewis, "Diophantine equations: p-adic methods", in: W.J. LeVeque, ed., *Studies in Number Theory*, Prentice-Hall, Englewood Cliffs (N.J.), 1969.

6. Roland Fraïssé, *Cours de Logique Mathématique II*, Gautier-Villars, Paris, 1972; English translation: *A Course in Mathematical Logic II*, Reidel, Dordrecht, 1974.

In fact, Robinson's own paper is itself worth looking into:

7. Julia Robinson, "Definability and decision problems in arithmetic", *J. Symbolic Logic* 14 (1949), 98 - 114.

Index of Names

There be of them, that have left a name behind them, that their praises might be reported.

—Ecclesiastes

Index of Subjects

So essential did I consider an Index to be to every book, that I proposed to bring a Bill into parliament to deprive an author who publishes a book without an Index of the privelege of copyright; and, moreover, to subject him, for his offence, to a pecuniary punishment.

—*Baron Campbell*

Notations

Local Notations

It does not seem supererogatory to inform the reader that the author is aware of an uniform misspelling of the name Leibniz throughout the present work. This was a genuine error brought on by the author's penchant for archaisms and his possession of a very nice print of "M. Leibnitz". This was not the antient Germanic spelling of the name, but a French one and, this not being a work in French, its use here must be deemed incorrect.

D. van Dalen, Rijksuniversiteit Utrecht

Logic and Structure

Universitext

2nd ed. 1983. Corr. 2nd printing 1988. X, 207 pp. Softcover DM 42,–
ISBN 3-540-12831-X

Contents: Introduction. – Propositional Logic. – Predicate Logic. – Completeness and Applications. – Second-Order Logic. – Intuitionistic Logic. – Appendix. – Bibliography. – Gothic Alphabet. – Index.

This book provides an efficient introduction to logic for students of mathematics. The central theme is that part of first order logic which can be handled directly on the basis of derivation and validity. Emphasis is placed on notions that play a role in every-day mathematics such as models, truth relativized quantifiers, consistency, Skolem functions, and extension by definition.

Following a self-contained presentation of propositional logic (including completeness), predicate logic – with applications to elementary algebra – is treated systematically, leading to an exposition of the first principles of model theory. A unique feature of this book is the systematic use of Fentzen's system of natural deduction, which is closer to natural informal reasoning than an axiomatic approach, enabling the student to devise derivations as a simple exercise.

Inductive definitions are employed in the book whenever appropriate. Model-theoretic topics include the main facts of compactness, non-standard models of arithmetic and the reals, and, in a special section, some of the properties of second-order logic. The material is illustrated by many exercises and demands of the reader a minimum background in mathematics.

Springer-Verlag
Berlin
Heidelberg
New York
London
Paris
Tokyo
Hong Kong
Barcelona

R. Mines, F. Richman, New Mexico State University, Las Cruces, NM;
W. Ruitenburg, Marquette University, Milwaukee, WI

A Course in Constructive Algebra

Universitext

1988. XI, 344 pp. Softcover DM 68,– ISBN 3-540-96640-4

Contents: Sets. – Basic Algebra. – Rings and Modules. – Divisibility in Discrete Domains. – Principal Ideal Domains. – Field Theory. – Factoring Polynomials. – Commutative Noetherian Rings. – Finite Dimensional Algebras. – Free Groups. – Abelian Groups. – Valuation Theory. – Dedekind Domains. – Bibliography. – Index.

P. J. McCarthy, University of Kansas, Lawrence, KS

Introduction to Arithmetical Functions

Universitext

1986. VII, 365 pp. 6 figs. Softcover DM 106,–
ISBN 3-540-96262-X

Contents: Multiplicative Functions. – Ramanujan Sums. – Counting Solutions of Congruences. – Generalizations of Dirichlet Convolution. – Dirichlet Series and Generating Functions. – Asymptotic Properties of Arithmetical Functions. – Generalized Arithmetical Functions. – References. – Bibliography. – Index.

Springer-Verlag
Berlin
Heidelberg
New York
London
Paris
Tokyo
Hong Kong
Barcelona

7842